Psychoneuroscience

Gerhard Roth · Andreas Heinz · Henrik Walter

Editors

Psychoneuroscience

 Springer

Editors
Gerhard Roth
Institut für Hirnforschung
Universität Bremen
Bremen, Germany

Andreas Heinz
Klinik für Psychiatrie und Psychotherapie
Charité Universitätsmedizin Berlin
Berlin, Germany

Henrik Walter
CharitéCentrum Psychatrie
Charité Universitätsmedizin Berlin
Berlin, Germany

ISBN 978-3-662-65776-8   ISBN 978-3-662-65774-4   (eBook)
https://doi.org/10.1007/978-3-662-65774-4

Illustrations by: Martin Lay

This Springer imprint is published by the registered company Springer-Verlag GmbH, DE, part of Springer Nature.
The registered company address is: Heidelberger Platz 3, 14197 Berlin, Germany

Preface

This textbook aims to build a bridge between the psychosciences, i.e., psychology, psychiatry, and psychotherapy and the neurosciences, i.e., neurobiology and neurology. Such bridging was already dreamed of by the most important neuroscientists, neurologists, and psychiatrists of the second half of the nineteenth century, such as Ernst von Brücke, Theodor Meynert, and Sigmund Exner, and also by the young Sigmund Freud, who, as is well known, began his career as a very talented neuroanatomist. However, such bridging, neglecting neither the humanities nor the natural sciences, could not be approached in a fruitful way until many decades later, towards the end of the last century, when findings in neuroanatomy, neurophysiology, developmental neurobiology, and neurogenetics greatly increased and consolidated.

While many psychoscientists and neuroscientists now emphasize the need for such a bridge and are actively working on it, there is still considerable resistance and tendencies towards isolation on both sides. This is often due to the fact that neurobiology and neurology see themselves as natural sciences and insist on the same scientific foundations and empirical evidence with regard to the theory and practice of the psychosciences, whereas many psychoscientists emphasize the independence of their humanistic orientation and take an aggressive stance against a "sovereignty of interpretation" of the neurosciences.

This often leads to considerable confusion in the respective training courses or advanced training events—after all, it is often a coincidence whether a student or an advanced trainee will be trained by a "hard" neurobiological reductionist, for whom mental illnesses are "nothing more" than malfunctions of synaptic transmission, or by a psychologist, psychiatrist, or psychotherapist who thinks and practices explicitly in terms of the humanities and who emphasizes the autonomy of the mind or at least the humanistic approaches to the relevant phenomena. While there are increasing calls for the abolition of such "camp thinking," there is usually a lack of clear concepts and overarching accounts of how such bridging can be accomplished. To accomplish this is the main intention of the textbook and its authors.

Accordingly, we intend to provide students and young scientists, doctoral students, and postdocs, as well as all those interested in further education and training in both psychosciences and neurosciences, with a compact and integrative overview of the current state of knowledge and findings that is not committed to any camp thinking. We therefore do not limit ourselves to talking about "mental correlates of neurobiological processes" or "neuronal correlates of mental processes," as this approach still implicitly suggests a traditional dualism of the mental and the neuronal. Rather, we assume that the "neuronal" and the "mental" form an inseparable unity in the sense that for every mental process there is a specific neuronal process and that the different levels of description can be represented within a common framework of scientific laws. What such a unity is ultimately based on, i.e., what the actual "nature of mind" is, remains the subject of philosophical discussions, which we will only touch upon in places. We are also aware that in many areas detailed proof and critical reflection of such a holistic explanation is still lacking. But wherever intensive research is carried out, this unity is emphasized.

We thus respect the autonomy of the phenomena and the domain-specific regularities associated with them, provided there is no contradiction between the neuronal and the psychological domains. In other words, psychology, psychiatry, and psychotherapy require a scientific-neurobiological basis, but the neurosciences, for their part, only provide a framework within which the psychosciences unfold in theory and practice. On a practical and methodological level, traditional boundaries are already disappearing, i.e., many neuroscientists deal with psychological-psychiatric-psychotherapeutic issues and vice versa, or are already qualified in both domains, and we assume that sharp demarcations are already obsolete today.

However, we are aware that we are only at the beginning with our bridge-building, especially since important new findings and methods are constantly emerging in the neurosciences and psychosciences that need to be integrated. Nevertheless, we believe that the direction we are taking with our textbook is irreversible.

The textbook is organized as follows: After a historical introduction by the three editors to the two-millennia-long search for the "nature of the soul," in the first part, the individual authors provide an overview of the state of neuroscientific research, which includes the functional neuroanatomy of the limbic system as a crucial carrier of psychological phenomena, the pharmacological and neurophysiological bases of neuronal excitation processing, brain development, the neurobiological bases of personality traits, emotions, and motivation, the neurobiological consequences of early stress experience, and the psychological and neurobiological bases of consciousness.

This is followed by the second part, which deals with questions of psychiatry and psychotherapy against the background of the neuroscientific part. It deals with the definition, classification, and diagnosis of mental disorders, followed by chapters on substance use disorders, psychotic disorders, affective disorders, and anxiety disorders as exemplary mental disorders. In addition, the effect of psychotherapeutic procedures and their neurobiological basis are discussed within the concept "neuropsychotherapy." The book concludes with the editors' consideration of psychoneurobiology and its relevance to psychiatric practice.

We would like to thank all the co-authors of this book for the effort they have taken to present their expertise in a form that is hopefully understandable to the other "side," without giving up the highest standards of scientific correctness and topicality.

We would also like to thank Stephanie Preuß and Bettina Saglio from Springer Verlag for their professional and patient supervision of this project.

Gerhard Roth
Bremen, Deutschland

Andreas Heinz
Berlin, Deutschland

Henrik Walter
Berlin, Deutschland
September 2019

Contents

Contributors

Ursula Dicke Brain Research Institute, Universität Bremen, Bremen, Germany

Stefan Gutwinski Charité – Universitätsmedizin Berlin, Berlin, Germany

John-Dylan Haynes Bernstein Center for Computational Neuroscience, Charité – Universitätsmedizin Berlin, Berlin, Germany

Christine Heim Charité – Universitätsmedizin Berlin, corporate member of Freie Universität Berlin and Humboldt-Universität zu Berlin, NeuroCure Cluster of Excellence, Berlin, Germany
Center for Safe and Healthy Children, College of Health and Human Development, The Pennsylvania State University, University Park, PA, USA

Andreas Heinz Klinik für Psychatrie und Psychotherpie Charité, Berlin, Germany

Andrea J. J. Knop Charité – Universitätsmedizin Berlin, corporate member of Freie Universität Berlin and Humboldt-Universität zu Berlin, Institute of Medical Psychology, Berlin, Germany

Michael Koch Department of Neuropharmacology, Universität Bremen FB 2, Brain Research Institute, Bremen, Germany

Stephan Köhler Klinik für Psychiatrie und Psychotherapie, Charité – Universitätsmedizin Berlin, Berlin, Germany

Nora K. Moog Charité – Universitätsmedizin Berlin, corporate member of Freie Universität Berlin and Humboldt-Universität zu Berlin, Institute of Medical Psychology, Berlin, Germany

Jens Plag Klinik für Psychiatrie und Psychotherapie, Charité Campus Mitte, Charité – Universitätsmedizin Berlin, Berlin, Germany

Nina Romanczuk-Seiferth Charité – Universitätsmedizin Berlin, corporate member of Freie Universität Berlin and Humboldt-Universität zu Berlin, Department of Psychiatry and Neuroscience, Charité Campus Mitte, Berlin, Germany

Gerhard Roth Institut für Hirnforschung, Universität Bremen, Bremen, Germany
Brain Research Institute, Bremen, Germany

Florian Schlagenhauf Klinik für Psychiatrie und Psychotherapie, Charité Campus Mitte, Charité – Universitätsmedizin Berlin, Berlin, Germany

Philipp Sterzer Universitäre Psychiatrische Kliniken Basel, Basel, Switzerland

Andreas Ströhle Klinik für Psychiatrie und Psychotherapie, Charité Campus Mitte, Charité – Universitätsmedizin Berlin, Berlin, Germany

Nicole Strüber Roth Institut, Bremen, Germany

Jakob von Engelhardt Institut für Pathophysiologie, Universitätsmedizin Mainz, Mainz, Germany

Henrik Walter Neurochirugie und Psychiatrie CC 15, Charité Centrum Neurologie, Berlin, Germany

Charité – Universtitätsmedizin Berlin, Department of Psychiatry and Psychotherapy, Berlin, Germany

Klinik für Psychiatrie und Psychotherapie, Charité – Universitätsmedizin Berlin, Berlin, Germany

The Search for the Nature of the Soul

Gerhard Roth, Andreas Heinz, and Henrik Walter

Contents

G. Roth et al. (eds.), *Psychoneuroscience*, https://doi.org/10.1007/978-3-662-65774-4_1

1

Learning Objectives

After reading this chapter, the reader should be able to comprehend the two millennia of philosophical and scientific views on the structure and functions of the human brain and their relationship to mental and psychological states.

In the current textbooks, manuals and dictionaries of neuroscience, psychology, psychiatry and psychotherapy, the term "soul" usually occurs only in a historical context and has been more or less completely replaced by the term "psyche" or "mind" meaning the entire thinking and feeling of man. However, this has been accompanied by a considerable restriction of the historical concept of "soul", as we will briefly illustrate here.

The question of the definition and nature of the soul has long preoccupied theologians, philosophers, and scientists. All major religions have developed some kind of doctrine of the soul, and nearly all major philosophers have commented on the nature and properties of the soul. In the following, we refer to European traditions, knowing that other cultures and regions of the world have sometimes developed quite different ideas (Morris 1994).

For the European-influenced tradition, it is true that from antiquity until the twentieth century brain researchers were searching for the "seat" of the soul in the nervous system (cf. Breidbach 1997; Florey 1996), and without this intensive search the great advances in brain research would not have been possible. It was not until the second half of the twentieth century that this question went out of fashion, and the attempts of the neurophilosopher John Eccles, a Nobel Prize winner for physiology, to link the concept of mind and soul with modern brain research (cf. Popper and Eccles 1977/1982; Eccles 1994) were laughed at by neurobiologists, psychologists and also most philosophers.

The need to think about the soul arises for a number of reasons. One reason lies in traditional views about the structure of the world. For Aristotle (384–322), the world is composed of two basic parts, i.e. dead things and animated things. The animated things in turn are divided by Aristotle into:

1. plants that can live and grow,
2. animals, which live and grow, but in addition can move on their own and probably have sensations, and finally
3. human beings who live and grow, who can move on their own, have sensations, but who, in addition, possess a mind with which they can think.[1]

Accordingly, in philosophical anthropology, distinctions are found between plants as "open" and animals as "closed" life forms with a boundary that separats the inner and the outer world. In this view, humans, unlike animals, not only live in the centre of their environment, but can also position themselves eccentrically, laughing or thinking about themselves, for example (Plessner 1975).

Another reason for the doctrine of the soul is our self-experience: we experience the world in characteristically different ways, namely, first, ourselves, i.e., our perceiving, thinking, remembering, imagining, feeling, and willing, which is most strongly connected with the ego-feeling (cf. Frank 1991, pp. 246 ff.; Sartre 2014, p. 157); second, our body, in which we live as a body, but which we can also experience as an object among other objects (Plessner 1975); and third, the world outside our ego and body, i.e., all that we cannot directly feel or influence (Roth 1996). The idea that our ego, our conscious states, our thinking, remembering, and willing, are seemingly immaterial and non-spatial is prevalent, at least in the European tradition. Usually, we cannot imagine that our mind should be of the same kind as a stone or a piece of wood, and even our body seems somehow of a different nature. There

---

1   Aristotle, On the Soul, 413a.

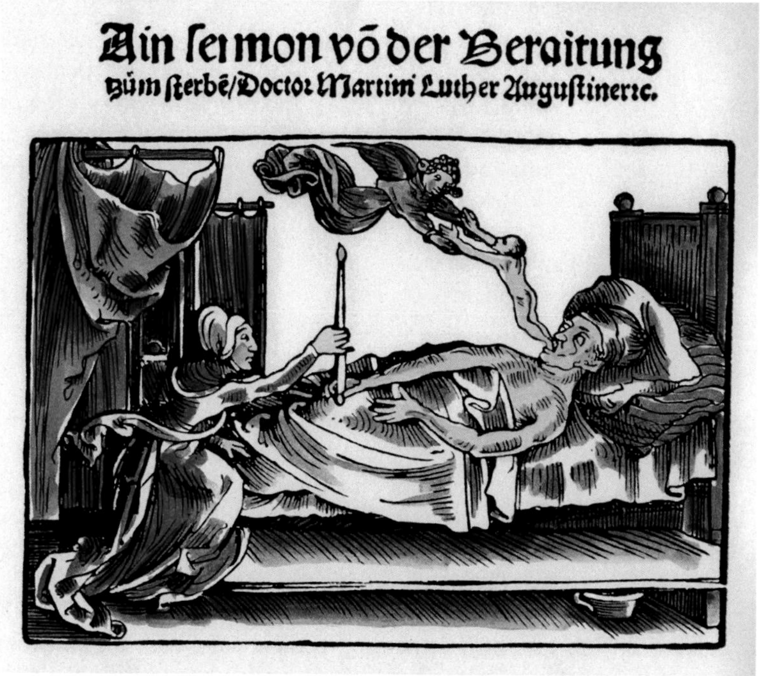

Ain sermon võ der Beraitung
zům sterbē/Doctoz Martini Luther Augustiner:c.

◘ Fig. 1.1 Popular representation from the sixteenth century showing how the soul escapes from the mouth of a dying person, leaving the dead body behind. The soul is received by an angel. The equation of soul and breath is of ancient origin. (From: Martin Luther: *Ain sermon von der Beraitung zum sterben.* Around 1520. © akg-images/picture alliance)

must, therefore, be something purely spiritual, purely soulful.[2] This difference is not as clear in the case of feelings, which on the one hand are pure sensations, but on the other hand are more or less strongly related to the body.

Another and usually little noticed reason comes from the experience of dreams. While I sleep and lie as if I were dead in bed or somewhere else, I have dreams in which I often move quite realistically in a dream world, experience things, say things, do things, have feelings, etc.; others also report this. This speaks for a separation of soul and body in dreams: According to this, there would have to be something that is often called the free soul, that animates the body, but can leave it temporarily (then the body is as if dead in deep sleep), and finally after death (◘ Fig. 1.1).

Already in antiquity there were discussions about the nature of these three aspects of the soul. The first two, *anima vegetativa* or *spiritus vegetativus* and *anima animalis* or *spiritus animalis*, were for the ancient philosophers indisputably of material nature, although (at least the *spiritus animalis*) of a very subtle nature like the spirit of wine, which is produced during the distillation of wine. The double meaning of "spiritus" as "spirit" and as "spirit of wine" has a tradition going back thousands of years. The nature of the third spirit has been controversial: While materialist philosophers also held it to be material-subtle (Lukrez 2017), for most other philosophers, especially Plato (428/427-348/347), the third soul, the *anima rationalis*, was immaterial, immortal, and

2   Scheler 1930, p. 82.

1

correspondingly detachable from the body.[3] Most philosophers up to the twentieth century were also of this opinion, as it was primarily represented by René Descartes in modern times.

Since ancient times, physicians and natural scientists have tried to locate these three souls or soul faculties somewhere in man. Plato states in his dialogue Timaeus[4] as well as in other dialogues that the soul faculty of desire is located in the abdomen, courage as the second soul faculty in the heart, and thinking or the mind as the third soul faculty in the brain. In this view, the brain is regarded as the central organ of the body; like reason directs our actions the brain direct our body. Mind/reason/psyche/soul are synonymous for Plato; they characterize an immortal, immaterial being of divine origin.

For Lucretius (99-55/53 BC), on the other hand, there was only matter (body) and emptiness. The subtly understood mind (*animus*) sits in the middle chest, while the soul (*anima*) is scattered throughout the body.[5] The middle chest or heart are also of central importance in other philosophical systems: for Aristotle, the heart was the seat of perceptions and sensations; it is from the heart that bodily functions are regulated.[6] Through the arteries filled with pneuma, the heart sends "life spirits" to the organs, and through the blood-filled warm veins, life-giving nutrients. Through its connections with the sense organs it is indirectly informed of the condition and processes of the outside world. This idea is still alive today in the popular conception of the heart as the seat of the emotions. The brain was for Aristotle a cooling organ for the warm blood; it was bloodless except for the meninges, and there-fore could not be the "soul organ" at all.[7] Aristotle, unlike his teacher Plato, did not regard the soul as an independent entity; for him it did not exist without the body, rather it was the inherent form of the respective substance and thus defined its concrete essence.[8]

The late antique physician and one of the first empirical brain researchers Claudius Galenos (129–199 A.D.) finally developed, on the basis of anatomical studies, a doctrine of the soul that was effective far into the seventeenth century, namely the doctrine of the three types of pneuma/spiritus/soul, which deviates somewhat from the aforementioned doctrine of the soul.[9] According to Galenos, in the liver the *spiritus naturalis/materialis* is formed from the *nutritive* substances, which is then transformed in the heart under the influence of the spiritus/pneuma inhaled with the air into the *spiritus vitalis* and released into the blood. In the brain, through which the blood flows, the *spiritus animalis* is formed by distillation via the miracle net (*rete mirabile*) from the *spiritus vitalis* (◻ Fig. 1.2). The three spirits are supplied to all organs of the body, the *spiritus naturalis* via the veins, the *spiritus vitalis* via the arteries and the *spiritus animalis* via the nerves. In short, veins nourish, arteries animate, nerves animate. Galenos is silent about a *spiritus rationalis*.[10]

The ambiguity found in Descartes as to whether signal transmission in the nerve tubes was mechanical via filaments, via vibrations of these filaments or hydraulic or gaseous continued well into the nineteenth century. Important natural scientists such as Albrecht von Haller (1708–1777) were of

---

3   Plato: Phaedrus 245, Timaeus 69c.
4   Plato: Timaeus 69c–d.
5   Lukrez 2017, pp. 51, 109.
6   Aristotle: On the Parts of Living Things, 652a–665 f.

7   Aristotle dto. 669.
8   Aristotle: Metaphysics 1036b–1037b, and On the Soul 413b.
9   Galenos: De placitis Hippocratis et Platonis. Opera omnia vol. V, 518.
10   in addition Julius Rocca (2003): Galen on the brain, Leiden, 34–38.

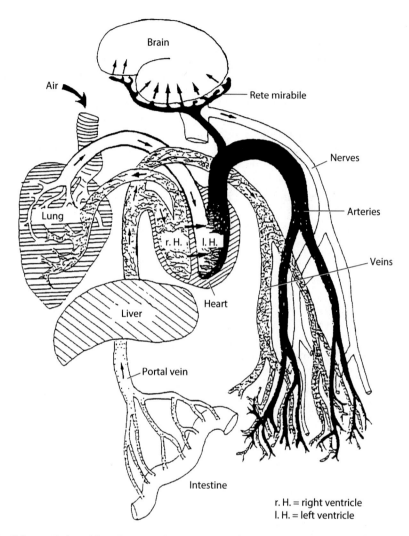

□ Fig. 1.2 Scheme of the spiritus doctrine of Galenos. From the nutrients taken from the food, the *spiritus naturalis* is formed in the liver, which is then transformed into the *spiritus vitalis in* the heart under the influence of the *spiritus/pneuma* inhaled with the air. In the brain the *spiritus animalis is* formed from this precursor. (Modified after Florey, from Roth and Prinz 1996)

the opinion that there is a fine fluid.[11] Isaac Newton, in his Principia mathematica, advocated the idea of signal transmission via vibrating elastic nerve filaments,[12] mentioning in the 2nd edition an *electric and elastic spirit* that supposedly propagates faster than the speed of light, but Newton did not elaborate further what role it played. Some Italian researchers from the eighteenth century, such as Leopoldo Caldani (1725–1813), following Newton, postulated that the "nerve fluid" was electric, and that electricity was the principle of life.[13] This

11 Haller: Elementa physiologiae corporis humanae, vol. 6: Cerebrum.
12 Newton, Principia mathematica III, 5.

13 Caldani: Institutiones. Physiologiae, 136.

triggered fierce opposition. The discoveries of Luigi Galvani (1737–1798) that muscle twitching could be triggered by electrical stimulation were the obvious proof.[14] Galvani, however, still adhered to the ancient pneuma or fluidum doctrine and believed that the electrical fluid in the cerebral cortex develops from the blood and flows into the nerves. Accordingly, the eminent anatomist Karl-Friedrich Burdach (1776–1847), in his three-volume work "Vom Baue und Leben des Gehirns" (1819–1826), held that the "nervous force" underlying soul activity was an energetic transformation of electricity. The *spiritus animalis* thus became an electric fluid, and the moving fluid was finally understood as an electric current.

Modern electrophysiology was founded by Emil Dubois-Reymond (1818–1896) and advanced by the invention of completely new devices and experiments. This development culminated in Dubois-Reymond's work "Untersuchungen über die thierische Elektrizität" (1848). Since the 60s of the nineteenth century, there was evidence that nerve excitation is pulse-like ("action potentials"). With the aid of very sensitive galvanometers, Richard Caton (1842–1926) demonstrated oscillating currents in the brains of rabbits, cats and monkeys, which occurred on the surface of the cerebral cortex during head turns, chewing movements or skin touches (Caton 1875). This became the basis of modern EEG, as developed in the 1920s by Hans Berger (1873–1941) (Berger 1929), and of modern electrophysiology.

Parallel to these investigations, functional neuroanatomy made progress. F. J. Gall (1758–1828) and J. C. Spurzheim (1775–1832) were important brain anatomists and the most prominent representatives of "phrenology", now considered unscientific, i.e. a detailed localization theory of brain functions and mental faculties derived from the form of the skull surface (Gall 1791), since the bulges of the brain were supposedly palpable on the skull bone. Gall and Spurzheim held a radically materialistic view: changes in the brain lead to changes in brain functions. In this view there was no longer any place for the "soul".

This history shows how within philosophy and brain research from antiquity to the present day, the soul in the proverbial "sense" evaporated. With Plato and the other ancient and medieval philosophers, physicians and natural scientists, the soul faculties were still located in the whole body, only the thinking soul had its seat in the brain. Since modern times, all the soul faculties have been transferred to the nervous system. In textbooks commonly used today, one still speaks of a *vegetative* nervous system (*anima vegetativa!*), which drives our bodily functions and our basic affects, and also of an *animal* nervous system (*anima animalis!*) or of an animal physiology, which drives all perceptions, sensations and movement control that we humans share with animals (called the "animal soul" by Dubois-Reymonds and Wundt, cf. Wundt 1863). This "animal soul" is then located in parts of the brain stem, the diencephalon, and the end brain, including some parts of the cerebral cortex. The third soul faculty, the *anima rationalis*, which is nowadays referred to by the term *cognitive* or *mental* states, has long been located in the so-called associative areas of the cerebral cortex. For a long time, one was rather uncertain about the seat of emotions; in the meantime, the neurobiological correlates of emotions are mostly observed in the brainstem, the diencephalon and parts of the cerebral cortex—often in centres that correspond to today's concept of the limbic system (► Chap. 2).

Modern neurobiology can link these states to structures and functions of the human (and to some extent the animal) brain, as illustrated in this textbook.

---

14 Caldani: De viribus electricitatis in motu muscularis commentarius.

Cognitive performance is tied to the activity of various brain regions, including the associative cerebral cortex, the allocortical memory system (hippocampus and surrounding cortex), and subcortical structures involved in the control of consciousness and attention (e.g., basal forebrain, basal ganglia, thalamic nuclei, reticular formation). Emotional states and moods are tied to the activity of the limbic system (for example, orbitofrontal and cingulate cortex, amygdala and mesolimbic system, hypothalamus, central cavernous grey and vegetative nuclei of the midbrain tegmentum, bridge and medulla oblongata). These are in turn influenced by hormonal systems from the whole body, so that a restriction of the "mental" to the brain is also mistaken from a neurobiological point of view.

In addition to the "fast" neurotransmitters such as glutamate, GABA and glycine, perhaps the most important function is performed by the so-called neuromodulatory systems, which are determined by the neurotransmitters noradrenaline, dopamine, serotonin and acetylcholine. These substances are produced in comparatively small centers of the brainstem (mainly the raphe nuclei, locus coeruleus, substantia nigra, ventral tegmental area) and the lower endbrain (basal forebrain) and are transported via an extensive fiber system throughout the brain, including the cerebral cortex, where they are released. These neuromodulatory substances, along with other neuroactive substances such as neuropeptides and neurohormones, are important components of mental functioning by mediating undirected attention and arousal (especially norepinephrine), happiness and well-being (e.g., serotonin), directed attention (such as acetylcholine), and drive, curiosity, and reward anticipation (especially dopamine). The dopaminergic system is closely coupled to the so-called endogenous opioids, which are the brain's actual pleasure and reward substances (Heinz et al. 2012). Other neuropeptides such as cholecystokinin and substance

P can trigger anxiety and malaise (cf. Roth and Strüber 2018). Whereas previously the neurosciences and cognitive sciences strongly separated the supposedly "higher" cognitive functions and the emotional/affective functions, there is now a growing realization that these functions are intimately connected in humans, indeed that cognitive performance is embedded in the much broader emotional and affective performance (Kaminski et al. 2018; Stephan and Walter 2003). This is the main reason why the concept of the mental can be considered quite useful in brain research, because it encompasses all states that affect our self-understanding, our thoughts, ideas, desires, feelings and affects, but also our drives and motivational states.

Historically speaking, there is a synthesis of the *anima animalis* and the *anima rationalis* of the old doctrine of the soul, i.e. everything that moves us motorically, emotionally-affectively and cognitively. The *anima vegetativa*, on the other hand, is realized by those biophysical, biochemical, and physiological processes that constitute the cellular activity of an organism. However, the neurobiological correlates of cognitive and affective (i.e. mental) states and processes accompanied by consciousness, cannot be located in specific brain regions, but rather involve an extensive network of brain centers that are also strongly influenced by bodily organs.

The characteristic of the "spiritual" that is often postulated in dualism, the *immaterial*, has become problematic because the understanding of matter has fundamentally changed in the modern natural sciences. When one thinks of the term "matter," one traditionally thinks of a piece of wood or brain tissue that can be touched, physiologically examined, sliced, and viewed under a microscope. If one has such a concept of matter, then it is of course puzzling how mind can arise from such material things as brain tissue. But modern physics teaches us that everything macrophysical is composed

1

of molecules, atoms, and finally subatomic particles which, like electrons, have only a tiny mass, or even no (rest) mass at all, as is the case with photons. Such elementary particles and their imputed building blocks, be they conceptualized as quarks or strings, create in the theories of many experts what we call matter, space and time.[15] Of whatever physical nature the soul may be, it need not be "material" at all in the sense of the crude nineteenth-century term, that is, something you can touch; you can't touch light, after all. It seems that the psychic must be regarded as a special physical principle to which the traditionally used distinction between the material and immaterial cannot be applied.

From a neurobiological point of view, nothing empirically verifiable can be said about the traditional aspect of the immortality of the soul: it is considered certain in brain research that everything spiritual, insofar as it is felt individually, is inseparably bound to brain structures and processes. A person stops thinking and feeling when his cerebral cortex is no longer supplied with sufficient oxygen and sugar; she then becomes unconscious, and finally cerebral death occurs. The unconscious-mental may continue to exist, but this part of our existence also ceases when the diencephalon and brain stem are no longer active. Then whole-brain death occurs. The question of whether there is or could be a supra-individual immortal soul is beyond the scope of neuroscience.

Summary
After about two thousand years of effort to solve the age-old mind-brain or body-soul problem, modern brain research is in a position to locate several of those processes that underlie our mental states of experience in the brain, and to answer, e.g., the question of how unconsciously occurring processes are related to these states of consciousness. The nature of stimulus conduction and stimulus propagation in the brain and nervous system can also be regarded as largely understood. From today's point of view, the mental is indispensably bound to brain processes. It is subject to the laws of nature (such as the conservation laws) and is influenced by processes that can be investigated and explained by means of the natural sciences. However, these findings have solved the mind-brain problem, brought it decisively closer to a solution or not touched it at all is still a hotly debated question.

Ancient Doctrine of the Soul from Plato to Galenos

Aristotle's reflections have their starting point in the doctrine of three soul faculties, which has dominated Western thought as well as the thought of manwpy other cultures for two and a half thousand years (Florey 1996). They all three derive from the universal world spirit, called God, pneuma or world ether, and are divided into the *anima vegetativa (spiritus vegetativus)*, which gives life to the living ("breath of life, breath"), the *anima animalis (spiritus animalis)*, which moves animals and gives them sensations, and the *anima rationalis (spiritus rationalis)*, which as intellect ("spirit") represents understanding and reason (Aristotle, ibid, 428a, 432a; 434a).

---

15 For an introduction, see Ernst and Heinz 2013, S. 46–54.

The Ventricular Gauges

Following Galenos, in the late fourth century A.D. Nemesios of Emesa developed the idea that the place of mental functions are the brain ventricles. This is followed by the medieval and early modern theory of the ventricles, which goes back to Albertus Magnus (c. 1200–1280): the three (in reality, with the two end-brain ventricles, there are four) ventricles or *cellulae* are arranged one behind the other and filled with pneuma/spiritus (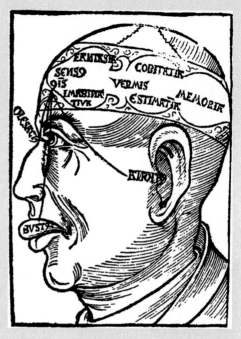 Fig. 1.3). The foremost "cell" is the site of common sense (*sensus communis*), perceptions (*phantasia*), and ideas generated by the soul (*imaginationes*). The second "cell" serves the evaluation of the perceived and the intellectual cognition (*cogitatio* and *aestimatio*), the third memory (*memoria*), in other authors also the will power (*voluntas, volitio*). Between the first and the second "cell" lies a worm-shaped clot which regulates the flow of information in the manner of a valve; anatomically, it corresponds to the pineal gland, which plays an important role in Descartes' work. Spirit in the narrower sense is always constructed as something immaterial and immortal that interacts with the brain in the ventricles (cf. Breidbach 1997). This doctrine persisted into the nineteenth century, as with the neuroanatomist Samuel Thomas Soemmerring (1755–1830), who in 1786 published the important book "On the Organ of the Soul," in which he regarded the ventricles as the "seat of the soul," i.e., the organ connecting body and spirit. The soul itself was also immaterial and immortal for Soemmerring.

◻ Fig. 1.3 A sixteenth century representation of the brain ventricles. Three of the five senses shown here, namely the sense of smell (*olfactus*), the sense of hearing (*auditus*) and the sense of taste (*gustus*), together with the sense of sight (*visus*) and the sense of touch (*tactus*), send excitations to the first ventricle, in which the common sense (*sensus communis*), the imagination (*fantasia*) and the faculty of seeing (*vis imaginativa*) have their seat. In the second cerebral ventricle are located the reasoning faculty (*vis cogitativa*) and the judgment faculty (*vis estimativa*); the first and second cerebral ventricles are separated by the worm (*vermis*) as a kind of valve. Memory (*memoria*) is located in the third cerebral ventricle. Today it is known that brain activity does not take place in the ventricles, but in the brain matter (grey and white matter). (From: Florey in Roth and Prinz 1996)

1

Mechanisms of Excitation Conduction in the Brain and the Beginning of Electrophysiology

An important question in modern philosophy and medicine was how excitation is transported from the sense organs to the brain, or more precisely to the ventricles or other mediating organs. For Descartes, as he detailed in his Traité de l'homme (published posthumously in 1742), the nerves were tubes containing a pith (Descartes 1969, 1993). The pineal gland projects into the ventricles. Via the sensory nerves, the filaments act mechanically on the surface of the pineal gland, causing it to vibrate. Its mirrored surface also serves to "represent" images. The deflection of the pineal gland pushes nerve fluid (*spiritus animalis*) in various ways into the motor tubular nerves, which then inflate the muscles and cause movement. Conscious perceptions and acts of will come about because the immaterial human soul can, on the one hand, perceive what is happening on the surface of the pineal gland and, on the other hand, move the pineal gland (cf. Métraux 1993).

The Current Conception of the Mental or Psychic and Its Neurobiological Basis

The soul, as mentioned at the beginning, is nowadays often replaced by the term mind or psyche, which is actually just a translation of the Greek word *psyché*, but sounds less problematic. Nevertheless, the concept of the mental (or psychological or psychic, in German: "psychisch") also remains ambiguous. In its most general sense, the mental includes everything about and within us that is not purely physical, such as the activity of the heart and liver, not purely motor, such as the lifting of the arm, or that concerns the elementary precursors of our perception, such as the processes in our retina or inner ear. Accordingly, the mental includes:

1. all conscious cognitive performances such as perceiving, recognising, thinking, imagining, remembering and planning;

2. all emotional states understood as intentional, such as joy/happiness, fear, anger, dread, disgust, etc., and

3. all moods, which can be understood as the affective background of experience and may also have an intentional character ("affective intentionality"), even if they are not directed at individual facts but at the general state of mind in the world (Gaebler et al. 2011; Heinz 2014).

The question of whether non-conscious processes also belong to the mental is controversial—most neurobiologists probably agree with the latter view.

The Question of the Nature of the Soul and the Relationship Between Brain and Mind

One classic question, however, remains unanswered: While it is now possible to state with reasonable precision what goes on in the brain when we have conscious mental states, or what neurobiological bases are required for this and how specific modifications of the neurobiological correlates (for example, in drug use) affect mental states such as the feeling of happiness, it has not yet been possible to state plausibly what the nature of these mental states is. The interactive dualism that is still widely popular today considers soul and mind to be something distinct in nature from the (materially understood) brain and, in principle, independent of it. This leads to the basic difficulty of explaining how then mind and soul can affect the brain, and vice versa. In addition to the reasons mentioned at the beginning, the influence of religious ideas (Boyer 2003) and particularily the influence of the major world religions, especially Christianity, is of continuing importance here. Even among theories that view the soul and mind as inseparable from the brain, there continues to be a wide range of positions (for a fuller account, see Walter 1997). These include, for example:

1. Eliminativism: Contrary to what everyday psychology of the soul would have us believe, the mental does not exist in itself, but is nothing more than the firing of certain neurons in certain brain regions (Bickl 2008; Churchland 1986; Crick 1998).

2. the identity theory (reductionism): the mental is identical with brain processes (e.g. Pauen 1999, 2016);

3. functionalism: the mental is a function of the neurobiological, but could also be realized by other physical structures (Fodor 1975)

4. neutral monism (two-aspect theory): neuronal and mental states are two aspects of a third, hitherto unknown state (cf. Vollmer 1975, evolutionary epistemology)

5. non-reductive materialism: the mental is always physically realized in or through the brain, but is in principle not reducible to physical properties. These include among others:

   - anomalous monism, which regards individual mental events (*tokens*) as identical with individual physical events, but does not regard types of mental events as identical with types of physical events (Davidson 1970),

   - property dualism, which views conscious properties as an intrinsic, elemental force within the physical world (Chalmers 1996),

   - various emergence theories, according to which complex mental properties arise or emerge from simple physical properties without being reducible to them (cf. on this Stephan 2016).

6. External materialism: the mental is not confined to the brain alone, but is constituted by processes that extend beyond the brain, namely into the body and the environment (Clark and Chalmers 1998; Walter 2018)

1

The Current State of the Debate

For today's neuroscientists dealing with mental or neuronal processes, the assumption that the mental is a part of nature, thus accessible to physical descriptions and subject to the laws of nature, does not contradict the fact that we will not yet, or perhaps never, know exactly how the mental comes about, what it actually is, and how exactly we should fit it into a scientific worldview. All of this was historically true of electricity and magnetism, as well as other observations that led to the development of relativity on the one hand and quantum physics on the other. Nor does it mean that we may reduce the mental to our understanding of its material, in this case its neurobiological, correlates. Any scientific explanation is in fact shaped by the prevailing paradigms of the time and, if it is to be amenable to experimental testing, complex phenomena must always be narrowed down to empirically testable questions and thus reduced (Ernst and Heinz 2013).

Also compatible with this position is a feature of conscious mental states which still causes trouble for mind-body philosophers today, namely the fact that these states can only introspectively be experience and therefore are more, than can ever be inferred from the observer's perspective ("from the out-

side"). This introspective privilege of conscious experience is also known as the qualia problem (Nagel 1974). Besides its intuitive plausibility, it is philosophically supported mainly by thought experiments such as the zombie argument, what Peter Bieri called the "Tibetan prayer wheel" (Bieri 1994): Thus, in any scientific explanation of conscious experience, the question can always be asked: Could not all this be the case without us having conscious experience? In other words, couldn't we all be zombies? However, there are also good arguments against these thought experiments (cf. Pauen 2016). Moreover, it is true that what can be experienced as different "from within" should, from a neuroscientific point of view, be associated with "externally" observable differences in its neurobiological correlates. The apparently exclusive accessibility of individual conscious experience can also be inferred from the extreme internal wiring and self-referentiality of the cortical and subcortical centers involved (Roth 2003). This topic is also the subject of current interdisciplinary research, such as in the research training group "Extrospection", which questions the epistemic privilege of introspection historically, philosophically and empirically (▶ http://www.mind-and-brain.de/rtg-2386/).

References

Aristoteles (1995) Philosophische Schriften in sechs Bänden. Felix Meiner, Hamburg

Berger H (1929) Über das Elektrenkephalogramm des Menschen. Arch Psychiatr Nervenkrankh 87(1):527–570

Bickl JW (2008) Psychoneural reduction. The new wave. MIT Press, Cambridge, MA

Bieri P (1994) Was macht Bewußtsein zu einem Rätsel? In: Singer W (ed) Gehirn und Bewusstsein. Spektrum, Heidelberg, pp 172–180

Boyer P (2003) Religious thought and behaviour as by-products of brain function. Trends Cognit Sci 7:119–124

Breidbach O (1997) Die Materialisierung des Ichs-. Zur Geschichte der Hirnforschung im 19. und 20. Jahrhundert. Suhrkamp, Frankfurt

Burdach KF (1822) Vom Baue und Leben des Gehirns. Dyk, Leipzig

Caldani LMA (1773) Institutiones physiologicae. Padua

Caton E (1875) The electric currents of the brain. Br Med J 765:278

Chalmers D (1996) The conscious mind: in search of a fundamental theory. Oxford University Press, Oxford

Churchland (1986) Neurophilosophy. MIT Press, Cambridge, MA

Clark A, Chalmers DJ (1998) The extended mind. Analysis 58:7–19

Crick (1998) Was die Seele wirklich ist. Rowohlt, Hamburg

Davidson D (1970) Mental events. In: Foster L, Swanson JW (eds) Essays on actions and events. Clarendon Press, Oxford, pp 207–224

Descartes R (1993) Meditationes de prima philosophia. Paris. Dt. Meditationen über die Grundlagen der Philosophie. Meiner, Hamburg (Erstveröffentlichung 1641)

Descartes R (1969) De Homine/Traité de l'Homme. Dt. Über den Menschen sowie Beschreibung des menschlichen Körpers, Heidelberg (Erstveröffentlichung 1662)

Dubois-Reymond (1848) Untersuchungen über die thierische Elektrizität (1848–1884). Reimer, Berlin

Eccles JC (1994) Wie das Selbst sein Gehirn steuert. Piper, München

Ernst G, Heinz A (2013) Die widerspenstige Materie. Neues aus der Naturwissenschaft und Konsequenzen für linke Theorie und Praxis. Schmetterling, Stuttgart

Florey E (1996) Geist – Gehirn – Seele: Eine kurze Ideengeschichte der Hirnforschung. In: Roth G, Prinz W (eds) Kopfarbeit. Kognitive Leistungen und ihre neuronalen Grundlagen. Spektrum Akademischer, Heidelberg, pp 37–86

Fodor J (1975) The language of thought. Harvard University Press, Cambridge, MA

Frank M (1991) Selbstbewußtsein und Selbsterkenntnis. Reclam, Stuttgart

Gaebler M, Daniels J, Walter H (2011) Affektive Intentionalität und existenzielle Gefühle aus Sicht der systemischen Neurowissenschaft. In: Slaby S, Stephan A, Walter H, Walter S (eds) Affektive Intentionalität. Beiträge zur welterschließenden Funktion der menschlichen Gefühle. Mentis, Paderborn, pp 321–339

Galenos Opera omnia (n.d.) I–XX. Hrsg. von K. G. Kühn. Cnobloch, pp 1821–1833

Gall FJ (1791) Philosophisch-Medicinische Untersuchungen über Natur und Kunst im kranken und gesunden Zustand des Menschen. Wien, Gräffer

Galvani L (1791) De viribus electricitatis in motu musculari. Commentarius, Bologna, Accademia delle Scienze

Haller A (1759–1776) Elementa physiologiae corporis humani. 8 Bde. Dt. Anfangsgründe der Phisiologie des menschlichen Körpers, Bd 8. Berlin

Heinz A (2014) Der Begriff psychischer Krankheit. Suhrkamp, Berlin

Heinz A, Batra A, Scherbaum N, Gouzoulis-Mayfrank E (2012) Neurobiologie der Abhängigkeit. Schattauer, Stuttgart

Kaminski JA, Schlagenhauf F, Rapp M, Awasthi S, Ruggeri B, Deserno L, Banaschewski T, Bokde ALW, Bromberg U, Büchel C, Quinlan EB, Desrivières S, Flor H, Frouin V, Garavan H, Gowland P, Ittermann B, Martinot JL, Martinot MP, Nees F, Orfanos DP, Paus T, Poustka L, Smolka MN, Fröhner JH, Walter H, Whelan R, Ripke S, Schumann G, Heinz A, IMAGEN Consortium (2018) Epigenetic variance in dopamine D2 receptor: a marker of IQ malleability? Transl Psychiatry (epub ahead of press)

Lukrez (2017) Über die Natur der Dinge. DTV, München

Métraux A (1993) Die Mikrophysik der Wahrnehmung und des Gedächtnisses in der französischen Aufklärung. In: Florey E, Breidbach O (eds) Das Gehirn – Organ der Seele? Akademie, Berlin

Morris B (1994) Anthropology of the self. Pluto Press, London

Nagel T (1974) What it is like to be a bat? Philos Rev 83:435–450

Newton I (1872) Philosophiae naturalis principia mathematica. Dt.: Sir Isaac Newton's Mathematische Principien der Naturlehre – Mit Bemerkungen und Erläuterungen herausgegeben von J. Ph. Wolfers. Berlin (Unveränderter Nachdruck Minerva, 1992)

Pauen M (1999) Das Rätsel des Bewusstseins. Eine Erklärungsstrategie. Mentis, Paderborn

Pauen M (2016) Die Natur des Geistes. Fischer, Frankfurt, S. Fischer

Platon Sämtliche Dialoge (2004) Hrsg. von O. Apelt, Bd 7. Meiner, Hamburg

Plessner H (1975) Die Stufen des Organischen und der Mensch. Walter de Gruyter, Berlin

Popper KR, Eccles JC (1982) Das Ich und sein Gehirn. Piper, München

Rocca J (2003) Galen on the brain. Anatomical knowledge and physiological speculation in the second century AD. Brill Academy Press, Leiden, S 34–38

Roth G (1996) Das Gehirn und seine Wirklichkeit, 2. veränderte Aufl. Suhrkamp, Frankfurt

Roth G (2003) Fühlen, Denken, Handeln. Wie das Gehirn unser Verhalten steuert. Suhrkamp, Frankfurt

Roth G, Prinz W (1996) Kopfarbeit. Kognitive Leistungen und ihre neuronalen Grundlagen. Spektrum Akademischer, Heidelberg

Roth G, Strüber N (2018) Wie das Gehirn die Seele macht. Klett-Cotta, Stuttgart

Sartre JP (2014) Das Sein und das Nichts, 18. Aufl. Rowohlt, Reinbek bei Hamburg

Scheler M (1930) Die Stellung des Menschen im Kosmos. Otto Reichl, Darmstadt

Soemmerring GT (1786) Über das Organ der Seele. F. Nicolovius, Königsberg

Stephan A (2016) Emergenz: Von der Unvorhersagbarkeit zur Selbstorganisation. Mentis, Paderborn

Stephan und Walter (Hrsg) (2003) Natur und Theorie der Emotion. (Nature and Theory of Emotions). Mentis, Paderborn

Vollmer G (1975) Evolutionäre Erkenntnistheorie. Hirzel, Stuttgart

Walter H (1997/1998) Neurophilosophie der Willensfreiheit. Von libertarischen Illusionen zum Konzept natürlicher Autonomie. Mentis, Paderborn

Walter H (2018) Über das Gehirn hinaus. Aktiver Externalismus und die Natur des Mentalen. Nervenheilkunde 7(8):479–486

Wundt W (1863) Vorlesungen über die Menschen- und Thierseele. Leopold Voß, Leipzig, S 1863

The Functional Neuroanatomy of the Limbic System

Ursula Dicke

Contents

G. Roth et al. (eds.), *Psychoneuroscience*, https://doi.org/10.1007/978-3-662-65774-4_2

2

In this chapter, we will review the anatomical components of the limbic system and its basic functions. The main structures of the brain will be presented in their morphology and neurochemistry as well as the connections of the limbic structures to each other and their links with motor and sensory-cognitive brain regions. The aim is to get to know the main functions of each limbic region, but also the overlapping and complementary functions with other limbic structures. Feelings, moods and affects are generated in limbic structures, and in complex cooperation with the other brain systems, emotional states, physical reactions, mental states and behavioural expressions are experienced as a unified whole. An important principle here is that of multiple networking and the formation of different networks of the limbic structures with the other brain systems, which act depending on the personal state of mind, the situational context and the external circumstances and requirements.

Learning Objectives

After reading this chapter, readers should know the main structures of the limbic system, be familiar with the respective functions of the limbic regions, understand the connections of limbic networks with the motor and sensory-cognitive systems and the associated control of emotions and behavior.

2.1 The Limbic System

The limbic system was considered the main centre for emotions by the American neurologist James Papez (1937). The reason for this view was the observation that diseases of this system often lead to severe emotional and psychological disorders. Papez included in this system the hypothalamus, including the mammillary bodies, the anterior thalamic nuclei, the cingulate gyrus, and the hippocampus. He considered these structures to be connected in a circle by powerful pathways and thus conceived of what is now called the "Papez circuit." The idea at the time that this circle was self-contained and closed off from the cerebral cortex is now considered to be refuted, even if the basic neuroanatomical features are correct.

The modern conception of the limbic system developed through the contributions of neuroanatomist Walle Nauta, who expanded the limbic system to include areas of the midbrain in the 1950s, and especially through the work of neuroanatomist Rudolf Nieuwenhuys, who included nuclei or areas of the pons and medulla oblongata (Nieuwenhuys 1985; Nieuwenhuys et al. 1991). Nieuwenhuys developed the fruitful concept of the "central limbic continuum" (Nieuwenhuys et al. 1991), which extends from the septum through the preoptic region and hypothalamus to the limbic centers of the ventral midbrain. At the subcortical level, the olfactory and vomeronasal systems, the amygdaloid complex, the pituitary gland, the habenula, and the limbic thalamic nuclei, and at the cortical level, the cingulate gyrus, the hippocampus, the parahippocampal gyrus, and the prepiriform cortex are directly associated with this central continuum (◘ Fig. 2.1).

Terminology and Anatomical Methods
Neurons are grouped together as a nuclear group or nucleus (abbreviated Ncl.) if their cell bodies are close together and if they have the same connections from or to other brain areas, possess the same transmitters, or have other common characteristics. The axonal connections of a nuclear group to another brain structure are called projections. An *efferent* represents output from one nucleus to another; inputs to a nucleus are *afferents*. The location of neurons or nuclear groups, their positional relationships, and directional designations are described in anatomy relative to the body. Thus, neu-

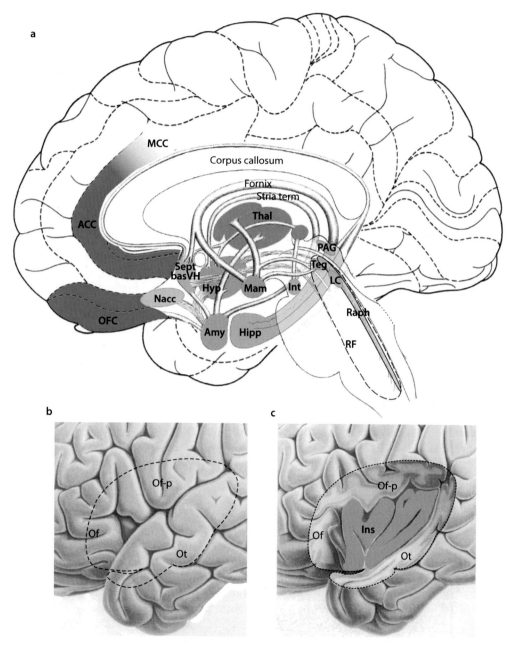

◻ Fig. 2.1 **a** Main structures of the limbic system with fiber connections in a medial view of the right hemisphere (modified after Nieuwenhuys et al. 1988). Rostral is on the left. Internal structures include nuclei in the subcortical telencephalon (nucleus accumbens Nacc, septum and basal forebrain Sept/bas VH, amygdaloid complex Amy, hippocampus Hipp), diencephalon (thalamus Thal with habenula Hab, hypothalamus Hyp with mammillary body Mam), and brainstem (periaqueductal gray PAG, tegmentum Teg, ncl. interpeduncularis Int, locus coeruleus LC, raphe nuclei Raph, reticular formation RF; stria term = stria terminalis). The orbitofrontal cortex (OFC), anterior cingulate cortex (ACC), and insular cortex **b, c** are limbic cortices. Adjacent to the ACC is the midcingular cortex (MCC). **b** Detail of the lateral view of the left temporal, frontal, and fronto-parietal cortex. The dashed line indicates the area showing the underlying insular cortex in **c**. The temporal operculum (Ot), frontal (Ot), and fronto-parietal operculum (Of-p) cover the insular cortex (Ins)

rons are referred to as cells or groups of cells located *dorsally*, to the back or upper side, or *ventrally*, to the front or lower side. Structures located on the midline are median; they are *medial* if located close to the midline or *lateral* if located sideward. Structures located toward the tip of the frontal lobe are called *rostral* or, in humans, cranial, and those located to the back of the brain are called *caudal*. Within a nuclear group, subdivisions are often made representing a front or *anterior* portion, a hind or posterior portion, and a lower or *inferior* portion and an upper or *superior* portion.

In the following, the limbic structures in the rostrocaudal order of the subcortical structures of the end brain (telencephalon), the interbrain (diencephalon), the midbrain (mesencephalon), and the medulla oblongata, followed by the major cortical limbic centers of the prefrontal cortex including the orbitofrontal cortex (OFC), the anterior cingulate cortex (ACC), and the insular cortex. The term brainstem subsumes the midbrain, pons, and medulla oblongata.

The anatomical connectivity (tracer) studies mentioned originate predominantly from experiments on macaque monkeys, only in a few cases data from rodents are cited. Cytological or immunohistochemical studies, as well as connectivity studies using imaging techniques, represent to a lesser extent the findings in humans. The taxon "primate" is used when homologous brain structures of monkeys and humans are presented.

2.1.1 Septal Region

The septum is a thin membrane in the middle of the brain between the two forebrain ventricles; the septal region refers to the adjacent nervous tissue on either side of this wall and borders the left and right ventricles. It is located in the subcortical telencephalon ventral to the corpus callosum and dorsal to the ncl. accumbens.

The septal region consists of different nuclei, which are divided into a *medial* and a *lateral septum*. The medial septum and the Ncl. of Broca's diagonal band (NDB) with a vertical and horizontal "limb" (abbreviated vertical and horizontal NDB, respectively) form the medial septal region. Neurons of the medial septum are cytoarchitecturally similar; they bear few or no *spines*. The largest group of neurons has the transmitter acetylcholine (ACh), others GABA or ACh and GABA, or GABA and the calcium-binding protein parvalbumin; still other neurons glutamate (Frotscher and Léránth 1985; Kiss et al. 1990a, b; Jakab and Leranth 1995; Hajszan et al. 2004). These neurons form a local network.

The *medial* septum receives afferents from the CA1-CA3 region of the hippocampus (▶ Sect. 2.1.3). The medial septal neurons project to the CA1 region of the hippocampus via the powerful fiber tract of the fimbria/fornix. GABAergic and glutamatergic neurons contact GABAergic hippocampal neurons, and cholinergic neurons contact pyramidal cells in CA1. This direct *septo-hippocampal loop* is critically involved in learning and memory; neurons in both areas show rhythmic activity in the theta frequency band (4–12 Hz); theta oscillations accompany voluntary movements, REM sleep, and states of arousal and attention. The cholinergic connection adjusts the excitability of hippocampal neurons in novel or familiar environments, glutamatergic septal neurons are involved in the initiation of locomotion, while GABAergic neurons influence or form theta rhythm and also provide information about the intensity of sensory stimuli (reviewed in Müller and Remy 2018). GABAergic and cholinergic neurons with projection to the hippocampus modulate different aspects of contextual fear as well as pain-related (nociceptive) information. The affective nociceptive component is processed via the projection of the medial septum to the limbic cortex and then experienced as pain. Efferents of the medial septal region exist to the medial prefrontal cortex (mPFC) and to the anterior cingulate cortex (ACC), furthermore to the amygdala, to the Ncl. accumbens and to the spinal cord.

Reciprocal relationships exist between medial septum and lateral, posterior and medial hypothalamus, preoptic region and supramammillary nucleus. Vegetative and endocrine functions are regulated by this axis. Finally, the medial septum has reciprocal connections with dopaminergic midbrain structures as well as with cholinergic, serotonergic and noradrenergic nuclei of the brainstem (▶ Sect. 2.1.5) and receives afferents from the spinal cord.

The *lateral* septum is divided into a dorsal, intermediate and ventral part. The neurons are GABAergic and, in contrast to those of the medial septum, are covered with spines. A variety of neuropeptides as well as steroid hormones are present in the neurons. The lateral septal region, like the medial, is closely associated with the hippocampus and entorhinal cortex; however, these afferents are purely glutamatergic. In addition, the bed nucleus of the stria terminalis (BNST) and the amygdala, hypothalamus, and limbic nuclei of the midbrain and pons project to the lateral septum. Its efferents terminate in limbic cortical and subcortical areas. Projections to the medial and lateral hypothalamus are strongly developed. Efferents also run to the preoptic region, to limbic thalamic nuclei and to nuclei of the brainstem, especially to the central gray, also called periaqueductal gray (PAG).

The lateral septum is involved in emotional-motivational behavior; for example, it controls emotional-cognitive aspects of food intake via its connections to the hypothalamus (Sweeney and Yang 2015; Carus-Cadavieco et al. 2017) as well as exploration and territorial behavior including aggressive responses (Toth et al. 2010; Oldfield et al. 2015). Social fear conditioning (Zoicas et al. 2014) and social interaction in substance use disorders (Zernig and Pinheiro 2015) are regulated by a network between the ncl. accumbens, amygdala, midbrain dopaminergic nuclei, medial and lateral septum. Maternal caregiving behavior is also controlled by the lateral septum (Zhao and Gammie 2014); sexual and reproductive behavior is influenced via a projection of the lateral septum to the PAG (Tsukahara and Yamanouchi 2001; Veening et al. 2014).

The *basal forebrain is* located ventral to the septal region and the BNST and is a cholinergic cell group that extends ventrally to the Ncl. accumbens and dorsally to the amygdala from rostral to caudal. In the basal forebrain, four groups of cholinergic neurons are distinguished: in addition to the Ncl. basalis Meynert (CH4 group), cholinergic neurons are found in the medial septal nucleus (CH1 group) and the NDB (vertical NDB CH2 and horizontal NDB CH3 group). The Ncl. basalis Meynert receives input from limbic frontal, insular and temporal cortical areas, from the septal nuclei, the Ncl. accumbens and ventral pallidum, the amygdala, the hypothalamus and the parabrachial nucleus in the brainstem. In addition to projecting to the hippocampus, CH1 cells project to the interpeduncular nucleus, CH2 cells also project to the hypothalamus and dopaminergic midbrain. The projection of CH3 cells is to the olfactory bulb, that of CH4 cells to the basal amygdala as well as to the isocortex.

Functions of cholinergic projections involve learning and extinction of contextual or stimulus-associated fear responses in cortico-amygdalar and cortico-hippocampal circuits (Knox 2016; Wilson and Fadel 2017), as well as regulation of sleep-wake rhythms (Yang et al. 2017). Cholinergic modulation of attentional processes occurs via "top-down" control of the PFC over sensory cortical areas. In Alzheimer's disease or Parkinson's disease with dementia, degeneration of cholinergic neurons, especially of the CH2 and CH4 groups, leads to attentional deficits, memory loss, language impairment, and, as degeneration progresses, further emotional-cognitive dysfunction (Liu et al. 2015; Ballinger et al. 2016).

2.1.2   Amygdala

The amygdala has historically been considered part of the limbic system, with predominant connections to the hypothalamus and brainstem. However, neuroanatomical studies from the last three decades show that the amygdala forms an extensive network with a variety of cortical and subcortical brain regions. The concept of the *extended* amygdala was developed by neuroanatomists Alheid and Heimer (1988) and includes, in addition to the classical amygdala nuclei, the aforementioned BNST and other nuclei lying between the amygdala and BNST. The extended amygdala complex is a heterogeneous group of nuclei located in the medial temporal lobe rostral to the hippocampal formation. There is little information on the anatomical connectivities of the amygdala in humans. However, the anatomical structure and connectivities of the macaque monkey amygdala are considered homologous to humans.

The following account of the classification of the nuclei of the amygdaloid complex is largely based on work in primates and follows the nomenclature of Freese and Amaral (2009). Up to 13 nuclei and cortical regions belong to the amygdaloid complex. They are divided into a deep and a superficial nuclear group. The deep lying group includes the lateral, basal, accessory basal, and paralaminar ncl; collectively, the lateral, basal, and accessory basal nuclei of the deep group are referred to as the *basolateral group*. The superficial group includes the medial, anterior, and posterior cortical ncl. as well as the ncl. of the lateral olfactory tract and the periamygdaloid ncl.; this nuclear group is also called the *cortical nucleus* without the medial

ncl. or the *corticomedial nucleus* with it. The remaining nuclei are the anterior amygdaloid area, the central ncl. also called the central amygdala, the amygdalo-hippocampal area, and the nuclei intercalares (Freese and Amaral 2009). Central and medial nuclei are also grouped together as the *centromedial group*, depending on the author and species studied.

2.1.2.1   Deep Nuclei (Basolateral Group)

The *lateral nucleus* is subdivided into a dorsal, an intermediate and a ventral part due to the cell density and size of the neurons and its immunoreactivity for the acetylcholine-degrading enzyme AChE. The dorsal lateral nucleus is considered a polysensory part of the lateral nucleus because of its inputs from the sensory cortex. Its neurons project to the more densely located and more AchE-immunoreactive neurons of the ventral lateral nucleus. The *basal nucleus* is divided into a dorsal and caudal magnocellular part with large neurons, an intermediate part, and a small-cell, so-called parvicellular part, which is located most ventrally and rostrally. The flow of information within the basal nucleus is from the dorsally to the ventrally located neurons. The *accessory basal nucleus is* located most medially of all four nuclei of the deep group, and here, too, a more AChE-reactive magnocellular neuronal group and densely packed, strongly AChE-reactive neurons are found in the ventromedial part. The *paralaminar nucleus is* located at the ventral and rostral edges of the amygdaloid complex and is connected to the lateral and basal nuclei.

2.1.2.2   Superficial Nuclei (Corticomedial Group)

The *medial nucleus* lies within the amygdaloid complex caudally. It has a larger proportion of GABAergic neurons. The *posterior cortical nucleus* is also located caudally within the amygdaloid complex. The *anterior cortical nucleus* is rostral to the medial nucleus and is demarcated from the medial nucleus, which has a demarcated layer II, because of its fusion of layers II and III. The *nucleus of the lateral olfactory tract* lies in the rostral part of the amygdaloid complex and is characterized by intense immunoreactivity for AChE. The *periamygdaloid nucleus*, also known as the periamygdaloid cortex, is located superficially medial and extends almost completely from rostral to caudal in the amygdaloid complex.

2.1.2.3 Central Amygdala and the Intercalated Nuclei

The *central nucleus is* located in the caudal half and is divided into medial and lateral divisions based on cytoarchitecture. The medial division is heterogeneous in terms of cell size and density, whereas the lateral division is more uniform in cell size and more densely packed. The prominent feature of the central nucleus is the presence of GABAergic neurons; accordingly, the projections of the central nucleus act predominantly inhibitory. In primates, the *nuclei intercalares* form a continuous inhibitory network of GABAergic neurons that lies between the basal nuclei and extends to the dorsally located anterior nuclei and the central and medial nuclei of the amygdaloid complex. *Spiny* neurons are found in greater numbers than *smooth* neurons.

The *BNST*, together with parts of the amygdaloid complex, is considered to be the "extended amygdala" and shares some similarities in terms of connections and chemoarchitecture (especially the presence of GABAergic neurons) with the central and medial nuclei of the amygdala. In humans, the BNST is divided into lateral, medial, central, and ventral portions and exhibits sexual dimorphism with up to 2.5-fold higher volume in males.

2.1.2.4 Intrinsic Connectivities of the Amygdalar Nuclei

The nuclei of the amygdaloid complex are closely interconnected. The lateral nucleus has projections to all other amygdalar nuclei; those to the basal, accessory basal, and periamygdaloid nuclei are particularly pronounced. The connection of the lateral to the central nucleus of the amygdala is weaker than that of the basal. In addition, the basal nucleus projects primarily to the medial and anterior cortical nuclei. Within the amygdala, the flow of information is generally from lateral to medial. The amygdala is connected to a variety of subcortical and cortical regions via the amygdalofugal fiber tract and stria terminalis. The fibers of the amygdaloid complex gather to form the ventral amygdalofugal fiber tract from rostral to caudal at the dorsomedial edge of the amygdala, while the stria terminalis is formed by fibers ventromedially in the caudal amygdala.

2.1.2.5 Extrinsic Connectivities and Functions of the Amygdalar Nuclei

The connections of the nuclei of the amygdaloid complex are primarily to the limbic cortical areas, to the other limbic subcortical brain structures, and to the thalamus and brainstem. The amydalar complex is involved in a variety of cognitive, emotional and affective-vegetative functions (◨ Fig. 2.2).

- **Olfactory System**
The olfactory bulb and also the piriform cortex send axons to the anterior cortical nucleus, the nucleus of the lateral olfactory tract, and the periamygdaloid nucleus (Turner et al. 1978); the latter two nuclei also project to the olfactory bulb.

- **Connections to Areas of the Cortex**
The majority of the basolateral group receives inputs from numerous areas in the frontal, insular, cingulate, and temporal cortex; extensive projections to a larger number of cortical areas (Brodmann areas, abbreviated BA) also originate from it. In general, the connections of the rostral areas of the cortical areas with the amygdalar nuclei are weaker. The cortical projections reach mainly the basolateral group of nuclei.

2

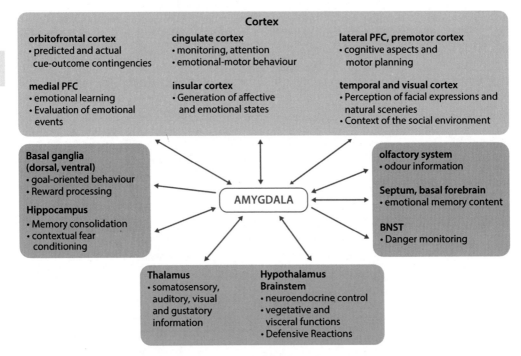

◻ Fig. 2.2 Functional relationships of the amygdala to cortical areas (light green), subcortical telencephalic structures (dark green) and other brain parts (gray-blue). Reciprocal projections are indicated by arrows with two tips; the line width of the arrows indicates the strength of the connection

Inputs from the caudal OFC run stronger and more widely branched in the caudal than in the rostral amygdala. Projections from the OFC (BA 11, 13 and parts of 10, 12, 14, and 24) and from the medial prefrontal cortex (mPFC; BA 32, parts of 9, 10, 14, and 24) extend to the medial nucleus, cortical nuclei, periamygdalar nucleus, and central nucleus in addition to the basolateral group. Laterally located prefrontal areas (BA 8, 45, 46, parts of 9 and 12) as well as the premotor cortex (BA 6) project—although less strongly—to the basal nucleus. In terms of efferents, the basal nucleus of the amygdala has the strongest projection to the OFC and mediolateral cortex and a weaker one to the dorsolateral PFC; again, the terminations are more pronounced caudally in the OFC but also reach the frontal pole. The projection to the mPFC also emanates most strongly from the basal nucleus and less so from the accessory basal and

medial nuclei and from the cortical nuclei. Only the basal nucleus has a projection, albeit small, to the dorsolateral PFC (Amaral and Insausti 1992; Carmichael and Price 1995a; Stefanacci and Amaral 2000; Ghashghaei and Barbas 2002).

The amygdala, together with the OFC and mediodorsal thalamus (▶ Sect. 2.1.5), encodes the specific identity of a predicted outcome (stimuli and/or actions) given an individual's current state (Rudebeck and Murray 2014). The basolateral amygdala thereby provides information about current stimulus-outcome contingencies, while the OFC forms a larger network of past and current associations that the basolateral amygdala can use for future learning episodes (Sharpe and Schoenbaum 2016).

One of the most important inputs to the amygdala is from the insular cortex. Projections from the rostral part of the insula (▶ Sect. 2.2.3) are stronger and reach mainly the lateral, basal, and central nuclei (Freese and Amaral 2009). Except for the central nucleus, these projections are reciprocal. Caudal portions of insular cortex project to a lesser extent to the lateral and central nuclei; a moderately pronounced reciprocal projection originates from the nuclei of the basolateral group. Reciprocal, more pronounced projections also originate from the basolateral group of nuclei of the amygdala with the rostral cingulate cortex (BA 24 and 25). However, inputs also run to a lesser extent to the anterior amygdaloid area and central nucleus.

Reciprocal relationships also exist between the temporal cortex and the basolateral group. Regions around the superior temporal gyrus and sulcus project primarily to the lateral and basal nucleus and, to a lesser extent, to the corticomedial group. Inputs from region TE in the inferior and middle temporal gyrus and the caudally located region TEO of the posterior middle temporal gyrus also reach the basolateral group (Stefanacci and Amaral 2002). Regions TE, TEO, V4, V2, and V1, which form the ventral pathway of visual cortex from occipital to temporal cortex, receive projections from the basal nucleus of the amygdala. Thereby, a rostrocaudal topography exists between the amygdala nucleus and the visual regions (Freese and Amaral 2005). Emotional facial expressions and the perception of natural sceneries activate the amygdala as well as limbic cortex areas (Sabatinelli et al. 2011). Sensory processing of emotional stimuli occurs early in the amygdala and thus may influence subsequent sensory processing in other brain areas (Pourtois et al. 2013). The context of the social environment can also be modulated via this amygdala connection for regulation of social behavior (Adolphs and Spezio 2006).

■ **Connections with the Basal Forebrain**
The basolateral group projects to the lateral regions of the cholinergic Ncl. basalis

Meynert and NDB of the basal forebrain; this projection also runs onward to other subcortical structures. The basolateral nuclei conversely receive powerful projections from the Ncl. basalis Meynert and are functionally involved in memory for contextual fear and extinction of fear. Cholinergic signaling is important in the generation of activity-dependent LTP in the amygdala and contributes to the maintenance of emotional memory content (Ballinger et al. 2016). Furthermore, the central nucleus of the amygdala projects to the Ncl. basalis Meynert, both structures have influence on fear conditioning processes (Knox 2016).

■ **Connections with the Basal Ganglia**
The amygdaloid complex, remarkably, receives no inputs from the striatum (Aggleton et al. 1980), but projects there. The basolateral group sends topographically ordered fibers to the caudate and putamen nuclei and to the ventral striatum and ncl. accumbens, respectively. Neurons of the small-cell portions of the two basal nuclei send their axons to the ncl. accumbens, while the magnocellular portions project to the caudate nucleus and rostroventral putamen, and the lateral nucleus projects to the caudoventral putamen and also to the tail of the caudate nucleus (Russchen et al. 1985).

Cho et al. (2013) distinguish three circuits that run from different cortical areas to the basal and accessory basal nuclei of the amygdala and from there to different regions of the striatum. A "primitive" pathway extends from BA 25 and 32 of the mPFC and from the agranular insula to the basal nuclei and thence to the rostral ventral striatum. This circuit presumably aligns internal emotional states with internal physiology and with motivation via attentional processes. An "intermediate" pathway runs from BA 24 and 14 of the mPFC and from the dysgranular as well as granular insula (▶ Sect. 2.2.3) via the basal nuclei also to the caudoventral striatum, the rostral body of the ventromedial striatum and the caudoventral putamen. This pathway may control responses to social events, as BA 24 processes social contacts, the insula processes tactile stimulation, and the striatum processes facial and eye movements. A "developed" pathway extends from the OFC and BA 10 of the mPFC to the dorsal parts of the basal nuclei. The projection then proceeds to the dorsolateral and caudal body, knee, and tail of the striatum. This latter pathway may be responsible for sensory-guided changes in behavior, as the OFC and BA 10 process more complex cognitive functions such as updating and temporal aspects of behavior, and the striatal portions reached by this pathway also receive information from auditory and visual association cortices.

The amygdala's projection to the ventral striatum is also activated during reward processing, especially when previously rewarded stimuli are attenuated (devalued), or in contexts involving the threat of punishment. Other research finds that salient rather than rewarding events lead to amygdala activation. Responses to rewarding stimuli quickly attenuate in amygdala neurons (as they do to emotional stimuli in general), in contrast to those in the nucleus accumbens (reviewed in Haber and Knutson 2010). Motivational aspects are controlled via a network that includes direct amygdalo-ventrostriatal projections as well as cortico-striato-pallido-thalamic and hippocampo-striatal circuits. Actions and predictive stimuli are associated with the value of subsequent events; this larger network thus ensures adaptive behavior (Zorrilla and Koob 2013).

■ **Connections with the Hippocampal Formation**

Inputs from the hippocampus proper (the hippocampus in the strict sense, ► Sect. 2.1.3) originate primarily in the CA1 region and travel to the basal and cortical nuclei and to the paralaminar and periamygdaloid nuclei. The dentate gyrus does not appear to have direct connections with the amygdala.

The projections from the amygdala to the CA1, CA2, and CA3 regions of the hippocampus are much stronger than those to the amygdala and originate from basal and cortical nuclei, while a projection from the basal and periamygdaloid nuclei also runs to the border region of the subiculum and CA1.

The entorhinal cortex (► Sect. 2.1.3) also projects to the lateral, basal, and periamygdaloid nuclei. In particular, the basal nucleus sends robust projections to the subiculum, para- and presubiculum of the hippocampal formation. Similarly, the lateral nucleus sends efferents to the parasubiculum, but influences the hippocampus proper primarily via a robust projection to the entorhinal cortex. Consolidation of memory content and reinforcement of declarative content of emotional events as well as (contextual) fear conditioning and extinction are important functions based on amygdalo-hippocampal interactions (McDonald and Mott 2017).

■ **Connections to the Thalamus**

Inputs from the thalamus to the basolateral group, medial nucleus, and central nucleus of the amygdala originate from the midline thalamic nuclei, e.g., the paraventricular and paratenial nuclei, which are activated during stressful situations, anxiety, and other affective behaviors. Projections of the thalamus from the Ncl. reuniens, the largest nucleus of the midline thalamic nuclei, and the intralaminar thalamic nuclei run to the medial, cortical, and central nuclei of the amygdala in addition to the basal nuclei. Strong inputs to the amygdala also arise from the nucleus centralis complex of the intralaminar thalamic nuclei (Aggleton et al. 1980; Mehler 1980).

The medial and central nuclei of the amygdala send projections to the reuniens nucleus. The latter has a massive connection to the hippocampus and limbic cortex areas, especially the mPFC (Vertes et al. 2015). The amygdala also projects strongly to the mediodorsal nucleus of the thalamus (► Sect.

2.1.5); the axons of the different amygdala nuclei terminate there separately (Russchen et al. 1987). The mediodorsal thalamic nucleus in turn has strong reciprocal relationships with limbic cortex areas, especially the mPFC, OFC, and insula. Similarly, the medial nucleus of the amygdala sends fibers to the central nuclei of the intralaminar thalamic nuclei, which are involved in attention and sensorimotor functions.

Reciprocal relationships exist between the amygdala and the pulvinar. The pulvinar is located in the caudal thalamus and has a strong connection to the visual cortex and is part of the visual attention system. The central nucleus projects to the pulvinar (Price and Amaral 1981), and the lateral nucleus of the amygdala receives a projection from the medial pulvinar (Aggleton et al. 1980).

■ **Connections with the Hypothalamus**
Strong reciprocal relationships exist between the ventromedial nucleus of the hypothalamus and the basal nuclei, central nucleus, and medial nucleus of the amygdala. The lateral hypothalamic region projects to the medial and central nuclei and to the cortical nuclei. The lateral mammillary nucleus of the hypothalamus innervates the central nucleus, and the supramammillary region innervates the medial nucleus.

The medial nucleus and parts of the cortical nuclei of the amygdala project to the preoptic region, the anterior hypothalamus, and to pre- and supramammillary regions of the hypothalamus. A strong projection of the central nucleus runs to the lateral hypothalamus and mammillary body. Neuroendocrine control, autonomic and visceral control, and defensive responses are major functions regulated by the amygdalo-hypothalamic axis.

■ **Connections with the Brain Stem**
In the central nucleus of the amygdala, dopaminergic projections terminate from the substantia nigra and the ventral tegmental area (VTA) of the midbrain, while the lateral and medial nuclei receive inputs from the peripe-

duncular nucleus. Inputs to the amygdaloid complex also originate from the serotonergic dorsal raphe nucleus and the PAG.

The central nucleus of the amygdala in turn sends projections to the dopaminergic midbrain nuclei, to the PAG and to the dorsal raphe nucleus and the ncl. raphe magnus. Reciprocal projections are also found between the central nucleus, the parabrachial nucleus, and the noradrenergic locus coeruleus. The central nucleus also projects to the reticular formation and to regions of the medulla oblongata up to the cervical spinal cord (Price and Amaral 1981; Amaral et al. 1982; Price 2003). Because the central nucleus has many GABAergic neurons, many of these long-descending projections to the brainstem likely have inhibitory effects. The connections of the central amygdala with the brainstem serve to regulate autonomic, visceral, and motor functions such as respiration, blood circulation, and defensive, avoidance, and escape behaviors.

2.1.2.6 Summary of the Functional Aspects of the Amygdaloid Complex

The amygdala is globally involved in emotional processing and is involved in motivation and memory. Information about external stimuli and the internal state of the organism is integrated by it and exerts an emotional influence on behavior via projection to other subcortical structures. This concerns olfactory signals directly mediated by the olfactory system and sensory information indirectly interconnected via the thalamus, such as taste, sight, hearing, touch. Vegetative centers, cardiovascular and respiratory centers from the brainstem as well as the hypothalamic neuroendocrine centers, visceral states and defensive reactions inform the amygdala about the internal state. The amygdala receives information about aspects of aggression and motivation processed in the septum and basal forebrain. These events and states influence actions, motivation, and reward behavior via projections to the dorsal and ventral striatum, anxiety via projections to the BNST, and learning and memory via those to the hippocampal formation.

The amygdala is strongly associated with the OFC and the mediodorsal PFC. These areas associate changes in the sensory environment relative to an individual's predicted and current state, mediating social signals to the amygdala, which influences social behavior via its projection to these cortices. The connections between the amygdala and insular cortex serve to detect and avoid danger in the environment as well as to regulate vegetative information. The recognition of emotional facial expressions and natural sceneries are integrated into behavior via the interaction between temporal visual areas and the amygdala. The amygdala is active in negative *and* positive emotional processing as well as in aversive and appetitive learning. However, it is unclear which nuclear groups and connections of the amygdala are relevant for appetitive, i.e. reward-oriented, signal processing (Correia and Goosens 2016; Kolada et al. 2017).

2.1.2.7 The Role of the BNST

The BNST is considered an important structure for fear responses in the presence of danger and for danger monitoring; it initiates the stress-relevant HPA (pituitary-hypothalamic-adrenal) axis in stress responses involving the medial PFC. In this context, the BNST appears to be activated in both imagined and actual danger, whereas the amygdala is activated only in the latter situation (Lebow and Chen 2016). The BNST is involved in an extensive limbic network via its connections, so that it is involved in many other functions such as mood, attention, sleep, appetite, but also in social interaction and reproductive behaviour (Lebow and Chen 2016).

2.1.3 Hippocampal Formation

The hippocampus proper is an elongated structure deep in the medial temporal lobe

that resembles a seahorse (Greek hippocampus) in cross-section. It consists of four morphologically distinct subregions: the dentate gyrus (DG), the ammonic horn, Latin cornu ammonis, (CA) with four fields CA1 to CA4, presubiculum, and subiculum (Amaral and Lavenex 2006). The CA4 region is located in the inner curvature of the DG and is also referred to as CA4/DG in the primate brain. Based on neurochemical features, in primates the subiculum complex is divided into a prosubiculum and an actual subiculum (collectively referred to as the subiculum), a presubiculum and postsubiculum (collectively referred to as the presubiculum), and a parasubiculum. The subiculum complex is located between the hippocampus and entorhinal cortex. The three main subicular parts are characterized by different connections and functions. While the subiculum is the main output structure of the hippocampal formation and is involved in encoding and retrieval of long-term memory content, the presubiculum has functions in spatial orientation ("landmark navigation") and, together with the anterior thalamic nucleus, lateral mammillary nucleus, and retrosplenial cortex (BA 29, 30), is a main structure of the head orientation system. The parasubiculum has strong connections to the entorhinal cortex and generates theta EEG activity (reviewed in Ding 2013), i.e., the cells fire at a frequency of 4–12 Hz. Theta frequency oscillation is generated locally in the hippocampus and by the septo-hippocampal circuit. In the EEG, it is measurable in REM sleep and during exploratory behavior in rats.

Based on the size and morphology of glutamatergic pyramidal cells as the major cell types of hippocampal circuits, two main regions, CA1 and CA3, can be distinguished. A trisynaptic pathway extends from the DG to CA1 and CA3; axons of the DG pass through the CA4 region to CA3. The entorhinal cortex, which is properly located upstream of the hippocampus, sends axons via the so-called *perforant pathway* to the granule cells of the DG, whose axons in turn terminate on the CA3 pyramidal cells via the so-called *mossy fiber tract*. The axons of the latter neurons form the *Schaffer collateral pathway*, which runs back to the subiculum and then to the entorhinal cortex. In addition to CA1, the axons of CA3 cells project to other CA3 neurons via axon collaterals, forming a recurrent collateral pathway. The CA3 region is therefore also considered to be an autoassociative, i.e., self-referral, memory system (Yau et al. 2015). CA3 cells also project back to the DG dentatus via excitatory mossy cells, so the pathways are not exclusively unidirectionally organized. In addition, recent data show that the CA2 region, previously considered a transition zone, represents a distinct functional unit equivalent to the CA1 and CA3 regions (Ding et al. 2010). CA2 and CA4 neurons are preferentially affected by degeneration processes in diseases such as chronic traumatic encephalopathy (McKee et al. 2016).

- **Connections of the Hippocampal Formation**

The parahippocampal region adjacent to the hippocampus proper comprises the entorhinal (BA 28), the parahippocampal (temporal areas TH and TF according to von Economo 1929) and the perirhinal cortex (BA 35, 36). The input structure for the hippocampus proper is the entorhinal cortex, which in turn has reciprocal connections with the perirhinal and parahippocampal cortex. The latter two cortices also project to each other. The perirhinal cortex is reciprocally connected with visual (associative) areas such as TE, TEO, and V4; however, it also has direct connections with the CA1 region and the subiculum. The parahippocampal cortex has reciprocal connections with the aforementioned visual cortical areas as well as with parietal and cingulate cortices. The parahippocampal cortex also has a direct input to the CA1 region and the subiculum.

The strongest hippocampal projections originate from the CA1 region and the

subiculum complex; only the projections to the septum and ncl. accumbens also contain information from the CA3 region. Four groups of different efferents can be distinguished. Projections to the retrosplenial cortex, anterior, lateral dorsal, and midline nuclei of the thalamus, and mammillary bodies run almost exclusively from the subiculum; pre- and parasubiculum contribute partially. A second projection originates equally from the subiculum and CA1 and to a lesser extent from the pre- and parasubiculum; it reaches the OFC (BA 11, 13) and the mPFC (BA 14, 25, 32) in the PFC, the amygdala as well as areas TE and TG in the temporal cortex. Efferents of the third group originate from the CA1 region and the subiculum and extend to the entorhinal, perirhinal, and parahippocampal cortex. The fourth group of projections from the CA3 and CA1 regions reach the septum, the vertical limb of Broca's diagonal band, and the ncl. accumbens (Friedman et al. 2002; Aggleton 2012; Aggleton et al. 2012).

The discovery of *place cells* in the hippocampus and their associated *grid cells* in the entorhinal cortex of rats provided a picture of the origin of spatial orientation and the formation of spatial memory and explained spatial orientation deficits following hippocampal lesions. Functionally, studies in humans and animals revealed that the anterior hippocampus is involved within a larger network in *non-spatial functions* such as context coding, attention or reward expectation and the posterior (dorsal in rodents) part in *spatial navigation* or memory of spatial arrangements of a scene (Viard et al. 2011; Nadel et al. 2013).

2.1.4 Basal Ganglia and Mesolimbic System

In the telencephalon, the basal ganglia include the corpus striatum, which is composed of the caudate nucleus, the putamen and the ventral striatum including the accumbens nucleus, and the globus pallidus. In the diencephalon, this includes the subthalamic nucleus and in the midbrain the substantia nigra, which in turn forms a functional unit with the ventral tegmental area. The basal ganglia are divided into a dorsal part with executive and sensorimotor functions, which prepares and controls motor actions, and a ventral part, which has emotional and motivational functions and belongs to the limbic system (for an overview, see Haber et al. 2012).

▪ Nucleus Caudatus/Putamen

The caudate nucleus and the putamen together form a large subcortical structure and are separated from each other by the fiber tracts of the internal capsule and are directly contiguous only in the rostral part. The caudate nucleus lies medial to the putamen, is divided into a head, body and tail, and extends around the putamen from dorsal to ventrolateral.

The majority of neurons in the caudate nucleus and putamen are medium *spiny* cells, densely *spiny* in the middle and distal regions of the dendritic tree, and are therefore referred to as *medium spiny cells*. They possess the transmitter GABA and project inhibitory to the dorsal pallidum, substantia nigra and VTA. The interneurons of the striatum are cholinergic or GABAergic and immunoreactive for a number of neuropeptides and proteins such as neurotensin, enkephalin, somatostatin, substance P, vasoactive intestinal peptide (VIP), neuropeptide Y, calbindin and parvalbumin, and for NADPH diaphorase.

The striatum in primates exhibits compartmentalization into striosomes (also called *patches*) and a matrix (Graybiel and Ragsdale Jr 1978). The striosomes account for 10–20% of the cell mass and are 300–600 μm wide islands with low density of the enzyme acetylcholine esterase (AChE) and high density of opioid receptors, as well as high immunoreactivity for GABA, enkephalin, substance P and neurotensin. In contrast, the intervening matrix has a high density of AChE and strong calbindin and somatostatin immunoreactivity. This subdivision is distinct in the head of the caudate nucleus, where associative and limbic afferents enter, but only weakly evident in the posterior part and in the putamen with sensorimotor inputs. Striosomes specifically receive projections from the OFC, ACC, and insular cortex, while the matrix receives afferents from the entire frontal areas (Eblen and Graybiel 1995). The projections originating from the limbic cortical areas terminate insularly within the striatum; therefore, so-called microcircuits are thought to exist between the different cortical areas and striatum. At the same time, the insular projections of interconnected PFC areas overlap in the striatum, so that intra-cortical projections may also be represented in the striatum.

▪ Ncl. Accumbens/Ventral Striatum

The ncl. accumbens comprises the rostral ventromedial part of the striatum and, together with the rostrally located olfactory tubercle and the ventrally located parts of the ncl. caudatus and putamen, is also considered the limbic part of the basal ganglia. It is anatomically and immunohistochemically subdivided into a ventromedial part, called shell, and a *core*, where the shell forms the dorsal and central part of the ncl. accumbens. The latter region is connected to the ventromedial part of the caudate nucleus and has similar histochemical features. The division into nucleus and shell is evident in primates only by the detection of histochemical markers (Meredith et al. 1996; Holt et al. 1997; Brauer et al. 2000). *Patches of* immunoreactivity for enkephalin or opioid receptors are found throughout the ncl. accumbens/ventral striatum. The shell shows stronger immunoreactivity for neurotensin and AChE and moderate calbindin and strong calretinin immunoreactivity, as well as weak presence of opioid receptors, whereas the core shows strong calbindin and only weak calretinin immunoreactivity, as well as strong presence of opioid receptors. The core region of the ncl. accumbens receives afferents from the OFC (BA 12, 13), whereas the shell region receives projections from the subgenual and pregenual cingulate cortex (BA 25) (Haber et al. 1995). Based on the cortical projections, there is thus a separation, albeit not complete, into an *associative* territory in the medial part of the caudate nucleus and putamen with input from the lateral OFC and dorsal PFC, and a *limbic* territory in the ventral striatum with input from the limbic cortex areas (Buot and Yelnik 2012).

Subcortical inputs to the striatum originate from the basolateral amygdala primarily to the head of the ncl. caudatus and to a lesser extent to the anterior and ventral parts of the putamen; these inputs did not differ with respect to their termination in the striosome matrix portions of the striatum. In the ncl. accumbens, projections from the anterior part of the basolateral amygdala reach the core region, and projections from the posterior part together with those from the central amygdala reach the shell region of the ncl. accumbens. Again, an insular distribution of afferents is found. The hippocampal formation innervates the ncl. accumbens excitatory, further afferents originate from the limbic midline nuclei and intralaminar nuclei of the thalamus.

The Ncl. accumbens-ventral striatum complex is involved in dopaminergic circuits with the midbrain nuclei. Striosomes receiving limbic information project to the dopaminergic neurons of the substantia nigra pars compacta (SNc); the latter form afferents to the striosomes and matrix. Through a series of circuits between the striatum and SNc, ventral striatal regions can influence dorsal striatal regions and relay information between limbic medially located neurons and motor laterally located neurons (Haber et al. 2000). Substantia nigra and VTA connect the motor and limbic parts of the basal ganglia, which, remarkably, have no direct connections to each other.

■ **Globus Pallidus**

The globus pallidus, also known as the pallidum for short, consists of an external and internal segment, which, like the striatum, is divided into a dorsal motor and a ventral limbic part. The latter, together with the ventral striatum/ncl. accumbens complex, forms the ventral striato-pallidal system. The primate ventral pallidum is a crescent-shaped structure whose external segment is rich in enkephalin immunoreactivity and lies ventral to the commissura anterior. The globus pallidus contains predominantly GABAergic neurons that also express calretinin, calbindin, parvalbumin, neuropeptide Y, or somatostatin. The GABAergic neurons are strongly occupied by GABAergic boutons—they are therefore inhibited in turn.

The projections of the ncl. accumbens to the ventral pallidum are topographically ordered. The limbic portion of the pallidum has a ventromedial portion that is rich in neurotensin and receives afferents from the shell region of the ncl. accumbens. This projection from the ncl. accumbens continues to the extended amygdala and lateral hypothalamus. A projection from the lateral part of the shell region and the olfactory tubercle runs to a ventrolateral subregion that is devoid of neurotensin. The core region of the ncl. accumbens projects to a ventrolateral part of the ventral pallidum with strong calbindin immunoreactivity. Cholinergic neurons in the ventral pallidum also receive GABAergic input from the ncl. accumbens, are locally interconnected, and project to the basolateral amygdala and PFC. Efferents from the ventral pallidum to the ncl. accumbens originate equally from the dorsolateral and ventromedial portions of the ventral pallidum and reach the shell or core of the ncl. accumbens. Efferents to the basolateral amygdala originate mainly from cholinergic neurons in the ventral pallidum regulated by *mu, kappa*, and *delta opioid receptors*. Some of these cholinergic neurons also innervate the PFC and entorhinal cortex; however, these do not appear to possess *kappa* or *delta opioid receptors*, so separate cholinergic ventropallidal projections to them appear to exist.

The neurons of the ventral pallidum have dopaminergic D1, D2, and D3 receptors. The dopaminergic inputs to the ventral pallidum are topographically ordered; a projection of the lateral VTA runs to the rostral, ventromedial, dorsolateral, and ventrolateral portions of the ventral pallidum, and that of the midline portion of the VTA to the medial ventral pallidum. Afferents from the substantia nigra are sparse.

The striosome and matrix neurons of the dorsal striatum form a so-called direct pathway through their projection to the internal segment of the dorsal pallidum to the substantia nigra pars reticulata (SNr) and an indirect pathway via the external segment of the dorsal pallidum to the subthalamic thalamus, which from there runs to the internal segment of the globus pallidus and further to the SNr. The majority of neurons projecting to the dorsal pallidum show immunoreactivity for GABA, neurotensin, and enkephalin and possess dopaminergic D2 receptors, whereas the majority of cells projecting to the SNc and SNr show immunoreactivity for GABA, dynorphin, neurotensin, and substance P and have dopaminergic D1 receptors. It is as yet unclear whether the strong efferent projection of the nucleus and core region of the ncl. accumbens also exhibits organization into a direct and indirect pathway.

Further subcortical connections exist to the thalamus and brainstem. Efferents of the ventral pallidum also reach the reticular nucleus of the thalamus. Another projection leads to the lateral habenula, which in turn projects to the mesopontine rostromedial tegmental nucleus (formerly also called the *tail of* the VTA), which is reciprocally connected to the ventral pallidum (Zahm and Root 2017). A projection to the lateral hypothalamus is topographically organized from medial to lateral; the ventral pallidum is also involved in the circuits of the medial preoptic nucleus. Projections exist to the pedunculopontine tegmental nucleus.

■ **Functions of Limbic Circuits**

The different functions within the ncl. accumbens/ventral pallidum complex are outlined in a review by Root et al. (2015) based on pharmacological microinjections into the ventral pallidum. The ventral pallidum is involved in a variety of motor behaviors such as unconscious reflexes, but also volitional actions, learning and memory, or reward-motivated actions. Consummatory behaviors such as food intake, food preference and taste responses, but also caring behaviors are also influenced by the ventral pallidum. Cognitive aspects during sensorimotor filtering mechanisms in the startle reflex, working memory and associative learning are affected in pharmacological microlesion, as well as reward mechanisms during self-stimulation and aversive behavior are regulated by ventral pallidal circuits.

The ventral striatum/ncl. accumbens appears to be more active in impulsive choice behavior than in inhibition of actions to be performed. However, the ncl. accumbens is not solely responsible for the regulation of impulsive behavior. In general, the pattern of neuronal activation in the ncl. accumbens appears to be partly genetically determined, partly learned, and has a strong influence on differences in impulsivity among individuals. Dopamine release by the VTA influences the strength of the neuronal representation and selection of the fronto-temporal limbic input to the ncl. accumbens and, via it, promotes either impulsive or controlled behavior depending on the relationship (contingency) of stimulus and reward, response and outcome (Basar et al. 2010).

Berridge and Kringelbach (2013) localize so-called hotspots in the ncl. accumbens and ventral pallidum, which are active for certain *hedonic aspects* such as liking and pleasure, while *wanting and craving for rewards* is represented in a larger and distributed dopaminergic network. Activation for valence (appetence and aversion) is also localized in the ncl. accumbens, which generates corresponding intense motivation in the presence of positively valenced wanting or negatively valenced fear. Graded mixtures of affective wanting-fear activations are found in microstimulation along a rostrocaudal axis in the medial shell region; namely, from rostral beginning with desire for food to caudal with fearful fear responses. Subjective, consciously experienced pleasure, on the other hand, is encoded in the OFC, where there are also hedonic hotspots (Berridge and Kringelbach 2015).

■ **Functions of the Basal Ganglia**

The dorsal and ventral parts of the basal ganglia and their circuits are involved in motor and limbic functions, respectively. Initial models of cortical projections to the basal ganglia assumed five separate circuits that exist between frontal, oculomotor, dorsolateral frontal, lateral orbitofrontal, and anterior cingulate cortex and distinct regions in the striatum, pallidum/substantia nigra, and thalamus before returning to the cortical output area (e.g., Alexander and Crutcher 1990). However, the ventral striatum also receives inputs from auditory and visual associative parietal areas and from the temporal gyrus, so that a division into a limbic, an associative, and a sensorimotor territory is assumed (Parent and Hazrati 1995). The basal ganglia play a role in response and selection processes of declarative and procedural memory. The ncl. accumbens and olfactory tubercle are involved in unconditioned and conditioned responses. The ventral striatum supports selection and response during instrumental learning. Goal-directed actions and habits are controlled by the dorsal striatum (da Cunha et al. 2012; Liljeholm and O'Doherty 2012).

2.1.5 **Thalamus**

In the thalamus, three major cell masses are classically distinguished, namely so-called relay nuclei, limbic midline and intralaminar nuclei, and associative nuclei (Price 1995; Jones 1998; Groenewegen and Witter 2004). The relay nuclei receive sensory and motor information via ascending modality-specific pathways and project to distinct regions in the cortex. They are also referred to as *specific* nuclei. They include the lateral and medial geniculate complex (LGN, MGN), ventral posteromedial and posterolateral nuclei, a posterior nucleus, a ventral lateral, ventral anterior, and ventral medial nucleus. The *associative* nuclei consist of the mediodorsal, anterior, submedial, and lateral nuclei. They receive inputs from the somatosensory cortex and mediate these to associative cortical areas.

The *limbic* nuclei areas were originally called "nonspecific" because they have wide-ranging projections and could not be assigned a specific function. However, midline and intralaminar nuclei each receive specific afferent projections from the brainstem, and these nuclei in turn project to specific and poorly overlapping regions of the cortex and striatum (Pereira de Vasconcelos and Cassel 2015). The reticular nucleus of the thalamus is a separate complex that inhibits and modulates the thalamic relay nuclei and is under the control of topographically organized afferents from the cortex and thalamus and disseminated afferents from the basal forebrain and brainstem (Guillery and Harting 2003).

The midline nuclei and the intralaminar nuclei are spatially separated, the former along the midline and the latter within a medullary lamina. The midline nuclei consist of the rhomboid nucleus, a periventricular area, the intermediodorsal nucleus, paraventricular nucleus, nucleus reuniens, and paratenial nucleus. The intralaminar nuclei include various central and parafascicular nuclei.

■ **Nucleus Reuniens and Nucleus Rhomboideus**

The largest midline nucleus is the ncl. reuniens. Inputs originate from the medial PFC (mPFC), anterior cingulate cortex, insular cortex, hippocampal formation, medial and anterior amygdala, lateral septum, and parts of the basal forebrain. Thalamic inputs from the reticular nucleus, corpus geniculatum laterale, and hypothalamic nuclei, as well as the brainstem (dopaminergic, serotonergic, and noradrenergic inputs; superior colliculus, periaqueductal gray, reticular formation, and other afferents) also reach this nucleus. Inputs to the rhomboid nucleus from the brainstem also originate from transmitter-specific nuclei and the reticular formation; from the cortex, projections originate from the mPFC and cingulate cortex as well as motor cortices and primary somatosensory cortex (Vertes et al. 2015).

Projections from the ncl. reuniens run to the rostral forebrain, especially to limbic cortices, most strongly to the mPFC. A topographically ordered projection runs to the CA1 region of the hippocampus and to the cortex surrounding it; a smaller proportion of neurons project parallel to the mPFC and to CA1. Efferents from the rhomboid nucleus also reach the ventral limbic frontal cortices, the cingulate cortex, and by extension, parts of the dorsal and ventral striatum, the lateral septum, and the core region of the ncl. accumbens. A dense projection terminates in the CA1 region of the dorsal hippocampus and in the surrounding cortex (Vertes et al. 2015).

The two aforementioned midline nuclei modulate circadian rhythms, eating behavior, and arousal states and are involved in stress and anxiety networks. According to Cassel et al. (2013), there is evidence for the involvement of the two nuclei in cognitive functions. These performances include attention, impulsivity, avoidance memory, (spatial) working memory as well as strategy switching and behavioral flexibility. These cognitive processes are influenced by the strong reciprocal connections of the two intralaminar nuclei with the medial PFC and the hippocampus. Thus, they may be a link between the mPFC and the hippocampus. Together with the ros-

tral intralaminar nucleus, the Ncl. reuniens and Ncl. rhomboideus form a hippocampo-cortico-thalamic network for the consolidation of persistent declarative memory at the systemic level (Pereira de Vasconcelos and Cassel 2015).

Dorsally located midline nuclei such as the paraventricular and paratenial nuclei have strong reciprocal connections to a medial prefrontal network consisting of BA 25, 32, and parts of BA 14 as well as adjacent regions of BA 13 (Hsu and Price 2007). The paratenial nucleus, along with the central intermedial nucleus, is also strongly connected to BA 13 and 12, that is, the orbital and medial PFC. The medial prefrontal network, particularly the subgenual cortex, controls visceral and emotional states and is also involved in anxiety disorders. The strong connection between paraventricular nuclei and subgenual cortex provides a pathway for processing stress signals in prefrontal circuits.

The paraventricular nucleus is the only thalamic nucleus with strong projections to limbic centers such as the amygdala, BNST, and cingulate cortex, which play important roles in fear, anxiety, and reward behavior. Reciprocal connections also exist between the suprachiasmatic nucleus of the hypothalamus, as the brain's circadian pacemaker, and the paraventricular nucleus, which plays a role in arousal states via its dense innervation of orexin-containing neurons (Colavito et al. 2015). Through its connections, the paraventricular nucleus influences important limbic structures that control motivation and mood (reviewed in Hsu et al. 2014), as well as modulating functions related to chronic stress, addictive and reward behaviors through its connections to the medial PFC, ncl. accumbens and amygdala (Colavito et al. 2015). In the limbic thalamus, pain processing is modulated and emotional motor behavior is controlled (Vogt et al. 2008).

In humans, the intralaminar nuclei have been studied in their connectivities with other brain centers using diffusion tensor imaging techniques (Jang et al. 2014). In particular, the PFC and the caudate nucleus of the basal ganglia, the primary motor and premotor cortex, the posterior parietal cortex, the globus pallidus, the basal forebrain and the hypothalamus as well as the reticular formation and the pedunculopontine nucleus in the brainstem are connected to the intralaminar nuclei; the cingulate cortex, however, has only minor connections. The connections are grouped by the authors into so-called "arousal" functions to control arousal states (PFC, brainstem, basal forebrain, and hypothalamus) and, for the PFC, also into an attentional function and into sensorimotor functions (motor cortices, parietal cortex, and basal ganglia).

2.1.6 Hypothalamus

At the base of the endbrain and diencephalon, the hypothalamus, including the preoptic region, forms the middle zone, which is bounded rostrally and medially by the anterior commissure and extends caudally to the ventral tegmental area and periaqueductal gray. The hypothalamus is a bilateral collection of nuclei divided into three longitudinal zones, periventricular, medial, and lateral. In the transverse plane, the medial zone is subdivided from rostral to caudal into a preoptic, supraoptic, tuberal, and mammillary zone (Mai et al. 2016). The hypothalamus is centrally located in the brain and connects to the cerebral cortex via the medial forebrain bundle, to the hippocampus via the fornix, and to the amygdala via the stria terminalis. The thalamus connects to the hypothalamus via the mammillo-thalamic tract, and the brainstem connects via the fasciculus longitudinalis dorsalis; the retino-hypothalamic tract connects the retina and hypothalamus (Bear and Bollu 2018).

The longitudinal periventricular zone has a close relationship to the pituitary gland, and accordingly neurons that secrete "releasing factors" are found in this zone. At the rostral pole of the preoptic region, at the midline, lies the *Ncl. periventricularis*, which extends caudally along the supraoptic and tuberal regions. The periventricular ncl. controls the cardiovascular system and fluid balance and projects to the supraoptic-paraventricular nucleus complex. Attached to the periventricular nucleus laterally is a *preoptic nucleus group*. Ventral to it is the supraoptic region; within it are a *paraventricular nucleus* adjacent to the ventricle and an anterior *lateral hypothalamic nucleus*. A *supraoptic and retrochiasmatic ncl.* join ventrally.

The tuberal region contains a *juxtaparaventricular hypothalamic area*, a *dorsal hypothalamic area*, a *medial hypothalamic group*, an *arcuate ncl.*, a *lateral tuberal ncl.* and a ventrally located *median eminence*. The mammillary region consists of the dorsally located *posterior hypothalamic nucleus*, the posterior *lateral hypothalamic ncl.* and the ventrally located *supramammillary and mammillary nucleus*. Laterally, a *tuberomammillary ncl.* and a *lateral mammillary ncl.* are adjacent. Further caudally, the *retromammillary area* is adjacent (Ding et al. 2016).

In the anterior and tuberal hypothalamus, neuroendocrine neurons are located in the para- and periventricular ncl, the supraoptic ncl and the arcuate ncl. The hormones produced, except for the transmitter dopamine, are peptides such as oxytocin, vasopressin, releasing hormones (corticotropin, thyrotropin, growth hormone, gonadotropin), and somatostatin; these neurohormones are released into the bloodstream. The production of neuroendocrine hormones as well as the production of other non-neuroendocrine neuropeptides in other hypothalamic nuclei is regulated by a complex network of transcription factors (Alvarez-Bolado 2019).

In the preoptic region, temperature regulation and endocrine regulation of sexual behavior are carried out by the preoptic nucleus, while neuroendocrine and autonomic stress responses and secretion of vasopressin and oxytocin are regulated in the hypothalamic paraventricular nucleus. The suprachiasmatic nucleus (SCN) regulates circadian rhythms, and the supraoptic nucleus secretes neurohypophyseal hormones. The tuberal region includes the arcuate nucleus, which monitors food intake, cardiovascular functions, and body adipose tissue, and the dorsomedial nucleus, with control over daily food timing, emotional stress responses, and libido. The ventromedial ncl. of the tuberal region is also involved in the control of food intake, weight loss and gain, fat digestion, and sexual behavior. In the mammillary region, the posterior nucleus organizes sympathetic nervous system responses and defensive and aggressive behaviors, while the tuberomammillary nucleus controls motivated behaviors related to food, fluid, sex, and intoxicants as well as wakefulness, and the mammillary nucleus is also active in encoding episodic memory content (Barbosa et al. 2017).

■ Eating Behavior

The control of the energy balance as well as the storage, use and conversion of nutrients is regulated by the *arcuate nucleus* (Joly-Amado et al. 2014). This is located close to the blood-brain barrier and controls hunger and satiety states via signals circulating in the blood (leptin, insulin, ghrelin). POMC (proopiomelanocortin) neurons of the Ncl. arcuatus reduce food intake and increase energy expenditure, while NPY/AgRP (neuropeptide Y/agouti-related protein) neurons are appetite-stimulating and anabolic. An extensive network controls eating behavior. The Ncl. arcuatus projects to other hypothalamic nuclei, the parabrachial nucleus, VTA and Ncl. solitarius in the brainstem. The latter are connected to limbic forebrain structures such as the ncl. accumbens, which acts as a guardian for the hedonic, or pleasure, value of food (Ferrario et al. 2016). The ncl. accumbens in turn projects to the lateral hypothalamus, which translates food signal-induced motivation into eating behavior;

however, it also appears to be critical for the acquisition of signal-food associations and the retrieval of corresponding memory (Petrovich 2018). The PFC processes cognitive-emotional aspects of food signals for decision making, and dorsal parts of the basal ganglia control motor eating behavior. The amygdala and insular cortex integrate homeostatic, cognitive, and visceral inputs to modulate eating behavior (Andermann and Lowell 2017; Sweeney and Yang 2017). This network controls metabolic rate, endocrine hormone release, and ultimately food intake via the autonomic nervous system, and the hypothalamus also interfaces with motivational and cognitive aspects of eating behavior.

■ Sleep-Wake Rhythm

The lateral hypothalamus regulates the sleep-wake rhythm via neuron populations that produce the peptide orexin/hypocretin (Ox) or the neuropeptide melanin-concentrating hormone (MCH). Ox neurons are active during wakefulness and cause a rapid shift from non-REM (*rapid eye movement*) to REM sleep or from REM sleep to wakefulness during sleep. Loss of Ox neurons (normal is about 70,000 neurons in humans) produces narcolepsy, an excessive sleepiness. Ox neurons have extensive projections in the brain and spinal cord and activate, for example, the dopaminergic VTA, the noradrenergic locus coeruleus, and histaminergic neurons of the tuberomammillary nucleus. Activation of brainstem and basal forebrain cholinergic neurons are also critical for maintaining wakefulness (Schwartz and Kilduff 2015). MCH neurons are also involved in the regulation of eating behavior and increase the duration of REM sleep. MCH neurons also project to all parts of the brain; they appear to inhibit "waking" brain structures. Descending connections affect the generators of REM sleep located in the pons. Together with GABAergic neurons of the preoptic region and brainstem structures, MCH neurons regulate sleep.

The SCN is a circadian zeitgeber (timer) for many physiological and biochemical factors and also for the 24-h sleep-wake rhythm. The sleep-wake rhythm is organized indirectly by the SCN via multiple pathways (Fuller et al. 2006); these reach the lateral hypothalamus via an indirect projection to the dorsomedial hypothalamic nucleus. GABAergic afferents to the lateral hypothalamus come from the preoptic nucleus, lateral septum, basal forebrain, and ncl. accumbens, whereas the ventromedial hypothalamus and PFC act glutamatergically on the lateral hypothalamus. These inputs can have depolarizing or hyperpolarizing effects on Ox and MCH neurons, depending on the synaptic connection (Yamashita and Yamanaka 2017). Functional connectivity between the hypothalamic structures for food and eating behavior and sleep and wakefulness rhythms is also evident in the fact that, for example, leptin, a hormone for satiety signaling, can inhibit Ox neurons, while ghrelin, a hormone signaling hunger, activates Ox neurons.

2.1.7 Limbic Brainstem

In this section, we describe those structures of the limbic brainstem that have important connections with the limbic brain structures already shown (◘ Fig. 2.3).

- **Periaqueductal Gray (Central Gray)**
The periaqueductal gray (PAG) is located periventricularly and extends rostrocaudally from the third to the fourth ventricle in the pons. It is divided into a dorsomedial, dorsolateral, lateral, and ventrolateral longitudinal column (Bandler and Shipley 1994). Like almost all periventricularly located structures, the PAG is characterized by the presence of numerous peptidergic systems; these include fibers and/or receptors for opioids, substance P, neurotensin, somatostatin, neurophysin, oxytocin, vasopressin, VIP, CGRP, neuropeptide Y, and other peptides (reviewed in Carrive and Morgan 2012).

Afferents to the PAG originate from all regions of the PFC except the medial anterior part of the OFC, from the ACC, and from the insular cortex. The projections are topographically ordered; those from the medial PFC run to the dorsolateral column, those from the posterior OFC and anterior insular cortex to the ventrolateral column, and those from the ACC to the lateral, ventrolateral, and dorsomedial columns (An et al. 1998). According to an imaging study in humans, the ventrolateral region is predominantly connected to pain-modulating centers such as the ACC, and the lateral and dorsolateral regions are connected to executive centers such as the PFC and striatum (Coulombe et al. 2016), The network consisting of the PFC and PAG is consequently differentially organized for processing the various motivational and emotional aspects of behavior. Other strong connections come from amygdalar, limbic-thalamic, and hypothalamic nuclei; the latter are predominantly reciprocally connected to the PAG. Projections from the brainstem originate from sensory midbrain structures, from the dopaminergic, noradrenergic, and serotonergic nuclei, the ncl. solitarius, and the parabrachial nucleus. Inputs from the trigeminal nucleus and the dorsal horn of the spinal cord are somatotopically ordered, i.e., projections from the trigeminal system terminate in the rostral PAG, those from the cervical spinal cord in the intermediate and the lumbar spinal cord in the caudal PAG. This arrangement reflects the importance of processing visceral, somatic, and nociceptive inputs in the PAG (Carrive and Morgan 2012).

The efferents of the PAG do not run directly to the cortex, but terminate in the limbic thalamic nuclei, which serve as a relay station to the medial PFC, amygdala, and basal ganglia (Hsu and Price 2009). The reticular ncl. of the thalamus also receives inputs through which the influence of the

2

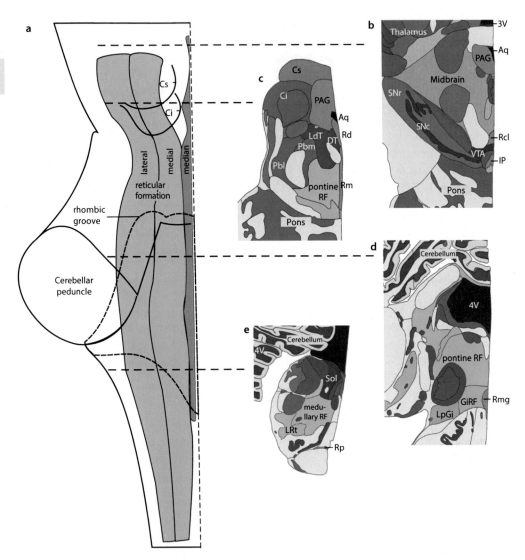

◻ Fig. 2.3   a Schematic dorsal view of the left hemisphere of the brainstem. The overlying cerebellum is not shown. The midline is on the right; top is rostral. **b–e** Semi-schematic cross-sections through the brainstem from rostral to caudal show the gray and white matter. Major nuclei of the limbic brainstem are colored (modified from Ding et al. 2016). The serotonergic raphe nuclei (Rcl caudal linear raphe nucleus, Rd raphe dorsalis, Rm raphe medianis, Rmg raphe magnus, Rp raphe pallidus; orange) are located in the median reticular formation. Rostrally in the basal midbrain are dopaminergic nuclei (SNc substantia nigra pars compacta, VTA ventral tegmental area). 3V/4V third/fourth ventricle, Aq aqueduct, Cs/Ci superior/inferior colliculus, GiRF gigantocellular reticular formation, IP interpeduncular nucleus, LdT/DT laterodorsal and dorsal tegmental nuclei, LpGi lateral paragigantocellular nucleus, LRt lateral reticular nucleus, PAG periaqueductal gray, Pbm/Pbl medial and lateral parabrachial nucleus, RF reticular formation, SNr substantia nigra pars reticulata, Sol nucleus solitarius

thalamic nuclei on the cortex can be modulated. The hypothalamus is an important input structure for the efferents of the PAG as are the reticular formation, the transmitter-specific brainstem nuclei, numerous premotor nucleus areas, and also the spinal cord (Carrive and Morgan 2012).

Due to its connections to the emotion-related forebrain system and to vegetative centers, the PAG is a significant integration point for an ascending pain/body feeling system as well as for a descending limbic-emotional motor system. Overall, the PAG is involved in pain modulation, cardiovascular and other autonomic processes, emotional affective behavior including defensive and panic behavior as well as sexual behavior, emotional vocalization and micturition.

- ## Substantia Nigra and Ventral Tegmental Area

The largest accumulation of dopaminergic neurons is found in the substantia nigra (SN), the ventral tegmental area (VTA), and the retrorubral field. In humans, approximately 600,000 dopaminergic neurons are found in the midbrain, with smaller numbers in the PAG, zona incerta, hypothalamus, olfactory bulb, and retina (group A9–A17; Dahlström and Fuxe 1964). The SN is located ventrally in the midbrain and has an elongated laminar architecture containing GABAergic neurons in a ventral row (pars reticulata) and overlying dopaminergic neurons in a ventral and dorsal row (pars compacta) (McRitchie et al. 1995). The dopaminergic neurons possess the namesake black pigment neuromelanin (Double et al. 2008). The VTA is located in the reticular formation of the midbrain, dorsal and medial to the substantia nigra in the human brain. The VTA is divided into ventromedial and dorsolateral zones; the former zone merges with the retrorubal field. The neurons of the VTA are less dense and smaller than those of the substantia nigra; 50% of the

neurons of the VTA have neuromelanin.

Dopaminergic connections to the caudate ncl. and/or putamen originate from dorsal and ventral SN neurons. A mesostriatal projection originates from dorsolateral neurons of the VTA and from the retrorubral field. The latter also modulates the interaction between midbrain dopaminergic neurons via dopamine. The strong mesolimbic projection of the VTA to the Ncl. accumbens, amygdala, lateral hypothalamus and subgenual limbic PFC as well as to the dentate gyrus of the hippocampus is 75% dopaminergic and 25% GABAergic in humans; a smaller projection also extends to the septum, hippocampus and entorhinal cortex. The mesocortical projection of the VTA is 50% dopaminergic and 50% GABAergic and in primates extends to the dorsolateral PFC, motor, parietal, and temporal cortices in addition to all limbic cortices (Berger et al. 1991).

These compounds are involved in the regulation of a variety of motor and cognitive functions and promote learning and reward mechanisms. Dopamine modulates the membrane states of neurons via different receptor types (▶ Chap. 3). The function of the nigrostriatal projection is primarily movement control, while that of the mesolimbic/cortical projection is the control of emotional behavior and motivation (Halliday et al. 2012). Reinforcement learning, reward seeking, working memory performance, addictive behavior, action drive and motivation as well as hippocampal plasticity in learning processes are processes in which the transmitter dopamine plays an important role. Electrophysiological studies in monkeys show that dopaminergic neurons encode both positive and negative reward expectations (Satoh et al. 2003). Accordingly, they are active not only during reward and positive motivation, but also in non-reward situations such as aversive and alarming events (Bromberg-Martin et al. 2010).

■ **Raphe Nuclei and Locus Coeruleus**

The raphe nuclei are part of the median reticular formation and extend bilaterally along the midline of the tegmentum from the rostral midbrain to caudally just before the junction of the pyramidal tract. The raphe nuclei contain many serotonin (5-HT)-containing neurons. In the primate brain, some of the serotonergic neurons are also located laterally from the median reticular formation. Serotonin is synthesized by approximately 80% of neurons in the rostrally located ncl. raphe dorsalis and up to 10–20% of neurons in the caudal medullary raphe nuclei. The projections of the raphe nuclei have serotonergic and non-serotonergic components (GABA, glutamate, other monoamines such as dopamine). In neurons of the dorsal raphe nucleus, colocalizations of serotonin with substance P, the neuropeptide CRF or galanin are found. The serotonergic nuclei are divided into a rostral (caudal linear ncl, raphe dorsalis, raphe medianus, oral pontine ncl, supralemniscal ncl) and a caudal group (raphe magnus, raphe obscurus, raphe pallidus and neurons of the medullary reticular formation). The dorsal raphe nucleus contains the largest number, approximately 170,000 serotonergic neurons.

The connections of the rostral raphe nucleus group are wide ranging and predominantly serotonergic. The dorsal and median raphe nuclei project to the cortex, striatum, amygdala, BNST, lateral septum, hippocampus, entorhinal cortex, thalamus, SN, and various brainstem nuclei. The median raphe nucleus influences the basal ganglia and hippocampus via direct projections as well as indirect projections via the SN, septum, and mammillary nuclei. The caudal raphe nucleus projects to the caudal brainstem and spinal cord. Afferents to the dorsal and median raphe nuclei originate from the medial PFC, central amygdala, medial septum, NDB, ventral pallidum, lateral habenula, hypothalamic regions, from the PAG and other brainstem nuclei, and from the reticular formation (Hornung 2012).

Raphe nuclei and serotonergic neurons are involved in a variety of functions. These include synaptic maturation and migration of neurons in early brain development, modulation of sleep-wake rhythm, central modulation of pain stimuli along with the PAG, control of motor activity and emotional behavior when processing appetitive or aversive information (Hornung 2012; Hayashi et al. 2015; Luo et al. 2016). Pain-relieving effects are produced serotonergically (tonically activated during stress) or non-serotonergically (during non-stress states) depending on the stress state (Mitchell et al. 1998). The serotonergic system is implicated in affective disorders such as depression, in coping with stress, and also in drug addiction; important control of socio-affective behavior occurs through the ventromedial PFC and the serotonergic system, in part via direct circuits between them (Challis and Berton 2015). The axis between lateral habenula and rostral raphe nuclei is also affected in depression; an important factor in the pathophysiology appears to be impaired serotonin-dependent modulation with hyperactivation of the lateral habenula (Metzger et al. 2017). In general, the synaptic effects in the different circuits are complex due to the multitude of different serotonin receptors (5-HT) and partly also occur via the modulation of other transmitters.

A complex chain of excitation and inhibition involving serotonin is also found in stress processing. For example, in the presence of stressors, corticotropin release factors generated in the hypothalamus can inhibit or excite serotonergic neurons in the raphe nucleus, depending on the activated receptor type CRF1 or CRF2, which then excites or inhibits glutamatergic neurons in the medial PFC via 5-HT1A receptors. The mPFC neurons project back to the raphe nucleus and excite or inhibit GABAergic interneurons there. These in turn excite or inhibit serotonergic neurons, which in turn project to the basolateral amygdala and inhibit or excite GABAergic neurons there.

Via the amygdala, active or passive behaviors to cope with stress can be initiated in this way (Puglisi-Allegra and Andolina 2015). The dorsal serotonergic raphe nucleus is also instrumental in the suppression or execution of aggressive behavior via similarly complex neuronal interactions (Miczek et al. 2015).

■ **Locus Coeruleus**

The locus coeruleus (LC, Latin "blue place") is a highly pigmented nucleus in the dorsal wall of the rostral pons in the lateral floor of the fourth ventricle. It consists predominantly of monoaminergic neurons that synthesize the neurotransmitter/neuromodulator norepinephrine. Polymerization of norepinephrine leads to the formation of neuromelanin, which causes the bluish coloration. This nucleus represents the main source of noradrenergic projections to the other parts of the brain. The LC is involved with attentional and perceptual processing, memory performance, and motivation.

Afferents to the LC originate from the dorsolateral and dorsomedial PFC, the ACC, and the amygdala; the latter terminate on LC neurons expressing corticotropin and thyrotropin-secreting hormones. Hypocretin- and orexin-containing neurons of the posterior hypothalamus innervate the LC, as do neurons of the nociception-processing lamina of the spinal cord and processes of vasopressin-, somatostatin-, and neuropeptide Y-containing neurons. In part, the latter originate from the central amygdala. Serotonin, angiotensin, acetylcholine, and dopamine receptors are located on LC neurons, so it can be assumed that a wide range of subcortical and cortical afferents regulate LC activity, and modulations occur in part via feedback loops (Counts and Mufson 2012). In addition to noradrenaline, LC neurons express a number of neuropeptides such as neuropeptide Y, calcitonin gene-related peptides, cholecystokinin and somatostatin.

Approximately 45,000 neurons in the rostral LC have widely branching axonal projections to the forebrain (Counts and Mufson 2012). Projections run across two axonal systems. A dorsal tegmental bundle extends to the forebrain and a smaller portion of it to local brainstem structures and the spinal cord. Via a rostral periventricular bundle, ascending projections run to the rostral diencephalon, and descending tracts run to sensory-recipient brainstem structures such as the ncl. solitarius. Telencephalic projections of the LC reach the frontal, dorsal, and lateral cortex. The PFC and parietal cortex are moderately innervated, while somatosensory and motor cortices have a dense network of noradrenergic fibers. In the subcortical telencephalon, there are projections to the amygdala, entorhinal cortex, hippocampus, septum, NDB and Ncl. basalis Meynert. Norepinephrine can be released via synaptic contacts as well as extrasynaptically via volume transmission (Aoki et al. 1998). Receptors for norepinephrine (and simultaneously for epinephrine) form different classes of G protein-coupled receptors (▶ Chap. 3).

LC neurons exhibit a tonic and a phasic mode of activity that depend on an individual's activity state. Phasic activation is driven by sensory inputs and is important for alertness and efficient processing of salient information. Accordingly, the LC is considered a monitor for important events that require immediate attention (Berridge and Waterhouse 2003; Sara 2009). In this regard, the LC modulates the formation and experience-dependent changes in memory through cortical and subcortical attentional and memory circuits. The LC also plays an important role in stress and emotional memory. An intact emotional memory requires the LC to interact with the amygdala and hippocampus. Normal aging processes include a loss of noradrenergic cells with a rostrocaudal gradient and averages 30–50% in 70-year-olds. The decreased number of LC neurons with forebrain projection may also have an explanation for decreased attention and memory performance in old age. In Alzheimer's disease, the loss is up to 80% of LC neurons. The reduction of LC neurons is associated with the occurrence of increased cortical plaque formation and neurofibrillary tangles; these phenomena correlate with the onset and progression of AD better than the loss of cholinergic neurons (Förstl et al. 1994; Zarow et al. 2003).

■ **The Reticular Formation**

The reticular formation (RF) is an extensive neuronal network along the rostrocaudal axis from the midbrain to the caudal medulla oblongata. It forms the inner core of the brainstem without conspicuous cytoarchitectonic boundaries; clearly delineated nuclei lie embedded within it. A median zone containing the raphe nuclei is distinguished from a medial large-cell and a lateral small-cell zone (Nieuwenhuys et al. 1991, 2008).

In the *medial RF*, due to the polarity of the cell bodies along a dorsomedial-ventrolateral axis and their main dendrites, an intermediate reticular zone is delineated by gigantocellular and parvicellular reticular nuclei with neurons of different orientations. The intermediate reticular zone contains catecholaminergic cells that have sympathetic cardiovascular and cardiorespiratory functions and control vasopressin release. Neuropeptide Y- and substance P-containing neurons as well as serotonergic neurons are present in this zone. The intermediate zone has ascending connections to the parabrachial region and descending connections to the ncl. solitarius and motor neurons of the phrenic nerve in the cervical medulla (Paxinos et al. 2012).

The gigantocellular ncl. and paragigantocellular ncl. of the medial RF contain serotonergic neurons that are larger than those in the intermediate reticular zone. They are involved in the inhibition of baroreflexes triggered by noxious stimuli in the nociceptive system. Adrenergic and noradrenergic neurons are also located in this zone. Unlike in rats or monkeys, no giant cells are found in the human gigantocellular nucleus. Descending projections of gigantocellular and paragigantocellular subnuclei to the spinal cord control structures of the autonomic system such as parasympathetic nuclei and neurons involved in the organization of reflexes of the pelvic floor (Hermann et al. 2003).

In the *lateral RF* is the parvicellular reticular nucleus with different subnuclei whose neurons are differentially immunoreactive for acetylcholine esterase (AChE). One subnucleus of this reticular nucleus has autonomic respiratory functions, as does the caudal medullary reticular nucleus. The parvocellular part with small compact cells and dense AChE reactivity contains dense neurokinin-containing fibers (Coveñas et al. 2003). In the caudal region lies the medullary reticular nucleus with a dorsal and ventral part, respectively. The dorsal medullary reticular nucleus is activated in response to noxious stimuli and is immunoreactive for opioids, neuropeptides, and monaminergic, catecholaminergic, and serotonergic transmitters/modulators. It is an important part of the pain control system and integrates excitatory and inhibitory inputs in nociceptive processing (Lima and Almeida 2002).

Inputs to the RF originate from many regions of the central nervous system. Cortical inputs originate from premotor, supplementary motor, and primary motor areas and innervate the RF ipsi- and contralaterally (Fregosi et al. 2017). Dorsal medial prefrontal areas and parietal areas of the cortex project to the pontine RF, and a projection from the central amygdala runs to the lateral RF (Price and Amaral 1981; Leichnetz et al. 1984). The cerebellum as well as the ascending spinoreticular tract from the spinal cord project to the RF.

The RF gives rise to descending tracts to the spinal cord and ascending tracts to brainstem and midbrain structures, the hypothalamus, the limbic thalamus, and the striatum (Angeles Fernández-Gil et al. 2010). The *ascending reticular activating sys-*

tem (ARAS) originates in the RF; major components of the ARAS consist of cholinergic nuclei in the brainstem and basal forebrain, noradrenergic nuclei, especially of the locus coeruleus, histaminergic and hypocretinergic hypothalamic projections, and dopaminergic and serotonergic projections from the brainstem and RF (Hobson and Pace-Schott 2002). The ARAS proceeds in a dorsal pathway via the limbic thalamic nuclei (intralaminar and midline nuclei and reticular thalamic nucleus) and in a ventral pathway via cholinergic nuclei in the basal forebrain. Via the ARAS originating from the thalamus, various projections extend to the ventromedial, dorso- and ventrolateral PFC, the OFC, the premotor cortex, the primary motor and somatosensory cortex, and the posterior parietal cortex. The most important function of the ARAS is the regulation of states of consciousness and wakefulness (Paus 2000; Jang and Kwak 2017); the ARAS also regulates related functions such as mood, motivation, attention, learning, memory, movement, and autonomic functions (Zeman and Coebergh 2013).

2.2 Limbic Cortical Areas

The orbitofrontal, cingulate and insular cortex are limbic structures in the frontal brain. They are named below according to the nomenclature of Ding et al. (2016). This study uses imaging, high-resolution cytoarchitectonic, and chemoarchitectonic data mapped to each other in a human brain. Brodmann's (1909) nomenclature of cortical areas is used as the primary reference and modified when necessary using nomenclature from a number of other studies of the human brain.

The entire prefrontal cortex (PFC) is divided into a frontopolar, dorsolateral prefrontal, ventrolateral prefrontal, orbitofrontal and a posterior frontal cortex. The frontopolar cortex (Brodmann area; BA 10) has a medial, lateral, and orbital area. The dorsolateral PFC has a rostral (BA 9), caudal (BA 8), intermediate (BA 9, 46), and rostroventral portion (BA 46). The ventrolateral PFC consists of a rostral (BA 45) and a caudal (BA 44) portion. The posterior frontal cortex is the motor cortex and contains the primary motor area (BA 4, also called area MI or area FA), the premotor cortex (BA 6 or area FB), which has laterodorsal, lateroventral, and medial (area MII) subdivisions, and a transitional area of premotor to cingulate cortex BA 6/32.

2.2.1 Orbitofrontal Cortex

According to Ding et al. (2016), the human orbitofrontal cortex (OFC) is divided into a medial, intermediate, and lateral area. The medial OFC (BA 14) has a rostral and caudal portion, while the intermediate OFC has a rostral, medial, and lateral portion (BA 11) and a caudal portion (BA 13). The lateral OFC (BA 12, 47) again has a medial and a lateral subdivision. The majority of the OFC connectivity studies presented here were conducted in macaque monkeys, whose OFC is divided into a frontopolar region, BA 10, and a lateral and medial OFC with different proportions of BA 11–14 (for a comparison, see Henssen et al. 2016).

The OFC has strong bidirectional connections within the cortex, especially with the other limbic cortical areas. Within the PFC, orbital, lateral, and medial prefrontal areas are intensely connected (Barbas and Pandya 1989; Carmichael and Price 1996). Limbic cortical areas connected to the OFC include the insular cortex, temporopolar cortex, and the medial-temporal hippocampal formation (Morecraft et al. 1992; Barbas 1993). This is indirectly connected to the OFC via the entorhinal and perirhinal cortex and the posteriorly located parahippocampal areas TF and TH and directly via the hippocampus proper (CA1, subiculum complex) (Cavada et al. 2000). The entorhinal cortex and the parahippocampal region, i.e., the temporal pole (TP), the perirhinal cortex, and the posterior parahippocampal cortex, project to the caudal orbitofrontal part of BA 12 and to the caudal half of BA 13 of the OFC (Muñoz and Insausti 2005).

The OFC receives direct olfactory input via the primary olfactory cortex, and indirect input via the anterior insula and entorhinal cortex. Gustatory information enters the OFC via the orbitofrontal operculum and insula (Insausti et al. 1987; Morecraft et al. 1992). These areas receive gustatory and visceral information via thalamic relay nuclei (Carmichael and Price 1995b). Somatosensory inputs originate from primary areas BA1 and 2 and secondary S2 cortex as well as from the insula and parts of parietal area BA 7, where somatosensory inputs from the face and hand are processed. Axons enter the OFC from primary auditory areas and especially from secondary areas, the auditory associative areas and the superior-temporal polysensory area STP (Hackett et al. 1999).

Projections of the OFC run to the anterior temporal lobe (ATL), which includes the temporal polar cortex, the rostral part of the perirhinal cortex (BA 35 and 36), area TE to the tip of the superior temporal sulcus including the anterior border of the superior temporal gyrus. Area BA 13 of the OFC projects to the entire ATL with a weaker projection to BA 35 and 36, and the temporal polar cortex receives a stronger projection from BA 11 (Markowitsch et al. 1985; Moran et al. 1987; Mohedano-Moriano et al. 2015). The influence of the OFC on the ATL is direct, whereas the dorsolateral and ventrolateral PFC appear to influence the ATL via indirect pathways (Mohedano-Moriano et al. 2015).

Subcortical afferents of the OFC originate from the nucleus basalis Meynert, which innervates the OFC as well as other cortical regions in a cholinergic manner (Mesulam et al. 1983). Limbic cortices with projections to the Ncl. basalis Meynert such as the insular cortex or parahippocampal cortices are connected to the OFC (Mesulam and Mufson 1984; Öngür et al. 1998). Intermediate dopaminergic innervation of the OFC originates primarily from the VTA; noradrenergic bilateral innervation also reaches it from the locus coeruleus.

The OFC, as well as the entire medial prefrontal region, is strongly connected to the amygdala. The basolateral, cortical, medial groups, and periamygdaloid cortex project to the OFC, and these projections originate primarily from numerous neurons in the basal and accessory basal nuclei (Porrino et al. 1981; Amaral and Price 1984; Barbas and De Olmos 1990). The OFC projects back to the aforementioned nuclei of the amygdala and also to additional nuclei, such as the paralaminar and central nuclei of the amygdala (Cavada et al. 2000). In particular, the posterior OFC exhibits strong efferents to the amygdala (Ghashghaei et al. 2007), which in turn has a back-projection that excites the posterior OFC. The OFC and amygdala both project to the mediodorsal nucleus of the thalamus. The latter nucleus in turn innervates the posterior OFC (Aggleton and Mishkin 1984; McFarland and Haber 2002; Timbie and Barbas 2015). This tripartite network consisting of the OFC, amygdala, and mediodorsal thalamus can provide information about emotionally significant events and influence higher-order cortex areas that integrate emotional cognitive processes for decision-making and flexible behavior.

Dense projections of the OFC reach the caudate ncl. and putamen, especially ventromedial areas, and extend along a considerable longitudinal extent to the head, body, and tail of the caudate ncl. (Selemon and Goldman-Rakic 1985). The OFC projections are topographically ordered in different areas. Axons of BA 13 project to the central ventral striatum, whereas those of BA 13a, 13b, and 14 terminate primarily in the medial ventral striatum, those of BA 12 in the nuclear region of the ncl. accumbens, and those of BA 11 in both of the above structures. In contrast, projections to the shell region of the ncl. accumbens originate predominantly in the anterior cingulate cortex (BA 32, 25; Haber et al. 1995). Projections from the OFC, as well as those from the other prefrontal areas also examined, appear to terminate predominantly in the striosomes. The topographic order of the projections is also maintained in the projections of the striatal areas to the medial and central globus pallidus, whereas the striatal projection to the substantia nigra is not topographically ordered.

Projections of the OFC to the thalamus, mostly reciprocal, reach the ipsilateral midline nuclei and intralaminar nuclei and, as already described, the mediodorsal nucleus of the medial nucleus group; but also parts of the pulvinar. The OFC has connections to autonomic centers in the hypothalamus and to the lateral and medial preoptic areas, as well as to the zona incerta and brainstem, here especially to the periaqueductal gray of the midbrain, the dopaminergic nuclei, and the interpeduncular nucleus (An et al. 1998; Rempel-Clower and Barbas 1998). The frontal cortex, like the amygdala, can exert rapid influence on autonomic systems responsible for the execution and expression of emotions through these connections (Barbas et al. 2003).

The OFC, with this multitude of connections to other limbic centers and to memory-processing and sensory structures, appears to form a special node where relevant past and present experiences, including their affective and social meanings, are collected and monitored. The OFC occupies an important role in the personality of individuals in the integration of complex memory contents and social adaptations. It involves comparing and processing personal experiences and intentions with contextual external stimuli to produce adapted and rational behavior. In general, the OFC is important in selecting appropriate behavior for the current context (Cavada et al. 2000; Wilkenheiser and Schoenbaum 2016).

The OFC integrates current information to make predictions or estimates about future outcomes. According to the group around Schoenbaum (Schoenbaum and Esber 2010; Schoenbaum et al. 2011), an important significance of the function of the OFC is that it is primarily important for outcome-guided and less for value-guided behavior. According to this view, in collaboration with the amygdala and hippocampus (and other brain structures), the OFC forms an extended network for processing past and present associations and forms new complex multidimensional associations specific to the current state. The amygdala updates information about signal-outcome conditions and accesses the associations of the OFC for use in future learning episodes (Sharpe and Schoenbaum 2016). The hippocampus, in turn, is a highly flexible system for rapidly grasping complex (directly experienced or deduced) features of the environment and encoding important information in such a way that higher-level spatial and relational information is retained. The hippocampus and OFC process parallel, but interactive, cognitive maps that capture the complex relationships between signals, actions, outcomes, and other environmental features. While the hippocampus provides abstract associations, information processing in the OFC concerns the direct biological relevance of events and objects (Wilkenheiser and Schoenbaum 2016).

The OFC, insula, and ACC, as well as subcortical regions such as the ncl. accumbens, ventral pallidum, and amygdala, form the brain's reward network. The anterior OFC seems to particularly register the feeling of subjective pleasure in different contexts (Gottfried et al. 2003; Grabenhorst and Rolls 2011; Rolls 2012) and is also active during sexual pleasure, the euphoric effects of drugs or music. An increase in the feeling of pleasure is regulated by opioid or orexin receptors in the hotspots of the OFC (Castro et al. 2014).

Koelsch et al. (2015) distinguish a total of four affective systems, namely a brainstem, a diencephalon, a hippocampus and an OFC-centered affective system. The OFC holds a number of functions, such as the integration of sensory information with stored memory content, decision-making, and preferences, which serve to motivate or inhibit a particular behavior. The OFC ensures that "somatic" markers in the sense of Damasio (1994), i.e. physical accompanying states, are used in this process. It modulates endocrine and vegetative processes in the hypothalamus and brainstem that contribute to the subjective feeling. Furthermore, the OFC flexibly processes rewards and punishments and generates "moral" affects based on representations of social norms and conventions. According to the authors' theory, the functions of the OFC in this context occur rapidly, automatically, and unconsciously. In contrast, the specific role of the OFC in states of consciousness is unclear. EEG findings in risk decision-making in humans show that OFC neurons encode a higher reward probability after 400–600 ms in response to cue stimuli, the OFC 1000–2000 ms after the decision is made produces a risk signal in the reward expectancy phase, and an evaluation signal 0–800 ms after the reward (Li et al. 2016). Conscious information processing is also likely to occur in this time window.

Lesion studies in macaque monkeys revealed that the lateral and medial OFC play different roles in choice behavior in uncertain and ambiguous situations, respectively. While the medial OFC tends to focus attention on the relevant decision variables for achieving a goal, the lateral OFC is required for rapid learning in fluctuating environments by associating a particular outcome with a particular decision (Walton et al. 2011). People with damage to the orbitofrontal and ventromedial cortex show weaknesses in decision making toward food, objects, object features, landscapes, or living things. Damage in the OFC appears to impair the accuracy of affective, value-based decision-making, without a large effect in the reaction time of a decision—nor impulsive action (Fellows 2011).

The functions of the OFC proposed by researchers were evaluated by Stalnaker et al. (2015), taking into account competing ideas and contradictory findings. Accordingly, response suppression, flexible associative coding, somatic markers for emotional states as well as value-based signal processing are *not* adequate explanations of the function of the OFC as a sole process. However, some of these functions such as response suppression, value calculation, error indication, or even their assignment to specific causes may be necessary in the formation of a so-called cognitive map in the current decision situation. A cognitive map contains the essential features of the current information and labels the current task state (Wilson et al. 2014). This shows that functions of the OFC clearly overlap with those of other limbic cortical areas.

2.2.2 Cingulate Cortex

The cingulate cortex is located in the medial wall of the cerebral hemisphere adjacent to the corpus callosum and is divided into anterior, middle, and posterior regions. These are divided into rostral and caudal as well as dorsal and ventral subregions.

In the *anterior* cingulate *cortex* (ACC), ventrodorsal BA 24 and a subcallosal or subgenual part BA 25 (lying below the corpus callosum or its "knee" or "genu") and dorsorostral BA 32 are distinguished. A ventral limbic series occupies the surface of the cingulate gyrus and contains BA 24a and 24b, subcallosal BA 25, and callosal part BA 33. A dorsal paralimbic row lies deep in the cingulate sulcus and corresponds to BA 24c and 32.

The *middle cingulate cortex* (MCC) is divided into an anterior part, again with a dorsal anterior (BA 32' and 24c'), a ventral anterior (BA 24a' and 24b') and a callosal part (BA 33'). The posterior portion of the MCC includes BA 24c' and 24d located in the cingulate sulcus (Vogt et al. 2006; Vogt 2016). The cingulate motor areas (CMA) include rostral BA 24c, dorsal BA 6c, and ventral BA 23c.

The *posterior cingulate cortex* (PCC) includes BA 23a-c, the dorsally located BA 31. Ventral to these areas and closely associated with them is the retrosplenial cortex (BA 29 and 30; Vogt and Palomero-Gallagher 2012).

- **Inputs of the Cingulate Cortex**

The anterior cingulate areas BA 25 and 24 are reciprocally connected and also receive inputs from the PCC. Both are innervated by the dorsolateral PFC and OFC and receive projections from the amygdala, hippocampus, and superior temporal sulcus. While BA 25 is also innervated by the superior temporal gyrus, BA 24 receives inputs from the temporal pole, parahippocampal areas, and the insula. Posterior BA 23 is connected to BA 24, is innervated by the dorsolateral PFC and OFC like the latter, but also receives inputs from the parietal and occipital cortex. The temporal inputs also originate from the superior temporal sulcus and parahippocampal areas (Vogt and Pandya 1987).

- **Outputs of the Cingulate Cortex**

The most rostrally located BA 32 projects to the lateral PFC, the middle OFC, and the rostral part of the superior temporal gyrus. The anteriorly located area BA 24 innervates premotor cortical regions, the OFC, the rostral inferior parietal lobe, the anterior insula, the perirhinal cortex and the basolateral amygdala. The posteriorly located area BA 23 sends efferents to the dorsal PFC, rostral orbital cortex, parieto-temporal cortex (posterior inferior parietal lobe and superior temporal sulcus), and parahippocampal areas (Pandya et al. 1981). Medial and rostral areas 25 and 32 of the ACC send strong projections to all areas of the hypothalamus (Öngür et al. 1998).

BA 24c (also referred to as M3 in monkeys), located rostrally in the ACC, and BA23c (M4), located ventrally in the PCC, are characterized by strongly different patterns of connections from thalamic and intracortical inputs (Hatanaka et al. 2003). The efferents of both areas are equally somatotopically ordered to the primary and supplementary motor cortex (M1 and M2 in the macaque brain). Neurons located anteriorly in BA 24c and in BA 23c project to the face area in M1 and M2, neurons located in the middle part send their axons to the anterior limb area, and posteriorly located neurons to the posterior limb area (Morecraft and Van Hoesen 1993). Consequently, frontal associative and limbic areas have direct access to the part of the corticospinal projection arising from the middle cingulate (motor) cortex via the cingulate cortex. Rostral regions of BA 23c and 24c project not only to the face area in M1 and M2, but also parallel to the facial nucleus in the pons. BA 23c and 24c each contain their own face representations and directly influence facial expressions via efferents to cortical and subcortical centers (Morecraft et al. 1996). The efferents of the cingulate motor cortex partly extend into the cervical medulla (Dum and Strick 1996), and at the same time these areas receive information from the spinothalamic tract via thalamic relay nuclei (Dum et al. 2009), so that these areas are involved not only in sensorimotor functions but also in pain processing.

The basolateral group and cortical nucleus of the amygdala project to BA 24c (M3). The medial temporal lobe influences facial expressions and, to a lesser extent, arm movements through this amygdalar pathway to the cingulate cortex (Morecraft et al. 2007). The primary motor cortex, ventral lateral premotor cortex, and supplementary motor area control *voluntary* facial expressions, while the cingulate cortical areas regulate *involuntary* emotional facial expressions due to their connection with limbic brain structures (Müri 2016).

Kunishio and Haber (1994) examined the projections of the cingulate cortex to the basal ganglia. Overall, medial regions of the cingulate cortex (BA 24c, 24c', 23c) project to the ventral striatum and to the shell region of the ncl. accumbens, whereas the lateral regions and the fundus of these areas project to the dorsal sensorimotor striatum. Thus, the fundus of the cingulate sulcus is involved in skeleto-motor functions through its connection to the dorsal striatum, whereas the medial region of the cingulate sulcus is involved with limbic and associative cortical functions.

Overall, the cingulate cortex has very extensive connections with limbic, parieto-temporal, and frontal associative areas (Morecraft et al. 2004). However, the ACC is also strongly connected to auditory associative areas and has a distinct connection to the amygdala as well as to a number of autonomic motor systems (Barbas et al. 1999; Öngür et al. 1998; Ghashghaei et al. 2007; García-Cabezas and Barbas 2017). ACC and posterior OFC are also strongly associated (Barbas and Pandya 1989; Cavada et al. 2000). A comparison of the strength of inputs and outputs reveals that the ACC has stronger projections to the amygdala than it receives from it; this is reversed for the posterior OFC (Ghashghaei et al. 2007). Due to its connections, the ACC shows strong involvement in the regulation of emotions and autonomic responses, whereas the functions of the MCC are in decision-making and skeleto-motor control (Vogt 2016). The ACC is active in attentional control and task-switching, and is particularly active when cognitive demands are high (Bush et al. 2000; Botvinick 2007). The ACC and area BA 10 are activated during mental tracking tasks (Burgess et al. 2007), so the network may be involved in maintaining concentration and, via it, supporting working memory and problem-solving.

The above findings are based primarily on studies in macaques. In humans, fMRI connectivity studies assume rostral and caudal as well as dorsal and ventral subregions of the ACC; these studies suggest functionally distinct networks (Margulies et al. 2007). The anterior MCC is active during the planning and execution of motor functions and also evaluates the outcome of an action via feedback detection (Picard and Strick 2001; Amiez and Petrides 2014; Procyk et al. 2016). Affective pain perception also occurs in the aMCC (Büchel et al. 2002; Rainville 2002). Increased activation of the aMCC and supplementary motor areas is also found when motor control and pain processing occur simultaneously (Misra and Coombes 2015). The cingulate cortex is also seen as an interface for the translation of intention and motivation into action: This explains the diversity of functions of the cingulate cortex (Paus 2001).

2.2.3 Insular Cortex

The insular cortex (also called insula) is located in the lateral wall of the cerebral hemisphere. It is deeply recessed between the frontal, parietal and temporal cortex and is externally overlaid by the operculum (◘ Fig. 2.4). It is divided into dysgranular (Idg) and granular (Ig) insular cortex, each with rostral and caudal parts, and agranular insular cortex (Iag), which is divided into frontal (FI) and temporal (TI) parts. In primates, from ventral to dorsal, the insular cortex consists of a rostroventral agranular, a caudodorsal granular, and a broader intermediate dysgranular zone. Granular and agranular refers to the presence or absence of an internal granular layer IV characterized by the presence of small cells called granule cells. The intermediate dysgranular zone has fewer granular neurons and incomplete laminar differentiation (Mesulam and Mufson 1985; Friedman et al. 1986). In a subregion of the agranular insula (FI) and dorsal to the anterior dysgranular zone are the often-cited, spindle-shaped large "von

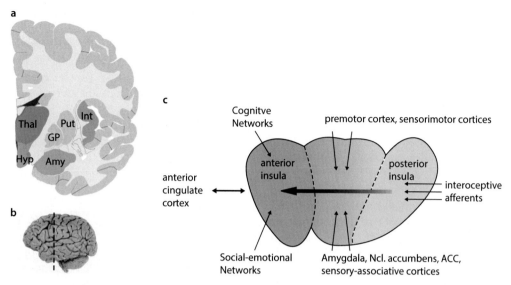

◻ Fig. 2.4 a Semi-schematic representation of the location of the insula in one hemisphere of the human brain (modified after Ding et al. 2016). The insular cortex (Ins) is covered by the parietal and temporal operculum. Amygdala; GP globus pallidus; hyp hypothalamus; putamen; thal thalamus. Gray lines in the surrounding cortex separate functionally distinct cortical areas. **b** Lateral view of the human brain. The dashed line indicates the location of the cross-section in **a**. Rostral is left and caudal is right. **c** Schematic representation of the insular cortex in a lateral view after removal of the opercula (modified after Nieuwenhuys 2012). According to Craig's (2010) model, information processing within the insula progresses from posterior to anterior. Visceral states from the body are primarily processed in the posterior insula. Limbic, sensory, and cognitive influences are integrated in the interoceptive processing pathway to the anterior insula. The anterior insula encodes a "global emotional awareness" and works closely with the anterior cingulate cortex (ACC). The insular cortex generates emotions and (physical) awareness and the ACC motivation and self-efficacy. Together they form the core network for adaptive homeostatic control of the body and brain

Economo" cells (named after their discoverer von Economo), which are thought to be an important part of circuits for cognitive, complex social functions, and consciousness. However, these cells, e.g. also described in the ACC and dorsolateral PFC, have been found not only in primates with larger brains, but also in brains of other larger and smaller mammals with very different cognitive abilities, so that the actual function remains unclear.

The insular cortex receives inputs from the rostral, orbital and dorsolateral PFC, from regions in the parietal and temporal lobes, from the ACC, the olfactory system, the entorhinal cortex and the basal ganglia. Limbic structures such as the entorhinal cortex (BA 28), perirhinal areas (BA 35, 36), posterior orbitofrontal areas (BA 13, 14), temporopolar (BA 38) and cingulate (BA 23, 24) cortex, and the amygdaloid complex are strongly and reciprocally connected to the anterobasal region of the insula (Mesulam and Mufson 1982a, b; Mufson and Mesulam 1982; Augustine 1996). This limbic part also receives inputs from the superiorly located insular somatic associative cortex. Therefore, the limbic insula is considered a somatolimbic integration site where events in the extrapersonal environment are linked to motivational states. The efferents of the insular cortex to the amygdala as well as those to the limbic areas of the perihippocampal cortex are more pronounced than vice versa. Overall, a wide range of cortical and subcortical limbic connections of the insula are found, as evidenced in macaque monkeys by tracing studies and in humans by imaging techniques (Ghaziri et al. 2017).

Efferents of the insular cortex run to the premotor cortex and the ventral and dorsal striatum. The agranular and ventral dysgranular zones project to the shell of the ncl. accumbens and to the medial ventral striatum, while the more dorsal and posterior parts of the agranular and dysgranular insula innervate the central ventral striatum. The dorsal dysgranular and granular insula project predominantly to the dorsolateral striatum (Chikama et al. 1997). Generally speaking, somatosensory information from the dorsal granular and dysgranular zone reaches the dorsolateral striatum, and limbic information about reward and memory content reaches the ventral striatum for feeding behavior. The agranular and rostral dysgranular parts of the anterior insula integrate sensory and amygdalar inputs and project to the caudal ventral striatum, which also receives projections from the amygdala. The posteromedial, lateral, and posterolateral portions of the agranular insula, which processes olfactory, gustatory, and visceral information, show particularly dense innervation of this striatal region (Fudge et al. 2005). After lesions of the insula, for example, conditioned feeding aversions are no longer maintained. Responses of the insula to visceral negative and positive stimuli can thus influence behavior organized by the caudal ventral striatum.

- **Sensory Processing**

Inputs entering the insula transmit gustatory, visceral, nociceptive, thermoreceptive, vestibular, somatosensory, and olfactory information/sensory stimuli. A *primary gustatory cortex (GI)* in the granular anterosuperior part of the insula is distinguished from a *secondary gustatory cortex* underlying the GI in the dysgranular part of the insula. Posterior to the GI is the *insular viscerosensory cortex*, which processes general

visceral information. Pain and temperature pathways terminate in the postero-superior insula, the *insular nociceptive and thermoreceptive cortex* (Brooks et al. 2005; Craig and Zhang 2006). Caudal to the latter, the postero-superior part contains the *parieto-insular vestibular cortex*, which receives information from the vestibular system. In the postero-superior insula, afferents from somatosensory, vestibular, and auditory cortices terminate in the *insular somatic associative cortex* (Nieuwenhuys 2012). The primary olfactory cortex, gustatory and viscerosensory insula project to the agranular anterior zone, which processes food-related information (Carmichael and Price 1996).

The visceral area of the insula processes taste information, regulates eating behavior, and has visceral motor autonomic functions (Augustine 1996). The somatosensory area of the insula plays a role in tactile perception, the pain-asymbolia syndrome that develops after stroke or trauma, i.e. the absence of normal pain-related motor and emotional responses and a sense of suffering despite pain recognition, and the thalamic pain syndrome, in which burning pain or cold pain occurs due to the disruption of thalamic connections.

In humans, four qualitatively and topographically different functional regions were described after electrical stimulation of the insula. A somatosensory representation was found in the posterior part, a representation of temperature and pain in a posterior superior part, and one for viscerosensory sensations anterior to the somatosensory representation. In a centrally located part of the insula, taste representations were localized, whereas no sensations were elicited upon stimulation of the anterior insula (Ostrowsky et al. 2002; Stephani et al. 2011).

- **Emotional and Cognitive Processing in the Insula**

The anterior insula and especially its anterior basal region are involved in the processing of emotions and empathy (Kurth et al.

2010; Nieuwenhuys 2012). This involves the recognition of the emotional meaning of stimuli such as the (re)recognition of emotion in facial expressions and the subsequent generation of an affective state based on this (Phillips et al. 2003). These processes are regulated by a ventral system consisting of the amygdala, insula, ventral striatum, ventral ACC, and PFC. A dorsal system consisting of the hippocampus, dorsal ACC, and PFC regulates and modulates the affective state through cognitive aspects.

The empathic feeling for the pain of others is primarily accompanied by an activation of the anterior insula. Pain experiences are segregated in sensory-discriminative and vegetative-affective attributes (Singer et al. 2004). The activation of a core network consisting of the anterior insula and ACC reflects the emotional component that gives rise to our responses to pain and provides the neural basis for recognizing the feelings of others and our own feelings (Lamm et al. 2011). In this context, the insula is also activated in the prediction of emotional states, it enables error learning in emotional states and emotional uncertainty, and it modulates individual preferences in risk avoidance and contextual appraisals (Singer et al. 2009; Lamm and Singer 2010; Bernhardt and Singer 2012).

Decety and Michalska (2010) used imaging techniques to study children and adults while they observed a subject either experiencing an involuntary, self-inflicted, painful minor accident in everyday life (sympathy situation), behaving in a non-painful everyday situation, or experiencing pain intentionally induced by another (empathy situation). In all subjects, the perception of pain in the sympathy situation produced similar activations in the ACC, somatosensory cortex, periaqueductal gray, and insula, whereas the empathy situation activated different PFC regions. In the empathy situation, a stronger negative arousal was found in children than in young adults. This suggests a developmental aspect.

The insular cortex is also involved in reward processing. Thus, when listening to music, there is an activation of the mesolimbic system (Ncl. accumbens, VTA), the OFC, the hypothalamus and the insula. Vegetative and physiological adaptations in pleasant, rewarding and emotional situations are regulated by the latter two structures (Menon and Levitin 2005).

- **The Functional Network of the Insula**

The insula is considered a major integrative cortex where multimodal information from somatosensory-limbic, insular-limbic, insular-orbital-temporal networks and from the axis between the PFC, basal ganglia, and basal forebrain converge (Shelley and Trimble 2004). Within the insula, homeostatic states appear to be connected to information from the sensory environment and to motivational, hedonic, and social information from different brain regions in a stepwise manner from posterior to anterior (Craig 2010).

The anterior insula is also a functionally complex area in humans, in which dorsal regions are involved with auditory-motor integration, while the ventral region is connected to the amygdala for the regulation of physiological parameters of emotional states (Mutschler et al. 2009). Cauda et al. (2011) used imaging techniques to investigate the connectivity of the insula in humans at rest and describe two complementary networks. In the first network, the ventral anterior insula has preferential connectivity to the middle and inferior temporal cortex and the ACC. This network controls arousal and attentional states that play a role in processing emotional stimuli and situations. In the second network, the middle posterior insula is connected to premotor, sensorimotor, supplementary motor cortices and the posterior cingulate cortex, has a more pronounced right-sided connection with the superior temporal and occipital cortex, and is functionally involved with sensorimotor integration.

2

Summary

In the endbrain, the limbic system comprises on the one hand the orbitofrontal cortex (OFC), the anterior cingulate cortex (ACC) and the insular cortex (insula), and on the other hand subcortical parts such as the septal region, the amygdaloid complex, the hippocampal formation, the habenula, parts of the basal ganglia and the mesolimbic system. The diencephalon contains limbic thalamic nuclei and the hypothalamus, and the midbrain contains the periaqueductal gray (PAG) with important limbic functions. The limbic brainstem contains transmitter-specific nuclei such as the dopaminergic ventral tegmental area (VTA), the serotonergic raphe nuclei, and the noradrenergic locus coeruleus. Other important limbic regions of the brainstem include the parabrachial nucleus, the nucleus solitarius, the reticular formation, and other smaller nuclei.

Functions of the limbic system involve emotional perception, evaluation, and behavioral control, which influence the cognitive and executive performances of the brain. Error detection and control, recognition of the emotional components of gestures, facial expressions, posture and language, learning and memory formation, as well as problem solving, action planning and attentional control, are substantially regulated by a limbic extensive network in the telencephalon. This consists of the OFC, ACC, medial septum, hippocampus, basal forebrain, amygdala, lateral habenula, ventral pallidum, and ncl. accumbens, and is closely intertwined with subcortical structures in the hypothalamus, transmitter-specific nuclei, and the reticular formation.

Emotional conditioning of behavior and motivational behavior control are also controlled by a functional unit consisting of the limbic cortical areas and especially the amygdala, ventral pallidum, and basal forebrain. These are interconnected in part by limbic thalamic nuclei and modulated by dopaminergic and serotonergic brainstem nuclei.

Affective states controlled by the limbic system include flight and avoidance behavior, defense, and attack, which are regulated by the amygdala, lateral septum, hypothalamus, PAG, and several nuclei in the brainstem, including transmitter-specific ones.

Control of feeding, reproduction, or sexual and caring behavior is by the lateral septum and ventral pallidum, hypothalamus, and PAG, whereas stress regulation and pain processing are controlled by the insula, ACC, hippocampus, limbic thalamus, and PAG; again, modulation by transmitter-specific and other brainstem nuclei.

The basic vegetative functions of the body, which include respiration, circulation, metabolism, digestion, hormonal balance, states of consciousness and arousal, and sleep-wake, are regulated by the hypothalamus and limbic brainstem nuclei and monitored and modulated by a telencephalic network of insula, ACC, and amygdala in cooperation with the limbic thalamus.

The limbic system exerts a direct influence (MCC) to a lesser extent and an indirect influence on the motor-executive and sensory-cognitive systems to a greater extent. According to current knowledge, these three systems are considered to be strongly interconnected. Intersections are e.g. the hippocampus, the (also cortical) far-reaching connections of the amygdala as well as the linkage of limbic cortical areas with PFC areas and thalamic regions, which establish direct and indirect connections to the sensory-cognitive system. Thus, the interaction of the three systems seems to be more dynamic than previously assumed. This is also increasingly reflected in the concepts and models of limbic networks.

The limbic system is the place of origin of affects and emotions and is considered the seat of the psyche. Mental illnesses may lie in disturbed functions of a particular limbic structure or in a disturbance of the balance between limbic centers. In many mental illnesses, alterations often also affect the transmitter/neuromodulator systems and the action of transmitters in their target areas.

References

Adolphs R, Spezio M (2006) Role of the amygdala in processing visual social stimuli. Prog Brain Res 156:363–378

Aggleton JP (2012) Multiple anatomical systems embedded within the primate medial temporal lobe: implications for hippocampal function. Neurosci Biobehav Rev 36:1579–1596

Aggleton JP, Mishkin M (1984) Projections of the amygdala to the thalamus in the cynomolgus monkey. J Comp Neurol 222:56–68

Aggleton JP, Burton MJ, Passingham RE (1980) Cortical and subcortical afferents to the amygdala of the rhesus monkey (Macaca mulatta). Brain Res 190:347–368

Aggleton JP, Wright NF, Vann SD, Saunders RC (2012) Medial temporal lobe projections to the retrosplenial cortex of the macaque monkey. Hippocampus 22:1883–1900

Alexander GE, Crutcher MD (1990) Functional architecture of basal ganglia circuits: neural substrates of parallel processing. Trends Neurosci 13:266–271

Alheid GF, Heimer L (1988) New perspectives in basal forebrain organization of special relevance for neuropsychiatric disorders: the striatopallidal, amygdaloid, and corticopetal components of substantia innominata. Neuroscience 27:1–39

Alvarez-Bolado G (2019) Development of neuroendocrine neurons in the mammalian hypothalamus. Cell Tissue Res 375:23–39

Amaral DG, Insausti R (1992) Retrograde transport of D-[3H]-aspartate injected into the monkey amygdaloid complex. Exp Brain Res 88:375–388

Amaral D, Lavenex P (2006) Hippocampal neuroanatomy. In: Andersen P, Morris R, Amaral D, Bliss T, O'Keefe J (eds) The hippocampus book. Oxford University Press, New York, pp 37–114

Amaral DG, Price JL (1984) Amygdalo-cortical projections in the monkey (Macaca fascicularis). J Comp Neurol 230:465–496

Amaral DG, Veazey RB, Cowan WM (1982) Some observations on hypothalamo-amygdaloid connections in the monkey. Brain Res 252:13–27

Amiez C, Petrides M (2014) Neuroimaging evidence of the anatomo-functional organization of the human cingulate motor areas. Cereb Cortex 24:563–578

An X, Bandler R, Öngür D, Price JL (1998) Prefrontal cortical projections to longitudinal columns in the midbrain periaqueductal gray in macaque monkeys. J Comp Neurol 401:455–479

Andermann ML, Lowell BB (2017) Toward a wiring diagram understanding of appetite control. Neuron 95:757–778

Angeles Fernández-Gil M, Palacios-Bote R, Leo-Barahona M, Mora-Encinas JP (2010) Anatomy of the brainstem: a gaze into the stem of life. Semin Ultrasound CT MR 31:196–219

Aoki C, Venkatesan C, Go CG, Forman R, Kurose H (1998) Cellular and subcellular sites for noradrenergic action in the monkey dorsolateral prefrontal cortex as revealed by the immunocytochemical localization of noradrenergic receptors and axons. Cereb Cortex 8:269–277

Augustine JR (1996) Circuitry and functional aspects of the insular lobe in primates including humans. Brain Res Brain Res Rev 22:229–244

Ballinger EC, Ananth M, Talmage DA, Role LW (2016) Basal forebrain cholinergic circuits and signaling in cognition and cognitive decline. Neuron 91:1199–1218

Bandler R, Shipley MT (1994) Columnar organization in the midbrain periaqueductal gray: modules for emotional expression? Trends Neurosci 17:379–389

Barbas H (1993) Organization of cortical afferent input to orbitofrontal areas in the rhesus monkey. Neuroscience 56:841–864

Barbas H, De Olmos J (1990) Projections from the amygdala to basoventral and mediodorsal prefrontal regions in the rhesus monkey. J Comp Neurol 300:549–571

Barbas H, Pandya DN (1989) Architecture and intrinsic connections of the prefrontal cortex in the rhesus monkey. J Comp Neurol 286:353–375

Barbas H, Ghashghaei H, Dombrowski SM, Rempel-Clower NL (1999) Medial prefrontal cortices are unified by common connections with superior temporal cortices and distinguished by input from memory-related areas in the rhesus monkey. J Comp Neurol 410:343–367

Barbas H, Saha S, Rempel-Clower N, Ghashghaei T (2003) Serial pathways from primate prefrontal cortex to autonomic areas may influence emotional expression. BMC Neurosci 10(4):25

Barbosa DAN, de Oliveira-Souza R, Monte Santo F, de Oliveira Faria AC, Gorgulho AA, De Salles AAF (2017) The hypothalamus at the crossroads of psychopathology and neurosurgery. Neurosurg Focus 43:E15

Basar K, Sesia T, Groenewegen H, Steinbusch HW, Visser-Vandewalle V, Temel Y (2010) Nucleus accumbens and impulsivity. Prog Neurobiol 92:533–557

Bear MH, Bollu PC (2018) Neuroanatomy, hypothalamus. In: StatPearls. StatPearls Publishing, Treasure Island, FL

Berger B, Gaspar P, Verney C (1991) Dopaminergic innervation of the cerebral cortex: unexpected differences between rodents and primates. Trends Neurosci 14:21–27

Bernhardt BC, Singer T (2012) The neural basis of empathy. Annu Rev Neurosci 35:1–23

Berridge CW, Waterhouse BD (2003) The locus coeruleus-noradrenergic system: modulation of behavioral state and state-dependent cognitive processes. Brain Res Rev 42:33–84

Berridge KC, Kringelbach ML (2013) Neuroscience of affect: brain mechanisms of pleasure and displeasure. Curr Opin Neurobiol 23:294–303

Berridge KC, Kringelbach ML (2015) Pleasure systems in the brain. Neuron 86:646–664

Botvinick MM (2007) Conflict monitoring and decision making: reconciling two perspectives on anterior cingulate function. Cogn Affect Behav Neurosci 7:356–366

Brauer K, Häusser M, Härtig W, Arendt T (2000) The core-shell dichotomy of nucleus accumbens in the rhesus monkey as revealed by double-immunofluorescence and morphology of cholinergic interneurons. Brain Res 858:151–162

Brodmann K (1909) Vergleichende Lokalisationslehre der Grosshirnrinde in ihren Prinzipien dargestellt auf Grund des Zellenbaues. Johann Ambrosius Barth, Leipzig

Bromberg-Martin ES, Matsumoto M, Hikosaka O (2010) Dopamine in motivational control: rewarding, aversive, and alerting. Neuron 68:815–834

Brooks JC, Zambreanu L, Godinez A, Craig AD, Tracey I (2005) Somatotopic organisation of the human insula to painful heat studied with high resolution functional imaging. NeuroImage 27:201–209

Büchel C, Bornhovd K, Quante M, Glauche V, Bromm B, Weiller C (2002) Dissociable neural responses related to pain intensity, stimulus intensity, and stimulus awareness within the anterior cingulate cortex: a parametric single-trial laser functional magnetic resonance imaging study. J Neurosci 22:970–976

Buot A, Yelnik J (2012) Functional anatomy of the basal ganglia: limbic aspects. Rev Neurol (Paris) 168:569–575

Burgess PW, Gilbert SJ, Dumontheil I (2007) Function and localization within rostral prefrontal cortex (area 10). Philos Trans R Soc Lond Ser B Biol Sci 362:887–899

Bush G, Luu P, Posner MI (2000) Cognitive and emotional influences in anterior cingulate cortex. Trends Cogn Sci 4:215–222

Carmichael ST, Price JL (1995a) Limbic connections of the orbital and medial prefrontal cortex in macaque monkeys. J Comp Neurol 363:615–641

Carmichael ST, Price JL (1995b) Sensory and premotor connections of the orbital and medial prefrontal cortex of macaque monkeys. J Comp Neurol 363:642–664

Carmichael ST, Price JL (1996) Connectional networks within the orbital and medial prefrontal cortex of macaque monkeys. J Comp Neurol 371:179–207

Carrive P, Morgan MM (2012) Periaqueductal gray. In: Mai JK, Paxinos G (eds) The human nervous system. Academic, London, pp 367–400

Carus-Cadavieco M, Gorbati M, Ye L, Bender F, van der Veldt S, Kosse C, Börgers C, Lee SY, Ramakrishnan C, Hu Y, Denisova N, Ramm F, Volitaki E, Burdakov D, Deisseroth K, Ponomarenko A, Korotkova T (2017) Gamma oscillations organize top-down signalling to hypothalamus and enable food seeking. Nature 542:232–236

Cassel JC, Pereira de Vasconcelos A, Loureiro M, Cholvin T, Dalrymple-Alford JC, Vertes RP (2013) The reuniens and rhomboid nuclei: neuroanatomy, electrophysiological characteristics and behavioral implications. Prog Neurobiol 111:34–52

Castro DC, Chesterman NS, Wu MKH, Berridge KC (2014) Two cortical hedonic hotspots: orbitofrontal and insular sites of sucrose 'liking' enhancement. In: Society for neuroscience conference, Washington, DC

Cauda F, D'Agata F, Sacco K, Duca S, Geminiani G, Vercelli A (2011) Functional connectivity of the insula in the resting brain. NeuroImage 55:8–23

Cavada C, Compañy T, Tejedor J, Cruz-Rizzolo RJ, Reinoso-Suárez F (2000) The anatomical connections of the macaque monkey orbitofrontal cortex. A review. Cereb Cortex 10:220–242

Challis C, Berton O (2015) Top-down control of serotonin systems by the prefrontal cortex: a path toward restored socioemotional function in depression. ACS Chem Neurosci 6:1040–1054

Chikama M, McFarland NR, Amaral DG, Haber SN (1997) Insular cortical projections to functional regions of the striatum correlate with cortical cytoarchitectonic organization in the primate. J Neurosci 17:9686–9705

Cho YT, Ernst M, Fudge JL (2013) Cortico-amygdala-striatal circuits are organized as hierarchical subsystems through the primate amygdala. J Neurosci 33:14017–14030

Colavito V, Tesoriero C, Wirtu AT, Grassi-Zucconi G, Bentivoglio M (2015) Limbic thalamus and state-dependent behavior: the paraventricular nucleus of the thalamic midline as a node in circadian timing and sleep/wake-regulatory networks. Neurosci Biobehav Rev 54:3–17

Correia SS, Goosens KA (2016) Input-specific contributions to valence processing in the amygdala. Learn Mem 23:534–543

Coulombe MA, Erpelding N, Kucyi A, Davis KD (2016) Intrinsic functional connectivity of periaqueductal gray subregions in humans. Hum Brain Mapp 37:1514–1530

Counts SE, Mufson EJ (2012) Locus coeruleus. In: Mai JK, Paxinos G (eds) The human nervous system. Academic, London, pp 425–438

Coveñas R, Martin F, Belda M, Smith V, Salinas P, Rivada E, Diaz-Cabiale Z, Narvaez JA, Marcos P, Tramu G, Gonzalez-Baron S (2003) Mapping of neurokinin-like immunoreactivity in the human brainstem. BMC Neurosci 4:3

Craig AD (2010) The sentient self. Brain Struct Funct 214:563–577

Craig AD, Zhang ET (2006) Retrograde analyses of spinothalamic projections in the macaque monkey: input to posterolateral thalamus. J Comp Neurol 499:953–964

Da Cunha C, Gomez-A A, Blaha CD (2012) The role of the basal ganglia in motivated behavior. Rev Neurosci 23:747–767

Dahlström A, Fuxe K (1964) Evidence for the existence of monoamine-containing neurons in the central nervous system. I. Demonstration of monoamines in the cell bodies of brain stem neurons. Acta Physiol Scand Suppl 62:1–55

Damasio AR (1994) Descartes' Irrtum. Fühlen, Denken und das menschliche Gehirn. Paul List Verlag, München

Decety J, Michalska KJ (2010) Neurodevelopmental changes in the circuits underlying empathy and sympathy from childhood to adulthood. Dev Sci 13:886–899

Ding SL (2013) Comparative anatomy of the prosubiculum, subiculum, presubiculum, postsubiculum, and parasubiculum in human, monkey, and rodent. J Comp Neurol 521:4145–4162

Ding SL, Haber SN, Van Hoesen GW (2010) Stratum radiatum of CA2 is an additional target of the perforant path in humans and monkeys. Neuroreport 21:245–249

Ding SL, Royall JJ, Sunkin SM, Ng L, Facer BA, Lesnar P, Guillozet-Bongaarts A, McMurray B, Szafer A, Dolbeare TA, Stevens A, Tirrell L, Benner T, Caldejon S, Dalley RA, Dee N, Lau C, Nyhus J, Reding M, Riley ZL, Sandman D, Shen E, van der Kouwe A, Varjabedian A, Wright M, Zöllei L, Dang C, Knowles JA, Koch C, Phillips JW, Sestan N, Wohnoutka P, Zielke HR, Hohmann JG, Jones AR, Bernard A, Hawrylycz MJ, Hof PR, Fischl B, Lein ES (2016) Comprehensive cellular-resolution atlas of the adult human brain. J Comp Neurol 524:3127–3481

Double KL, Dedov VN, Fedorow H, Kettle E, Halliday GM, Garner B, Brunk UT (2008) The comparative biology of neuromelanin and lipofuscin in the human brain. Cell Mol Life Sci 65:1669–1682

Dum RP, Strick PL (1996) Spinal cord terminations of the medial wall motor areas in macaque monkeys. J Neurosci 16:6513–6525

Dum RP, Levinthal DJ, Strick PL (2009) The spinothalamic system targets motor and sensory areas in the cerebral cortex of monkeys. J Neurosci 29:14223–14235

Eblen F, Graybiel AM (1995) Highly restricted origin of prefrontal cortical inputs to striosomes in the macaque monkey. J Neurosci 15:5999–6013

Fellows LK (2011) Orbitofrontal contributions to value-based decision making: evidence from humans with frontal lobe damage. Ann N Y Acad Sci 1239:51–58

Ferrario CR, Labouèbe G, Liu S, Nieh EH, Routh VH, Xu S, O'Connor EC (2016) Homeostasis meets motivation in the battle to control food intake. J Neurosci 36:11469–11481

Förstl H, Levy R, Burns A, Luthert P, Cairns N (1994) Disproportionate loss of noradrenergic and cholinergic neurons as cause of depression in Alzheimer's disease—a hypothesis. Pharmacopsychiatry 27:11–15

Freese JL, Amaral DG (2005) The organization of projections from the amygdala to visual cortical areas TE and V1 in the macaque monkey. J Comp Neurol 486:295–317

Freese JL, Amaral DG (2009) Neuroanatomy of the primate amygdala. In: Whalen PJ, Phelps EA (eds) The human amygdala. The Guilford Press, New York, pp 3–42

Fregosi M, Contestabile A, Hamadjida A, Rouiller EM (2017) Corticobulbar projections from distinct motor cortical areas to the reticular formation in macaque monkeys. Eur J Neurosci 45:1379–1395

Friedman DP, Murray EA, O'Neill JB, Mishkin M (1986) Cortical connections of the somatosensory fields of the lateral sulcus of macaques: evidence for a corticolimbic pathway for touch. J Comp Neurol 252:323–347

Friedman DP, Aggleton JP, Saunders RC (2002) Comparison of hippocampal, amygdala, and perirhinal projections to the nucleus accumbens: combined anterograde and retrograde tracing study in the Macaque brain. J Comp Neurol 450:345–365

Frotscher M, Léránth C (1985) Cholinergic innervation of the rat hippocampus as revealed by choline acetyltransferase immunocytochemistry: a combined light and electron microscopic study. J Comp Neurol 239:237–246

Fudge JL, Breitbart MA, Danish M, Pannoni V (2005) Insular and gustatory inputs to the caudal ventral striatum in primates. J Comp Neurol 490:101–118

Fuller PM, Gooley JJ, Saper CB (2006) Neurobiology of the sleep-wake cycle: sleep architecture, circadian regulation, and regulatory feedback. J Biol Rhythm 21:482–493

García-Cabezas MÁ, Barbas H (2017) Anterior cingulate pathways may affect emotions through orbitofrontal cortex. Cereb Cortex 27:4891–4910

Ghashghaei HT, Barbas H (2002) Pathways for emotion: interactions of prefrontal and anterior temporal pathways in the amygdala of the rhesus monkey. Neuroscience 115:1261–1279

Ghashghaei HT, Hilgetag CC, Barbas H (2007) Sequence of information processing for emotions based on the anatomic dialogue between prefrontal cortex and amygdala. NeuroImage 34:905–923

Ghaziri J, Tucholka A, Girard G, Houde JC, Boucher O, Gilbert G, Descoteaux M, Lippé S, Rainville P, Nguyen DK (2017) The corticocortical structural connectivity of the human insula. Cereb Cortex 27:1216–1228

Gottfried JA, O'Doherty J, Dolan RJ (2003) Encoding predictive reward value in human amygdala and orbitofrontal cortex. Science 301:1104–1107

Grabenhorst F, Rolls ET (2011) Value, pleasure and choice in the ventral prefrontal cortex. Trends Cog Sci 15:56–67

Graybiel AM, Ragsdale CW Jr (1978) Histochemically distinct compartments in the striatum of human, monkeys, and cat demonstrated by acetylthiocholinesterase staining. Proc Natl Acad Sci U S A 75:5723–5726

Groenewegen HJ, Witter MP (2004) Thalamus. In: Paxinos G (ed) The rat nervous system. Academic, San Diego, pp 408–441

Guillery RW, Harting JK (2003) Structure and connections of the thalamic reticular nucleus: advancing views over half a century. J Comp Neurol 463:360–371

Haber SN, Knutson B (2010) The reward circuit: linking primate anatomy and human imaging. Neuropsychopharmacology 35:4–26

Haber SN, Kunishio K, Mizobuchi M, Lynd-Balta E (1995) The orbital and medial prefrontal circuit through the primate basal ganglia. J Neurosci 15:4851–4867

Haber SN, Fudge JL, McFarland NR (2000) Striatonigrostriatal pathways in primates form an ascending spiral from the shell to the dorsolateral striatum. J Neurosci 20:2369–2382

Haber SN, Adler A, Bergman H (2012) The basal ganglia. In: Mai JK, Paxinos G (eds) The human nervous system. Academic, London, pp 678–738

Hackett TA, Stepniewska I, Kaas JH (1999) Prefrontal connections of the parabelt auditory cortex in macaque monkeys. Brain Res 817:45–58

Hajszan T, Alreja M, Leranth C (2004) Intrinsic vesicular glutamate transporter 2-immunoreactive input to septohippocampal parvalbumin-containing neurons: novel glutamatergic local circuit cells. Hippocampus 14:499–509

Halliday G, Reyes S, Double K (2012) Substantia nigra, ventral tegmental area, and retrorubral fields. In: Mai JK, Paxinos G (eds) The human nervous system. Academic, London, pp 439–454

Hatanaka N, Tokuno H, Hamada I, Inase M, Ito Y, Imanishi M, Hasegawa N, Akazawa T, Nambu A, Takada M (2003) Thalamocortical and intracortical connections of monkey cingulate motor areas. J Comp Neurol 462:121–138

Hayashi K, Nakao K, Nakamura K (2015) Appetitive and aversive information coding in the primate dorsal raphé nucleus. J Neurosci 35:6195–6208

Henssen A, Zilles K, Palomero-Gallagher N, Schleicher A, Mohlberg H, Gerboga F, Eickhoff SB, Bludau S, Amunts K (2016) Cytoarchitecture and probability maps of the human medial orbitofrontal cortex. Cortex 75:87–112

Hermann GE, Holmes GM, Rogers RC, Beattie MS, Bresnahan JC (2003) Descending spinal projections from the rostral gigantocellular reticular nuclei complex. J Comp Neurol 455:210–221

Hobson JA, Pace-Schott EF (2002) The cognitive neuroscience of sleep: neuronal systems, consciousness and learning. Nat Rev Neurosci 3:679–693

Holt DJ, Graybiel AM, Saper CB (1997) Neurochemical architecture of the human striatum. J Comp Neurol 384:1–25

Hornung JP (2012) Raphe Nuclei. In: Mai JK, Paxinos G (eds) The human nervous system. Academic, London, pp 401–424

Hsu DT, Price JL (2007) Midline and intralaminar thalamic connections with the orbital and medial prefrontal networks in macaque monkeys. J Comp Neurol 504:89–111

Hsu DT, Price JL (2009) Paraventricular thalamic nucleus: subcortical connections and innervation by serotonin, orexin, and corticotropin-releasing hormone in macaque monkeys. J Comp Neurol 512:825–848

Hsu DT, Kirouac GJ, Zubieta JK, Bhatnagar S (2014) Contributions of the paraventricular thalamic nucleus in the regulation of stress, motivation, and mood. Front Behav Neurosci 8:73

Insausti R, Amaral DG, Cowan WM (1987) The entorhinal cortex of the monkey: II. Cortical afferents. J Comp Neurol 264:356–395

Jakab RL, Leranth C (1995) Chapter 20—Septum. In: Paxinos G (ed) The rat nervous system, 2. Aufl. Academic, San Diego, pp 405–442

Jang S, Kwak S (2017) The upper ascending reticular activating system between intralaminar thalamic nuclei and cerebral cortex in the human brain. J Korean Phys Ther 29:109–114

Jang SH, Lim HW, Yeo SS (2014) The neural connectivity of the intralaminar thalamic nuclei in the human brain: a diffusion tensor tractography study. Neurosci Lett 579:140–144

Joly-Amado A, Cansell C, Denis RG, Delbes AS, Castel J, Martinez S, Luquet S (2014) The hypothalamic arcuate nucleus and the control of peripheral substrates. Best Pract Res Clin Endocrinol Metab 28:725–737

Jones EG (1998) The thalamus of primates. In: Bloom FE, Björklund A, Hökfelt T (eds) The primate nervous system. Part II. Handbook of chemical neuroanatomy, Bd 14. Elsevier, Amsterdam, pp 1–298

Kiss J, Patel AJ, Baimbridge KG, Freund TF (1990a) Topographical localization of neurons containing parvalbumin and choline acetyl-transferase in the medial septum-diagonal band region of the rat. Neuroscience 36:61–72

Kiss J, Patel AJ, Freund TF (1990b) Distribution of septohippocampal neurons containing parvalbumin or choline acetyltransferase in the rat brain. J Comp Neurol 298:362–372

Knox D (2016) The role of basal forebrain cholinergic neurons in fear and extinction memory. Neurobiol Learn Mem 133:39–52

Koelsch S, Jacobs AM, Menninghaus W, Liebal K, Klann-Delius G, von Scheve C, Gebauer G (2015) The quartet theory of human emotions: an integrative and neurofunctional model. Phys Life Rev 13:1–17

Kolada E, Bielski K, Falkiewicz M, Szatkowska I (2017) Functional organization of the human amygdala in appetitive learning. Acta Neurobiol Exp (Wars) 77:118–127

Kunishio K, Haber SN (1994) Primate cingulostriatal projection: limbic striatal versus sensorimotor striatal input. J Comp Neurol 350:337–356

Kurth F, Zilles K, Fox PT, Laird AR, Eickhoff SB (2010) A link between the systems: functional differentiation and integration within the human insula revealed by meta-analysis. Brain Struct Funct 214:519–534

Lamm C, Singer T (2010) The role of anterior insular cortex in social emotions. Brain Struct Funct 214:579–591

Lamm C, Decety J, Singer T (2011) Meta-analytic evidence for common and distinct neural networks associated with directly experienced pain and empathy for pain. NeuroImage 54:2492–2502

Lebow MA, Chen A (2016) Overshadowed by the amygdala: the bed nucleus of the stria terminalis emerges as key to psychiatric disorders. Mol Psychiatry 21:450–463

Leichnetz GR, Smith DJ, Spencer RF (1984) Cortical projections to the paramedian tegmental and basilar pons in the monkey. J Comp Neurol 228:388–408

Liljeholm M, O'Doherty JP (2012) Contributions of the striatum to learning, motivation, and performance: an associative account. Trends Cognit Sci 16:467–475

Lima D, Almeida A (2002) The medullary dorsal reticular nucleus as a proprioceptive centre of the pain control system. Prog Neurobiol 66:81–108

Li Y, Vanni-Mercier G, Isnard J, Mauguière F, Dreher JC (2016) The neural dynamics of reward value and risk coding in the human orbitofrontal cortex. Brain 139:1295–1309

Liu AK, Chang RC, Pearce RK, Gentleman SM (2015) Nucleus basalis of Meynert revisited: anatomy, history and differential involvement in Alzheimer's and Parkinson's disease. Acta Neuropathol 129:527–540

Luo M, Li Y, Zhong W (2016) Do dorsal raphe 5-HT neurons encode "beneficialness"? Neurobiol Learn Mem 135:40–49

Mai JK, Majtanik M, Paxinos G (2016) Atlas of the human brain, 4. Aufl. Academic, London

Margulies DS, Kelly AM, Uddin LQ, Biswal BB, Castellanos FX, Milham MP (2007) Mapping the functional connectivity of anterior cingulate cortex. NeuroImage 37:579–588

Markowitsch HJ, Emmans D, Irle E, Streicher M, Preilowski B (1985) Cortical and subcortical afferent connections of the primate's temporal pole: a study of rhesus monkeys, squirrel monkeys, and marmosets. J Comp Neurol 242:425–458

McDonald AJ, Mott DD (2017) Functional neuroanatomy of amygdalohippocampal interconnections and their role in learning and memory. J Neurosci Res 95:797–820

McFarland NR, Haber SN (2002) Thalamic relay nuclei of the basal ganglia form both reciprocal and nonreciprocal cortical connections, linking multiple frontal cortical areas. J Neurosci 22:8117–8132

McKee AC, Cairns NJ, Dickson DW, Folkerth RD, Keene CD, Litvan I, Perl DP, Stein TD, Vonsattel JP, Stewart W, Tripodis Y, Crary JF, Bieniek KF, Dams-O'Connor K, Alvarez VE, Gordon WA, TBI/CTE Group (2016) The first NINDS/NIBIB consensus meeting to define neuropathological criteria for the diagnosis of chronic traumatic encephalopathy. Acta Neuropathol 131:75–86

McRitchie DA, Halliday GM, Cartwright H (1995) Quantitative analysis of the variability of substantia nigra cell clusters in the human. Neuroscience 68:539–551

Mehler WR (1980) Subcortical afferent connections of the amygdala in the monkey. J Comp Neurol 190:733–762

Menon V, Levitin DJ (2005) The rewards of music listening: response and physiological connectivity of the mesolimbic system. NeuroImage 28:175–184

Meredith GE, Pattiselanno A, Groenewegen HJ, Haber SN (1996) Shell and core in monkey and human nucleus accumbens identified with antibodies to calbindin-D28k. J Comp Neurol 365:628–639

Mesulam MM, Mufson EJ (1982a) Insula of the old world monkey. I. Architectonics in the insulo-orbito-temporal component of the paralimbic brain. J Comp Neurol 212:1–22

Mesulam MM, Mufson EJ (1982b) Insula of the old world monkey. III: Efferent cortical output and comments on function. J Comp Neurol 212:38–52

Mesulam MM, Mufson EJ (1984) Neural inputs into the nucleus basalis of the substantia innominata (Ch4) in the rhesus monkey. Brain 107:253–274

Mesulam MM, Mufson EJ (1985) The insula of Reil in man and monkey. In: Peters A, Jones EG (eds) Association and auditory cortices. Plenum, New York, pp 179–226

Mesulam MM, Mufson EJ, Levey AI, Wainer BH (1983) Cholinergic innervation of cortex by the basal forebrain: cytochemistry and cortical connections of the septal area, diagonal band nuclei, nucleus basalis (substantia innominata), and hypothalamus in the rhesus monkey. J Comp Neurol 214:170–197

Metzger M, Bueno D, Lima LB (2017) The lateral habenula and the serotonergic system. Pharmacol Biochem Behav 162:22–28

Miczek KA, DeBold JF, Hwa LS, Newman EL, de Almeida RM (2015) Alcohol and violence: neuropeptidergic modulation of monoamine systems. Ann N Y Acad Sci 1349:96–118

Misra G, Coombes SA (2015) Neuroimaging evidence of motor control and pain processing in the human midcingulate cortex. Cereb Cortex 25:1906–1919

Mitchell JM, Lowe D, Fields HL (1998) The contribution of the rostral ventromedial medulla to the antinociceptive effects of systemic morphine in restrained and unrestrained rats. Neuroscience 87:123–133

Mohedano-Moriano A, Muñoz-López M, Sanz-Arigita E, Pró-Sistiaga P, Martínez-Marcos A, Legidos-Garcia ME, Insausti AM, Cebada-Sánchez S, Arroyo-Jiménez Mdel M, Marcos P, Artacho-Pérula E, Insausti R (2015) Prefrontal cortex afferents to the anterior temporal lobe in the Macaca fascicularis monkey. J Comp Neurol 523:2570–2598

Moran MA, Mufson EJ, Mesulam MM (1987) Neural inputs into the temporopolar cortex of the rhesus monkey. J Comp Neurol 256:88–103

Morecraft RJ, Van Hoesen GW (1993) Frontal granular cortex input to the cingulate (M3), supplementary (M2) and primary (M1) motor cortices in the rhesus monkey. J Comp Neurol 337:669–689

Morecraft RJ, Geula C, Mesulam MM (1992) Cytoarchitecture and neural afferents of orbitofrontal cortex in the brain of the monkey. J Comp Neurol 323:341–358

Morecraft RJ, Schroeder CM, Keifer J (1996) Organization of face representation in the cingulate cortex of the rhesus monkey. Neuroreport 7:1343–1348

Morecraft RJ, Cipolloni PB, Stilwell-Morecraft KS, Gedney MT, Pandya DN (2004) Cytoarchitecture and cortical connections of the posterior cingulate and adjacent somatosensory fields in the rhesus monkey. J Comp Neurol 469:37–69

Morecraft RJ, McNeal DW, Stilwell-Morecraft KS, Gedney M, Ge J, Schroeder CM, van Hoesen GW (2007) Amygdala interconnections with the cingulate motor cortex in the rhesus monkey. J Comp Neurol 500:134–165

Mufson EJ, Mesulam MM (1982) Insula of the old world monkey. II: Afferent cortical input and comments on the claustrum. J Comp Neurol 212:23–37

Müller C, Remy S (2018) Septo-hippocampal interaction. Cell Tissue Res 373:565–575

Muñoz M, Insausti R (2005) Cortical efferents of the entorhinal cortex and the adjacent parahippocampal region in the monkey (Macaca fascicularis). Eur J Neurosci 22:1368–1388

Müri RM (2016) Cortical control of facial expression. J Comp Neurol 524:1578–1585

Mutschler I, Wieckhorst B, Kowalevski S, Derix J, Wentlandt J, Schulze-Bonhage A, Ball T (2009) Functional organization of the human anterior insular cortex. Neurosci Lett 457:66–70

Nadel L, Hoscheidt S, Ryan LR (2013) Spatial cognition and the hippocampus: the anterior–posterior axis. J Cognit Neurosci 25:22–28

Nieuwenhuys R (1985) Chemoarchitecture of the brain. Springer, Berlin

Nieuwenhuys R (2012) The insular cortex: a review. Prog Brain Res 195:123–163

Nieuwenhuys R, Voogd J, van Huijzen C (1988) The human central nervous system. Springer, Berlin

Nieuwenhuys R, Voogd J, van Huijzen C (1991) Das Zentralnervensystem des Menschen. Springer, Berlin

Nieuwenhuys R, Voogd J, Van Huijzen C (2008) The human central nervous system. Springer, Berlin

Oldfield RG, Harris RM, Hofmann HA (2015) Integrating resource defence theory with a neural nonapeptide pathway to explain territory-based mating systems. Front Zool 12(Suppl 1):S16

Öngür D, An X, Price JL (1998) Prefrontal cortical projections to the hypothalamus in macaque monkeys. J Comp Neurol 401:480–505

Ostrowsky K, Magnin M, Ryvlin P, Isnard J, Guenot M, Mauguière F (2002) Representation of pain and somatic sensation in the human insula: a study of responses to direct electrical cortical stimulation. Cereb Cortex 12:376–385

Pandya DN, Van Hoesen GW, Mesulam MM (1981) Efferent connections of the cingulate gyrus in the rhesus monkey. Exp Brain Res 42:319–330

Papez JW (1937) A proposed mechanism of emotion. Arch Neurol Psychiatr 38:725–743

Parent A, Hazrati LN (1995) Functional anatomy of the basal ganglia. I. The cortico-basal ganglia-thalamo-cortical loop. Brain Res Rev 20:91–127

Paus T (2000) Functional anatomy of arousal and attention systems in the human brain. Prog Brain Res 126:65–77

Paus T (2001) Primate anterior cingulate cortex: where motor control, drive and cognition interface. Nat Rev Neurosci 2:417–424

Paxinos G, Xu-Feng H, Sengul G, Watson C (2012) Organization of brainstem nuclei. In: Mai JK, Paxinos G (eds) The human nervous system. Academic, London, pp 260–327

Pereira de Vasconcelos A, Cassel JC (2015) The non-specific thalamus: a place in a wedding bed for making memories last? Neurosci Biobehav Rev 54:175–196

Petrovich GD (2018) Lateral hypothalamus as a motivation-cognition interface in the control of feeding behavior. Front Syst Neurosci 12:14

Phillips ML, Drevets WC, Rauch SL, Lane R (2003) Neurobiology of emotion perception I: the neural basis of normal emotion perception. Biol Psychiatry 54:504–514

Picard N, Strick PL (2001) Imaging the premotor areas. Curr Opin Neurobiol 11:663–672

Porrino LJ, Crane AM, Goldman-Rakic PS (1981) Direct and indirect pathways from the amygdala to the frontal lobe in rhesus monkeys. J Comp Neurol 198:121–136

Pourtois G, Schettino A, Vuilleumier P (2013) Brain mechanisms for emotional influences on perception and attention: what is magic and what is not. Biol Psychol 92:492–512

Price JL (1995) Thalamus. In: Paxinos G (ed) The rat nervous system, 2nd Aufl. Academic, San Diego, pp 629–648

Price JL (2003) Comparative aspects of amygdala connectivity. Ann N Y Acad Sci 985:50–58

Price JL, Amaral DG (1981) An autoradiographic study of the projections of the central nucleus of the monkey amygdala. J Neurosci 1:1242–1259

Procyk E, Wilson CR, Stoll FM, Faraut MC, Petrides M, Amiez C (2016) Midcingulate motor map and feedback detection: converging data from humans and monkeys. Cereb Cortex 26:467–476

Puglisi-Allegra S, Andolina D (2015) Serotonin and stress coping. Behav Brain Res 277:58–67

Rainville P (2002) Brain mechanisms of pain affect and pain modulation. Curr Opin Neurobiol 12:195–204

Rempel-Clower NL, Barbas H (1998) Topographic organization of connections between the hypo-thalamus and prefrontal cortex in the rhesus monkey. J Comp Neurol 398:393–419

Rolls ET (2012) The emotional systems. In: Mai JK, Paxinos G (eds) The human nervous system. Academic, London, pp 1328–1350

Root DH, Melendez RI, Zaborszky L, Napier TC (2015) The ventral pallidum: subregion-specific functional anatomy and roles in motivated behaviors. Prog Neurobiol 130:29–70

Rudebeck PH, Murray EA (2014) The orbitofrontal oracle: cortical mechanisms for the prediction and evaluation of specific behavioral outcomes. Neuron 84:1143–1156

Russchen FT, Amaral DG, Price JL (1985) The afferent connections of the substantia innominata in the monkey, Macaca fascicularis. J Comp Neurol 242:1–27

Russchen FT, Amaral DG, Price JL (1987) The afferent input to the magnocellular division of the mediodorsal thalamic nucleus in the monkey, Macaca fascicularis. J Comp Neurol 256:175–210

Sabatinelli D, Fortune EE, Li Q, Siddiqui A, Krafft C, Oliver WT, Beck S, Jeffries J (2011) Emotional perception: meta-analyses of face and natural scene processing. NeuroImage 54:2524–2533

Sara SJ (2009) The locus coeruleus and noradrenergic modulation of cognition. Nat Rev Neurosci 10:211–223

Satoh T, Nakai S, Sato T, Kimura M (2003) Correlated coding of motivation and outcome of decision by dopamine neurons. J Neurosci 23:9913–9923

Schoenbaum G, Esber GR (2010) How do you (estimate you will) like them apples? Integration as a defining trait of orbitofrontal function. Curr Opin Neurobiol 20:205–211

Schoenbaum G, Takahashi Y, Tzu-Lan L, McDannald MA (2011) Does the orbitofrontal cortex signal value? Ann N Y Acad Sci 1239:87–99

Schwartz MD, Kilduff TS (2015) The neurobiology of sleep and wakefulness. Psychiatr Clin North Am 38:615–644

Selemon LD, Goldman-Rakic PS (1985) Longitudinal topography and interdigitation of corticostriatal projections in the rhesus monkey. J Neurosci 5:776–794

Sharpe MJ, Schoenbaum G (2016) Back to basics: making predictions in the orbitofrontal-amygdala circuit. Neurobiol Learn Mem 131:201–206

Shelley BP, Trimble MR (2004) The insular lobe of Reil—its anatamico-functional, behavioural and neuropsychiatric attributes in humans—a review. World J Biol Psychiatry 5:176–200

Singer T, Seymour B, O'Doherty J, Kaube H, Dolan RJ, Frith CD (2004) Empathy for pain involves the affective but not sensory components of pain. Science 303:1157–1162

Singer T, Critchley HD, Preuschoff K (2009) A common role of insula in feelings, empathy and uncertainty. Trends Cogn Sci 13:334–340

Stalnaker TA, Cooch NK, Schoenbaum G (2015) What the orbitofrontal cortex does not do. Nat Neurosci 18:620–627

Stefanacci L, Amaral DG (2000) Topographic organization of cortical inputs to the lateral nucleus of the macaque monkey amygdala: a retrograde tracing study. J Comp Neurol 421:52–79

Stefanacci L, Amaral DG (2002) Some observations on cortical inputs to the macaque monkey amygdala: an anterograde tracing study. J Comp Neurol 451:301–323

Stephani C, Fernandez-Baca Vaca G, Maciunas R, Koubeissi M, Lüders HO (2011) Functional neuroanatomy of the insular lobe. Brain Struct Funct 216:137–149

Sweeney P, Yang Y (2015) An excitatory ventral hippocampus to lateral septum circuit that suppresses feeding. Nat Commun 6:10188

Sweeney P, Yang Y (2017) Neural circuit mechanisms underlying emotional regulation of homeostatic feeding. Trends Endocrinol Metab 28:437–448

Timbie C, Barbas H (2015) Pathways for emotions: specializations in the amygdalar, mediodorsal thalamic, and posterior orbitofrontal network. J Neurosci 35:11976–11987

Toth M, Fuzesi T, Halasz J, Tulogdi A, Haller J (2010) Neural inputs of the hypothalamic "aggression area" in the rat. Behav Brain Res 215:7–20

Tsukahara S, Yamanouchi K (2001) Neurohistological and behavioral evidence for lordosis-inhibiting tract from lateral septum to periaqueductal gray in male rats. J Comp Neurol 431:293–310

Turner BH, Gupta KC, Mishkin M (1978) The locus and cytoarchitecture of the projection areas of the olfactory bulb in Macaca mulatta. J Comp Neurol 177:381–396

Veening JG, Coolen LM, Gerrits PO (2014) Neural mechanisms of female sexual behavior in the rat; comparison with male ejaculatory control. Pharmacol Biochem Behav 121:16–30

Vertes RP, Linley SB, Hoover WB (2015) Limbic circuitry of the midline thalamus. Neurosci Biobehav Rev 54:89–107

Viard A, Doeller CF, Hartley T, Bird CM, Burgess N (2011) Anterior hippocampus and goal-directed spatial decision making. J Neurosci 31:4613–4621

Vogt BA (2016) Midcingulate cortex: structure, connections, homologies, functions and diseases. J Chem Neuroanat 74:28–46

Vogt BA, Pandya DN (1987) Cingulate cortex of the rhesus monkey: II. Cortical afferents. J Comp Neurol 262:271–289

Vogt BA, Palomero-Gallagher N (2012) Cingulate cortex. In: Mai JK, Paxinos G (eds) The human nervous system. Academic, London, pp 943–987

Vogt BA, Vogt L, Laureys S (2006) Cytology and functionally correlated circuits of human posterior cingulate areas. NeuroImage 29:452–466

Vogt BA, Hof PR, Friedman DP, Sikes RW, Vogt LJ (2008) Norepinephrinergic afferents and cytology of the macaque monkey midline, mediodorsal, and intralaminar thalamic nuclei. Brain Struct Funct 212:465–479

Von Economo C (1929) The cytoarchitectonics of the human cerebral cortex. Oxford University Press, London

Walton ME, Behrens TE, Noonan MP, Rushworth MF (2011) Giving credit where credit is due: orbitofrontal cortex and valuation in an uncertain world. Ann N Y Acad Sci 1239:14–24

Wilkenheiser AM, Schoenbaum G (2016) Over the river, through the woods: cognitive maps in the hippocampus and orbitofrontal cortex. Nat Rev Neurosci 17:513–523

Wilson MA, Fadel JR (2017) Cholinergic regulation of fear learning and extinction. J Neurosci Res 95:836–852

Wilson RC, Takahashi YK, Schoenbaum G, Niv Y (2014) Orbitofrontal cortex as a cognitive map of task space. Neuron 81:267–279

Yamashita T, Yamanaka A (2017) Lateral hypothalamic circuits for sleep-wake control. Curr Opin Neurobiol 44:94–100

Yang C, Thankachan S, McCarley RW, Brown RE (2017) The menagerie of the basal forebrain: how many (neural) species are there, what do they look like, how do they behave and who talks to whom? Curr Opin Neurobiol 44:159–166

Yau SY, Li A, So KF (2015) Involvement of adult hippocampal neurogenesis in learning and forgetting. Neural Plast 2015:717958

Zahm DS, Root DH (2017) Review of the cytology and connections of the lateral habenula, an avatar of adaptive behaving. Pharmacol Biochem Behav 162:3–21

Zarow C, Lyness SA, Mortimer JA, Chui HC (2003) Neuronal loss is greater in the locus coeruleus than nucleus basalis and substantia nigra in Alzheimer and Parkinson diseases. Arch Neurol 60:337–341

Zeman A, Coebergh JA (2013) The nature of consciousness. Handb Clin Neurol 118:373–407

Zernig G, Pinheiro BS (2015) Dyadic social interaction inhibits cocaine-conditioned place preference and the associated activation of the accumbens corridor. Behav Pharmacol 26:580–594

Zhao C, Gammie SC (2014) Glutamate, GABA, and glutamine are synchronously upregulated in the mouse lateral septum during the postpartum period. Brain Res 1591:53–62

Zoicas I, Slattery DA, Neumann ID (2014) Brain oxytocin in social fear conditioning and its extinction: involvement of the lateral septum. Neuropsychopharmacology 39:3027–3035

Zorrilla EP, Koob GF (2013) Amygdalostriatal projections in the neurocircuitry for motivation: a neuroanatomical thread through the career of Ann Kelley. Neurosci Biobehav Rev 37:1932–1945

Neuro- and Psychopharmacology

Michael Koch

Contents

The aim of neuro- and psychopharmacology is the elucidation and targeted manipulation of chemical signal transmission in the central nervous system (CNS) for scientific or therapeutic purposes. The basis for this is the fact that signal transmission between nerve cells essentially takes place via neurotransmitters and neuropeptides, i.e. via chemical messengers that are synthesized and released by a cell and cause physiological changes (increase or decrease of membrane potential, regulation of the activity of intracellular signalling cascades or gene expression) at postsynaptic cells. Neuromodulators (e.g., endocannabinoids) may regulate these processes. Some neurotransmitters (e.g. glutamate and γ-aminobutyric acid, GABA) are ubiquitously distributed in the CNS, whereas other neurotransmitters (e.g. acetylcholine, dopamine, norepinephrine, and serotonin) are found in neuroanatomically defined "systems." Neuropeptides are mostly cotransmitters of the "classical" neurotransmitters. For each of these signaling molecules, there are relatively specific receptors, metabolizing enzymes, and transporters. Neuromodulators, such as endocannabinoids, regulate the activity of the classical transmitters. A variety of neurological and psychiatric diseases are characterized by pathological changes in the chemical communication between neurons. Therefore, neurotransmitters represent an important starting point for the pharmacotherapy of these diseases. Considering that ultimately all cognitive and emotional processes in the brain are based on communication in neuronal networks, one recognizes the great importance of neuropharmacology not only for the neurosciences in the narrower sense, but also for psychotherapy and psychology.

The transmission of information between nerve cells (neurons) occurs primarily through a combination of electrical and chemical processes, i.e. the release of a neurotransmitter and/or a neuropeptide from the synapse of a neuron due to incoming action potentials, the diffusion of these signal molecules to another neuron, the binding to specific receptors and the short- or long-term change in the activity of the recipient neuron (Cooper et al. 2003; Koch 2006; Siegel et al. 2006). That certain plant extracts can alter mental states is an experience that dates back thousands of years in human history, when hemp extracts were used in Chinese medicine. The discovery and scientific consideration of chemical signaling in the peripheral nervous system and CNS is over a hundred years old. The mechanisms of action of signaling molecules in the brain are essential for understanding neurological and psychiatric disorders and form the basis of psychopharmacotherapy (Davis et al. 2002; Tretter and Albus 2004; Gründer and Benkert 2012).

3.1 Structure of the Chemical Synapse and Transmitter Release

A historically important landmark of brain research is the "neuron theory" formulated around 1910 by the Spanish neuroanatomist Santiago Ramón y Cajal, which states that neurons represent distinct units and that the brain consists of a network of interconnected neurons.[1] The points of contact where signal transmission occurs between neurons through chemical messengers are

Learning Objectives

This chapter familiarizes readers with the basic principles of chemical signal transduction in the brain and highlights its pharmacotherapeutic possibilities.

---

1   Ramón y Cajal received the Nobel Prize together with Camillo Golgi in 1906. The essential anatomical discoveries that formed the basis of the neuron theory can be traced back to the use of the method developed by Golgi for staining individual nerve cells.

called synapses. This term goes back to the British physiologist Sherrington. Synapses are specialized structures formed by the axon of the presynaptic neuron at various sites in the postsynaptic neuron (cell body, axon, dendrites). For simplicity, this chapter will not further discuss the fact that postsynaptic targets of neurons can also be muscle cells, glial cells, glandular cells, or endothelial cells (◻ Fig. 3.1). It is historically interesting that the fact of synaptic information transmission by chemical messengers was

doubted for a long time because the idea of purely electrical transmission dominated (Valenstein 2005).

The release of the transmitter occurs after the arrival of an action potential at the synapse through Ca^{2+}-dependent exocytosis. A number of Ca^{2+}-dependent enzymes and other proteins play an important role in this process, ultimately leading to the formation of a fusion pore between the vesicle membrane and the presynaptic membrane. Previously, the vesicles in the presynapse

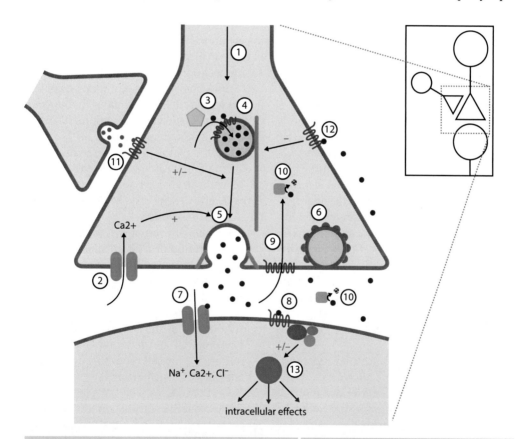

Ca2+

Na⁺, Ca2+, Cl⁻

intracellular effects

1	Action potential	7	postsynaptic ionotropic receptors
2	Calcium channel	8	postsynaptic metabotropic receptors
3	Synthesis enzyme	9	cytoplasmic transporter
4	Transmitter vesicle with vesicular transporter is displaced to the active zone on a cytoskeletal filament	10	Degradation enzymes
5	Transmitter release; calcium-dependent formation of the fusion pore	11	presynaptic heteroreceptors
		12	presynaptic autoreceptors
6	Vesicle recycling by clathrin	13	intracellular signalling cascades (second messenger)

◻ **Fig. 3.1** Schematic of a chemical synapse with the essential processes of signal transmission

were filled with several thousand transmitter molecules by transport proteins. This transport is driven by an electrochemical gradient generated by an ATP-dependent proton pump (▶ Chap. 4). The filled vesicles are transported to the presynaptic membrane by cytoskeletal proteins (e.g. actin and microtubules). Upon arrival of an action potential at the presynapse, voltage-gated Ca^{2+} channels are opened and Ca^{2+} flows into the presynapse. This triggers the formation of the fusion pore, which establishes a direct connection between the vesicle membrane and the presynaptic membrane through which the transmitter is released. After fusion and transmitter release, the piece of vesicle membrane inserted into the membrane is pinched off and reabsorbed by endocytosis. This "recycling" process is tightly coupled to the release process and is regulated by the protein clathrin. Transmitter molecules diffuse through the approximately 20 nm wide synaptic cleft to the postsynaptic neuron. Pre- and postsynaptic neurons are connected by various cell adhesion molecules, which stabilize the synapses. In addition, various so-called extracellular matrix proteins have modulatory effects on transmitter release. All these individual processes of transmitter release occur within only 1–2 ms. The termination of the transmitter effect occurs through enzymatic degradation, diffusion and reuptake via transporter proteins.

In addition to the transmitter receptors, cytoskeletal proteins and enzymes for the degradation of the transmitter are inserted in the *postsynaptic* membrane, which can be seen under the microscope as thickened areas (*postsynaptic densities*). The presynaptic membrane also shows microscopically conspicuous accumulations, which are referred to as the "active zone" because the already filled vesicles, Ca^{2+} channels necessary for exocytosis and fusion proteins are present there in particularly high density. The cytoskeletal proteins by which the transmitter receptors are anchored in the post-synaptic membrane can move the receptors. Thus, receptors can be transported from the extrasynaptic zone of the membrane to the synaptic zone and are thus activatable by the transmitter. The reverse, i.e., removal of receptors from the synaptic area, is also observed. These dynamics of receptor distribution at the postsynapse play a major role in synaptic plasticity in the context of learning and memory. Lateral movement of postsynaptic receptors also plays an important role in some pathological processes. For example, a reduced number of synaptic receptors for the transmitter GABA is found in certain parts of the brain in epilepsy and in chronic alcohol abuse (Jacob et al. 2008). In connection with neurotoxic processes, it was found that synaptic glutamate receptors activate different intracellular signalling cascades than extrasynaptic glutamate receptors (Hardingham and Bading 2010).

Receptor synthesis and degradation rates are in the minute range, while the duration of receptor transport into or out of the active zone of the postsynaptic membrane is fractions of a second.

The following criteria were established for the definition of a neurotransmitter:
- Synthesis by specific enzymes and synaptic release
- Physiological effects of the transmitter on the postsynaptic cell are similar to the stimulation of the afferent neuron
- Degradation and reuptake of the transmitter into the cell occur through specific enzymes or transporters.

Some neuroactive substances do not fully meet these criteria and are therefore called *neuromodulators*, which enhance or attenuate the function of the actual transmitters. Examples of such neuromodulators are endocannabinoids, adenosine, neuroactive steroid hormones and nitric oxide (NO). In addition, there are a large number of *neuropeptides* that show some differences (especially with regard to synthesis and transport) to the neurotransmitters (Bohlen et al. 2002).

3.2 Transmitter Receptors

Neurotransmitters and -peptides are mostly water-soluble, hydrophilic molecules, which due to this property cannot penetrate the cell membrane consisting of a lipid bilayer. Therefore, transmitter receptors are usually transmembrane proteins with extracellular binding sites for the messenger, which influence the activity of the postsynaptic neuron by conformational changes and thus transmit the signal.

Basically, two types of transmitter receptors are distinguished: *ionotropic* and *metabotropic* receptors. Ionotropic receptors are ligand-gated ion channels, i.e. the receptor itself is an ion channel (e.g. a sodium or a chloride channel). Ionotropic receptors are quasi-sealable membrane pores that allow direct connection between the extracellular and intracellular environments. They are usually composed of four or five protein subunits that can interact with each other. Pore formation after binding of the transmitter (or an agonist) occurs through conformational changes of the protein subunits. Interestingly, most ionotropic receptors are heterotetramers or heteropentamers, meaning that they are composed of slightly different subunits. This leads to the fact that different combinations of subunits result in somewhat different receptor properties (e.g. different opening kinetics). This phenomenon will be discussed in more detail in the specific sections on individual receptors.

Metabotropic receptors are not ion channels but act on the physiological properties of the postsynaptic cell via second messenger systems (e.g. G protein-coupled receptors that modulate the opening kinetics of ion channels or gene activity via enzymes such as adenylate cyclase and various protein kinases by phosphorylation).

G Protein-Coupled Receptors (GPCR)
The coupling of metabotropic receptors to GTP-binding proteins (G-proteins) is one of the most common and effective mechanisms for the modulation of intracellular processes. For their discovery, Gilman and Rodbell received the Nobel Prize in 1994. G-proteins are heterotrimeric proteins, i.e. they consist of three different protein subunits: α, β and γ, of which the α-subunit can bind guanosine diphosphate (GDP) or guanosine triphosphate (GTP). In the inactive state, GDP is bound. A conformational change of the transmitter receptor activates the G protein—as a result of ligand binding—in that GTP now binds to the α-subunit instead of GDP. As a result, the heterotrimeric G protein dissociates and influences various intracellular signalling pathways, e.g. the activity of enzymes such as adenylate cyclase or phospholipases. Simplified, there are three subfamilies of G-proteins:

1. G_s proteins have an α-subunit that *stimulates* the enzyme adenylate cyclase. This converts adenosine triphosphate (ATP) into cyclic adenosine monophosphate (cAMP). cAMP is an intracellular messenger that can activate a number of protein kinases. Protein kinases can regulate the activity of numerous proteins (ion channels, transcription factors, synthesis and degradation enzymes, etc.) by transferring phosphate residues (phosphorylation).

2. G_i proteins have an α-subunit that *inhibits* adenylate cyclase and does not result in the re-synthesis of cAMP. Since cAMP is rapidly degraded by constitutively expressed phosphodiesterases, activation of a G_s protein results in a decrease in cAMP levels.

3. G_q proteins have an α-subunit that stimulates phospholipases. Activated phospholipases cleave phospholipids and generate further intracellular messenger molecules, e.g. inositol trisphosphate and diacylglycerol, which can release Ca^{2+} ions from intracellular stores.

Since all these biochemical signalling cascades are enzymatic, i.e. catalytic processes, this type of intracellular signal processing enables a considerable amplification of the input signal—i.e. the transmitter binding to the receptor. Another advantage of signal processing by metabotropic receptors is the *cross-talk* of different receptors via the regulation of common second messenger systems. Besides GPCR, there are other metabotropic receptor systems, e.g. receptor tyrosine kinases.

The interaction of a neurotransmitter or an agonist or antagonist[2] with a receptor occurs

---

2   Agonists usually have the same physiological effect as the natural ligand. Antagonists reduce or block the action of the natural ligand and agonists. There are competitive antagonists, which compete with the transmitter or its agonists for the same receptor binding site, and non-competitive antagonists, which bind to a different site on the receptor and decrease its activity.

through electrostatic allosteric interactions, i.e. the attachment to specific areas of the receptor protein and its conformational change caused by this. The binding is described by the mass action law, where the binding or dissociation constant (K_d) is the ratio of the concentrations of free ligands and receptors to the concentration of the ligand-receptor complex. Thus, the K_d value is an important measure of the affinity between ligand and receptor. The smaller the K_d value, the higher the affinity (or specificity) of the ligand for the receptor. The rule of thumb is that hydrophilic (water-soluble) substances usually have a high affinity and specificity, while lipophilic (fat-soluble) substances have a rather non-specific effect.

For the physiological effect of a ligand on the postsynaptic cell, however, not only the affinity to the receptor is important, but also the so-called "intrinsic activity". This refers to the functional coupling strength of the activated receptor (ligand-receptor complex) with the downstream intracellular effectors (G-proteins or second messenger systems) or the type and extent of the conformational change of a receptor.

Substances that bind to a transmitter receptor with high affinity but activate the receptor only to a weak cellular response are called *partial agonists*. Partial agonists play an important role in psychopharmacology because their high affinity for the receptor with weak intrinsic activity causes them to act as antagonists in disorders resulting from overactivity of a transmitter. In contrast to full antagonists, however, they usually have fewer side effects due to their weak physiological activity.

3.3 Regulation of Transmitter Release by Presynaptic Autoreceptors and Heteroreceptors

The synthesis and release of a transmitter is regulated by presynaptic receptors through feedback inhibition. These so-called autoreceptors are mostly metabotropic receptors coupled to a G_i-protein. Activation of autoreceptors inhibits transmitter synthesis and release. The action of autoreceptors leads to an interesting pharmacological phenomenon: namely, blockade of feedback inhibition by an antagonist of the autoreceptor leads to an *increase* in transmitter release. This allows an antagonist to functionally act like an agonist. Although the same receptor type is then blocked at the postsynaptic neuron, the effect of the transmitter at other postsynaptic receptor types is enhanced. A good example of this is the effect of the aphrodisiac yohimbine, which increases the release of the monoamine transmitter norepinephrine by blocking α2-adrenergic autoreceptors, thereby causing the activation of postsynaptic α1- and β-adrenergic receptors.

Autoreceptors usually have a higher affinity for the transmitter than the postsynaptic receptors, so that low concentrations of a transmitter or its agonist reduce transmitter release without activating the postsynaptic receptors. An agonist can thus functionally act like an antagonist—a principle of action that is also used in psychopharmacology. In the 1970s, for example, attempts were made to activate dopaminergic autoreceptors relatively selectively by low doses of the dopamine receptor agonist apomorphine in order to reduce the pathologically increased dopamine release in patients with schizophrenia.

In addition to autoreceptors, presynaptic heteroreceptors are also known, whereby a transmitter A regulates the release of transmitter B via presynaptic axon terminals. Important examples are glutamatergic afferents on dopaminergic axon terminals in the basal ganglia, which promote the release of dopamine via glutamate heteroreceptors, and cannabinoid receptors on glutamatergic and GABAergic synapses.

Interestingly, in recent years it has been shown that complex formation and functional coupling of postsynaptic heteroreceptors can apparently occur in some brain areas. For example, it has been demonstrated that metabotropic serotonin receptors and metabotropic glutamate receptors can occur as functionally antagonistically coupled dimers and that this coupling can be disturbed in schizophrenic patients (Wischhof and Koch 2016).

3.4 Acetylcholine

For historical reasons, the transmitter acetylcholine (ACh) is discussed here first, because ACh is the first chemical messenger to be scientifically described and physiologically characterized. Around 1921, Otto Loewi studied the regulation of the heartbeat by the vagus nerve. Stimulation of the vagus nerve in frogs reduced the beating force and frequency of the heart. This effect was also achieved by extracts from vagus nerves, so that Loewi called the hitherto unknown transmitter "vagus substance". A little later it was shown by Henry Dale that the transmitter was ACh.[3] ACh acts as a transmitter at postganglionic neurons of the parasympathetic nervous system, at the neuromuscular end plate on muscles and in the brain.

■ Synthesis and Degradation

ACh is synthesized from choline (formed from lecithin = phosphatidylcholine) and acetyl-CoA ("activated acetic acid") under the catalytic action of the enzyme choline acetyltransferase (ChAT) in the presynaptic axon terminals. ACh is degraded by the enzyme acetylcholine esterase (AChE). AChE inhibitors play an important role in the pharmacotherapy of dementias (e.g. donepezil), as neurotoxins (e.g. sarin and nowitchok) and in pest control (parathion, E605).

■ Cholinergic Systems

ACh is the transmitter at the neuromuscular endplate, i.e. at the synapses that motoneurons form on muscles. In the brain, ACh plays an important role as a transmitter in neuroanatomically relatively well-defined systems. The mapping of cholinergic neurons and their projections are essentially based on the immunohistochemical detection of the synthesis enzyme ChAT and on the histochemical detection of AChE.

1. Basal forebrain: medial septum (CH1 cell group), nucleus of the diagonal, or horizontal, limb of Broca's band (CH2 and CH3), and nucleus basalis (CH4). These nuclei supply ACh to almost the entire forebrain, with the medial septum essentially projecting to the hippocampus and the nucleus basalis to the cortex and amygdala. The nuclei of Broca's ligament project to the cortex and to the olfactory bulb. The functions of ACh in the brain have been deduced from pharmacological studies with agonists and antagonists, and from failure symptoms in transgenic knock-out animals and after specific cholinergic lesions. The cholinergic forebrain areas play an important role in sensory processing, attention, learning and memory. Dysfunctions of the cholinergic systems are significantly involved in the symptomatology of senile dementias. Therefore, AChE inhibitors (e.g. donepezil) are used pharmacotherapeutically.

2. Ponto-mesencephalic cholinergic nuclei: The pedunculopontine (CH5) and laterodorsal (CH6) tegmental nuclei project to the thalamus and substantia nigra, as well as to the formatio reticularis and cerebellum. The cholinergic brainstem systems are important for the general state of arousal and wakefulness (arousal), for the reward system and the sleep-wake rhythm. They essentially form the anatomical substrate for the *Ascending Reticular Activation System* ("ARAS") described by Moruzzi and Magoun

---

3 Loewi and Dale received the Nobel Prize for their work in 1936.

3

around 1950. Functional disturbances of the CH5 and -6 cell groups may be involved in narcolepsy. In addition, ACh is also present as a transmitter in the habenula and parabigeminal nucleus and in interneurons of the striatum.

ACh affects the postsynaptic neuron via two classes of receptors, each named after its most potent agonist:

1. Muscarinic receptors[4]: Five different G protein-coupled muscarinic receptor types (M1–M5) have now been characterized: M1, M3 and M5 receptors are G_q -protein-coupled receptors that increase the formation of diacylglycerol and inositol triphosphate, whereas M2 and M4 receptors inhibit cAMP synthesis via coupling to G_i-proteins. Activation of muscarinic receptors opens or closes K^+, Ca^{2+}, or Cl^- channels, allowing ACh at muscarinic receptors to both depolarize and hyperpolarize the cell. Agonists for muscarinic receptors are muscarinic and oxotremorine, whereas scopolamine and atropine are antagonists.

2. Nicotinic receptors[5]: The nicotinic ACh receptor is considered the prototypical ligand-gated ion channel. It is a cation channel consisting of five subunits (mostly 2 x α, β, γ and δ) that allows cations such as Na^+, Ca^{2+} and K^+ to pass. Agonists at nicotinic ACh receptors in the CNS are nicotine and anatoxin A, antagonists are mecamylamine and α-bungarotoxin (a venom of the poisonous snake *Bungarus*), as well as some α-conotoxins (i.e. venoms of marine cone snails of the family *Conidae*, which are also used therapeutically, e.g. in pain treatment).

3.5   Amino Acid Transmitters

3.5.1   Glycine

Glycine is the simplest amino acid and occurs in all cells of the brain as a product of protein metabolism. As a messenger, glycine has two different functions in the CNS: as an inhibitory transmitter and as a modulator of the glutamatergic N-methyl-d-aspartate (NMDA) receptor (▶ Sect. 3.5.3).

Glycine is formed from the amino acid serine by *trans-hydroxymethylase*. Glial and neuronal transporters are responsible for the uptake from the synaptic cleft. Glycine is degraded by glycine dehydrogenase and aminomethyl transferase.

The glycine receptor is a chloride channel consisting of five subunits (α1–4- and β-subunits), its activation leads to hyperpolarization (inhibition) of the cell. The distribution of glycine receptors in the CNS shows that the importance of glycine as an inhibitory transmitter increases along the neuraxis from the forebrain to the pons to the medulla oblongata and spinal cord. The glycine receptor is blocked by the plant toxin strychnine, which is why it is also called the strychnine-sensitive glycine receptor, to distinguish it from the strychnine-insensitive glycine binding site on the NMDA receptor. From the symptomatology of strychnine intoxication (convulsions and muscle spasms), it appears that glycine acts primarily on motor and premotor neurons as an inhibitory transmitter. The rare neurological disease hyperekplexia (*startle disease*), which results from a point mutation in the gene of the α1-subunit and correspondingly reduced glycine binding, is characterized by explosive motor behavior.

Glycine also acts as a positive allosteric modulator at the NMDA glutamate receptor (▶ Sect. 3.5.3), where it prolongs the opening time of the ion channel. Glycine is necessary for NMDA receptor function, as is known from studies with the blocker of

---

4   Muscarin belongs, together with muscimol and ibotenic acid, to the poisons of the fly agaric (*Amanita muscaria*).

5   Nicotine is an alkaloid found in the tobacco plant (*Nicotiana tabacum*).

this binding site (7-chlorokynurenic acid), which can act like an NMDA receptor antagonist. In contrast, agonists of this glycine binding site, such as the antibiotic D-cycloserine, exert a stimulatory effect on the NMDA receptor and therefore act as *cognitive enhancers* (▶ Sect. 3.5.3).

3.5.2 γ-Aminobutyric Acid

γ-Aminobutyric acid (GABA) is a neuron-specific amino acid formed from glutamate (▶ Sect. 3.5.3) by decarboxylation using glutamate decarboxylase (GAD). GAD occurs in two versions. GAD_{65}[6] is found primarily in the synapse, whereas GAD_{67} is found throughout the cytoplasm. Immunohistochemical detection of $GAD_{65/67}$ in the CNS is one of the most important methods for studying the distribution of GABAergic neurons. GABAergic *interneurons* are commonly found in the cerebellum, thalamus, striatum, throughout the cortex, hippocampus, and amygdala. GABAergic interneurons occur in a variety of morphological-physiological variations. GABAergic *projection neurons* are found in the striatum (caudate nucleus, putamen, and nucleus accumbens), reticular thalamic nucleus, substantia nigra pars reticulata, globus pallidus, and cerebellum. GABA is removed from the synaptic cleft by high-affinity GABA transporters (GAT) and transported back to either the presynapse (GAT-1 and -4) or glial cells (GAT-2 and -3). The degradation of GABA to succinate semialdehyde is carried out by a transaminase.

Over 60 years ago, it was shown by Eugene Roberts and Ernst Florey that GABA is an inhibitory transmitter (*Factor I*). GABA can act through two different receptor subtypes: $GABA_A$- and $GABA_B$-receptors. More than 20 different subunit

variants have been described for the ligand-activated $GABA_A$ receptor. The combinatorics of these subunits is crucial for the physiological and pharmacological properties of the pentameric chloride channel. Binding of GABA between the α- and the β-subunit leads directly to an increase in chloride permeability and thus, under normal conditions (high extracellular chloride concentration), to a hyperpolarization of the cell.[7]

Approximately 75% of $GABA_A$ receptors contain binding sites for benzodiazepines, which prolong the opening time of the channel. Benzodiazepines have been used clinically since the 1960s—chlordiazepoxide (Librium®), diazepam (Valium®)—as anxiolytics and sedatives, although it should be noted that dependence can occur with chronic treatment. $GABA_A$ receptors are also the site of action of some sleep and narcotic drugs (e.g., pentobarbital and propofol), which also act as allosteric modulators of channel opening kinetics. Alcohol (ethanol), barbiturates, anesthetics, and neuroactive steroids (e.g., allopregnanolol, a breakdown product of the pregnancy hormone progesterone) bind to the β-subunits.

$GABA_B$ receptors are dimeric metabotropic G_i protein-coupled receptors that inhibit adenylate cyclase, decrease the conductance of Ca^{2+} channels and increase the permeability of K^+-channels. Binding of GABA to $GABA_B$-receptors leads to a slow onset and long lasting hyperpolarization of the cell. $GABA_B$-receptors can also regulate the release of GABA as presynaptic autore-

---

6   65 and 67 represent the atomic masses of the enzymes in kilodaltons (kDa).

7   In a strict sense, it is incorrect to speak of excitatory or inhibitory transmitters, because the effects of opening ion channels on the membrane potential of the cell depend on the ion distribution at the membrane. For example, in the embryonic brain, a high intracellular chloride concentration is sometimes present, so that GABA receptors have an excitatory effect on the cell in the early stages of CNS development by an efflux of chloride from the cell.

ceptors and the release of acetylcholine, glutamate and monoamines as well as neuropeptides as heteroreceptors.

GABA transporters are found in glial cells and in vesicles in the presynaptic terminals of GABAergic neurons. Four different variants of these transporter proteins have been identified. The binding of GABA to the transporter and the uptake of the transmitter into the vesicles are regulated by an electrochemical gradient of sodium and chloride ions.

GABA is the most important inhibitory transmitter in the brain. Therefore, deficits in GABAergic transmission result in severe neurological and psychiatric disorders. Prominent disorders in which defects of the GABAergic system play an important role are anxiety and fear disorders, schizophrenia, impulse control disorders, epilepsy and Huntington's disease.

3.5.3 Glutamate

The amino acid L-glutamate is the most important excitatory neurotransmitter in the CNS. Glutamate is formed by transamination from 2-oxoglutarate or oxaloacetate (metabolites of the citrate cycle). Glutamate is present in relatively high concentrations throughout the brain, both in local circuits and in projection neurons. The glutamatergic projections from the cortex to subcortical areas (hippocampus, amygdala, basal ganglia and via the pyramidal tract into the spinal cord) as well as the thalamo-cortical and cerebellar connections are particularly important for cognitive, emotional, sensory and motor functions.

Selective transporters in the membrane of neurons and glial cells play an essential role in regulating the transmitter action of glutamate. Glutamate uptake by these transporters is driven by the electrochemical gradients of Na^+ and K^+. Currently, five different glutamate transporters (*Excitatory Amino Acid Transporters*, EAAT1-5) are known, which differ in their structure, localization and pharmacology. Deficits in glutamate uptake resulting from disturbances in EAAT function during cellular stress or energy deficiency are involved in the development of neurodegenerative diseases.

Ionotropic and metabotropic receptors mediate the cellular effects of glutamate. The three different ionotropic glutamate receptors are named after their specific agonists: NMDA (*N-methyl-d-aspartate*), AMPA (α-amino-3-hydroxy-5-methylisoxazole-propionic acid), and kainate. The metabotropic glutamate receptors (mGluR) are also subdivided into three classes and further into different subtypes (▶ Sect. 3.5.3.1).

The NMDA receptor is a ligand *and* voltage-gated ion channel composed of four transmembrane protein subunits and is permeable primarily to Na^+ and Ca^{2+}. The permeability of the NMDA receptor to cations can be regulated by several other factors; the binding of glycine as a cofactor at the so-called strychnine-insensitive glycine binding site (so called because it is not blocked by the glycine receptor antagonist strychnine) is a necessary prerequisite for NMDA receptor function. In addition, protons (pH), Zn^{2+} and polyamines (e.g., spermidine) regulate the receptor's channel properties. Interestingly, the ion permeability of the NMDA receptor is voltage dependent and blocked at a membrane potential up to about −35 mV by Mg^{2+}. Only when this threshold is exceeded is the Mg^{2+} block removed and the channel becomes permeable to Na^+ and Ca^{2+}. The ligand *and* voltage dependence of the NMDA receptor accounts for its special property as a "coincidence detector" in *long-term potentiation* (LTP) as a cellular basis of learning. When a neuron is strongly excited, e.g. by rapid repetitive ("tetanic") stimulation or simultaneous activation by two afferent inputs of the neuron, the channel becomes active and allows the influx of Ca^{2+} ions, which contribute to the activation of numerous intracellular processes (e.g. activation of protein kinases and transcription factors). Because of these properties, the NMDA receptor is critically involved in learning, memory, and developmental processes. NMDA receptor antagonists such as ketamine (short-term narcotic) and phencyclidine ("angel dust") impair learning and lead to alterations in consciousness and even loss of consciousness (anesthesia). Agonists of the NMDA receptor (e.g., D-cycloserine) can improve learning and memory functions as *cognitive enhancers* and can be used in the treatment of dementia and in exposure therapy for anxiety disorders (Ressler et al. 2004). The most important antagonists are the (noncompetitive) channel blockers ketamine, phencyclidine and dizocilpine (MK-801) and the competitive antagonists AP-5 (2-amino-5-phosphonopentanoic acid) and CGS19755.

NMDA receptors are widely distributed in the brain, especially in the cortex and the limbic system (e.g. in the amygdala and the hippocampus). They are also involved in neurotoxic processes. A number of endogenous and exogenous neurotoxins are NMDA receptor agonists. Quinolinic acid is a metabolite of the amino acid tryptophan. Domoic acid is produced by algae and ingested by molluscs, which can then cause intoxication with hippocampal damage in humans after consumption. Ibotenic acid is a toxin and neurotoxin found in fly agarics.

AMPA and kainate receptors are differentiated from NMDA receptors as non-NMDA receptors. They are ligand-gated ion channels with preferential permeability to sodium ions. AMPA receptors consist of four subunits, of which there are eight subtypes and several splice variants.[8] Here, the combination of subunits determines the properties (affinity for glutamate, opening kinetics, etc.) of the channel. AMPA receptors are activated by AMPA, quisqualate and glutamate and inhibited by CNQX. AMPA receptors are distributed throughout the brain and are significantly involved in the generation of action potentials in glutamatergic pathways. During long-term potentiation (LTP), new synthesis and postsynaptic incorporation of AMPA receptors occurs.

Kainate receptors are ligand-gated sodium channels consisting of four subunits, of which five different types and several splice variants have been characterized so far. Kainate receptors are responsible for long-lasting excitatory postsynaptic potentials and play a role in the temporal integration of neuronal signals in the context of synaptic plasticity.

The density of postsynaptic transmitter receptors and thus the efficacy of a transmitter are highly variable. Receptors can be removed from the active zone of the postsynapse by different cytoskeletal proteins (e.g. by actin), or additional receptors can be incorporated. Thus, the "physiological strength" of a synapse can be regulated (Collingridge et al. 2004).

3.5.3.1 Metabotropic Glutamate Receptors (mGluR)

In addition to ligand-gated ion channels, glutamate can also activate metabotropic, i.e. G-protein-coupled receptors. The mGluR are divided into three groups: Group I increases the formation of inositol triphosphate by activating phospholipase C and has an activating effect. Group II and group

---

8   When a gene (DNA sequence) is transcribed into the corresponding messenger RNA, a pre-mRNA is first formed, which consists of coding and non-coding nucleic acid sequences. Slightly different mature mRNA molecules can then be formed from the same gene by so-called "*alternative splicing*", depending on the position at which the splicing enzymes cut out the non-coding sequences. If these mRNAs are then translated into the corresponding protein in ribosomes, these differences in amino acid sequence may result in somewhat different biochemical and physiological properties. This increases the diversity of the protein encoded by a gene.

III, on the other hand, inhibit adenylate cyclase. Group I mGluR are mainly localized in the cortex, limbic system, striatum and cerebellum and play important roles in learning and memory. Behavioral studies in mGluR knock-out mice and pharmacological tests with specific mGluR antagonists showed that mGluR are involved in a variety of motor and cognitive functions.

Processes at the level of glutamate receptors play an important role in neurodegenerative diseases and stroke. The uncontrolled activation of NMDA receptors and the resulting influx of Ca^{2+} into the cell leads to osmotic and metabolic cell stress, which results in the death of neurons.

3.5.4   Monoamines

The monoamines represent an important class of neurotransmitters. They include the catecholamines adrenaline (epinephrine), noradrenaline (norepinephrine) and dopamine, and the indoleamine serotonin. The catecholamines are synthesized from amino acids:

L-Tyrosine is formed from the amino acid L-phenylalanine and hydroxylated by the enzyme tyrosine hydroxylase to dihydroxyphenylalanine (L-DOPA), which is converted to dopamine by DOPA decarboxylase. Under the action of dopamine-β-hydroxylase, this becomes noradrenaline, which in turn becomes adrenaline under enzymatic cleavage of a methyl group. Here it should be emphasized again that ultimately the presence or absence of the corresponding enzymes determines the transmitter equipment of a neuron.

The first detection of catecholamines in the brain was achieved around 1960 by the Swedes Falck and Hillarp using histochemical methods. At that time, they already created "maps" of the distribution of catecholaminergic neurons in the peripheral nervous system and in the CNS, which are essentially still valid today.

The control of the effect of the monoamine transmitters takes place on the one

hand via their reuptake and on the other hand via their enzymatic degradation.

The reuptake of transmitters from the synaptic cleft into the presynaptic terminals occurs through transporter proteins. The pharmacological manipulation (mostly inhibition) of monoamine transporters plays an important role in the treatment of depression and ADHD. The addictive effects of amphetamine and cocaine are due to the inhibition of the dopamine transporter.

The enzymes monoamine oxidase (MAO) and catechol ortho-methyltransferase (COMT) are responsible for the degradation of monoamines. By blocking these degradation enzymes, the effect of the transmitters can be prolonged. This is exploited in psychopharmacology: Numerous antidepressants act by inhibiting the degradation or reuptake of norepinephrine and serotonin; similar effects are achieved in the therapy of Parkinson's disease by COMT inhibitors and the thereby mediated reduced degradation of dopamine in the synaptic cleft. However, the delayed onset of the therapeutic effect of antidepressants (usually several weeks) suggests that the effect is not due to the acute pharmacological action of the reuptake inhibitors, but rather to plastic processes at the monoaminergic synapses.

3.5.4.1 Noradrenaline and Adrenaline

These two monoamines were initially known as messengers in the autonomic nervous system. In the CNS, their transmitter function has only been described for about 60 years, whereby adrenaline has a significantly lower significance in the brain than noradrenaline. Therefore, the effects of norepinephrine will be primarily discussed here. As mentioned above, norepinephrine is formed from dopamine under the action of dopamine-β-hydroxylase and enters synaptic vesicles via the vesicular monoamine transporter. This vesicular transporter can be inhibited by the alkaloid reserpine. Reserpine used to play an important role in the pharmacotherapy of hypertension and schizophrenia as well as in experimental neuropharmacology (especially in the establishment of an animal model for Parkinson's disease).

Noradrenaline acts through different classes and subtypes of metabotropic receptors:

- $\alpha1_{A-D}$ receptors are coupled to a G_q protein and stimulate phospholipase C activity, ultimately increasing intracellular Ca^{2+} concentration.
- $\alpha2_{A-C}$ receptors activate a G_i protein and inhibit adenylate cyclase, and
- β-Receptors increase the activity of adenylate cyclase via a G_s protein. Central nervous β-receptors are stimulated by isoproterenol and inhibited by propranolol (a prototypical "beta-blocker").

Noradrenergic neuron groups are found primarily in the pontine and medullary brainstem. The most prominent noradrenergic nucleus is the locus coeruleus, which in humans consists of only about 60,000 neurons and supplies almost the entire forebrain with noradrenaline via a dorsal and a ventral projection bundle. The nucleus of the tractus solitarius projects mainly to the hypothalamus, while the noradrenergic cells of the lateral tegmental field innervate the spinal cord in particular.

Behavioral pharmacological studies in humans and experimental animals have shown that norepinephrine plays an important role in defensive responses (stress, fear, and anxiety), in attention and arousal, in the signal-to-noise ratio of sensory systems, and in the consolidation of aversive memory content. Beta-blockers may therefore prevent the storage of negative experiences. The drug atomoxetine, used for the treatment of ADHD, increases the effect of noradrenaline by decreasing its reuptake by transporters. The tricyclic antidepressants such as imipramine and desipramine also act by inhibiting noradrenaline transporters.

3.5.4.2 Dopamine

Dopamine, too, was initially regarded only as a synthesis precursor of adrenaline, before the Swedish pharmacologist Carlsson[9] in particular demonstrated in the 1950s the importance of dopamine in behavioural control, especially in the initiation of movements in the basal ganglia and the control of executive functions in the frontal brain.

Dopamine is synthesized by decarboxylation of L-DOPA and packaged into vesicles by transporters. The degradation enzymes MAO and COMT as well as reuptake by cytoplasmic transporters contribute to the termination of its effect. Cocaine and amphetamine inhibit this transporter, thereby enhancing the effects of dopamine. The substance methylphenidate (e.g. "Ritalin") used for the treatment of ADHD also acts as an inhibitor of the cytoplasmic monoamine transporters.

An anatomical distinction is made between the following dopaminergic systems (cell groups and their projections):

- The retrorubral nucleus projects to the dorsal and ventral striatum (caudate nucleus, putamen, and nucleus accumbens), the perirhinal and piriform cortex, and the amygdala.
- The substantia nigra pars compacta projects primarily to the dorsal striatum (caudate nucleus and putamen). This projection is called the *nigrostriatal* system.
- The ventral tegmental area (VTA) innervates various cortex areas, especially the prefrontal cortex (*mesocortical* system), the nucleus accumbens (*mesoaccumbal* system), the amygdala and the hippocampus (*mesolimbic* system).
- The *tuberoinfundibular* system projects from the hypothalamus to the pituitary

---

9   Arvid Carlsson received the Nobel Prize in Medicine for his discoveries in 2000, together with Eric Kandel and Paul Greengard.

gland, where dopamine inhibits prolactin release. This projection is also relevant in psychiatry, since classical neuroleptics act as dopamine receptor antagonists and block the inhibitory effect of dopamine on prolactin release leading to undesirable side effects (breast growth, loss of libido), especially in male patients, via hyperprolactinaemia triggered in this way.

In addition to the dopaminergic projection systems, there are also local dopaminergic cell groups in the bulbus olfactorius and in the retina of the eye.

Dopamine acts through G protein-coupled receptors that occur both postsynaptically and presynaptically (autoreceptors) and are divided into two groups (D1 and D2).

The group of D1 receptors includes D1 and D5 receptors, which are linked to adenylate cyclase via a G_s-protein, promote the formation of cAMP, and are mostly *excitatory* to the postsynaptic neuron. D1 receptors are widely distributed in the dopaminergic termination areas (basal ganglia, cortex), in the retina and on vascular smooth muscle. Selective D1 receptor agonists are SKF38393 and dihydrexidine, and SCH23390 acts as an antagonist. D5 receptors are mainly found in the thalamus and hippocampus.

The group of D2 receptors includes D2, D3 and D4 receptors, which *inhibit* cAMP formation via a G_s-protein and the inhibition of adenylate cyclase. Since cAMP is constantly degraded by the enzyme phosphodiesterase, this decreases its intracellular concentration. D2 receptors are also common postsynaptically in the basal ganglia, cortex, limbic system, retina, and pituitary gland, and also occur as presynaptic autoreceptors, especially in the frontal cortex. Binding to autoreceptors decreases the synthesis and release of dopamine. D3 and D4 receptors are particularly abundant in the limbic system and frontal cortex. Selective

agonists at D2 receptors are quinpirole and PHNO (9-hydroxynaphthoxazine); antagonists (which are important mainly because of their possible action as neuroleptics) are sulpiride, raclopride and haloperidol.

For the experimental investigation of the function of dopamine, genetic manipulations (knock-out mice) and selective lesions by neurotoxins in experimental animals have proved to be helpful methods, in addition to psychopharmacological studies in humans and experimental animals.[10] The functions of dopamine are diverse and must be considered separately for the anatomically different systems:

— *Nigrostriatal system:* Dopaminergic neurons in the substantia nigra pars compacta project to the caudate nucleus and the putamen. In the dorsal striatum, dopamine plays an important role in the initiation of movement (locomotion, reaching movements) and in the selection of alternative behavioral programs. Deficiency of dopamine in the nigrostriatal system due to degeneration of dopaminergic neurons in the substantia nigra is responsible for the characteristic motor symptoms of PD (rigor, akinesia, postural instability). Since dopamine cannot cross the blood-brain barrier, it is not suitable for the treatment of dopamine deficiency. Instead, the synthesis precursor of dopamine, L-DOPA

---

10 Recently, optogenetic methods have also become available as further molecular biological methods. In this process, ion channels are expressed by introducing light-sensitive channel rhodopsins into the genome of nerve cells by viral vectors. These ion-channels can be opened by targeted illumination via intracerebral light probes. In this way, these cells can be activated or inhibited with the highest temporal precision. Another modern method is the "DREADDs" (*Designer Receptors Exclusively Activated by Designer Drugs*). Here, artificially produced G protein-coupled receptors are also specifically expressed in certain nerve cells, which can only be activated by artificially produced ligands (e.g. clozapine *N-oxide*).

(levodopa),[11] which can cross the blood-brain barrier, is used. To delay the breakdown of the newly synthesized dopamine, levodopa is usually combined with COMT- or decarboxylase-inhibitors.

– *Mesolimbic and mesoaccumbal systems:* Dopaminergic neurons in the ventral tegmental area (VTA) project to the ventral striatum/nucleus accumbens and other limbic areas such as the amygdala. In the ventral striatum (nucleus accumbens), dopamine is involved in behavioral control in the context of reward. In particular, the driving (appetitive) motor component (e.g., approaching palatable food, sexual partners, etc.) is controlled by dopaminergic neurons. Based on numerous human and animal studies over the past 60 years, the mesolimbic-mesoaccumbal dopamine system is referred to as the brain's "reward system".[12] Consumption of common stimulants (e.g., coffee, tea, tobacco/nicotine, alcohol) as well as life-sustaining behaviors such as food intake, drinking, sex, and brood care are dependent on activation of the mesoaccumbal dopamine system in humans and animals. Addictive drugs and psychostimulants such as cocaine and amphetamine, as well as the opioids, also act preferentially via the release of dopamine in the nucleus accumbens. The non-substance addictions (e.g., pathological gambling) have also been shown to be due to dysfunction of the mesoaccumbal dopamine system. The reduction of the effect of mesoaccumbal dopamine (e.g. through the administration of neuroleptics) leads to the inability to feel pleasure and joy (anhedonia), because the blockade of the dopamine receptors prevents the expectation of reward.

– *Mesocortical system:* Another ascending projection of dopaminergic neurons in the VTA innervates the frontal cortex. D1 and D2 receptors are present in the cortex on both glutamatergic and GABAergic interneurons. Dopamine plays an important role in the prefrontal cortex in working memory and executive functions such as action planning, action preparation and impulse control. In this context, the studies by Goldman-Rakic in particular showed that there is an optimum of activity of the mesocortical dopamine system according to the respective cognitive requirements of the test task, i.e. that both too much and too little dopamine can impair the performance of the subjects.

The function of the dopaminergic forebrain systems is relevant for the understanding of addiction, as well as for depression, ADHD and schizophrenia. The "dopamine hypothesis" of schizophrenia assumes, in simplified terms, an overfunction of mesoaccumbal and mesolimbic systems, whereas there is a dopamine deficiency in the mesocortical system. The neuroleptics developed in the 1950s for the treatment of schizophrenia are primarily dopamine D2 receptor antagonists, alleviate acute psychotic symptoms, but have numerous side effects due to the blockade of striatal and pituitary dopamine receptors, such as Parkinsonian-like movement disorders (extrapyramidal symptoms) and an increase in prolactin release (hyperprolactinaemia). Atypical antipsychotics

---

11 The prefixes "L-" or "D-" in the case of natural substances go back to their mirror image isomerism (chirality or handedness). Mirror image isomers are similar in all physicochemical properties (solubility, melting point, etc.), but differ in their ability to turn polarized light either to the right (D = *dexter*, Latin right) or to the left (L = *laevus*, Latin left). Amino acids occur in nature only in the L-form.

12 The discovery of the reward system essentially goes back to brain stimulation experiments on rats conducted by James Olds and Peter Milner in the 1950s. Together with the subsequent animal experiments on intracranial self-administration of drugs by Bartley Hoebel, they laid the foundation for modern research on the reward system.

(clozapine, risperidone, olanzapine) show a significantly weaker D2 receptor blockade and even increase dopamine release in the frontal brain after chronic administration, resulting in an improvement of the patients' cognitive performance.

Third-generation antipsychotics include partial agonists of D2 receptors, which have a high affinity for the receptor but result in only a weak cellular response (about 30% of full agonist). The property of high affinity combined with weak coupling of receptor and G protein results in the drug (e.g., aripiprazole) acting as an agonist when dopamine levels are low, whereas an antagonistic effect occurs when dopamine levels are in excess. This mechanism of action largely avoids complete blockade of the dopamine receptors and thus the side effects described above, while still achieving therapeutic attenuation of dopaminergic hyperfunction (Koch 2007).

3.5.4.3 Serotonin (5-Hydroxytryptamine Serotonin, 5-HT)

The indolamine 5-HT was discovered around 1950 as a vasoconstrictive agent in the viscera and blood serum (hence the name "sero-tonin"). A short time later, the structural similarity of 5-HT with the psychedelically acting ergot alkaloid lysergic acid diethylamide (LSD) was pointed out, whereupon the role of 5-HT as a neurotransmitter in the brain was increasingly investigated.

5-HT is formed from the essential amino acid tryptophan, which enters the brain through an amino acid transporter across the blood-brain barrier.[13] The degradation of 5-HT takes place through the MAO as well as through aldehyde dehydrogenase to 5-hydroxyindoleacetic acid, which can be detected in the blood and urine and thus provides information about the 5-HT metabolism.

In addition to the enzymatic degradation of 5-HT, reuptake from the synaptic cleft into the presynapse is the most important mechanism for terminating the effect of this transmitter. Selective cytoplasmic serotonin transporters, which have gained considerable therapeutic importance for the treatment of depression, ensure reuptake.

5-HT is produced by neurons of the midbrain, pons and medulla, which send serotonergic projections to many parts of the brain. Innervation of the forebrain (cortex, limbic system, striatum, bed nucleus of the stria terminalis) with 5-HT is mainly provided by the dorsal raphé nuclei of the midbrain. The serotonergic nucleus groups in the medulla project into the ventral horn of the spinal cord, while those of the pons supply the thalamus with 5-HT.

5-HT receptors show a considerable diversity of subtypes. In addition to the metabotropic receptors of the 5-HT1 family with seven different subtypes, the 5-HT2 family (subtypes 5-HT2A-C) and the 5-HT4, 5, 6, 7, there is also a class of ionotropic 5-HT3 receptors. 5-HT1 receptors frequently occur as autoreceptors that inhibit 5-HT release. 5-HT2 receptors couple to G_q-proteins and activate phospholipase C. This ultimately results in excitation of the neuron

---

13 The fact that the starting material for serotonin is an *essential* amino acid has led to interesting dietary considerations in the past. Essential amino acids are needed by the organism, but cannot be synthesized in the body, but must be ingested through food. Therefore, there have been considerations that by supplying nutrients rich in trypto-phan, the serotonin level in the brain can be increased. However, since the two synthesis enzymes of serotonin (tryptophan hydroxylase and amino acid decarboxylase) are already saturated in a normal diet, such dietary measures have little effect (just as a car with a full tank of gas does not reach a higher speed than one with an almost empty tank). However, a diet low in tryptophan can indeed lead to serotonin deficiency. This is also exploited experimentally, for example to make test subjects more aggressive in the short term.

via inhibition of K^+ conductance and increase in intracellular Ca^{2+} concentration. 5-HT3 receptors are ligand-gated cation channels and are involved in the excitation of neurons. 5-HT4 receptors are coupled to adenylate cyclase via a G_s-protein and have an excitatory effect on the neuron by lowering the cAMP concentration via a reduction in K^+ conductance. The intracellular action of 5-HT5 receptors is unknown. 5-HT6 and -7 receptors activate the adenylate cyclase via a G_s-protein and have a depolarizing effect.

5-HT1 receptors are located as autoreceptors in the raphé nuclei and the hippocampus. They are also found postsynaptically in the basal ganglia. 5-HT2, -4, and -6 receptors are found in the cortex, hippocampus, and amygdala.[14] 5-HT3 receptors have been detected in the entorhinal cortex, area postrema, and peripheral nervous system. 5-HT7 receptors are found in the thalamus and hypothalamus and in the limbic system.

The great functional diversity of 5-HT receptor systems and their broad distribution in the brain is the basis for the fact that 5-HT is involved in many different physiological and cognitive functions, namely learning, memory formation, pain processing, sexual behavior, food intake, mood and affect, sleep-wake rhythm and aggression. Disorders of the various serotonergic systems are accordingly involved in a wide range of disorders, e.g. depression, anxiety and fear disorders, eating and sleeping problems, schizophrenia, aggression, migraine. Disorders resulting from a deficiency of 5-HT (e.g. depression) can be treated with selective serotonin reuptake inhibitors (SSRIs) such as fluoxetine (Prozac®), paroxetine, sertraline or citalopram. SSRIs are also used in obsessive-compulsive disorders

and in the treatment of anxiety and fear disorders, although, as noted above, the onset of action is usually delayed. Overall, the effect of SSRIs has not yet been fully elucidated.

The hallucinogenic effect of LSD, which was first synthesized around 1938 by the Swiss chemist Albert Hofmann from lysergic acid (an alkaloid of ergot mushrooms), as well as of mescaline and other psychedelics, is due to partial agonism at the 5-HT2A receptor. MDMA (3,4-methylenedioxy-N-methylamphetamine, the essential ingredient in "ecstasy") also acts at postsynaptic 5-HT2A receptors, among others, and also increases the release of the monoamines 5-HT, norepinephrine, and dopamine. MDMA causes euphoria, increases self-esteem, has an entactogenic effect, and produces a feeling of social closeness. However, repeated use of MDMA can cause opposite effects (dysphoria, listlessness, feelings of loneliness). Long-term studies in humans and animal experiments have also demonstrated neurotoxic effects of MDMA.

In summary, the interplay of the three monoamine transmitters, dopamine, norepinephrine, and serotonin, is important for a variety of cognitive performances (attention, working memory, executive functions) and of emotions. Accordingly, these transmitter systems (receptors, metabolizing enzymes, transporters) represent the sites of action of numerous drugs and psychotropic medications.

3.5.4.4 Histamine

Histamine is a phylogenetically ancient messenger substance that occurs not only in vertebrates but also in insects and molluscs. In addition to its inflammation-mediating effect in the blood and skin and functions in the gastrointestinal tract, the monoamine histamine[15] also acts as a neurotransmitter in the brain. It is formed from the amino acid histidine and degraded by the enzyme MAO.

---

14 Interestingly, 5-HT2A receptors also occur as so-called heterodimers—functionally antagonistic—coupled to metabotropic glutamate receptors (mGluR 2/3).

---

15 *Histos*, ancient Greek tissue.

The histaminergic system consists of only one group of neurons that projects from the tuberomammillary nucleus of the posterior hypothalamus via the medial forebrain bundle into the forebrain, where it mainly innervates the striatum, amygdala, septum, cortex and hippocampus. There is also a descending histaminergic projection into nuclei of the brainstem and spinal cord.

Histamine acts on the postsynaptic cell via three different G-protein coupled receptors H1–H3. Activation of H1 receptors leads to excitation of the cell via increased Ca^{2+} influx. H1 receptor antagonists (mepyramine), which are used as anti-allergic drugs and penetrate the blood-brain barrier, have sedative effects. H2 receptors are coupled to adenylate cyclase via a G_s protein. H3 receptors act as inhibitory autoreceptors at histaminergic synapses, but also regulate the release of other transmitters (NA, DA, 5-HT) and neuropeptides as heteroreceptors.

The effect of histamine as a neurotransmitter in the brain is not yet precisely known. It is likely that histamine locally promotes blood flow and the permeability of blood vessels in the brain. Furthermore, the involvement of histamine as a neurotransmitter in hormone release (ACTH) in the pituitary gland, in the control of the vestibular system and in thermoregulation was shown. It is also thought to play a role in cortical arousal and in learning processes in the hippocampus. Some of the side effects of antipsychotics (e.g. weight gain and sedation after taking clozapine) are attributed to the antagonism of histamine at the H1 receptors.

3.5.5   Adenosine and ATP

Adenosine and also the ubiquitous biochemical "energy carrier" adenosine triphosphate (ATP) act as neuroactive substances in addition to their role as general energy transmitters and their role in membrane potential (▶ Chap. 4), but are usually referred to as *neuromodulators* because they do not quite fulfil the criteria for classical neurotransmitters (▶ Sect. 3.1).

Adenosine and ATP act in a para- or autocrine manner after release into neuronal tissue and diffusion to neighboring cells or to the releasing neuron itself. ATP acts in the nervous system through P2 purine receptors, which are divided into subtypes X and Y. P2X receptors are cation channels found in the spinal cord and hippocampus. P2Y receptors are coupled to G_q proteins, are relatively abundant in the brain, and increase intracellular Ca^{2+} concentration. The most important function of extracellular ATP in the brain is probably the activation of repair mechanisms after damage to neuronal tissue. ATP is released from injured neurons and, after binding to P2 receptors on glial cells, leads to the release of cytokines and the activation of phagocytosis processes. The cytokines (e.g. interleukins) activate microglial cells in particular, which then migrate amoeboid to the damaged site and ingest and resorb destroyed tissue. Astrocytes are also involved in these ATP-initiated repair processes.

Adenosine acts through four subtypes of P1 receptors (A1, A2A, A2B, and A3), all of which are G protein-coupled metabotropic receptors. A1 receptors are widely distributed in the brain and inhibit adenylate cyclase via a G_i protein. A2 receptors are also common in the brain and stimulate adenylate cyclase via a G_s protein. The A3 receptor, on the other hand, plays no role in the brain. Important antagonists of adenosine at A1 and A2 receptors are the stimulants caffeine and theophylline from the substance class of methylxanthines.

Interestingly, adenosine receptors are colocalized with dopamine receptors, especially in the basal ganglia, and are coupled antagonistically. Therefore, A1 and A2 receptor agonists act like dopamine D1 and D2 antagonists, respectively. Conversely, the adenosine receptor antagonists (theophylline and caffeine) act like dopamine receptor agonists. Presumably, some of the stimulant, excitatory, and euphoric effects of tea and coffee result from stimulation of the dopamine system.[16] These interactions are being intensively researched to support pharmacotherapy of disorders of the dopaminergic system (such as Parkinson's disease or schizophrenia) via adenosine receptors (Agnati et al. 2003).

3.5.6 Cannabinoids

Parts of the hemp plant (*Cannabis sativa*) have been used as a drug for almost 5000 years. The resin and leaves of this plant contain numerous psychoactive compounds, of which Δ^9-tetrahydrocannabinol (THC) is the most potent. THC was characterized by Mechoulam and co-workers around 1960, but the discovery of cannabinoid receptors and their endogenous ligands (endocannabinoids) was not achieved until around 1990 (Murray et al. 2007).

The most important endocannabinoids are the neurotransmitters anandamide and 2-arachidonylglycerol, synthesized from the cell membrane component arachidonic acid. These act as retrograde neuromodulators, meaning they are synthesized in the postsynaptic membrane and then diffuse to cannabinoid receptors anchored in the presynaptic membrane. There are two different cannabinoid receptors, namely CB1 and CB2 receptors. CB1 receptors are commonly found in the brain and other parts of the CNS, while CB2 receptors are predominant in the body. CB2 receptors play an important role in immune system activity. Both receptors are G_i-protein-coupled receptor proteins, which means that the CB1 receptor, which is common in the brain, inhibits adenylate cyclase in neurons. CB1 receptors are mostly localized as presynaptic heteroreceptors on synapses of various transmitter systems (dopamine, GABA, glutamate) and inhibit the release of these transmitters. A particularly high density of CB1 receptors is found in the cortex, hippocampus and basal ganglia. Endocannabinoids play an important role in the regulation of food intake, thermoregulation, pain perception, and learning and memory. They also have an antiemetic effect, i.e. they inhibit nausea.

In addition, endocannabinoids play a role in numerous ontogenetic processes, including synaptogenesis and myelination of fiber tracts, during both early CNS development (embryogenesis) and late development (adolescence). All these natural neuromodulatory effects of endocannabinoids are influenced by exocannabinoids (e.g. THC or dronabinol). Some findings suggest an involvement of a dysfunction of the endocannabinoid system by chronic intake of THC in the development of schizophrenic psychosis. In particular, cannabis use during adolescence appears to lead to permanent, adverse brain changes (Meier et al. 2012).

3.5.7 Neuropeptides

The number of neuroactive peptides (i.e. amino acid chains that are linked to each other via so-called peptide bonds[17]) significantly exceeds that of the classical transmitters. In the meantime, far more than 40 neuropeptides are known and their effects have been characterised. The discovery of the

---

16 In addition, the methylxanthines also act as inhibitors of the enzyme phosphodiesterase, which degrades the intracellular messenger substances cAMP and cGMP to AMP and GMP, respectively. By inhibiting phosphodiesterase, the increase in cAMP concentration caused by stimulation of adenylate cyclase is maintained.

17 The formation of a peptide bond is a condensation reaction in which the amino group of one amino acid is covalently linked to the carboxyl group of a second amino acid with the elimination of water. In this way, long peptide chains and ultimately proteins can be formed.

first neuropeptide, substance P, about 90 years ago by Ulf von Euler and John Gaddum, which coexists with the classical transmitter glutamate, disproved Dale's principle, which states that each neuron releases only one neurotransmitter at its synapses. In fact, very many classical neurotransmitters are colocalized with neuropeptides.[18]

There are some fundamental differences between neuropeptides and classical transmitters. The concentration of neuropeptides in the vesicles is usually lower than that of classical transmitters, and neuropeptides are always formed in the soma of the neuron and conducted to the synapse in transport vesicles via axonal transport along cytoskeletal molecules. Synthesis does not occur via specific synthesis enzymes, but rather according to the principles of gene transcription and translation on ribosomes. Neuropeptides are always synthesized as precursor peptides, which are subsequently degraded by specific proteases to yield the active neuropeptide. This construction principle allows different neuropeptides to be formed from the same precursor peptide, depending on the position in the amino acid sequence at which the proteases cleave the peptide bond. The length of neuropeptides, i.e., the number of amino acid residues from which they are constructed, varies widely (from four for cholecystokinin to 41 for corticotropin-releasing factor). The effect of neuropeptides on the postsynaptic cell usually lasts much longer than that of classical transmitters. The postsynaptic neuropeptide receptors are always G-protein-coupled metabotropic receptors.

Due to the large number of neuropeptides, only a few prominent representatives are discussed here.

3.5.7.1 Substance P (SP)

SP consists of 11–13 amino acid residues and belongs to the tachykinin family together with the neurokinins. It was isolated from brain and visceral tissue about 90 years ago by Von Euler and Gaddum and is involved in a variety of central nervous functions as a messenger substance. In neurons of the spinal cord, neurons that produce SP as well as SP receptors are abundant. SP is often colocalized with glutamate and is responsible for pain processing. In the CNS, SP is colocalized with serotonin in neurons of raphé nuclei and with acetylcholine in neurons of mesopontine nuclei and controls aversive behaviors. Importantly, SP also colocalizes with GABA in neurons of the striatum, which represent the "direct pathway" of the striatopallido-thalamic loop that is impaired in Parkinson's disease.

Receptors for SP (NK1 receptors) in the brain are located in the amygdala, septum, hippocampus, central gray, pontine reticular formation, cortex, and hypothalamus. This broad receptor distribution suggests that SP is involved in a variety of brain functions. SP plays a role in stress and anxiety, the central nervous aspects of pain, and generally increases the sensitivity of sensory systems.

3.5.7.2 Oxytocin and Vasopressin

Both neuropeptides consist of nine amino acid residues each and are formed in neurons of the paraventricular hypothalamic nucleus (PVN). The PVN consists of a magnocellular and a parvocellular part. Via the axons of magnocellular neurons, oxytocin and vasopressin are transported to the neurohypophysis, where they are released into the bloodstream, from where they reach peripheral target areas (e.g. smooth muscle in blood vessels, uterus and kidney). The parvocellular neurons of the PVN also produce both neuropeptides and, after release into the portal system of the pituitary gland, control the production of ACTH in the adenohypophysis. Receptors for both neuropeptides are found primarily in the hypothalamus, amygdala and ventral striatum.

Oxytocin—and to a somewhat lesser extent vasopressin—have received consider-

---

18 The coexistence principle was introduced by the Swedish neuroscientist Tomas Hökfelt.

able attention from neuroscientific research in recent years, as they appear to play an important role in social behavior. In particular, oxytocin appears to have a promoting effect in the emotional bonding of partners, between parents and children, and within social groups. Scientific evidence that single nucleotide polymorphisms[19] are present in the oxytocin receptor gene in patients with autistic spectrum disorder supports the assumption that autism and related disorders may be due, at least in part, to a deficit in the oxytocin system and that, accordingly, oxytocin may represent a promising treatment option (Meyer-Lindenberg et al. 2011).

3.5.7.3 Vasoactive Intestinal Peptide (VIP)

VIP consists of 28 amino acid residues and, as the name suggests, was first found in the intestine where it regulates local blood flow and secretion release. Later, VIP was also detected in the parasympathetic nervous system colocalized with acetylcholine. In addition, VIP-immunoreactive neurons and receptors are also found in the CNS, especially in the hypothalamus, nucleus accumbens and cortex. VIP influences hormonal processes in the context of social behavior via VPAC1 and 2 receptors.

3.5.7.4 Neuropeptide Y (NPY)

NPY is composed of 36 amino acid residues and is found widely in the peripheral nervous system and brain, particularly in the amygdala and the nucleus arcuatus of the hypothalamus, which projects to the PVN. NPY has a wide range of functions in the brain and controls day-night rhythms, sexual behavior, and food intake, among other functions. In the sympathetic nervous system, NPY is colocalized with norepinephrine and is involved in blood pressure elevation in the context of stress responses.

3.5.7.5 Corticotropin Releasing Factor (CRF) and Adrenocorticotropic Hormone (ACTH)

CRF was described by Vale and colleagues in 1983 as the neuropeptide that controls the hormonal stress response (formation of cortisol in the adrenal cortex) via the release of ACTH in the adenohypophysis. Thus, CRF is an elementary component of the so-called HPA axis.[20] CRF consists of 41 amino acid residues, is formed in the parvocellular neurons of the PVN from the precursor peptide prepro-CRF and is conducted to the adenohypophysis via the pituitary portal vein. ACTH consists of 39 amino acid residues, is formed in the adenohypophysis, and from there travels via the bloodstream to the adrenal cortex, where it controls the formation of gluco- and mineralocorticoids. These in turn (especially cortisol) are involved in the immediate stress reactions (stimulation of energy metabolism, suppression of the immune system, but also central nervous effects on learning and memory).

However, CRF not only controls the release of ACTH from the adenohypophysis, but is also found in neurons of the amygdala and hippocampus, where it is involved in defensive responses (fear as well as "fight-or-flight" responses). The postsynaptic action of CRF is mediated by two G protein-coupled receptors (CRF1 and 2). CRF1 receptors are commonly found in the adenohypophysis, cortex, hypothalamus, olfactory

---

19 *Single nucleotide polymorphisms* (SNPs) are relatively common changes in the nucleotide sequence of a particular gene in which one nucleotide of the DNA is replaced by another, e.g. cytosine can be replaced by thymine. Since the genetic code is redundant, this may be without consequence for the protein encoded by that gene. However, there is also the possibility that the SNP results in the substitution of an amino acid in the protein, which alters the function of that protein—for example, a particular transmitter receptor.

20 HPA: *hypothalamus-pituitary gland-adrenal cortex.*

bulb, cerebellum, and locus coeruleus, whereas CRF2 receptors are found in neurons of PVN, septum, amygdala, interstitial nucleus of stria terminalis (BNST), hippocampus, and pons. Activation of CRF1 receptors is responsible for the defensive response immediately elicited by the stressor and the accompanying aversive feeling states. Accordingly, CRF1 receptor antagonists are being considered as potential treatment options for anxiety and fear disorders and depression. The importance of CRF2 receptors is seen more in terminating the immediate stress response and initiating the recovery phase. The neuropeptide oxytocin (▶ Sect. 3.5.7) and the neurotransmitter serotonin (▶ Sect. 3.5.4) also appear to suppress the stress response.

Moreover, the formation and release of CRF and ACTH are under the feedback control of cortisol, so that stress reactions are terminated relatively quickly by this feedback. Here, glucocorticoid (GR) and mineralocorticoid (MR) receptors in the hippocampus play an important role, because their activation triggers an inhibition of CRF production in the PVN. Extremely severe or chronic stress can impair this feedback system by damaging hippocampal GR and MR such that depressive disorders and/or anxiety and fear disorders, among others, may develop as a result of overproduction of CRF, ACTH, and eventually cortisol (hypercortisolism) (▶ Chaps. 5 and 7). CRF receptor ligands are therefore an interesting starting point for the pharmacotherapy of depression and anxiety disorders (De Kloet et al. 2005).

3.5.7.6 Cholecystokinin (CCK)

CCK was first found in digestive secretions, which control the activity of bile and pancreas. It was not until around 1975 that this intestinal peptide was also detected in the brain. CCK is formed from a precursor peptide (115 amino acid residues) and post-translationally cleaved into various fragments that act as neuropeptides in the brain. Especially the short fragments CCK-4 and CCK-8 play an important role in the brain, for example in the reduction of appetite, thermoregulation, fear reactions and pain perception. Neurons producing CCK and CCK receptors are widely distributed in the brain, particularly in the amygdala, hypothalamus and entorhinal and piriform cortex, olfactory tubercle, septum, basal ganglia (substantia nigra, dorsal and ventral striatum), ventral tegmental area and central gray. CCK is often colocalized with GABA and dopamine. CCK is also discussed as a regulator of dopaminergic neurons in the ventral tegmentum in relation to neurological and psychiatric disorders (schizophrenia, Huntington's disease and Parkinson's disease).

3.5.7.7 Somatostatin

Somatostatin was isolated from bovine brains in 1973 and occurs in the brain in two variants (of 14 or 28 amino acids). Its action was discovered when pituitary extracts were studied for their effectiveness in inhibiting the release of growth and thyroid hormones from the pituitary gland. It is widely distributed in the central and peripheral nervous system and is also found in glands of the viscera. In the brain, somatostatin-producing cells are found in the hypothalamus, amygdala, hippocampus, somatosensory cortex, basal ganglia, and midbrain central gray, as well as in the nucleus tractus solitarius. Somatostatin acts through five different G-protein-coupled receptors. Its main action is the modulation of classical transmitters such as norepinephrine, dopamine and acetylcholine. In this context, somatostatin plays a role in motor function, sleep and also anxiety.

3.5.7.8 Orexins

The neuropeptides orexin A and B (or hypocretin 1 and 2) were first described in 1998. Orexin A consists of 33 and orexin B of 28 amino acids. They are formed from a pre-pro-orexin precursor peptide in neurons of the lateral hypothalamus, particularly in the posterior perifornical area, from where they influence various brain areas via long-range projections even beyond the hypothalamus. Thus, orexin-containing fibers are found in the locus coeruleus, VTA, septum, midbrain central gray, amygdala, hippocampus, cortex, and nucleus accumbens, as well as in autonomic brainstem centers (e.g., nucleus tractus solitarius). Orexins bind to orexin-1 and orexin-2 receptors, which are G_q-protein-coupled. Orexins play an important role in the regulation of food intake, reward behavior (ancient Greek *orexis*: appetite, desire), sleep-wake rhythms, stress and panic reactions, and general wakefulness ("arousal").

It is possible that disorders in the orexin system are involved in the development of narcolepsy and other forms of sleep disorders and would thus be a target system for the treatment of these diseases. Pathological changes in eating behavior (e.g., "binge-eating" disorders) could also be due to alterations in the orexin system. Because of the importance of orexins in panic behavior, orexin receptor antagonists could also represent a treatment option here (James et al. 2017).

3.5.8 Opioids

Extracts from poppy plants have been used by humans for thousands of years as opium for pain relief, relaxation and intoxication. Morphine[21] was isolated from opium poppy extracts by Sertürner in the 1800s and identified as the narcotic as well as analgesic component of opium. It was not until after this discovery that the *endogenous* opioid systems were identified. In the 1970s, the various opioid receptors and their endogenous ligands were characterized.

Endogenous opioids are encoded by three different genes and belong to three dif-

ferent families of peptides: Endorphins, Enkephalins and Dynorphins. All endogenous opioids bind with slightly different affinities to three different G_i/G_o-protein-coupled receptor subtypes, called μ-, δ- and κ-receptors. The physiological effects usually consist of hyperpolarization of the cell by inhibition of adenylate cyclase, a decrease in Ca^{2+} conductance, or an increase in K^+ conductance. All endogenous opioids are provided from precursor peptides (e.g. pro-opio-melanocortin, POMC).

Beta-endorphin consists of 31 amino acid residues, binds to μ-, δ- and κ-receptors, which are widely distributed in both the central and peripheral nervous system. Endorphin has similar effects to morphine (euphoria, analgesia, blood pressure lowering, and respiratory depression). Incidentally, POMC, the precursor peptide of beta-endorphin also contains the amino acid sequences of the signal peptides ACTH (see above) and α-melanocyte stimulating hormone, which are cut out of POMC via specific peptidases.

Enkephalins are pentapeptides that are distinguished by the fifth amino acid residue into methionine (met)enkephalin and leucine (leu)enkephalin. Both are widely distributed in the brain in neurons and nerve fibers, especially in the cortex, amygdala, basal ganglia, midbrain (central gray and ventral tegmental area), hypothalamus, and medulla. Importantly, enkephalins colocalize in GABAergic neurons of the striatum, which represent the "indirect pathway" of the striatopallido-thalamic basal ganglia loop. Enkephalins have a particularly high affinity for δ-receptors and show an antinociceptive, i.e. analgesic, effect. They probably attenuate pain at the spinal level following stress-induced activation of medullary neurons and thus mediate stress analgesia. The function of enkephalins in the midbrain and forebrain is seen in connection with the pleasurable (hedonic) aspects of reward.

Dynorphin A (8–17 amino acids) and dynorphin B (29 amino acids) are found in

---

21 First called "morphine" after Morpheus, the god of dreams in Greek mythology.

the cortex, striatum, globus pallidus, hippo-campus and hypothalamus, but also in the cerebellum and brainstem. Dynorphins have high affinity for κ- and relatively low affinity for μ- and δ-receptors. Dynorphins are also involved in pain processing, but also in food intake and respiration. In the hippocampus, dynorphins have been shown to be involved in learning.

Heroin is a semi-synthetic morphine derivative (diacetylmorphine) that induces strong euphoria and has extremely high addictive potential. Its analgesic effect is significantly stronger than that of morphine. Methadone is a synthetic opioid with high selectivity for the μ-receptor and strong analgesic as well as sedative effects, which is mainly used in drug substitution programs for the treatment of heroin addiction. Due to special pharmacokinetic properties, it has less addictive potential than heroin. The addictive effects of opioids are probably primarily due to their inhibition of GABAergic interneurons in the VTA, which increases accumbal dopamine release. Fentanyl is a synthetic opioid with potent analgesic properties (about 80 times more potent than morphine) used to treat chronic pain and pain occurring during surgical procedures. Buprenorphine and tilidine are also used as opioid receptor agonists for analgesia. Naloxone is a competitive opioid receptor antagonist that is administered in the event of opioid overdose (for information on the development and treatment of addiction, see ▶ Chap. 14).

Summary

Neuro- and psychopharmacology deals with the influence of drugs and psychotropic substances on chemical signal transduction in the brain. This, in addition to and in combination with electrical signal processing, forms the "language of the brain". As mentioned briefly and presented in more detail in further chapters of this book, both the mental well-being of a person and mental illness are directly related to the chemical processes at the synapses. To influence these processes, there are a variety of starting points at chemical synapses through which appropriate neuroactive substances can modulate the flow of information in neuronal networks and circuits. Since information processing in the brain determines our thinking and feeling, our behavior, our personality, this discipline provides important contributions to the understanding of the human psyche as well as options for the therapy of neurological and psychiatric disorders.

References

Agnati LF, Ferré S, Lluis C, Franco R, Fuxe K (2003) Molecular mechanisms and therapeutic implications of intramembrane receptor/receptor interactions among heptahelical receptors with examples from the striatopallidal GABA neurons. Pharmacol Rev 55:509–550

Collingridge GL, Isaac JTR, Wang YT (2004) Receptor trafficking and synaptic plasticity. Nat Rev Neurosci 5:953–962

Cooper JR, Bloom FE, Roth RH (2003) The biochemical basis of neuropharmacology, 8. Aufl. Oxford University Press, Oxford

Davis KL, Charney D, Coyle JT, Nemeroff C (eds) (2002) Neuropsychopharmacology. The fifth generation of progress. Lippincott Williams & Wilkins, Philadelphia

De Kloet ER, Joels M, Holsboer F (2005) Stress and the braun: from adaptation to disease. Nat Rev Neurosci 6:463–475

Gründer G, Benkert O (eds) (2012) Handbuch der Psychopharmakotherapie. Springer, Berlin

Hardingham GE, Bading H (2010) Synaptic versus extrasynaptic NMDA receptor signalling: implications for neurodegenerative disorders. Nat Rev Neurosci 11:682–696

Jacob TC, Moss SJ, Jurd R (2008) GABA(A) receptor trafficking and its role in the dynamic modulation of neuronal inhibition. Nat Rev Neurosci 9:331–343

James MH, Campbell EJ, Dayas CV (2017) Role of the orexin/hypocretin system in stress-related psychiatric diseases. Curr Top Behav Neurosci 33:197–220

Koch M (2006) Neuropharmakologie. In: Förstl H, Hautzinger M, Roth G (eds) Neurobiologie psychischer Störungen. Springer, Berlin, pp 178–219

Koch M (2007) On the effects of partial agonists of dopamine receptors for the treatment of schizophrenia. Pharmacopsychiatry 40:1–6

Meier MH, Caspi A, Ambler A, Harrington H, Houts R, Keefe RSE, McDonald K, Ward A, Poulton R, Moffitt TE (2012) Persistent cannabis users show neuropsychological decline from childhood to midlife. Proc Natl Acad Sci U S A 109:E2657–E2664

Meyer-Lindenberg A, Domes G, Kirsch P, Heinrichs M (2011) Oxytocin and vasopressin in the human brain: social neuropeptides for translational medicine. Nat Rev Neurosci 12:524–538

Murray RM, Morrison PD, Henquet C, Di Forti M (2007) Cannabis, the mind and society: the hash realities. Nat Rev Neurosci 8:885–895

Ressler KJ, Rothbaum BO, Tannenbaum L, Anderson P, Graap K, Zimand E, Hodges L, Davis M (2004) Cognitive enhancers as adjuncts to psychotherapy. Arch Gen Psychiatry 61:1136–1144

Siegel GJ, Albers RW, Brady ST, Price DL (eds) (2006) Basic neurochemistry, 7. Aufl. Elsevier Academic Press, Burlington

Tretter F, Albus M (2004) Einführung in die Psychopharmakotherapie. Georg Thieme, Stuttgart

Valenstein ES (2005) The war of the soups and the sparks. Columbia University Press, New York

Bohlen V, Halbach O, Dermietzel R (2002) Neurotransmitters and neuromodulators. Wiley-VCH, Weinheim

Wischhof L, Koch M (2016) 5-HT2A and mGlu2/3 receptor interactions: on their relevance to cognitive function and psychosis. Behav Pharmacol 27:1–11

Neurophysiology

Jakob von Engelhardt

Contents

G. Roth et al. (eds.), *Psychoneuroscience*, https://doi.org/10.1007/978-3-662-65774-4_4

The aim of this chapter is to introduce the interested reader to the neurophysiological processes that underlie the behavior of humans and most mammals. The communication of neurons is a basic prerequisite for us to perceive, process and react to environmental stimuli. If we touch a hot stove top, the stimulus is transmitted to the thalamus and cortex via nerve fibers in the peripheral and central nervous system. In these higher brain regions, the incoming information is processed and a conscious perception of the heat of the hot plate occurs, i.e. a sensation of pain in the case of a very hot plate. At the same time, a complex motor reaction is triggered by the activation of efferent nerve cells, and we pull our arm back. The association between touching the hot plate and the sensation of pain also sets in motion a learning process that is capable of changing our behavior in the long term. The sight of a hot plate, even one that is turned off, will signal danger to us in the future. Accordingly, we will exercise greater caution when dealing with hotplates and first check whether they are on or off before touching them. Motor skills, sensations and feelings involve the communication of neurons within one brain region, but also between different brain regions. The neurophysiological correlate for learning processes is a change in the strength of this communication.

Learning Objectives

After reading this chapter, the reader should understand the neurophysiological basis of neural communication and be aware of the major changes in this communication that underlie learning processes.

4.1 Membrane Potential

Membrane potential is the electrical voltage between the extracellular and intracellular space separated by the cell membrane (Hille 2001). All cells of an organism have a char-acteristic membrane potential. This is usually negative under resting conditions, meaning that the cell interior is more electrically negative than the extracellular space. Changes in this characteristic potential, which occur for example during the activity of an excitable cell, can be measured experimentally with the aid of microelectrodes. One measuring electrode is located in the extracellular space and a second in the cell or, in the *patch clamp technique*, directly on the cell membrane (◘ Fig. 4.1).

The characteristic membrane potential of excitable cells is mainly due to two conditions (Hille 2001):

1. Ions, i.e. electrically charged atoms or molecules, are unequally distributed intracellularly and extracellularly, and
2. the plasma membrane of the cell is permeable only for certain ions.

The most important charge carriers for the membrane potential of excitable cells are the positively charged cations sodium (Na^+) and potassium (K^+) and the negatively charged anion chloride (Cl^-). The concentration of positively charged Na^+ ions and negatively charged Cl^- ions is about 10–25 times higher extracellularly than intracellularly. In contrast, the intracellular concentration of K^+ ions is 30 times higher than extracellular (◘ Table 4.1). Intracellularly, anionic organic molecules are the most important carriers of negative charges.

The different distribution of ions is mainly due to the fact that proteins located in the cell membrane, so-called *pumps*, transport ions into or out of the cell against their concentration gradient. A prominent example of such a pump is the **sodium-potassium ATPase** (Skou 1957), for whose research the Scandinavian scientist Jens Christian Skou received the Nobel Prize in 1997. The sodium-potassium ATPase pumps three Na^+ ions out of the cell and two K^+ ions into the cell. It requires energy for this transport process, as it occurs in opposition to the concentration gradient of the two ions to be

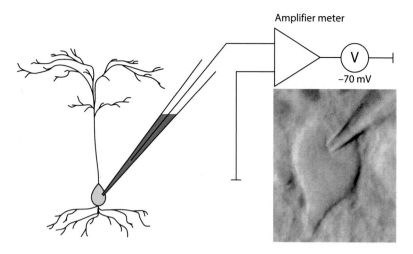

Fig. 4.1 Determination of the membrane potential of a neuron. To determine the potential of a cell membrane, a thin glass capillary with an opening of a few micrometers is positioned directly at the membrane. By sucking the cell membrane to the opening of the capillary, a high electrical seal is created. This is essential for measuring currents and potential changes in the pA or mV range that typically occur in nerve cells. Inside the glass capillary is a salt solution that is in contact with an electrode. Outside the cell there is another electrode. This setup allows the voltage to be measured. Due to the small size of the measured currents/voltages, they have to be electronically amplified before evaluation. In this example, the resting membrane potential of the neuron is −70 mV

Table 4.1 Intracellular and extracellular ion concentrations in humans

Ion	Concentration intracellular (mM)	Concentration extracellular (mM)	Equilibriumpotential (≈mV)
Na^+	15	145	+60
K^+	120	3	−95
Ca^{2+}	0.0001	1[a]	+120
Cl^-	7	120	−75
HCO_3^-	15	24	−13
A^{-b}	≈120–150	–	–

[a] The total Ca^{2+} ion concentration is about 2 mM, about half of which is unbound and ionized
[b] A^-: negatively charged organic molecules that cannot pass through the plasma membrane

transported. Due to the unequal ratios of this transport (the pump transports only two positively charged K^+ ions for three positively charged Na^+ ions), a slightly negative membrane potential is already generated by the activity of the sodium-potassium ATPase. Such potential-generating transport is called electrogenic (**Fig. 4.2**). As its name suggests, the sodium-potassium ATPase uses adenosine triphosphate (ATP) as an energy carrier. In this process, the hydrolysis of ATP to adenosine diphosphate (ADP) and phosphate by the phosphatate activity of the pump leads to the release of

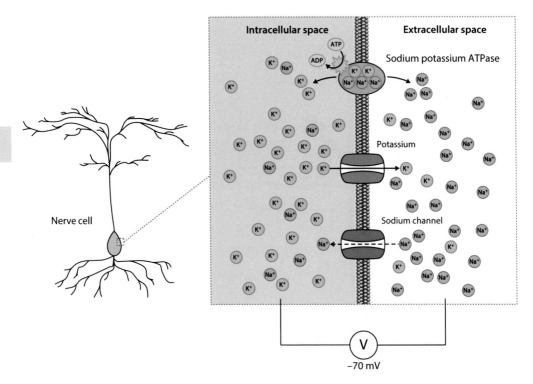

◘ Fig. 4.2 Development of the negative membrane potential of a neuron. In order for the resting membrane potential of a nerve cell to settle at −70 mV, it requires various preconditions. First, there must be an unequal distribution of sodium and potassium ions intracellularly and extracellularly. This unequal distribution is generated and maintained under energy consumption (ATP → ADP) by the activity of the sodium-potassium ATPase (light blue). Second, the cell membrane must be particularly permeable to potassium ions and only slightly permeable to sodium ions. The relatively high potassium permeability under conditions of resting membrane potential is mediated by open potassium channels (green). Potassium continuously flows out of the cell into the extracellular space through these potassium channels. This constant flow of potassium causes the membrane potential of the cell to approach the equilibrium potential of potassium (◘ Table 4.1). However, the equilibrium potential of potassium (about −90 mV) is usually lower than the resting membrane potential of a neuron (about −70 mV). This is because most neurons also have a low permeability to sodium ions under resting conditions. This permeability is mediated by a small number of open sodium channels (red). Sodium flux slightly shifts the membrane potential towards the sodium reversal potential (approx. +60 mV), causing the resting membrane potential to settle at values around −70 mV

the energy required for the transport process. This explains the observation that the negative resting membrane potential of excitable cells collapses when the energy carrier ATP is insufficiently available.

As described above, a basic requirement for the formation of the negative membrane potential of nerve cells is the unequal distribution of different ions between extracellular and intracellular space. In order to maintain this unequal distribution in a stable manner, an important property of the cell membrane is required. The cell membrane must have only a low *permeability* for ions, otherwise the extracellular and intracellular ion concentrations would equalize again over time via diffusion. The permeability of the lipid bilayer of the cell membrane itself is very low. However, there are channel proteins in the membrane of neurons that are permeable to certain ions. While under resting conditions the permeability for ions such as Na^+ and Cl^- is low, there are K^+ channels in the cell membrane of neurons that are open even at negative membrane potential. K^+ ions can therefore flow out of the cell into the extracellular space through these open channels in the direction of the concentration gradient established by the sodium-potassium ATPase, even under resting conditions. In contrast, Na^+ ions can hardly flow into the cell under resting conditions because most Na^+ channels are closed. The membrane potential becomes more negative due to this shift of the positive charge carrier K^+ from intracellular to extracellular (◘ Fig. 4.2). Accordingly, the electrogenic activity of sodium-potassium ATPase only partly explains the unequal charge distribution between the cell interior and the extracellular space. The negative resting membrane potential is additionally caused by a higher permeability for K^+ ions than, for example, for Na^+ or Cl^- ions. As we will see later, the permeability of the cell membrane to specific ions can change very rapidly. As a consequence of such a permeability change, the membrane potential shifts towards more positive or more negative values, depending on which ion flows more strongly across the membrane (Hille 2001).

Patch Clamp Technique

The technique is a measurement method used primarily in basic neuroscientific research, with the help of which scientists can trace the smallest currents and potential changes of living nerve cells. The technique was developed in the late 1970s and early 1980s by the two German electrophysiologists Bert Sakmann and Erwin Neher (Hamill et al. 1981). In order to be able to measure currents or potentials of a single nerve cell with as little interference as possible, a measuring electrode is brought into very close contact with the cell membrane of the cell. Neher and Sakmann succeeded in doing this by using a special glass electrode filled with a salt solution, which enabled them to suck the cell membrane onto the electrode by means of negative pressure. The currents or potential changes of the nerve cell can be derived via a wire in the glass electrode. The diameter of the tip of such a glass electrode is 1–2 μM, much smaller than the diameter of most neurons (about 20 μm for cortical pyramidal cells). Thus, only a small area (the eponymous *patch*) of the cell membrane is located under the glass electrode. In fact, the area of the cell membrane is so small that it often contains only one ion channel. The patch-clamp technique can therefore be used to study the activity of a single channel at very high resolution. One advantage of the patch-clamp technique is a strong reduction of background noise, which allows the derivation of even very small currents or voltage changes. This low noise of the patch clamp technique is a basic requirement when investigating the activity of individual channels. Thus, currents with amplitudes of a few picoamperes (1 pA $= 10^{-12}$ A) can be easily resolved. At the same time, the patch clamp technique allows either the membrane voltage to be held constant (the eponymous clamp) when measuring currents or the current to be held constant when measuring voltage changes. This offers a great advantage in the controlled analysis of ion channel properties. However, the membrane under the pipette

4

can also be disrupted, so that in the *whole cell configuration* one can measure currents and voltage changes caused by the opening of ion channels in the membrane of the whole neuron. Since it is also possible to "patch" neurons in acute brain slices and even in vivo, the patch-clamp technique allows the analysis of the activity of a neuron that is in a neuronal network and receives information from other neurons. With the development of the patch-clamp technique, Bert Sakmann and Erwin Neher decisively advanced the understanding of the function of ion channels, which was honored with the award of the Nobel Prize in 1991.

4.2 Equilibrium Potential of an Ion

In ▶ Sect. 4.1 we have already discussed the two forces that are responsible for the movement of ions from intracellular to extracellular. Firstly, ions move in the direction of the concentration gradient. This chemical driving force leads to a movement of K$^+$ ions from intracellular to extracellular. On the other hand, ions also move along an electrical potential gradient. The negative membrane potential (i.e. the intracellular space is more negative than the extracellular space) thus causes K$^+$ ions to flow into the cell. However, chemical and electrical driving forces cannot be considered separately. If we take the movement of K$^+$ ions at a negative resting membrane potential of −70 mV as an example, the chemical driving force opposes the electrical one. Since at this potential the chemical driving force is greater than the electrical driving force, more K$^+$ ions will flow out of the cell than

into it. Net, therefore, there is a K$^+$ efflux. In contrast, when Na$^+$ channels are opened, both driving forces will cause Na$^+$ ions to flow into the cell (Hille 2001).

The net movement of an ion is therefore dependent on the electrochemical driving force, in which both driving forces are combined. If the driving forces for an ion are opposite and equal, there is no net movement of the ion across the membrane. For example, this is the case for K$^+$ ions at about −95 mV. At this membrane potential, the chemical driving force is directed extracellularly and the electrical driving force is directed intracellularly. Since the driving forces are equal at about −95 mV, exactly as much K$^+$ ions flow out of the cell as into the cell. Thus, there is no net current flow. The potential of the cell at which this is true is therefore called the **equilibrium potential of a given ion** (◘ Fig. 4.3). If the intracellular and extracellular ion concentrations are known, the equilibrium potential of an ion can be calculated using the **Nernst equation** (Hille 2001):

$$E_{ion} = \frac{R \cdot T}{z \cdot F} * \ln \frac{\left[K^+\right]_a}{\left[K^+\right]_i} \qquad (4.1)$$

with E_{ion} = equilibrium potential, R = general gas constant, T = absolute temperature, F = Faraday constant, z = valence of ion.

In our body, the Na$^+$ ion concentration is usually much higher extracellularly than intracellularly (◘ Table 4.1). This explains why the equilibrium potential for Na$^+$ ions is positive (approx. +75 mV). The Cl$^-$ ion concentration is also much higher extracellularly than intracellularly, but since Cl$^-$ ions are negatively charged, the equilibrium potential is negative (approx. −50 mV).

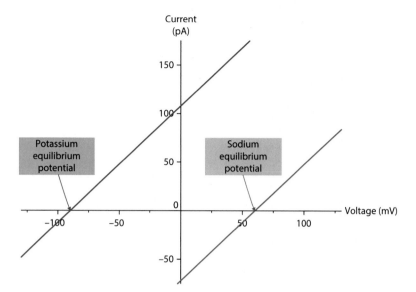

◻ Fig. 4.3 Current-voltage characteristic of an ion channel. The passage of ions across the cell membrane occurs through ion channels along the electrochemical gradient. The magnitude and direction of the ion current depend on the membrane voltage and the concentration gradient. The current is zero when all driving forces cancel out (equilibrium potential, see Nernst Eq. 4.1). If, for example, the membrane voltage changes the magnitude and direction of the ion current will also change as a result. This in turn can be measured as an electric current. Due to the different charge and concentration distributions there is a specific relationship between current and voltage for each ion. The current-voltage curves shown here for idealized sodium and potassium channels are strictly linear. In nature, however, most channels show a divergent current-voltage curve. The channels change their permeability as a function of membrane voltage, thereby affecting the amplitude of potential ionic currents. Due to these properties, they are then referred to as inward or outward rectifying channels (cf. NMDA receptor channel, ◻ Fig. 4.12)

4.3 Resting Membrane Potential

The resting membrane potential is the membrane potential of a neuron at rest, i.e. when the neuron is not firing an action potential. In the brains of mammals, including humans, the resting membrane potential for most neurons is about −70 mV, which is close to the equilibrium potential for K^+ ions (about −95 mV). This can be explained by the fact that the cell membrane has a high permeability for K^+ ions under resting membrane conditions. Na^+ ions and Cl^- ions can generally hardly flow under these conditions and therefore have little influence on the resting membrane potential. Thus, not only the extracellular and intracellular concentrations of ions are crucial for the membrane potential of a cell, but also the permeability of the membrane for these ions. The ion for which the membrane is particularly permeable sets the tone, so to speak. It "pulls" the membrane potential close to its equilibrium potential. Ions with lower permeabilities have less influence on the membrane potential (Hodgkin 1964). The **Goldman-Hodgkin-Katz equation** takes into account not only the concentration difference of the ions but also the permeability and allows the membrane potential to be calculated:

$$E_{M} = \frac{R \cdot T}{z \cdot F} * \ln \frac{P_{Na} \cdot \left[Na^{+} \right]_{a} + P_{K} \cdot \left[K^{+} \right]_{a} + P_{Cl} \cdot \left[Cl^{-} \right]_{i}}{P_{Na} \cdot \left[Na^{+} \right]_{i} + P_{K} \cdot \left[K^{+} \right]_{i} + P_{Cl} \cdot \left[Cl^{-} \right]_{a}}$$

with E_M = membrane potential, R, T and F as in the Nernst equation, P_{Na}, P_K, P_{Cl} = permeability for K^+ ions, Na^+ ions and Cl^- ions.

As can be seen, the Goldman-Hodgkin-Katz equation is an extension of the Nernst equation. All ions that play a role in the membrane potential are included in the calculation. Under resting conditions—as already mentioned—the permeability for K^+ ions is high, that for Na^+ ions and Cl^- ions is low. Accordingly, the resting membrane potential is close to the equilibrium potential for K^+ (generally somewhat more positive, since the permeability for Na^+ ions and Cl^- ions is not zero).

If the permeability for K^+ ions increases as additional K^+-channels open, the membrane potential approaches the equilibrium potential for K^+, i.e. it becomes more negative. This is referred to as **hyperpolarization**. As we will see later, hyperpolarization causes the excitability of the neuron to decrease. The equilibrium potential for Na^+ is about +60 mV. This means that when the resting membrane potential is, say, −70 mV, Na^+ ions will strive to flow into the cell. This happens, for example, when a neuron is activated in such a way that the permeability of the membrane to Na^+ ions increases. In this case, the membrane potential approaches the equilibrium potential for Na^+. The membrane potential thus becomes less negative or even positive. This is referred to as **depolarization**.

4.4 Action Potential

Neurons are among the excitable cells of our body. This means that they are capable of firing an action potential (Hille 2001). An action potential is understood to be a very rapid change in the membrane potential, due to a depolarization of the membrane potential to approx. −50. At this so-called **threshold potential**, voltage-gated Na^+ channels (Na_V channels) open. Voltage-gated ion channels change their ion permeability depending on the membrane potential (Armstrong and Hille 1998). They thus differ from ligand-gated ion channels (= ionotropic receptor ion channel, ligand-gated), in which the binding of a transmitter causes the opening of the channel. Voltage-gated Na^+ channels are closed at normal resting membrane potential and open at a membrane potential of about −60 mV. As shown in ▶ Sect. 4.3, the opening of channels that are selectively permeable to Na^+ ions leads to depolarization (the so-called **rising phase** the action potential). The membrane potential becomes positive (**overshoot**) and approaches the equilibrium potential for Na^+ ions at about +60 mV. However, voltage-gated Na^+ channels close again very quickly. Voltage-gated K^+ channels open somewhat later than Na^+ channels. This leads to a **repolarization** of the membrane potential (◻ Fig. 4.4). The permeability of the membrane to K^+ ions is usually greater at this stage than before the action potential. This explains why the membrane potential is hyperpolarized after an action potential, i.e. closer to the equilibrium potential for K^+ ions than before the action potential. Action potentials follow an all-or-nothing law. This means that reaching the threshold potential always triggers an action potential, which then proceeds in a very stereotyped manner with a similar amplitude (approximately 80–100 mV) and temporal duration (approximately 1 ms). The increasing depolarization during the upstroke leads to the explosive opening of all voltage-gated Na^+ channels. Accordingly, after an action potential all voltage-gated Na^+ channels are also inactivated. The cell is then in the absolute **refractory period** during which no further action potential can be triggered. In the subsequent relative refractory period, voltage-gated Na^+ channels can be partially reactivated so that a strong depolarization

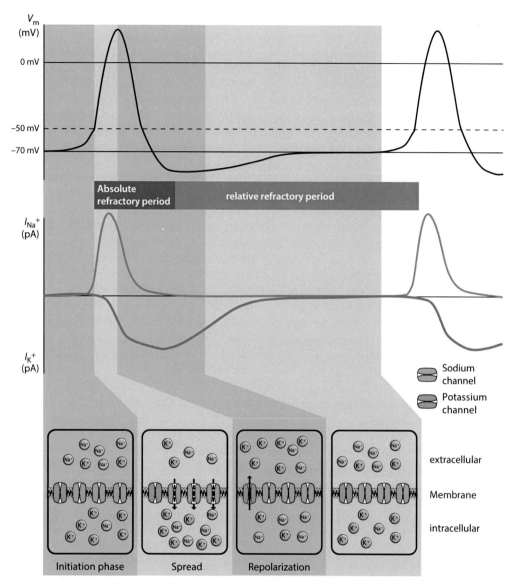

■ **Fig. 4.4** The action potential. The upper part of the figure shows the course of the membrane voltage over time during two action potentials. Excitatory postsynaptic potentials depolarize the membrane potential during the initiation phase (▶ Sects. 4.14 and 4.16). When this then reaches the threshold potential (dashed line), there is a rapid, strong depolarisation with overshoot (positive membrane potential). During repolarization, the cell is briefly hyperpolarized. A further action potential can be formed at the earliest in the relative refractory period

can again trigger an action potential. The refractory period plays an important role in controlling the excitability of neurons. Mutations that shorten the duration of inactivation of voltage-gated Na⁺ channels lead to hyperexcitability of neurons and are thus causative for certain forms of epilepsy (Catterall 2017).

4.5   Electrical Conduction

Changes in the membrane potential are passively and actively conducted in the membrane. To understand passive (*electrotonic*) conduction, it is helpful to think of the membrane as an electrical *capacitor-resistor element* (◘ Fig. 4.5a). The lipid bilayer of the membrane is a capacitor because it can store electric charge as an insulating layer between two conducting media. However, depending on the number of open ion channels, the membrane also has some electrical conductivity (as the reciprocal of resistance). K^+ channels, for example, which are open even in resting neurons, reduce the membrane resistance and cause a leakage current. If the membrane potential changes at one site of a neuron, this membrane potential change will passively propagate across the membrane of the neuron. The capacitor property of the membrane explains why electrotonic conduction is quite slow and the amplitude of the propagated membrane potential change decreases with distance from the site of origin (◘ Fig. 4.5). Accordingly, electrotonic conduction over long distances is not very effective. However, changes in membrane potential must sometimes be propagated over very long distances. The longest axons in humans, for example, can be over 2 m long (e.g., from neurons that conduct sensitive information from the big toe to the nucleus gracilis located in the brainstem). In the blue whale, axons can even be over 30 m long. Passive conduction would be much too slow here, and even with large changes in membrane potential at the axon hillock, there would be no measurable change in membrane potential at the synapses.

Electrical conduction over long distances must therefore be active. One speaks of active mechanisms when it comes to the opening of voltage-gated channels. Action potentials originate at the axon hillock. The density of voltage-gated Na^+ channels is particularly high in the membrane of this initial segment of the axon. From there, the action potential must be conducted to the synapses at the end of the axon. An axonal conduction of an action potential can be imagined as follows: An action potential arising at the axon hillock leads to depolarization of the neighboring axon membrane. This causes voltage-gated Na^+ channels to open. As in passive conduction, the action potential is thus continuously propagated. In contrast to passive conduction, however, the amplitude of the action potential does not reduce in active (*regenerative*) conduction (◘ Fig. 4.5). An action potential is generally propagated only in the *orthodromic* direction, that is, from the axon hillock along the axon toward the synapses. Only in the orthodromic direction does the action potential encounter closed and activatable voltage-gated Na^+ channels. The inactivation of Na^+ channels that have just been activated and the opening of voltage-gated K^+ channels prevents *antidromic* conduction (▶ Sect. 4.4).

However, action potentials do not only play a role at the axon. So-called *back-propagating action potentials* can propagate from the axon hillock via the cell soma into the dendrites. The back-propagating conduction into the dendrites is based on passive (electrotonic) as well as active (activation of voltage-dependent Na^+ or Ca^{2+}-channels) mechanisms. The exact function of the retrograde action potentials has not yet been elucidated. However, it is likely that they remove the Mg^{2+} block of NMDA receptors (▶ Sect. 4.17) by depolarizing the dendritic membrane and thus influence synaptic signal transmission.

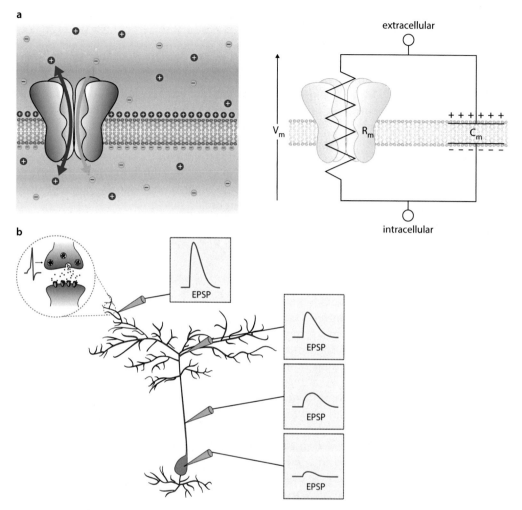

◻ Fig. 4.5 Electrical conduction. **a** The membrane of neurons has properties of a capacitor-resistor element. Charges can be stored at the lipid bilayer, as in a capacitor. At the same time, a current can flow through ion channels. Depending on the number of open channels, the cell membrane has a resistance. **b** In passive (electrotonic) conduction, a current flow (e.g. by opening excitatory receptors) causes the membrane potential to change locally. As a result, the membrane potential of neighboring membrane sec-

tions is also changed, leading to passive conduction. Leakage currents flowing through open channels reduce the membrane potential change. In electrotonic conduction, the amplitude of the membrane potential change decreases with distance from the origin. At the same time, a high capacitance of the membrane slows electrical conduction. EPSP potential, excitatory postsynaptic: excitatory postsynaptic potential

4.6 Electrical Conduction Along Myelinated Axons

Although active/regenerative conduction prevents the reduction of the amplitude of the action potential, it is quite time-consuming. Very fast conduction is therefore not

possible. In order for action potentials to be conducted not only over long distances but also very quickly, active mechanisms must be combined with passive mechanisms. This occurs in myelinated axons. **Myelination** is an electrically insulating layer that tightly surrounds the axon. It does not originate from the neuron itself, but is formed by glial cells (oligodendrocytes in the central nervous system, Schwann cells in the peripheral nervous system), whose membrane protrusions grow in a spiral around the axon of the neuron (◘ Fig. 4.6). As the degree of myelination increases, the number of lamellae enveloping the axon increases.

How does myelination affect conduction? In ▶ Sect. 4.5, we discussed that the membrane can be viewed as a capacitor-resistor element. Myelin, which has a high lipid content and a relatively low protein content, performs a similar function to the insulating plastic sheath around a power cable. It increases the membrane resistance and reduces the capacitance of the capacitor. The increased membrane resistance (reduction of leakage currents) and low electrical capacitance due to myelin reduce the amplitude drop of the electrotonically propagated action potentials and cause the membrane potential to change much more rapidly when current flows through it. As a result, myelinated axons conduct action potentials electrotonically much faster than unmyelinated axons. Since the thickness of the myelin layer increases the speed of conduction, long axons in particular, which have to conduct information very quickly, have a high degree of myelination, i.e. a thicker myelin sheath. This applies, for example, to axons of motor neurons in the cortex or spinal cord. Unmyelinated or poorly myelinated axons would not be able to initiate fast, coordinated movement patterns.

However, myelination cannot completely prevent the amplitude reduction of the action potentials. For this reason, there are regular interruptions of myelination in the course of the axon. These myelin-free axon regions are called **Nodes of Ranvier**. In the myelin-free regions are voltage-gated Na^+ channels in particularly high density. The opening of these channels leads to the regeneration of the action potential. While the electrotonic conduction velocity is very high in the myelinated region, there is a delay at the Nodes of Ranvier because the regeneration of the action potential by opening the Na^+ channels takes more time. The conduction is therefore not continuous, but the action potential appears to jump from one Node of Ranvier to the next Node of Ranvier. This is therefore referred to as saltatory conduction (◘ Fig. 4.6). The maximum conduction velocity can reach 130 m s^{-1} in myelinated axons. In contrast, non-myelinated axons only conduct at a speed of 1–3 m s^{-1}.

Multiple Sclerosis and Guillain-Barré Syndrome

Multiple sclerosis (encephalomyelitis dissiminata) is an inflammatory disease of the central nervous system that is associated with demyelination of axons. The resulting impaired conduction of excitation leads to deficits such as paralysis, sensory disturbances or visual disturbances. Since demyelination occurs in many different parts of the brain (hence dissiminata), the clinical picture is colourful. For example, patients may suffer from visual disturbance of the left eye, numbness of the right foot, and coordination problems and paralysis of both hands at the same time. As expected in a demyelinating disease, conduction is slowed, which can be documented by electroencephalogram (EEG). For example, a common examination in multiple sclerosis patients with visual disturbances is an EEG recording with visual stimulation. Shortly after the visual stimulus, a signal can be detected over the visual cortex, but because it is very small, it becomes visible only after averaging many EEG recordings. Signals that are independent of the stimulus are filtered out by averaging. The primary cortical visual evoked potential appears after about 100 ms in healthy individuals, and a latency prolongation is

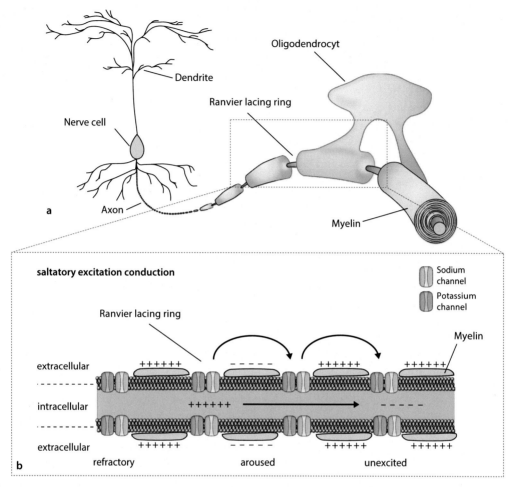

□ Fig. 4.6 Saltatory conduction. **a** The figure shows a neuron whose axon is covered by a myelin layer. The myelin layer, which is formed in the CNS by oligodendrocytes, is interrupted at regular intervals (Node of Ranvier). Myelination causes action potentials to be transmitted faster along the axon. **b** The high number of voltage gated sodium channels and the absence of myelin allow formation of action potentials at the Nodes of Ranvier. The depoplarisation is then passively transmitted from one Node of Ranvier to the next where a new regenerative action potential is formed. Condution of action potentials is usually unidirectional as the refractory period after sodium channel activation prevents backpropagation of the action potential

expected in patients with reduction in conduction. Demyelinating diseases are not unique to the central nervous system. Guillain-Barré syndrome, for example, is an inflammatory disease in which antibodies produced by the body (autoantibodies) are directed against the myelin sheath of the peripheral nerves. Here, too, the consequence of demyelination is slowed conduction, which leads to symptoms such as paralysis and sensory disturbances.

4.7 Synaptic Transmission

> So far as our present knowledge goes, we are led to think that the tip of a twig of the arborescence is not continuous with but merely in contact with the substance of the dendrite or cell body on which it impinges. Such a special connection of one nerve cell with another might be called a synapsis.
>
> Sir Charles Scott Sherrington 1897 (Foster 1897)

At the end of the nineteenth century, the Italian neuroanatomist Camillo Golgi developed a silver stain with which he could not stain the whole tissue, but only individual cells in the brain. This method, later further developed by the Spanish neuroanatomist Santiago Ramón y Cajal, made it possible for the first time to visualize the anatomy of neurons. Golgi and Ramón y Cajal thus laid the foundation for modern neuroscience and were awarded the Nobel Prize in 1906 for their achievements. However, both scientists drew very different conclusions from the observation of the stained cells. The different hypotheses were the cause of a bitter antagonism between the two scientists. Golgi's reticular theory stated that nerve tissue is a network, a "reticulum," of interconnected cells. Indeed, neurons can be directly connected to each other via *gap junctions*. However, such connections are the exception rather than the rule. Rather, neurons are separated from each other, as assumed by Ramón y Cajal, even if they are spatially in very close contact with each other. It was not until the 1950s that the dispute was settled in favor of the neuron theory by the development of electron microscopy. This microscopy is so high-resolution that it can be used to see the very small gap (30–50 nm) between the synapse **terminal** (**axon terminal** or *twig of the arborescence*) of one neuron and the dendrite or cell body of the second neuron. This point of contact between two neurons was named synapse by Charles Sherrington at the suggestion of his colleague Michael Foster and the philologist Arthur Verrall (Sherrington had initially preferred *syndesm as* a term, but was persuaded that synapse was the better choice, among other things because it allows a better adjective to be formed: synaptic versus syndesmic).

Today we know that neuronal "communication" between presynaptic and postsynaptic neuron takes place at the synapses. We speak of presynaptic and postsynaptic, since communication generally only goes in one direction: the presynaptic neuron "speaks", the postsynaptic neuron "listens". In this process, the electrical signal (the action potential) must be converted into a chemical signal. To do this, the presynaptic neuron releases transmitters that bind to receptors in the membrane of the postsynaptic neuron, causing it to be excited or inhibited. Presynaptic transmitter release is triggered by the action potential arriving at the axon terminal. In this process, voltage-gated Ca^{2+} channels are opened by depolarization of the cell membrane. Ca^{2+} entering the axon terminals binds to synaptotagmin—a protein that acts as a calcium sensor. The binding of Ca^{2+} to synaptotagmin in turn initiates the fusion of vesicles, i.e. small membrane-enveloped vesicles, with the presynaptic cell membrane via activation of proteins of the so-called SNARE complex (including synaptobrevin, syntaxin, SNAP25). The fusion results in the release (*exocytosis*) of transmitter molecules located in the vesicles into the synaptic cleft (◘ Fig. 4.7) (Jahn and Fasshauer 2012). Presynaptic vesicles contain several thousand neurotransmitter molecules. Diffusion of the neurotransmitter across the synaptic cleft to the postsynaptic membrane occurs in a fraction of a millisecond. The membrane of the postsynaptic neuron contains receptors to which the transmitter molecules bind and thereby exert their effect.

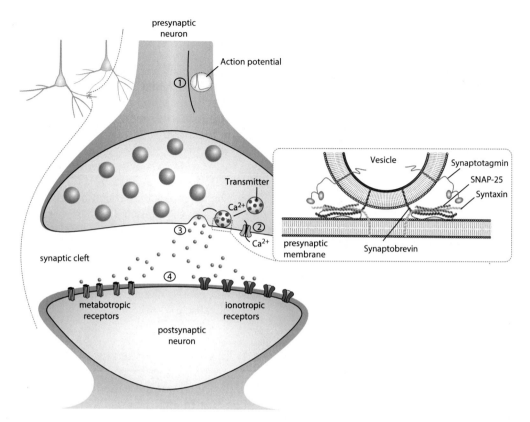

◻ Fig. 4.7 Structure of a synapse. Top left: two nerve cells connected by a synapse. Middle: simplified structure of the synapse consisting of the axon terminal of the presynaptic nerve cell and the so-called *spine* of the postsynaptic nerve cell (Yuste 2015). An incoming action potential (1) leads to Ca^{2+} influx via presynaptic Ca^{2+} channels (2), which in turn initiates fusion of synaptic vesicles with the presynaptic membrane and subsequent transmitter release (3). The transmitter molecules then bind specifically to ionotropic or metabotropic receptors (4) localized in the postsynaptic membrane. Right: structure of synaptic proteins involved in calcium-dependent exocytosis. The inflowing Ca^{2+} binds to the vesicle protein synaptotagmin. As a consequence, synaptotagmin binds to the presynaptic membrane and to the SNARE complex, which is formed by the vesicle protein synaptobrevin and the two synaptic membrane proteins, syntaxin and SNAP-25. The SNARE complex contributes to the fusion of the vesicle membrane with the adjacent presynaptic membrane. (Jahn and Fasshauer 2012)

Tetanus and Botulism

The proteins of the SNARE complex play a crucial role in the symptoms of tetanus (lockjaw) and botulism. In both cases, bacteria (*Clostridium tetani* and *Clostridium botulinum*) produce toxins (tetanus toxin and botulinus toxin) that cleave proteins of the SNARE complex. Tetanus bacteria occur ubiquitously (especially in the soil). Infection occurs, for example, after an injury to a splinter of wood. If the injury is deep, the bacteria can multiply well, especially in an anaerobic (i.e. low-oxygen) environment, and produce tetanus toxin. This is transported retrogradely via axons of motor neurons into the spinal cord, where it exerts its effect in inhibitory neurons. Via cleavage of synaptobrevin, it inhibits synaptic transmission between inhibitory neurons and motoneurons. The reduced inhibition of motoneurons leads to increased activation of muscles and thus to the eponymous convulsions. Acute treatment consists of wound sanitation, antibiotic treatment, and administration of antitoxin (antibodies to tetanus toxin). Good protection is provided by active vaccination, in which the administration of tetanus toxoid (inactivated tetanus toxin) induces the production of protective antibodies.

Botulism is generally caused by food poisoning, usually meat or sausage, in which botulinum bacteria multiply due to improper food hygiene. Typically, these are home-cooked canned foods that have not been sterilized with overpressure and temperatures above 100 °C. In these, the also anaerobic bacteria can grow and build botulinus toxin. This toxin is then ingested and enters the motor neurons where, like tetanus toxin in inhibitory neurons, it cleaves synaptobrevin and thus inhibits the transmission from motor neuron to muscle. Paralysis is the result. In addition to symptomatic therapy (e.g. ventilation in the case of respiratory paralysis), the administration of antitoxin (antibodies against botulinus toxin) is indicated. In aesthetic medicine, the use of botulinus toxin as an anti-aging substance is widespread. There it is known as Botox and is mainly injected in the facial area, which leads to the slackening of muscles and thus to the temporary disappearance of wrinkles.

4.8 Motor End Plate

The motor endplate is a special form of synapse, as it does not allow communication between two neurons, but between neuron and skeletal muscle cell. However, the principles of chemical transmission are similar to synapses between two neurons. The neurotransmitter of the motor endplate is acetylcholine, which binds postsynaptically to **nicotinic acetylcholine receptors** of the muscle cell. These are ionotropic receptors (as distinct from *muscarinic* acetylcholine receptors; ▶ Sect. 4.10), through which ions can flow when the transmitter binds, depolarizing the muscle cell. The resulting action potential leads to a release of Ca^{2+} ions from the sarcoplasmic reticulum, which in turn causes the muscle cell to contract (electromechanical coupling). Acetylcholine is cleaved in the synaptic cleft within a very short time by the enzyme acetylcholineesterase to choline and acetate, which limits the duration of the chemical signal and thus also the muscle contraction.

4.9 Electrical Synapses

Camillo Golgi was not entirely wrong when he postulated that neurons form a syncytium in which neurons are directly connected. Although the spatial separation of neurons that communicate with each other via chemical synapses is the rule, neurons can also form direct connections via **gap junctions** (◘ Fig. 4.8; Bennett and Zukin 2004). These are channels formed from connexins. Six connexins form a connexon, and the interconnected cells each have a connexon that together form a gap junction. Gap junctions are found in many tissues (heart, liver, intestine, vessels), where they enable, among other things, a direct exchange of substances between neighbouring cells. However, the diameter of the gap junctions limits the exchange of substances to small molecules up to a size of approx. 500–1000 Da. This allows, for example, the exchange of second messengers such as cAMP or Ca^{2+}. In the brain, glial cells in

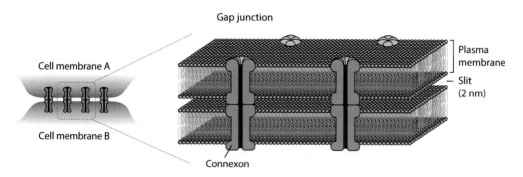

Fig. 4.8 Electrical synapses. Left: The plasma membranes of two cells are connected by gap junctions. Right: Schematic representation of a cross-section through the plasma membranes with connexons

particular express gap junctions. Direct connection between neurons via gap junctions is comparatively more common in invertebrates than in vertebrates. Gap junctions allow not only an exchange of substances, but also direct electrical communication between the two neurons. They are therefore also referred to as electrical synapses. In contrast to chemical synapses, the communication is bidirectional. Since the conversion of the electrical signal into a chemical signal is omitted, electrical synapses are also much faster. Thus, a depolarization in one neuron will cause a depolarization in a neuron connected by electrical synapses with minimal time delay. When many neurons are connected to each other via gap junctions, the activity of this neuron network can be synchronized by the electrical synapses. The synchronous activity of interneurons plays a role in the generation of oscillatory activity in the brain, for example.

4.10 Ionotropic Receptors

Ionotropic receptors are membrane proteins that not only bind neurotransmitters but also form an ion channel. The binding of the neurotransmitter on the extracellular membrane side leads to a structural change of the protein, which opens the ion channel. Depending on the membrane potential, ions flow in and out of the cell through the channel; thus, a current flows. However, ionotropic receptors are not equally permeable to all ions. Ionotropic **glutamate receptors** and **acetylcholine receptors** are examples of transmitter-gated cation channels. Binding of glutamate or acetylcholine opens channels that are permeable to Na^+ and K^+ ions. Depending on the membrane potential, there is an influx and efflux of Na^+ and K^+ ions (**Fig. 4.9**). At a resting membrane potential of -70 mV, the influx of Na^+ ions outweighs the efflux of K^+ ions resulting in a net inward current and thus a depolarization of the neuron. The membrane potential thus approaches the threshold potential for action potentials. Accordingly, glutamate and acetylcholine are excitatory neurotransmitters, and the neurons that release these transmitters are excitatory neurons. In contrast, the inhibitory neurotransmitters γ-aminobutyric acid (GABA) and glycine bind to transmitter-gated anion channels that are permeable to Cl^- ions. The influx of Cl^- ions into the cell through **$GABA_A$ receptors** or **glycine receptors** leads to hyperpolarization and thus to inhibition of the neuron.

a ionotropic receptors

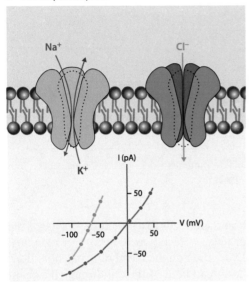

b metabotropic receptors

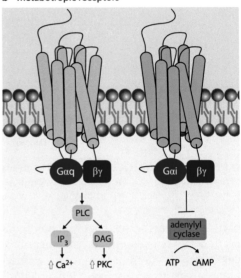

☐ **Fig. 4.9** Ionotropic and metabotropic receptors. **a** Structure and current voltage curves of ionotropic receptors. Ionotropic glutamate receptors and acetylcholine receptors (left) are (excitatory) cation channels that open after binding by a neurotransmitter and lead to membrane depolarization. The depolarization here is mainly carried by the inflowing sodium ions. As can be seen from the current-voltage curve, the reversal potentials of glutamate receptors are usually around 0 mV. At membrane voltages below 0 mV, the sodium influx predominates, i.e. the cell depolarizes; at membrane voltages above 0 mV, the potassium outflow predominates, and hyperpolarization occurs.

However, physiologically important here is mainly the depolarizing effect, since membrane voltages above 0 mV are not seldom reached at the synapse—$GABA_A$ receptors conduct Cl^- ions, which means that opening the channel usually leads to a hyperpolarization of the cell. **b** Structure and intracellular signaling cascades of metabotropic receptors. Metabotropic receptors differ in their action primarily by the G protein coupled to the receptor. G_q-coupled receptors increase intracellular Ca^{2+} levels via activation of phospholipase C and stimulate protein kinase C activity. G_i-coupled receptors inhibit the adenylate cyclase and thereby increase the intracellular cAMP level

4.11 Metabotropic Receptors

The effect of binding of a neurotransmitter to metabotropic receptors is more indirect than that to ionotropic receptors. Metabotropic receptors are not channels, but exert their effect by activating intracellular signalling cascades and changing the concentration of intracellular messengers (second messengers). Primarily, they activate proteins on the intracellular side. Many neurotransmitters bind to G protein-coupled metabotropic receptors. Binding of the neurotransmitter to these receptors leads to activation of the guanosine triphosphate (GTP)-binding protein (G protein for short), which itself is composed of three subunits (α-, β- and γ-subunit). The consequence of this activation depends on which G protein is coupled to the receptor. For example, metabotropic glutamate receptors, of which there are eight subtypes ($mGluR_1$–$mGluR_8$), are either G_q-coupled or G_i-coupled. Activation of G_q-coupled receptors (mGluR$_1$ and mGluR$_5$) leads to activation of a signaling cascade that ultimately results in activation of protein kinase C and release of Ca^{2+} ions from the endoplasmic reticulum. In contrast, activation of G$_i$-coupled receptors (mGluR$_{2-4}$ and mGluR$_{6-8}$) inhibits adenylate cyclase, thereby reducing the intracellular concentration of the second messenger cAMP. The consequences of Ca^{2+} or cAMP concentration change and activation of protein kinase C are complex and dependent on neuron type.

Another example of a metabotropic receptor is the GABA$_B$ receptor. In contrast to the GABA$_A$ receptor, which is an ionotropic channel that inhibits the neuron via a Cl$^-$ ion influx when opened, the GABA$_B$ receptor is a metabotropic receptor coupled to a G protein. Binding of GABA to the GABA$_B$ receptor leads to postsynaptic opening of K$^+$ channels via activation of the G protein. The efflux of K$^+$ ions from the cell hyperpolarises the cell (insofar as its membrane potential is less negative than the equilibrium potential for K$^+$), and an IPSP or IPSC (▶ Sects. 4.12, 4.13, and 4.14) can be measured. However, metabotropic receptors (the same is true to some extent for ionotropic receptors) are not only expressed postsynaptically. They are also found presynaptically, where, for example, activation of mGluRs or GABA$_B$ receptors reduces the likelihood that neurotransmitters will be released. Excessive transmitter release can thus be prevented via presynaptic receptors. The effect of metabotropic receptors lasts much longer than that of ionotropic receptors. For example, a GABA$_B$ receptor-mediated IPSP can last one second, whereas

a GABA$_A$ receptor-mediated IPSP, in contrast, has kinetics that are approximately 50-fold faster. Metabotropic receptors are therefore also thought to have a signal-modulating effect.

4.12 Excitatory and Inhibitory Postsynaptic Potentials

The influence that a presynaptic neuron exerts on a postsynaptic neuron depends on which neurotransmitter has been released and which receptors the postsynaptic neuron carries on its membrane. Excitatory transmitters, such as glutamate and acetylcholine, generally lead to membrane depolarization by binding to their specific ionotropic receptors, while inhibitory transmitters lead to hyperpolarization. Excitatory postsynaptic potentials (EPSPs) and inhibitory postsynaptic potentials (IPSPs) are thus potentials that either bring the membrane potential closer to or remove it from the threshold potential for triggering an action potential. The currents underlying an EPSP and IPSP are referred to as *excitatory postsynaptic current (EPSC)* and *inhibitory postsynaptic current (IPSC)*. EPSPs, IPSPs, EPSCs and IPSCs are short-lived (a few milliseconds) because the neurotransmitters very quickly unbinds from the receptors and diffuse out of the synapse (◘ Fig. 4.10). The duration of action of neurotransmitters is also limited by the fact that they are either degraded very rapidly (e.g. acetylcholine by acetylcholineesterase) or eliminated by uptake into neurons or glial cells.

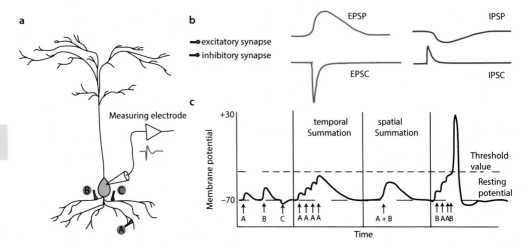

◻ Fig. 4.10 Excitatory and inhibitory currents and potentials. **a** Excitatory and inhibitory synapses are located throughout the dendritic tree; inhibitory synapses are often located at the cell body. **b** Excitatory and inhibitory postsynaptic currents (EPSC and IPSC) and excitatory and inhibitory postsynaptic potentials (EPSP and IPSP) result from the activation of excitatory and inhibitory synapses. **c** Activation of individual excitatory and inhibitory synapses usually changes the membrane potential only slightly. When a synapse is activated repeatedly over a short period of time, temporal summation results in a greater change in membrane potential. When multiple synapses are activated, summation of the individual potentials occurs (spatial summation). If many synapses are active in a short period of time, the spatial and temporal summation can result in a potential change that is suprathreshold, so that an action potential is generated

4.13   mEPSPs and mIPSPs

The amplitude of EPSPs and IPSPs depends on the neurotransmitter concentration in the vesicles, the number of vesicles exocytosed in response to an action potential, and the number of postsynaptic receptors. So-called miniature EPSPs (mEPSPs) and miniature IPSPs (mIPSPs) occur when only one vesicle fuses with the presynaptic membrane. The quantum hypothesis developed in this context states that when multiple vesicles are released, the amplitudes of EPSPs and IPSPs are integer multiples of the amplitudes of mEPSPs and mIPSPs.[1] Here, the mEPSPs and mIPSPs represent the so-called "quanta" as the smallest and indivisible responses. The quantum hypothesis was postulated based on observations of the motor endplate by

Bernard Katz. The motor endplate is the synapse between the motoneuron and muscle. The motoneuron releases acetylcholine as a transmitter, which binds to nicotinic actelycholine receptors in the membrane of the muscle cell. The depolarization in response to exocytosis of an acetylcholine vesicle is referred to here as miniature endplate potential (mEPSP). Bernard Katz noted that the amplitudes of muscle cell depolarization are integer multiples of this mEPSPs.

mEPSPs and mIPSPs as well as their underlying currents (mEPSCs and mIPSCs) can be visualized by a methodical trick. This involves blocking voltage-gated Na^+ channels with tetrodotoxin (TTX) derived from puffer fish of the family Tetraodontidae. This prevents action potential generation and thus action potential-induced exocytosis of vesicles. In rare cases, however, fusion of a vesicle with the presynaptic membrane still occurs. Since this fusion is not triggered by an action potential, there is no temporally coordinated exocytosis of multiple vesicles.

---

1   Despite the similarity of terms, however, the quantum hypothesis has primarily nothing to do with quantum physical phenomena.

The response of the transmitter molecules of one vesicle can then be measured as mEPSP, mIPSP, mEPSC or mIPSC in the postsynaptic neuron. The analysis of these smallest responses is interesting for two main reasons: First, the frequency with which they occur correlates with the number of synapses of the postsynaptic neuron, and second, the amplitude of the responses correlates with the number of receptors per synapse.

4.14 Integration of EPSPs and IPSPs

A single EPSP is normally not able to depolarize a neuron to the extent that the threshold potential is reached at the axon hillock and the neuron generates an action potential. Rather, an integration of many EPSPs and IPSPs results in a constant fluctuation of the membrane potential. A presynaptic neuron often forms multiple synapses with a postsynaptic neuron. Nevertheless, the activity of a single presynaptic neuron is usually not sufficient to trigger an action potential in a postsynaptic neuron. The more or less coordinated activity of several excitatory neurons is necessary for the postsynaptic neuron to reach the threshold potential. This is partly because an action potential does not result in the release of vesicles in every synapse (the presynapse is not particularly reliable, ▶ Sect. 4.16), and partly because a single EPSP changes the membrane potential very little. Locally at the synapse, an EPSP often already has a small amplitude. Due to the electrotonic (i.e. purely passive) signal transmission from the dendritic synapses towards the cell body, the amplitude reduces even further (▶ Sect. 4.6) and the resulting depolarization at the axon hillock is very small (in fact, it is around 1 mV). Synapses located farther from the soma are thus at a disadvantage and have less weight in generating action potentials than synapses located closer to the axon hillock. Distally arising EPSPs or IPSPs may

nevertheless play a relevant role because their amplitude at the site of origin is either relatively large or by activating dendritic voltage-gated Na^+ and Ca^{2+} channels.

Despite these amplification mechanisms, one can imagine that at a resting membrane potential of about -70 mV, a summation of many EPSPs is necessary to reach a threshold potential of about -50 mV (◨ Fig. 4.10). Here, the active synapses may be widely separated in different dendrites. The EPSPs are ultimately summed at the axon hillock. In addition to this **spatial summation**, there is also temporal summation (and combinations of spatial and temporal summation). **Temporal summation** of EPSPs occurs when they are not triggered by synaptic activity at the same time. For EPSPs to sum, they need only occur in a temporally overlapping time window; precisely coordinated synaptic activity is not required (◨ Fig. 4.10). An IPSP arriving in the same time window will again reduce the resulting depolarization and thus the probability that an action potential will be generated. Inhibitory synapses are found throughout the dendrites, but in particularly high density at the cell body, i.e., in strategic proximity to the axon hillock. There they are ideally positioned to effectively prevent the generation of action potentials.

Inhibitory transmitters such as GABA can inhibit a neuron even if the opening of their receptors does not lead to an IPSP. The reversal potential for Cl^- ions is quite close to the resting membrane potential. If the reversal potential is equal to the resting membrane potential, the opening of Cl^--permeable channels will lead to no IPSP. There are neurons whose resting membrane potential is even more negative than the reversal potential for Cl^- ions. In this case, the opening of Cl^--permeable channels even leads to a depolarization of the neuron (formally, one can then measure an EPSPs in electrophysiological studies). In general, the neuron is nevertheless inhibited by the opening of Cl^--permeable channels (e.g. $GABA_A$-receptors). This is partly because the reversal potential for Cl$^-$ ions is more negative than the threshold potential. Once the membrane potential through EPSPs approaches the threshold potential, and the membrane potential is then less negative than the reversal potential for Cl$^-$ ions, the opening of Cl$^-$-permeable channels will again trigger an IPSP. On the other hand, the opening of Cl$^-$-permeable channels reduces the membrane resistance. In ▶ Sect. 4.6 we mentioned that the amplitude of membrane potential changes depends on the membrane resistance. The lower this is, e.g. if many K$^+$ channels are open causing leakage currents, the smaller the membrane potential change will be. In a very similar way, open Cl$^-$-permeable channels reduce the amplitude of EPSP, even if the Cl$^-$-permeable channels themselves do not trigger IPSP or even depolarize the neuron. This is referred to as ***shunting inhibition***.

Electroencephalography

Electroencephalography (EEG) is used to graphically record electrical brain activity via electrodes attached to the scalp. In the process, potential fluctuations are derived from many neurons, especially cortical neurons (primarily potentials that result from synaptic activity). Depending on the state of wakefulness or consciousness, the activity of populations of neurons synchronizes. This synchronous activity is reflected in oscillatory EEG signals. The frequency spectrum of the EEG can be analyzed quantitatively by Fourier analysis. Most often, however, the predominant frequency is read directly from the EEG by the experienced examiner. α-waves (alpha waves; 8–13 Hz) are found in the EEG as the dominant frequency in awake and relaxed people with eyes closed (especially over the occipital cortex). When the eyes are opened or mentally tense, desynchronization occurs and faster β-waves (beta-waves; 13–30 Hz) replace the α-waves (α-blocking; arousal response).

During sleep, the dominant EEG frequency slows down, ϑ-waves (theta-waves; 4–7 Hz) and with increasing sleep depth δ-waves (delta-waves; 0.5–3 Hz) are observed. The slow frequencies are explained by the fact that as the state of wakefulness or consciousness decreases, the thalamus determines the rhythm of cortical activity, leading to the synchronous activity of large populations of neurons. As the number of synchronously active neurons increases so does the amplitude of the recorded potential fluctuations. δ-waves therefore generally have a higher amplitude than β-waves. γ-waves (gamma-waves; 30–80 Hz) occur in the awake and alert state, but have such small amplitudes that intracortical EEG recordings are usually required for their analysis. The sensitivity of intracerebral EEG is so good that action potentials of individual neurons can also be recorded. This means, for example, that the single-cell activity of over a hundred neurons can be recorded simultaneously and correlated with specific behavioral patterns. Intracerebral EEGs are not only used in animal experiments as part of basic research, but are also performed on epilepsy patients as part of preoperative diagnostics. EEGs are of particular importance in epilepsy diagnostics. In addition to the frequency spectrum, the neurologist primarily analyses the forms of the potential fluctuations. During an epileptic seizure, the synchronous activity of large neuron populations can be seen in the high-amplitude potential fluctuations. In the seizure-free interval, characteristic potential fluctuations (epilepsy-typical potentials; e.g. *spike-waves*) indicate the presence of the disease. The EEG is also used in brain death diagnostics. Here, the absence of potential fluctuations (zero-line EEG) indicates irreversible brain damage or brain death.

4.15 Synaptic Short-Term Plasticity

The strength of communication between presynaptic and postsynaptic neuron is variable. This variability is the basis for adaptive processes, learning and memory formation. Different mechanisms cause the amplitude of postsynaptic responses to increase or decrease for short but also very long periods of time. If the change only lasts for a short time (a few seconds), we speak of short-term plasticity.

Short-term plasticity refers to the change in synaptic responses when a synapse is active several times in a short period of time, e.g. when the presynaptic neuron fires at a frequency of 10 Hz. This can increase or decrease the amplitude of the responses (□ Fig. 4.11). This is referred to as synaptic **facilitation** or **depression, respectively**. There are presynaptic and postsynaptic mechanisms that lead to a short-lasting change in synaptic strength. On the presynaptic side, high frequency activity generally leads to a change in the probability that a transmitter-filled vesicle will be released. In synapses where the probability of release is very high, each subsequent action potential reduces this probability. The first action potential and the resulting Ca^{2+}-influx have a high probability to trigger a release if there are vesicles that are very close to the cell membrane and are just "waiting" for the signal to exocytose. When the presynaptic neuron is active again shortly afterwards, these easily released vesicles no longer exist. Each subsequent action potential releases fewer vesicles. The postsynaptic responses therefore become smaller (depression). The situation is quite different in synapses where the release probability is low. In these synapses, the presynaptic Ca^{2+} influx is often insufficient for a vesicle to be exocytosed at all. The repetitive activity of such synapses leads to an accumulation of Ca^{2+} in the axon terminal so that the release

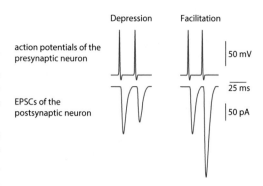

□ **Fig. 4.11** Synaptic short-term plasticity. If a presynaptic neuron is excited several times within a short period of time, synaptic short-term plasticity may occur. Short-term plasticity can occur as facilitation or depression. Facilitation means that EPSC amplitudes increase. Depression causes the opposite, i.e., the EPSC amplitude decreases. Both pre- and postsynaptic mechanisms can be responsible for both mechanisms

probability increases with each action potential. This is referred to as synaptic facilitation.

In fact, the release probability in synapses of the central nervous system is mostly smaller than 1. There are for example synapses where the release probability is 0.05. This means that only every 20th action potential triggers the exocytosis of a vesicle. The effectiveness of communication between two neurons is increased by the fact that a presynaptic neuron often has more than one synapse with the postsynaptic neuron. Nevertheless, a single action potential does not always lead to a postsynaptic response. Synaptic facilitation then strengthens these less effective connections.

In some synapses with high release probability, postsynaptic mechanisms play an important role in short-term plasticity. The binding of the transmitter to its receptor not only causes it to open but also to desensitize, i.e. to become less sensitive. Some receptors are slow to recover from this desensitization. Thus, in synapses with a high release probability the prolonged desensitization of receptors can reduce the current amplitude and thereby contribute to synaptic depression.

4.16 Synaptic Long-Term Plasticity

In contrast to short-term plasticity, changes in synaptic strength in long-term plasticity can last for many hours or days. It is likely that long-term plasticity is a fundamental mechanism for the long-lasting storage of learned material in memory. Communication between neurons can be strengthened and weakened. This is referred to as long-term potentiation and long-term depression, respectively. The Canadian psychologist Donald Hebb postulated in 1949 that the synaptic communication of two neurons is strengthened if the activity of one neuron repeatedly causes activity in the other neuron (Hebb 1949) (*what fires together, wires together*). Long-term potentiation can explain, for example, the synaptic changes in **classical conditioning**: When an animal repeatedly hears a tone just before or during an electric shock *fear conditioning* occurs. After conditioning, the animal will respond with a fear response when it hears the tone. Before conditioning, the tone, as a **neutral stimulus**, will not elicit a fear response. At the cellular level, we can imagine that before conditioning the auditory neurons elicit an EPSP but no action potential in neurons of the amygdala, a brain region that is important for the fear response. The electrical shock will elicit an EPSP large enough to form an action potential in the same cells. Thus, as an **unconditioned stimulus** (UCS), the electrical shock will always elicit a fear response. If sound and shock occur repeatedly together the auditory neurons will be active more or less simultaneously with the neurons of the amygdala. Thus, as predicted by Hebb, synaptic communication is strengthened and in our example in such a way that, after conditioning, activity in the auditory neurons triggers a suprathreshold EPSP in the neurons of the amygdala. The tone thus becomes a **conditioned stimulus** (CS). If the tone is later repeatedly presented without electrical shock the fear response will decrease (**extinction**). The influence of auditory neurons on neurons of the amygdala will weaken again. Long-term synaptic depression, in which there is a prolonged weakening of synaptic communication, appears to play a role in extinction.

The timing of presynaptic and postsynaptic neuron activity is important in determining whether potentiation or depression occurs (Markram et al. 1997). This is referred to as *spike-timing-dependent plasticity*. In general, long-term potentiation occurs when the presynaptic neuron is active just before the postsynaptic neuron (◘ Figs. 4.12 and 4.13). This temporal sequence suggests a causal link. In our fear conditioning example, the auditory neurons will be active before the neurons in the amygdala. With this apparent causal connection (tone triggers electrical shock) conditioning makes sense. The tone predicts the shock that is about to follow. If the tone is not followed by a shock, e.g., if the shock-tone sequence is reversed, then the tone has no predictive value for the occurrence of the tone. Synaptic connections in which the presynaptic neuron is repeatedly active after the postsynaptic neuron are generally weakened accordingly.

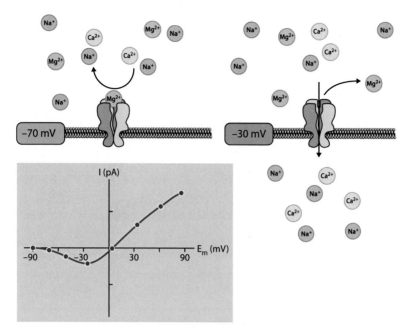

◻ Fig. 4.12 Magnesium block of NMDA receptors. NMDA receptors have a voltage dependence mediated by Mg^{2+} ions. If the receptors open after binding glutamate at values around the resting membrane potential of a nerve cell (approx. −70 mV), the Mg^{2+} ions attracted by the negative cell interior block the receptor because they cannot pass through it. This so-called Mg^{2+}-block also prevents the other ions (especially Na^+ and Ca^{2+} ions) from passing through the channel. However, if the membrane potential shifts towards more positive values the Mg^{2+} block is reduced. As a result, more and more Na^+ and Ca^{2+} ions can pass the receptor and an EPSP is formed. The dependence of the ion flux on the membrane voltage can be shown particularly well in a so-called current-voltage curve in which the current flow through the receptor is plotted against the membrane voltage of the cell. In the curve shown here for an NMDA receptor it can be seen that no current flows through the receptor at values below −70 mV. With depolarization of the membrane potential, the Mg^{2+} block described above increasingly decreases and a current flow occurs

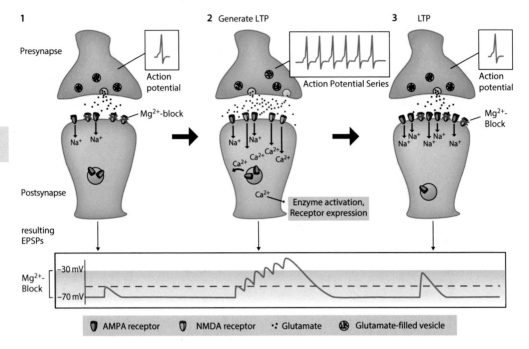

◻ Fig. 4.13 Development of long-term potentiation. At excitatory synapses of the brain, action potentials arriving at the presynapse lead to Ca^{2+}-mediated release of glutamate-filled vesicles and binding of glutamate to postsynaptic glutamate receptors. (1) At negative resting membrane potential, only AMPA receptors open because NMDA receptors are blocked by Mg^{2+} ions. The depolarization resulting from opening of fewer AMPA receptors is too small to remove the Mg^{2+}-block. (2) Upon higher frequency activation of the synapse and temporal summation of EPSPs, the Mg^{2+}-block of NMDA receptors is removed. Ca^{2+} ions flowing into the cell through NMDA receptors lead to increased AMPA receptor incorporation into the synapse via various mechanisms. (3) Subsequently, when another action potential encounters a synapse enhanced by the incorporation of more AMPA receptors it generates an EPSP that is larger than the initial EPSP. Since the amplification of the EPSP can persist over long periods of time it is referred to as long-term potentiation. (Bredt and Nicoll 2003)

4.17 Cellular Mechanisms of Synaptic Long-Term Plasticity

What happens at the synaptic level during long-term potentiation and long-term depression? As with short-term synaptic plasticity, there are in principle presynaptic and postsynaptic mechanisms that can contribute to changes in synaptic communication strength. In contrast to short-term plasticity, however, postsynaptic mechanisms play a greater role in most synapses.

Changes in presynaptic function were analyzed in particular detail by the American neurophysiologist Erik Kandel and his colleagues in the marine snail *Aplysia*. They investigated the cellular mechanisms underlying associative and non-associative memory. **Associative learning** was studied using **classical conditioning** in which an electric shock was used as an unconditioned stimulus and touching the snail's siphon was used as a neutral stimulus. After repeated pairing of both stimuli, touching the siphon elicits a contraction of the snail's gill. This results in a long-lasting strengthening of communication between sensory neurons which are active when the siphon is touched and postsynaptic motor neurons which are responsible for the contraction of the gill. The **induction** of presynaptic long-term potentiation (LTP) requires the activity of interneurons that are active when the electric shock is applied. These interneurons secrete the transmitter serotonin, which binds to metabotropic receptors of sensory siphon touch neurons. In the axon terminals of sensory neurons, second messenger signaling cascades increase the release probability for glutamate vesicles. The resulting long-term potentiation of synaptic communication explains why touching the siphon triggers gill contraction even days after conditioning. Again, the order of stimulation plays a role. A long-term potentiation and conditioned response is observed when touching the siphon occurs just before the electrical shock. If the stimulation is reversed, long-term depression (LTD) occurs.

The mechanisms that lead to the strengthening of the synaptic connection in classical conditioning are in principle very similar to the mechanisms that play a role in **sensitization**. Here, a repeatedly applied strong electrical stimulus enhances the transmitter release of sensory neurons that are active upon electrical stimulation. These presynaptic changes are also triggered via the activity of serotonergic neurons. The resulting long-term potentiation between sensory and motor neurons causes gill contraction to increase in response to weak electrical stimuli. Because sensitization occurs on repeated application of an electrical stimulus, i.e., does not require pairing with another stimulus, it is a form of **nonassociative learning**. In contrast, with primarily rather weak electrical stimuli **habituation** occurs another form of nonassociative learning. The serotonergic interneurons are not activated to a sufficient extent so that the strength of synaptic communication between sensory and motor neurons weakens. Consequently, gill contraction decreases with repeated application of a weak electrical stimulus.

In more complex organisms, presynaptic and postsynaptic mechanisms lead to the sustained strengthening of neuronal communication. A prominent example of presynaptic long-term potentiation is the long-lasting strengthening of communication between granule cells and CA3 neurons in the hippocampus. Granule cells of the dentate gyrus project with their axons known as mossy fibers to CA3 neurons. High-frequency stimulation of mossy fibers induces long-term potentiation in granule cell-to-CA3 synapses. Activation of second messenger signaling cascades thereby increases the release probability of glutamate vesicles in axon terminals of granule cells (similar to sensory neurons of *Aplysia*). This potentiates EPSP amplitudes in postsynaptic CA3 neurons.

In the hippocampus, CA3 neurons project to CA1 neurons. Here, high-frequency stimulation or even low-frequency stimulation in which a presynaptic action potential is paired with depolarization of the postsynaptic neuron induces a postsynaptic form of long-term potentiation. Crucial for the induction of potentiation are the activation of NMDA-type glutamate receptors and an influx of Ca^{2+} ions into the postsynaptic neuron. NMDA receptors have two properties that make them central players in the induction of synaptic plasticity (Figs. 4.12 and 4.13). First, at a resting membrane potential of, say, -70 mV, NMDA receptors are blocked by Mg^{2+} ions which sit on the extracellular side in the channel of NMDA receptors and prevent other ions from flowing through the channel (Fig. 4.12). Thus, binding of glutamate to NMDA receptors does not result in current flow in nondepolarized neurons. When the neuron is depolarized, Mg^{2+} ions leave the channel of NMDA receptors. At membrane potentials more positive than -30 mV, Mg^{2+} ions do not block NMDA receptors so that binding of glutamate can induce a current. Thus, NMDA receptors are **coincidence detectors** that open only in the presence of simultaneous activity of the pre-synaptic (release of glutamate) and postsyn-aptic neuron (depolarization). Let us recall Donald Hebb. He had postulated that neu-rons strengthen their connection when they are active together. Thus, the molecular basis for "Hebbian learning" is the NMDA recep-tor, since it can detect joint activity. This also explains why synaptic plasticity is generally restricted to the synapses where this simulta-neous activity occurs (**input specificity**). High-frequency presynaptic activity, for example, can lead to postsynaptic depolarization suffi-cient to open NMDA receptors. Less active synapses, where depolarization is insufficient, are not potentiated. However, if they are active at the same time as a synapse that leads to strong depolarization they too can be potentiated (**associativity**). Potentiation of "weak" synapses upon synchronous activa-tion of "strong" synapses is a form of synap-tic plasticity that likely underlies associative learning. In our example above, before condi-tioning, auditory stimuli activate weak syn-apses. Simultaneous application of electrical stimuli stimulates strong synapses such that depolarization is sufficient to open NMDA receptors in the weak synapses. These are thereby potentiated, the primarily neutral stimulus becomes a suprathreshold condi-tioned stimulus.

The second property of NMDA recep-tors that is crucial for the induction of synap-tic plasticity is their high permeability for Ca^{2+} ions. Ca^{2+} ions flowing into the post-synaptic neuron activate intracellular signal-ing cascades as second messengers, which ultimately lead to the expression of long-term potentiation. Kinases such as $Ca^{2+}/$Calmodulin kinase II (CaMKII), protein kinase A and C play important roles in this process (Raymond 2007). The subsequent phosphorylation of AMPA receptors increases their conductance. In addition, new AMPA receptors are incorporated into the synaptic membrane. Both mechanisms lead to a rapid potentiation of synaptic transmis-sion (◘ Fig. 4.13).

AMPA receptors are not expressed in every excitatory synapse. These synapses are also referred to as *silent synapses* since they do not form EPSPs at a resting potential of, for example, −70 mV. However, silent syn-apses may express NMDA receptors the activation of which leads to the incorpora-tion of AMPA receptors during the induc-tion of synaptic plasticity by the mechanisms mentioned above. This form of long-term potentiation is also referred to as *unsilencing* (Bredt and Nicoll 2003).

For a long-lasting potentiation, an increased protein synthesis is also necessary. A crucial step in this process is the transcrip-tion of DNA into RNA, which is induced by transcription activators such as CREB (*cAMP response element-binding protein*). This leads to increased synthesis of AMPA receptors and structural proteins. The result is not only an increased incorporation of AMPA receptors into the synapse, but also a restructuring of the synapse. However, the potentiation results not only from the strengthening of the existing synapses, which become larger and contain more AMPA receptors, but also from a new for-mation of additional synaptic connections.

However, the activation of NMDA receptors can induce not only long-term potentiation but also long-term depression. In this case, the activation of NMDA recep-tors is generally weaker than during the induction of long-term potentiation and consequently the increase in the intracellular Ca^{2+} ion concentration is also lower. The signalling cascades activated after induction of long-term depression thus also differ from those after induction of long-term potentiation. For example, whereas kinases play a special role in long-term potentiation, phosphatases, i.e. enzymes that dephosphor-ylate phosphorylated proteins, play a more important role in long-term depression. The depression of synaptic strength results in particular from a reduction in the number of synaptic AMPA receptors.

Summary

How does consciousness arise? How does our memory work? What causes diseases that influence our behavior? These are only three examples of current scientific questions, the answers to which are indispensably linked to knowledge of the basic neurophysiological processes in our brain, as we have described them in the present chapter. Although great progress has been made in the field of basic neurophysiological research due to the enormous technical progress since the establishment of the membrane theory by Julius Bernstein in the nineteenth century, we are still far from understanding the complexity of the human brain in all its facets. Above all, experiments on rodents, usually mice or rats, provided important insights into the understanding of physiological as well as pathophysiological processes in the brain. For example, optogenetic stimulation of individual nerve cells in the brains of living mice has been able to manipulate their memory leading to changes in their behaviour. The findings of basic neurophysiological research also play a major role in medical research. The autism spectrum disorders, for example, are associated with a pathological change in the number of synapses. Here, too, the basic principles of the disease were studied in rodents and the findings were transferred to humans. The investigation of pathophysiological changes in brain function is essential for the understanding of brain diseases such as stroke, migraine, Parkinson's disease, multiple sclerosis, epilepsies, Alzheimer's disease, schizophrenia, autism, depression, anxiety disorders and thus an important building block for the development of new therapeutic approaches.

References

Armstrong CM, Hille B (1998) Voltage-gated ion channels and electrical excitability. Neuron 20:371–380

Bennett MV, Zukin RS (2004) Electrical coupling and neuronal synchronization in the mammalian brain. Neuron 41:495–511

Bredt DS, Nicoll RA (2003) AMPA receptor trafficking at excitatory synapses. Neuron 40:361–379

Catterall WA (2017) Forty years of sodium channels: structure, function, pharmacology, and epilepsy. Neurochem Res 42(9):2495–2504

Foster M (1897) A textbook of physiology, part III, 7. Aufl. Macmillian, London

Hamill OP, Marty A, Neher E, Sakmann B, Sigworth FJ (1981) Improved patch-clamp techniques for high-resolution current recording from cells and cell-free membrane patches. Pflüg Arch 391:85–100

Hebb DO (1949) The organization of behavior. A neuropsychological theory. Wiley, New York

Hille B (2001) Ionic channels of excitable membranes, 3. Aufl. Sinauer, Sunderland

Hodgkin AL (1964) The ionic basis of nervous conduction. Science 145:1287

Jahn R, Fasshauer D (2012) Molecular machines governing exocytosis of synaptic vesicles. Nature 490(7419):201–207

Markram H, Lübke J, Frotscher M, Sakmann B (1997) Regulation of synaptic efficacy by coincidence of postsynaptic APs and EPSPs. Science 275:213–215

Raymond CR (2007) LTP forms 1, 2 and 3: different mechanisms for the 'long' in long-term potentiation. Trends Neurosci 30:167–175

Skou JC (1957) The influence of some cations on an adenosine triphosphatase from peripheral nerves. Biochim Biophys Acta 23:394–401

Yuste R (2015) The discovery of dendritic spines by Cajal. Front Neuroanat 9:18

Further Reading

Andersen P, Morris R, Amaral D, Bliss T, O'Keefe J (2007) The hippocampus book. Oxford University Press, New York

Jack JJB, Noble D, Tsien RW (1975) Electric current flow in excitable cells. Clarendon Press, Oxford

Kandel ER (2013) Principles of neural science, 5. Aufl. McGraw-Hill Professional, New York

Kreutz MR, Sala C (2012) Synaptic plasticity. Springer, Wien

Llinás RR (1988) The intrinsic electrophysiological properties of mammalian neurons: insights into central nervous system function. Science 242:1654–1664

Developmental Neurobiology

Nicole Strüber and Gerhard Roth

Contents

G. Roth et al. (eds.), *Psychoneuroscience*, https://doi.org/10.1007/978-3-662-65774-4_5

Trailer

The brain perceives its environment, it feels, compares, infers and initiates and controls behaviour and language—and much more. How does all this arise from a single fertilized egg?

Already early in the prenatal development, the first brain structures are formed. The development is initially genetically controlled. However, even before birth, the environment begins to influence the developing brain, for example when the mother-to-be experiences considerable stress. At birth, the morphological appearance of the brain is already very similar to that of the mature brain. Even to the naked eye, convolutions (gyri) and furrows (sulci) are clearly visible, and imaging studies show the typical structure of the brain with portions of grey and white matter. After birth, the influence of experiences is great. These can significantly influence brain development and are even required for normal development in many circuits. They refine the synaptic circuitry and basic psychoneural systems in the maturing brain and thus influence the way the brain perceives, feels, evaluates and controls behavior.

Learning Objectives

You will learn how the brain develops prenatally and early postnatally and how humans acquire individual tendencies of feeling, thinking and acting against the background of an interplay of genes, epigenetics and environment.

5.1 Prenatal Development of the Brain

Prenatal brain development initially follows a genetically determined plan that produces a functional nervous system if it is not disrupted by mutations or harmful environmental influences such as drugs, medications and nutrient deficiencies. In the first 8 weeks after fertilization, new structures, i.e. tissues and organs, are formed in a short time during the embryonic period. This is followed by a phase whose essential characteristics are growth and histological differentiation: the fetal period. The cerebral cortex becomes more complex and thicker, increasingly concealing underlying "subcortical" areas such as the diencephalon, the mesencephalon, and parts of the cerebellum. Sulci and gyri of the cortex also form. A highly complex central nervous system emerges that can process information and organize behavior (◘ Figs. 5.1 and 5.2). In the following, we will take a closer look at the individual steps of this development. We begin with the embryo.

5.1.1 Formation and Differentiation of Neuronal Tissue

In the days and weeks after fertilization, the cells of the young embryo divide over and over again. Already at the beginning of the third week, structures of cells with different destinations form, the so-called germ layers (cotyleda): endoderm, mesoderm and ectoderm. The ectoderm is particularly relevant for neurobiology, because it gives rise to the nervous system in addition to the outermost skin layer. The central area of this embryonic structure is important for this, the so-called neural plate, which appears on the 19th day of development. But how do the cells located in the middle of the ectoderm know that they should not form skin tissue like the cells located at the sides, but instead neural tissue? This happens because of signals from neighboring cells of a second cotyledon—the mesoderm. The genetic program of these cells makes them release specific factors that in turn tell the centrally located cells of the ectoderm to turn on different genes than the later skin cells located to the side. This is called **neural induction**. The neural plate elongates and

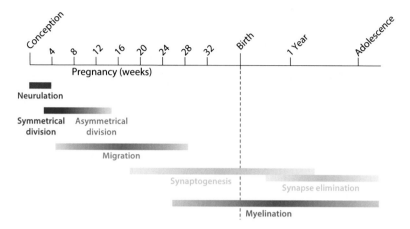

Fig. 5.1 Overview of the approximate time course of essential processes of brain development

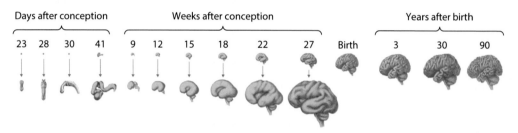

Fig. 5.2 Overview of the anatomical features of the developing human brain. The prenatal stages of brain development are shown again in magnification in the second row. (Modified after Silbereis et al. 2016)

deepens along the midline, and the so-called neural groove forms, the edges of which begin to fuse into the neural tube toward the end of the fourth week (**neurulation;** ■ Fig. 5.1). Some cells migrate out of the fusion area and form paired strands, the neural crests. They give rise to most parts of the peripheral nervous system as well as the adrenal medulla. The neural tube runs lengthwise through the embryo. In the direction of the rump (in animals in the direction of the tail, i. e., caudally) the central canal of the spinal cord arises from the cavity of this structure, while in the anterior, cephalic region (rostrally) it widens to form the ventricles of the brain. The walls of the internally hollow neural tube are composed of neural stem cells or progenitor cells. In this area, also called the ventricular zone, the stem cells divide by mitosis.

The resulting multiplication of cells, **proliferation**, takes place in two stages, namely symmetrical and asymmetrical proliferation. Initially, cells divide symmetrically. A progenitor cell gives rise to two identical cells, and the number of neural stem cells grows quadratically. At a regionally different time point, the stem cells switch to an asymmetric mode of cell division. One daughter cell remains a progenitor and continues to divide, and the other cell becomes either a glial cell or, in the process of prenatal neurogenesis, a neuron.

Neuronal Induction

Due to signals from other cells, the centrally located cells of the ectoderm differentiate into neuronal cells.

5

> **Neurulation**
>
> This process refers to the formation of the neural tube and, with it, the beginning of the development of the central nervous system. It begins with the formation of the neural plate from cells of the ectoderm and ends with the formation of a tube that will develop into the brain and spinal cord.

There are many different types of neurons in the brain. Accordingly, the cells must **differentiate**. Depending on where a cell is located within the tissue, it receives a specific pattern of chemical signals from the neighboring tissue. This leads to different gene expression in different cells, and a wide variety of neuronal cell types are generated, each type with a characteristic shape, with specific synaptic properties and neurotransmitters. In the posterior (caudal) region of the neural tube, the spinal cord is formed as a result of this differentiation; in the anterior region, the first anlagen of the future brain are already visible at the embryonic age of three and a half weeks. At the age of 5 weeks, the five brain structures can be identified by local bulging of the interior (vesicles) at the rostral end of the now closed neural tube: endbrain (cerebrum, telencephalon), diencephalon (dienencephalon), midbrain (mesencephalon), hindbrain (metencephalon) and afterbrain (myelencephalon).

> **Differentiation**
>
> The evolution of cells towards higher structural and functional specialization.

The hindbrain develops rapidly and its complexity already corresponds to that of a newborn at an embryonic age of 8 weeks. The first attachments of the various structures of the cerebrum are also formed at an early age. For example, the hippocampus begins to form at the age of 5 weeks. Nuclei of the amygdala complex can also be detected at this time. Fibrous connections between the nuclei of the amygdala with the septum, the hippocampus and also the diencephalon form the beginning of the limbic system before the end of the embryonic period.

The cerebral cortex is formed by **migration** of young neurons from their place of origin in the ventricular zone of the end brain towards the brain surface (◘ Fig. 5.1). To enable migration, some of the progenitor cells first develop into specialized cells, the radial fiber glia. Their cell bodies remain in the ventricular zone but send long projections both to the inner membrane separating the ventricular zone from the ventricle and to the membrane spanning the brain, called the pia mater. Thus, they form a scaffold along which neurons can migrate toward the brain surface. In humans, cortical migration begins around day 33 of embryonic development. After migration, the radial glial cells disappear. Many of them transform into astrocytes.

The mature cerebral cortex is divided into distinct areas that are anatomically distinguishable and functionally specialized based on the pattern of their connections. Where within the cortex a neuron settles depends on where it is generated within the ventricular zone and when this occurs within early development. The first out-migrating cells settle in the deepest, i.e. innermost, layer of the cortex (layer 6); later-generated cells migrate past the early-born neurons and colonize more superficial layers of the cortex. The peak of migration occurs between the third and fifth months of gestation. During the third trimester of pregnancy, migration is complete, and the neocortex is already divided into its characteristic six layers by the seventh month.

> **Migration**
>
> Emigration of newly formed neurons during brain development from their place of origin in the ventricular zone towards the brain surface.

5.1.2 Formation of Networks of Synaptic Connections

In a mature brain, about 90 billion neurons (Herculano-Houzel 2009) of different regions are precisely interconnected and form the basis of all behavior. The formation of such networks requires that the axons of the neurons grow out and find their correct path during early prenatal development. The possibilities for **axon growth** are limited here. The axon can grow in a certain direction, turn or stop. Its outgrowth is controlled by molecular signals. At the tip of the axon there is a growth cone which, with the help of specific receptors, is able to recognise and integrate the molecular cues in order to guide the axon to its destination. In the growth cone, the structure of the cytoskeleton changes, and this determines in which direction and at what speed the axon spreads.

> **Axon Growth**
>
> The targeted outgrowth of neuronal cell axons during early brain development.

Once the axons have found the adequate neighborhood for them, they have to recognize and contact their corresponding synaptic target from a multitude of potential partners. In this process, growth factors and other molecules released by the target neurons or surrounding glial cells stimulate the local spread of the processes and the further differentiation and maturation of the neuron. Once the partners have converged, the formation of synapses, **synaptogenesis**, is an interactive process. Neurons exchange numerous signals to coordinate their activities and stabilize sites of contact. Synaptogenesis begins during the second trimester of pregnancy and continues through the first years of life.

> **Synaptogenesis**
>
> The formation of contact points, synapses, between two nerve cells.

Due to the processes described here, each neuron has numerous synaptic partners, on average several thousand, sometimes more than 100,000. During the development of the nervous system, however, up to 50% of neurons die shortly after they have formed synaptic connections with their target cells. The tissue does not shrink, however, because for almost every cell that dies, another divides and replaces it. The cause of this "**programmed cell death**" could be a lack of supply of growth factors, which are essential for the survival of every neuron. In its search for synaptic partner cells, a neuron must take up the substances released by the neuron's target cells, via its presynaptic terminals. However, the substances are present only in a limited quantity, and this is not sufficient to supply all the neurons generated during development. The undersupply, however, is not a shortcoming of the developing brain, but a clever move of the brain's "developmental program": only neurons with active synaptic connections receive sufficient growth factors. For all others, an intracellular "suicide program" is released, and they die. This mechanism ensures that only cells with suitable synaptic targets survive.

> **Programmed Cell Death**
>
> Regulated elimination of cells, which ensures that suitable synaptic connections are formed in the brain.

5.1.3 Formation of Glial Cells and Myelin

From the same precursor cells that give rise to neurons, another class of cells develops: glial cells. Their formation begins four and a half weeks after fertilization and continues postnatally. Different subtypes of glial cells perform different functions in the brain. For example, astrocytes are involved in regulating the composition of the extracellular milieu, the reuptake of excess neurotransmitters, and synaptogenesis. They constitute almost half of all cells in the human brain. Oligodendrocytes, another class of glial cells, are responsible in the central nervous system for, among other things, the formation of myelin. Myelin is a lipid-rich membrane formed by glial cells that wraps around and insulates the axons of neurons, thereby promoting the rapid and efficient transmission of action potentials along the axons.

This process called **myelination** begins in the brain towards the end of the second trimester of fetal development and extends after birth well into the third decade of life and beyond. It progresses from caudal to rostral, following a basic principle of maturation in the central nervous system: caudal areas, such as those of the spinal cord responsible for simple reflexes, mature first, while rostral brain structures with more complicated functions mature later. Last in line are the cortical areas of the frontal lobe. Here, myelination begins only at the age of 7–11 months after birth.

The progressive coating of axons with myelin promotes the rapid transmission of action potentials. This can also increase the speed and efficiency with which sensory, cognitive, emotional and motor content can be processed. The delayed or temporally extended development of this process is accordingly partly responsible for the fact that many abilities are only stably developed with increasing age.

Myelination

The insulating coating of axons with a lipid-rich membrane to increase the transmission efficiency of action potentials.

Essential processes in prenatal development follow a developmental plan stored in the genes. Accordingly, they take place in a similar way in each individual. Prenatally, however, the environment also begins to influence the developing brain in an individual way, such as the child's stress system. We will return to this influence later, in a connection with the psycho-neuronal personality systems (▶ Sect. 5.3.2). First, let us look at the progression of neuronal connectivity after birth.

Prenatal Development

Prenatal development initially follows a genetically predetermined plan. From a small area of one of the embryonic germ layers, a functioning, highly complex central nervous system emerges via processes such as neuronal induction, neurulation, differentiation, migration, axon growth, synaptogenesis, programmed cell death, myelination and many other processes.

5.2 Postnatal Development

Postnatally, the number of connections between neurons is rapidly increased. The rapid and extensive synaptogenesis initially produces an overproduction of synaptic connections. Just as some of the neurons are degraded again, a large proportion of the synapses subsequently undergoes elimination. In this process, the initial pattern of connections is refined in an activity-dependent manner, and coordinated functional circuits are generated: those connections that are active, are stabilized, and unused synapses are eliminated (Changeux and Danchin 1976). This depends on the particular experiences a child has in his specific environment.

We will now explain this process and its importance for adaptation to an individual environment. Subsequently, we will deal with the importance of sensitive periods for these processes.

5.2.1 Activity-Dependent Modification of Neuronal Circuits

Synaptogenesis produces prenatally and early postnatally an initially overconnected network of neurons. Experiences now decide which synapses are stabilized by neuronal activity and which are eliminated: Those synapses that are activated by excitations from sensory or motor centers of the brain survive, the rest disappear due to disuse. The involved projections are retracted—a process also called pruning.

The fine-tuning of synaptic connections during brain development follows the general learning principle proposed by Donald O. Hebb (1949), according to which connections between cells are strengthened when one cell is repeatedly involved in exciting another cell. Following this rule, the activity-dependent modification seems to involve that

- synaptic contacts between synchronously active pre- and postsynaptic neurons are strengthened and
- synaptic contacts between non-synchronously active neurons are weakened or eliminated (Constantine-Paton et al. 1990; Singer 1990).

The activity-dependent change in the permanent strength of synaptic connections is called **synaptic plasticity**. Synaptic plasticity is the basis of all learning and allows the organism to adapt to its respective environment.

> **Synaptic Plasticity**
>
> The activity-dependent change in the permanent strength of synaptic connections. It is the basis of all learning and enables the organism to adapt to its respective environment.

At the molecular level, the NMDA receptor (▶ Chap. 4) plays an important role in associative learning processes, and its activation is also crucial for the stabilization of synaptic connections during early development. However, the early fine-tuning of synaptic connections is influenced by numerous other processes, such as the activity-dependent release of growth factors or the maturation of inhibitory circuits by GABAergic neurons. The latter is described in ▶ Sect. 5.2.2 in relation to sensitive periods.

The early overproduction of synapses is thus followed by a stabilization of those synapses that are used again and again due to specific experiences, and a degradation of those that remain unused. In this way, the brain is adapted to its environment. A "full wiring" between brain cells would be useless—it would correspond to an administrative operation in which all employees constantly talk to all others. Instead, a linkage structure is formed in the brain that has been shown to be optimal in network theory terms, namely a combination of intensive local ("local") linkage and fewer and fewer and increasingly selective "global" linkages over increasing distances. This is called the "small-world linkage model" (Watts and Strogatz 1998).

The brain must be able to respond appropriately to its respective environment. The genome determines a large part of the basic structure and function of the nervous system. However, the environment and the physical characteristics of the individual cannot be encoded by the genes. This information must be acquired through experience. Activity-dependent refinement of networks can allow the nervous system's functions to adapt to conditions in the outside world, while ensuring that metabolic energy is expended only on those processes that are needed. This will allow the brain to function quickly and efficiently in its particular environment. Because of the development designed in this way, the brain is shaped by the effects of early experiences. However, this also means that the consequences of negative experiences can be permanent.

For psychoneuroscience, the exact timing of these processes is of great importance, because it determines at what age the brain is particularly receptive to the influence of experiences and these have a major impact on the child's psychological development. It is around such periods of special receptivity that we will now turn.

5.2.2 Sensitive Periods of Brain Development

Many neural circuits are shaped by experiences during early sensory periods. The occurrence of such time windows has been studied primarily in animal sensory systems. Although we are primarily concerned with mental development in this textbook, the visual system is a good model for the importance of sensitive periods. Indeed, it has been shown that the visual system of mammals, including humans, requires visual information beginning with the first month of life in order to develop normal function. This is why it is called a critical period rather than a sensitive period.

Strictly speaking, a distinction must be made between **critical periods** and **sensitive periods**. Critical periods denote time windows within which certain experiences are absolutely necessary. The brain requires information from the environment that is the same for all members of the species and is universally present in the environment, such as basic elements of visual patterns. The brain waits for these experiences during this period (Greenough et al. 1987). If they are not made, the window of opportunity closes, and subsequent experiences can make little difference. In addition, there is the concept of sensitive periods. These are windows of time within which the brain is particularly sensitive to experience. It is influenced by experiences more than usual during this time. Every critical period is also a sensitive period, but not every sensitive period is also a critical period—the experiences don't always have to happen during that time. The

brain would just like to have the information during that time.

> **Definition**
>
> **Critical periods** denote time windows within which certain experiences are absolutely necessary for the brain to produce a function characteristic of the species.
>
> **Sensitive periods** refer to time windows within which the neuronal circuitry of the brain is particularly sensitive to the influence of experience.

In humans, the development of visual acuity has been studied in particular. The human visual system requires 10 years of structured visual information to form stable functional units of interconnected neurons in the brain areas responsible for visual processing and to enable high visual acuity. This period thus represents a critical period for visual acuity. The importance of early visual information for visual acuity has been studied, among other things, in people whose eyes only gained full functionality after surgery due to a congenital lens opacity (cataract). Although they may still be able to see spatially or distinguish between faces after an operation in later childhood or adolescence, as impressively demonstrated by the so-called Prakash Project, which aims to provide medical care for blind children in India, their visual acuity never develops normally. In fact, the brain needs the visual information in the first months of life so that the ability to resolve details can develop normally later on. This is especially remarkable because a child with healthy eyes cannot yet see fine details at all during this time. The information is therefore necessary at a time when the child is not yet able to use it (Maurer 2017).

Circuits of the brain that are responsible for other functions are also particularly receptive to experience during sensitive periods of development. However, one cannot define a general sensitive period, because each brain area and each function has its own time window. This can be illustrated by language development. There is a sensitive period for sound formation, another for sentence construction. Infants can basically distinguish all sounds in the first months of life, and this is why Japanese infants can distinguish the "r" and the "l". This ability is maintained until 10–12 months of age, after which infants can distinguish only the sounds of their native language(s)-the sensitive period for the acquisition of specific sounds is closed. The ability to learn sentence structure, on the other hand, does not decline steeply until after the age of seven. Vocabulary can be learned throughout life (Werker and Hensch 2015).

Thus, depending on the function in the brain, there are different time windows within which experiences are expected or the brain is particularly sensitive to experiences. It is assumed that at the beginning of this early sensitive or critical period, the brain is in a state of excessive synaptic interconnection. Experiences within the period now lead to the stabilization of certain neuronal connections. Connections that would have served other purposes (e.g., distinguishing "r" from "l" in a Japanese child) are broken down. In a further process, the active and stabilised synapses are now structurally strengthened by the use of certain molecules (▶ Sect. 5.2.3). They are then no longer susceptible to being eliminated (Knudsen 2004).

> **Overview**
> Three processes characterize what happens during sensitive periods:
> 1. the spreading of the processes and the formation of stable synapses
> 2. the elimination of processes and synapses
> 3. the consolidation or hardening of synapses

Of course, the human brain is also changed later, i.e. beyond sensitive periods, by learning experiences. And even then, associative learning processes lead to the stabilization of new synaptic connections of neurons, as described in ▶ Chap. 4. However, beyond the sensitive period, the molecular milieu in the brain differs from that of the maturing brain. Let us now ask ourselves why the brain is actually not always ready for change in the same way? Can this be influenced?

5.2.3 Regulation of Plasticity

The discussion about the significance of early sensitive periods has been enriched by another aspect in recent years. Nowadays it is assumed that the time windows of the sensitive periods are not rigid. This means: their period is not irretrievably genetically determined; rather, the opening and closing of the time windows are also influenced by experience. Accordingly, sensitive periods can be experimentally delayed or accelerated. Under certain conditions, it also appears that a molecular milieu can later be restored in which extensive changes are possible. The time windows are reopened. This, of course, has great significance for a person's ability to change, for example in the context of psychotherapy.

It is assumed that the increased plasticity during sensitive periods in early childhood is not an initial state in the child's brain. Accordingly, the time window of increased plasticity would initially still be closed. In order for the sensitive period to begin in the first place, certain molecular prerequisites must be fulfilled—and these are in turn influenced by experience. These prerequisites include a certain level of growth factor supply as well as the maturation of a subgroup of inhibitory GABAergic interneurons (the so-called parvalbumin-positive interneurons). Only when these have reached a certain degree of maturity—partly due to experience—is plasticity particularly pronounced. A further increase in maturity closes the sensitive period again (Hensch

2018). In certain states of maturity, inhibitory circuits can act like a **molecular brake** on cortex plasticity. They prevent neuronal connections from changing too much before and after the sensitive periods (◻ Fig. 5.3).

┌─ **Molecular Brakes** ─────────────
│ Molecular mechanisms for temporary
│ prevention of plasticity. These may limit
│ the plasticity of the developing brain
│ before and after the sensitive periods.
└──────────────────────────────────

The molecular brakes or preconditions for plasticity can be manipulated in animal experiments, again using the visual system as a model: If an animal grows up in complete darkness (i.e., if the eyes are not even stimulated by the light of a clouded lens, as in the children with a cataract mentioned above), then the window of increased receptivity initially remains closed. Without experience, few growth factors act on the cortex responsible for visual processing. In addition, inhibitory circuits are delayed in maturing. Under these conditions, the sensitive period begins later and sometimes remains open into the animals' adulthood (◻ Fig. 5.4). However, one can also trigger an early onset of the sensitive period by administering growth factors to animals growing up under normal light conditions or by using benzodiazepines to induce the inhibitory circuits to mature prematurely (Hensch 2018). Stimulating environmental conditions, on the other hand, can *prolong* the time period of increased susceptibility, as shown in a German study (Box: Prolonging the Time Period of Increased Susceptibility). According to this, individual experiences influence the molecular milieu, and this in turn affects how long the brain can be altered by environmental influences.

Extension of the Period of Increased Susceptibility Due to Stimulating Environmental Conditions
Depending on whether the studied animals grew up in a stimulating or a deprived environment, a sensitive period in the visual system, which is important for seeing with both eyes, was shorter or longer in them. If the animals grew up in typical laboratory cages, the

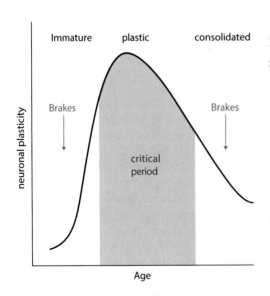

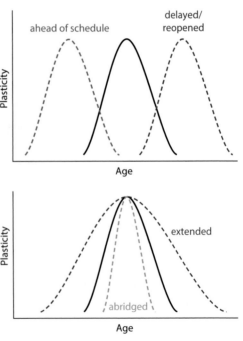

◻ **Fig. 5.3** Molecular brakes suppress the plasticity of the cerebral cortex. (After Werker and Hensch 2015)

◻ **Fig. 5.4** Changeable time course of critical periods. (After Werker and Hensch 2015)

sensitive period ended when they were young. If, on the other hand, the animals grew up in a stimulating environment and experienced social interaction, physical and also cognitive stimuli, then their visual cortex was very receptive to experiences well into adulthood. The animals had retained their youthful brains, so to speak. An electrophysiological analysis of activity in the brain suggested that a particular dominance of inhibitory circuits was responsible for the end of the sensitive period in the deprived animals (Greifzu et al. 2014).

Towards the end of the sensitive period, long-term wiring patterns are consolidated. This significantly shapes the later function of the cortex. Children in whom a one-sided visual impairment was not corrected early must now live with the visual impairment, even if the optical problem of the eye is corrected. Similarly, what sounds children can distinguish is determined, whether they can differentiate between the letters "r" and "l", for example. In the brain, in addition to the now mature inhibitory circuits, structural changes are also responsible for this consolidation. The latter include the so-called **perineuronal networks**. These are networks of extracellular matrix proteins that mature towards the end of the sensitive period, protect the synapses and thus also represent a molecular brake.

Perineuronal Networks

Networks of extracellular matrix proteins for the protection and consolidation of synaptic contacts.

These mechanisms of consolidation can, for example, lead to fear memories being permanently stored. If animals learn early after birth that certain stimuli (such as sounds) are coupled to pain, then they subsequently fear it (fear conditioning). However, if the coupling is subsequently absent (still early after birth), then this learned content is erased again (extinction). Such extinction is quite feasible early after birth, which corresponds to early childhood in humans. Later, extinction of the context is no longer possi-

ble, and the fear conditioning remains—even if it is learned that the stimulus is not so bad after all. Apparently, fear memories are actively protected at the level of the amygdala by the perineuronal networks (Gogolla et al. 2009).

If one wants to reopen time windows of increased susceptibility later, the molecular brakes must be released, e.g. pharmacologically. This allows the brain networks to become sensitive again. They are receptive to new things, and old learning contents can be better overlearned. This can be achieved, for example, by disturbing the activity of the inhibitory circuits of the cerebral cortex (Harauzov et al. 2010) or by acting on the perineuronal networks. If one pharmacologically induces a breakdown of these networks in fear-conditioned animals, subsequent fear conditioning can be better extinguished. Unfortunately, research in this area is still in its infancy, and it is not yet possible to treat anxiety disorders in humans, for example, using appropriate procedures. According to various studies, there are further possibilities to reopen the time windows of increased plasticity. According to these, both the rearing of animals in a complex and stimulating environment and the chronic administration of antidepressants could reopen time windows of increased changeability of the brain and enable the integration of new information (Box: Reopening the Time Period of Increased Receptivity). However, findings in this regard are also limited to the results of studies of the rodent visual system.

Reopening the Period of Increased Susceptibility Through Stimulating Environmental Conditions

If—as described in the example "Extension of the time period of increased receptivity"—the animals raised in the laboratory cage were kept in a stimulating environment in adulthood, this was able to restore the receptivity of the cerebral cortex. The sensitive period within which environmental stimuli can influence vision with both eyes had been reopened (Greifzu et al. 2014). In a similar manner, administration of antidepressants (serotonin reuptake inhibitors) was also able to reopen time windows of increased receptivity in the visual cortex of adult rats. The positive effects of the drug were accompanied by decreased activity of inhibitory circuits as well as increased formation of growth factors in the visual cortex. The authors of the study suggest that chronic administration of antidepressants could also have this effect in humans and underlie their therapeutic effect (Vetencourt et al. 2008).

It is not yet clear whether the findings presented here can be transferred in detail to the maturation of the areas of the human brain responsible for mental functions. Nevertheless, it can be assumed that a high degree of maturation is reached when the processes of synapse stabilization and elimination have been completed and the connections have been structurally consolidated. Experiences themselves seem to be able to promote these processes and thus also maturation.

Background Information

According to psychologist Mark H. Johnson, experiences or learning processes cause interconnected brain structures or cortical areas to specialize in certain functions, which in turn terminates sensitive periods.

Johnson opposes what he sees as a widespread view according to which brain regions mature as a result of genetic control and, with a sufficient degree of maturity, produce certain sensory, motor or even cognitive abilities (*maturational view*). According to this model, which he criticized, the high plasticity in sensitive periods would be determined by intrinsic factors of the cerebral cortex (e.g. certain maturation-dependent changes). The time windows of increased plasticity would be fixed in duration by maturation. Johnson, on the other hand, emphasizes that brain regions are not inherently specialized, and contrasts this view with his "interactive specialization" hypothesis. He assumes that the various areas of the cerebral cortex are initially only specialized to a limited extent for certain functions. Accordingly, they are active in a wide variety of situations. If these different brain areas are repeatedly activated together in the course of development, their functions become more and more sharpened, and their activity is increasingly reduced to circumscribed situations (e.g., a region that previously reacted to all kinds of visual objects will then react only to uprightly presented human faces). Assuming this view, a sensitive period will end when a brain region has gained full function as a result of its interaction with the other brain areas. The end of the sensitive period is thus triggered by the learning process itself (Johnson 2005).

5.2.4 Time Course of Synaptogenesis and Synapse Elimination

If we want to know at what age the networks relevant for mental functions are particularly receptive to experience, we have to ask ourselves when the processes of synapse stabilization and degradation described in ▶ Sects. 5.2.1 and 5.2.2 take place in the cerebral cortex, because, as explained, these produce stable wiring patterns that prepare the child for its respective environment.

Insights were initially provided here by synapse counts carried out postmortem. This showed an increase in synapse density in early childhood, followed by a plateau in childhood and a reduction of synapses in later childhood and adolescence. However, depending on the brain region and function, the processes took place at different ages. Nowadays, imaging techniques are generally used to gain insights into the course of development of the cerebral cortex. The thickness or the volume of the cerebral cortex is measured. Due to the grey colouring of the cerebral cortex caused by the numerous nerve cell bodies, this is usually referred to as grey matter. It is supposed that an increase in the thickness or volume of the gray matter is mainly the result of a spreading and remodeling of the axonal and dendritic processes of the nerve cells in the cortex, which takes place at this time. The decrease over time, on the other hand, is attributed to the elimination of unused synapses and the consequent retraction of the processes. However, the increasing insulation of neurons with myelin and the consequent ingrowth of myelin into the gray matter layer may also reduce cortex thickness. A cortex thinned in this way is considered mature. Gray matter volume and thickness are not synonymous. Indeed, what is important for gray matter volume is not only how thick the cortex is, but also how spread out it is. The latter property is given as surface area.

- **Synapse density** Number of synapses per spatial unit of the cerebral cortex. Synapse density first increases and then decreases during the development of the cerebral cortex.
- **Gray matter thickness** Measured thickness of the cerebral cortex. The thickness of gray matter first increases and then decreases. A thinned cortex is considered mature.
- **Gray matter volume** Measured volume of the cerebral cortex. The volume results from the thickness and the surface of the gray matter and first increases and then decreases.
- **surface of the gray matter** spread of the cerebral cortex. The surface first increases and then decreases, but the latter later than the thickness and volume.

There is still no consensus in science about the timing of these maturation processes. The first studies using imaging techniques (e.g. Sowell et al. 2004; Shaw et al. 2008) showed that there is still an increase in cortical thickness during the early school years. A peak occurring then should be followed by a decrease in cortical thickness in later childhood and adolescence. In recent years, however, many large studies have observed that both volume and thickness begin to decrease in early childhood and continue this process into adolescence (for a review, see Walhovd et al. 2017).

How could such different results come about? Various methodological problems are discussed here, such as the occurrence of motion artifacts. Slight movements in the course of imaging examinations can lead to the cortical thickness being underestimated.

It is assumed that children in the study presumably move more the younger they are, and that this led to—incorrect—results of a lower cortical thickness at a young age and the observed increase in the course of childhood (Walhovd et al. 2017).

Studies specifically investigating early childhood show that the thinning of the cerebral cortex begins in many regions as early as the first or second year of life, for example also in areas of the cerebral cortex (e.g. medial superior frontal cortex, orbitofrontal cortex) that are responsible for processing emotions and information about reward and punishment (Li et al. 2015; Croteau-Chonka et al. 2016). Based on these findings, it is hypothesized that a rapid build-up of synapses is followed by a rapid and extensive breakdown as early as infancy, with a lesser degree of breakdown continuing into adolescence (◘ Fig. 5.5). Based on the findings of this and other studies, it is concluded that experiences in the first 2 years of life in particular (▸ Chap. 7), including traumatic stressful experiences, significantly influence the networks in the child's brain and, in the case of negative experiences, promote the development of neuropsychiatric disorders (Lyall et al. 2015). This assumption is supported by various findings from psychological research, as we will now show.

5.2.5 Early Experiences and Psychological Development

Findings from psychological research repeatedly show that not only sensory systems or language development, but also psychological development is influenced in the long term by early experiences. This becomes clear when children develop attention deficits, mental illnesses, an increased sense of pain or behavioural disorders more frequently than others as a result of an early

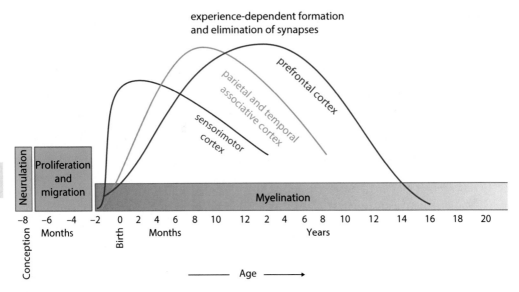

◻ Fig. 5.5 Approximate time course of processes of synapse formation and degradation in different regions of the cerebral cortex. (Modified after Casey et al. 2005)

experience of stress or neglect (▶ Chap. 7). However, because in most cases children continue to be exposed to similar living conditions after early childhood, it is difficult to determine whether it is really the early experiences that have left a lasting impact. To clarify this question, children whose environment has changed significantly after early childhood are studied, such as orphans (Box: Studies Investigating the Influence of Early Experiences).

Studies Investigating the Influence of Early Experiences

Several large research projects have addressed this task. Orphans who spent their early years in orphanages under significantly disadvantageous conditions and were subsequently adopted (e.g. English and Romanian Adoptees Study, ERA, Rutter et al. 2010; e.g. International Adoption Project Survey, Gunnar and van Dulmen 2007) or were cared for by a foster family (e.g. Bucharest Early Intervention Project, BEIP, Zeanah et al. 2017) or whose conditions improved significantly in the institution (e.g. The St. Petersburg-USA Orphanage Research Team 2008). All projects studied children whose living conditions changed dramatically at different times in their young lives.

The study of orphans yielded important insights into the role of early experiences for psychological development. In particular, studies of Romanian orphans were revealing in this regard. When the Romanian Ceauşescu regime came down in 1989, the public became aware of the deplorable living conditions of orphans housed in Romania, who received little individual attention or social affection and were significantly retarded in their physical, cognitive, and behavioral development. Many of the children were adopted by families in Western Europe and North America or placed in Romanian foster care. Since this occurred at an individually different age, it provided an opportunity for researchers to examine the consequences of early neglect in relation to age. This made it possible to examine whether the children's development was shaped primarily during sensitive periods of development.

The findings of the various research projects were similar. For example, it was found that children adopted after the age of two or later were more likely to develop mental disorders such as depression, anxiety disorders and social behaviour disorders later in life than children adopted earlier

(Gunnar and van Dulmen 2007). The results of the Bucharest Intervention Project's randomized controlled trials, in which groups of Romanian orphans of different ages were either assigned to available foster families or remained in orphanages after random selection, were particularly telling. According to this study, the children who entered foster care early did better in many areas than those who entered late. For example, good language skills were able to develop when children were assigned to foster care within the first 15 months of life. Adequate ability to respond to social stress required that children entered foster care before 24 months of age (for a review, see Zeanah et al. 2017).

Both the findings from brain research on the time course of synapse formation and elimination and the findings from the study of orphans suggest that experiences in the first 2 years of life can have a considerable influence on the development of social, emotional and cognitive abilities. According to this view, early childhood represents a *sensitive period* within which early experiences of affection and social as well as cognitive stimulation have a particularly strong influence on development. Other research approaches (for an overview, see Teicher et al. 2016; see also ▶ Chap. 7) show that *maltreatment* during this early sensitive period also has a lasting impact on the child's brain and thus its psyche.

For an assessment of the consequences of neglect, it is also important whether these first years of life also represent a *critical period* within which certain experiences are necessary for brain structures to develop and the corresponding abilities to unfold. This is not clearly answered by the psychological findings from the study of orphans. And also the findings of brain research on the maturation processes in the brain merely suggest that the brain is particularly strongly influenced in the first years of life, but do not prove that in the absence of experiences

in the context of neglect these experiences cannot be made up for.

The fact that psychological problems often follow when the life situation of a previously neglected child changes only after the second year of life may also be due to the fact that a recovery of the child would require a catching up of the experiences of reliable and sensitive provision missed in early childhood, but this catching up does not take place. This can happen, for example, when a child traumatized at an early age triggers ill-will with his or her behavior in a new environment and vicious circles develop. Or it may be the case if a child is *not* offered the opportunity to catch up on missed experiences within the framework of support measures appropriate to his or her current age. It is quite conceivable that support for the child that is not age-appropriate, but instead starts at the child's current stage of development, makes it possible to catch up on the relevant experiences and can thus have a profound and long-term positive influence on psychological development (for a detailed account, see Strüber 2019).

Based on all these findings presented in this subchapter about the maturation of brain areas relevant to mental functions, we will now explain how mental brain function emerges at different levels of the brain and how different functional systems are individually shaped.

Postnatal Development

Postnatally, neurons are initially connected to each other in abundance. Subsequently, the circuits are refined depending on activity: used connections are stabilized, unused ones are eliminated. In this way, the brain is adapted to its environment. In early childhood, periods occur in which the brain is particularly strongly influenced by experiences. According to new findings, the course of these periods is also influenced by experiences, and they can even be reopened under certain circumstances. If one wishes to understand which age is of crucial importance for psychological development, then it is useful to look both at the neurobiological processes, i.e. the time course of synaptogenesis and synapse elimination, and at the psychological findings from the study of children whose living conditions have changed drastically at a certain point in their lives. Both disciplines emphasize the importance of the first 2 years of life, so much so that they are thought to represent a sensitive period for brain functions that underlie psychological development. This also shows how a considerable gain in knowledge can result from bringing together the disciplines of psychology and neurobiology.

5.3 Maturation of Psychologically Relevant Brain Functions

We will now see how mental development occurs at four levels of brain function and that neuromodulatory molecules give rise to functional systems that influence our psyche quite significantly.

5.3.1 Four-Level Model

In this model, four anatomical and functional brain levels are distinguished in justifiable simplification, which arise in different developmental periods and produce different aspects of feeling and thinking. These include three limbic levels and one cognitive level (◘ Figs. 5.6 and 5.7). A detailed account of functional neuroanatomy can be found in ▶ Chap. 2, and a detailed account

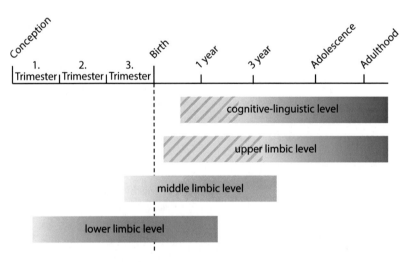

◘ **Fig. 5.6**   Overview of the approximate time course of maturation of the four anatomical and functional brain levels. In their development, the levels build on the abilities and characteristics imparted by the respective level below. Shaded areas mark times when the maturation of the level is already influenced by experience, but the level is not yet fully functional. The lower limbic level with its vegetative functions begins its development in the first weeks after conception. In the last trimester of pregnancy, environmental stimuli perceived in utero begin to shape the middle limbic level. The upper limbic level begins to develop as soon as the child starts to smile specifically at his parents or other attachment figures. At the age of 3–4 years, it enables the child to actively participate in social life. The cognitive-linguistic level is already influenced in the first year of life, but is only able to equip the child with grammatical-syntactical skills between the second and third years of life

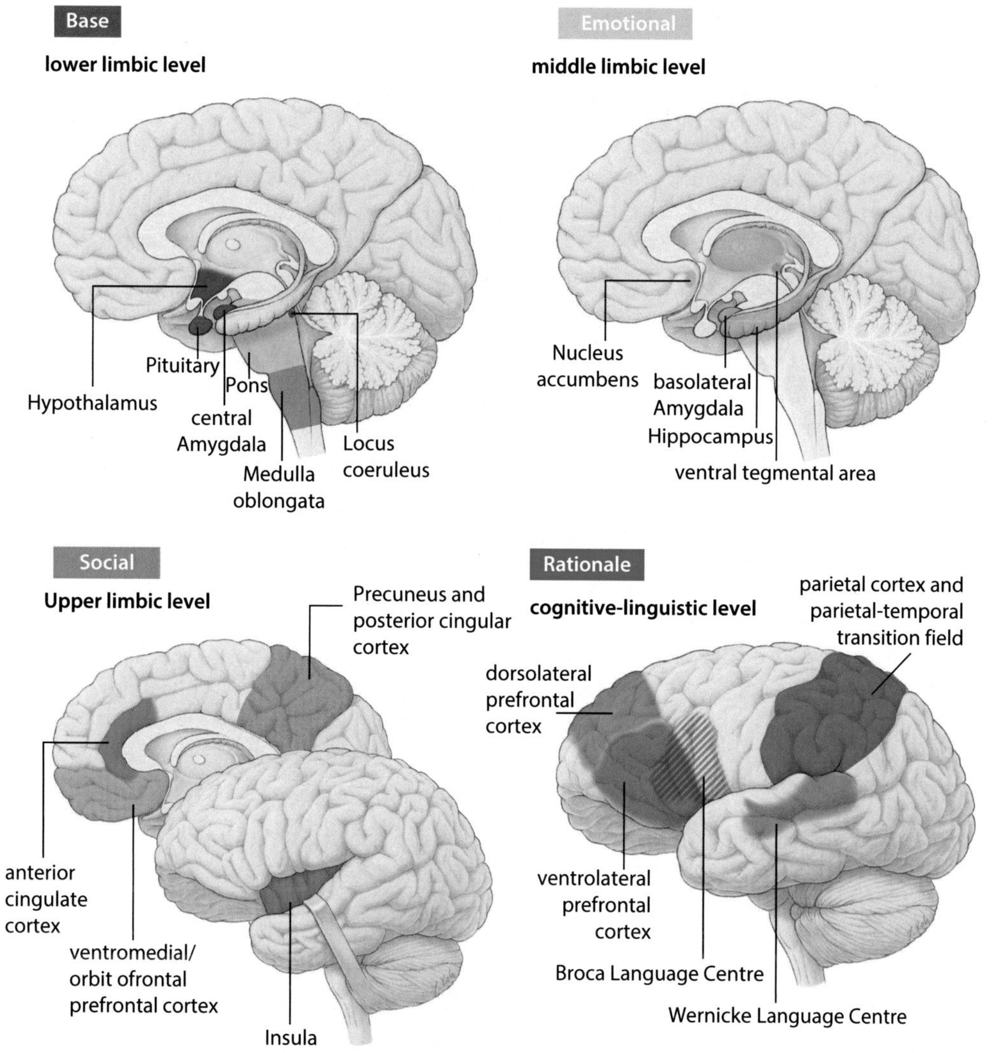
Base
lower limbic level
Pituitary
Pons
Hypothalamus
central
Amygdala
Medulla
oblongata
Locus
coeruleus
Emotional
middle limbic level
Nucleus
accumbens
basolateral
Amygdala
Hippocampus
ventral tegmental area
Social
Upper limbic level
Precuneus and
posterior cingular
cortex
anterior
cingulate
cortex
ventromedial/
orbit ofrontal
prefrontal cortex
Insula
Rationale
cognitive-linguistic level
parietal cortex and
parietal-temporal
transition field
dorsolateral
prefrontal
cortex
ventrolateral
prefrontal
cortex
Broca Language Centre
Wernicke Language Centre

Fig. 5.7 Four anatomical and functional levels of brain function. (© Youson Koh; from Strüber and Roth 2017)

of the four levels and their interaction can be found in Roth and Strüber (2018).

5.3.1.1 Lower Limbic Level

This level includes brain structures such as the hypothalamus, the central nucleus of the amygdala and autonomic-vegetative centers of the brainstem. The hypothalamus in particular is of great importance for the function of this level. It is the major control center for basic biological functions such as food and fluid intake, sleep and wakefulness, temperature and circulatory regulation, attack and defense behavior, and sexual behavior. Together with the pituitary gland (hypophysis) and the periaqueductal grey (PAG), it is hereby the unconscious place of origin of drive and affect states such as hunger, thirst, sexual desire, aggression, anger, but also positive feelings such as lust, attachment and love.

These structures are about survival and reproduction.

The development of this level begins early in pregnancy and is already partly completed by the time of birth. The individual function is influenced accordingly by genes, epigenetic factors and prenatal experiences and produces our **temperament** (▶ Chap. 6). For example, a certain tendency to react in a certain way to stress is stored here. Later experiences, for example in the context of upbringing, can change the function of the lower limbic level only with difficulty.

5.3.1.2 Middle Limbic Level

Middle limbic level brain structures deal with individual learning experiences. This level includes brain structures such as the basolateral amygdala, the basal ganglia and the hippocampus. At this level, experiences are unconsciously evaluated and the outcome is stored for future action. Threats experienced in childhood, for example, are firmly integrated into the circuitry of the amygdala, while positive experiences such as the experience of sensitive care find their way into the network of the basal ganglia, including the nucleus accumbens. Experiences during early childhood are particularly important here. The evaluations made during this time unconsciously guide future behaviour.

The middle limbic level also begins its development during pregnancy, but continues its maturation into early childhood. In adolescence or adulthood, this level can only be changed with the use of strong emotional or long-lasting influences.

One particular structure at this level is a good illustration of the connection between structural maturation and the emergence of specific mental functions, namely the hippocampus. The hippocampus is important for the storage and retrieval of memories, including autobiographical content. Paradoxically, although early experiences do significantly shape later mental functioning, most adult humans are virtually unable to recall memories of their first 2–3 years of life. This phenomenon is called **infantile amnesia**. One of the causes seems to be the incomplete maturation of the hippocampus in the first years of life. However, children aged 5–10 years can sometimes still remember their first years of life well, so that storage must have occurred initially and forgetting only took place later. Animal models suggest that a particularly pronounced formation of new neurons (**neurogenesis**) in early childhood, particularly in the dentate gyrus of the hippocampus, is responsible for early forgetting. If neurogenesis is experimentally stimulated, this increases forgetting even in adult animals. On the other hand, if new neuron formation is blocked, this can attenuate infantile amnesia and promote memory in young animals (Akers et al. 2014). It is thought that the integration of new neurons into existing hippocampal circuits can destabilize previously formed memories. Only when the extent of neurogenesis decreases as the hippocampus matures can lasting memories be formed (for a review, see Roth and Strüber 2018).

> **Infantile Amnesia**
>
> The inability to remember the experiences of the first years of life.

> **Neurogenesis**
>
> The formation of new nerve cells.

5.3.1.3 Upper Limbic Level

The upper limbic level contains the parts of the cerebral cortex that are important for conscious emotional processing, such as the orbitofrontal, ventromedial prefrontal and anterior and posterior cingulate cortex, but also the anterior insula and the precuneus. In the brain structures of the upper limbic level, the emotions generated in the lower two limbic levels are processed in an at least partially conscious and differentiated manner as "feelings" and checked for their situational appropriateness. This level is responsible for ensuring that people behave adequately in their social environment. Accordingly, it is not only concerned with simple emotions such as fear or aggression, but also with such complicated and often consciously experienced human sensitivities as pride, shame, disappointment and Schadenfreude. The upper limbic level evaluates one's social behavior in terms of its consequences and enables the regulation of one's emotions and the curbing of rash impulses. The behavioral tendencies of the two lower limbic levels are strengthened or weakened by this level depending on socialization.

Although the brain structures of the upper limbic level also begin to develop during pregnancy, they are not very mature after birth and continue to change into adolescence. If one wants to influence the contents stored in these circuits, then this requires emotional and social stimuli. By means of a purely rational and linguistic influence, hardly any changes in the circuits there are likely to be achieved.

5.3.1.4 Cognitive-Linguistic Level

The cognitive-linguistic level comprises prefrontal, temporal and parietal associative cortical areas, such as those for language comprehension and language production and also those which, as working memory, form the basis for ideas and conscious action planning. For the latter, the dorsolateral prefrontal cortex is particularly important as the seat of the anterior working memory. At this level, the feelings and motives developed in the limbic levels are verbalized, thoughts are structured and sequenced, behavioral goals are internally mapped and maintained, compared with each other and with models. One's motives do not always coincide with one's assumptions about correct behavior or with one's self-image. Then it is up to the cognitive-linguistic level to find justifications.

The cognitive-linguistic level matures only gradually in the course of childhood

and adolescence, and the development of its individual function is never complete. It changes constantly in linguistic interaction with others, but does not reflect the core of the personality, but rather how we would like to see ourselves and be seen. One reason for this is that the cortical associative areas of this level have no effective influence on the centers of the lower and middle limbic levels and only a limited influence on the areas of the upper limbic level (Ray and Zald 2012; cf. ▶ Chap. 6):

> **Overview**
> - **Lower limbic level:** Level of the vital regulation of vegetative functions and innate behaviours as well as temperament.
> - **Middle limbic level:** level of unconscious individual learning and emotional conditioning
> - **Upper limbic level:** level of social learning and consciously experienced feelings and goals
> - **Cognitive-linguistic level:** level of working memory, goal-directed action planning and grammatical-syntactical language.

5.3.2 Six Basic Psychoneural Systems

If we ask ourselves how the different levels communicate with each other and why the lower limbic level is so important for our temperament, we must direct our attention to the so-called **neuromodulators**. These are substances such as noradrenaline, serotonin, dopamine and acetylcholine, which, like neuropeptides and neurohormones, influence the communication between the nerve cells of the different levels and thus have a significant effect on our mental state. Most of these substances are produced in narrowly defined areas of the lower and middle limbic level and distributed from there.

Neuromodulators

Slow and rather globally acting transmitters that modulate the activity of fast and locally acting transmitters at the synapse, such as glutamate, GABA and glycine. These include the substances noradrenaline, serotonin, dopamine and acetylcholine. Depending on the definition, neuropeptides and neurohormones are also included.

The number of substances active in the human brain is large. In addition, there are different receptor types for many substances, which react to different concentrations of these substances and sometimes have opposite, i.e. excitatory or inhibitory, effects on the activity of the nerve cells. If one wishes to understand the effect of these substances on the development of the human psyche, it is useful to assign them to functional systems with the aim of reducing complexity. For this reason, we name here six basic "psychoneural" systems that underlie a person's personality and psychological state (see also ▶ Chap. 6; see also Roth 2019).

The genetic make-up of a person plays a major role in the effect of neuromodulatory substances and thus also of the basic psychoneural systems. The genes for receptors of the substances, but also for other molecules that influence their function, such as transport proteins or degradation enzymes, can exist in different variants (Gene Polymorphisms, box). Depending on the gene variant, the neuromodulatory substances then function more or less efficiently.

Gene Polymorphisms
In the context of genetic inheritance, parents pass on genes, i.e. specific DNA gene sequences, to their children. Different variants of a gene are called "alleles". Sequence variations are often *single nucleotide polymorphisms* (SNPs). In these SNPs, one nucleotide, i.e. one of the basic building blocks of the DNA or RNA

sequence consisting of the bases adenine, guanine, cytosine, thymine (in the case of DNA) or uracil (in the case of RNA), a monosaccharide and a phosphoric acid residue, is exchanged. Such an exchange can affect a regulatory sequence of the gene and thereby permanently alter its expression. If, on the other hand, the exchange affects the protein-coding region of a gene sequence, this can lead to the formation of a different amino acid and thus an altered protein, e.g. an enzyme or a receptor with altered properties.

However, early experiences also influence the function of neuromodulatory substances. Thus, early experiences of stress may be associated with reduced receptor function or other molecular changes in neuromodulatory systems. Many studies now show that experiences influence the organism via Epigenetic Changes (Box).

Epigenetic Changes
The term refers to changes in gene expression that, in contrast to SNPs (see gene polymorphisms), do not involve a change in the nucleotide sequence. They are often based on changes in the function of so-called promoter regions, which are responsible for the expression of the nucleotide sequences. Differences in the extent of epigenetic changes in a given tissue, or in the same tissue of different individuals, can then lead to differences arising at the phenotype level without genetic differences being present (Szyf 2015). Epigenetic changes often occur via environmental influences. This can reprogram gene expression with the goal of adapting an organism to its environment. If these epigenetic changes occur during life as a result of environmental influences within germ cells, then they can also influence the characteristics of subsequent generations (Bale 2015). Epigenetic changes include, for example, the methylation of certain nucleotides of DNA by the enzyme DNA methyltransferase. This usually involves the attachment of methyl groups to cytosine nucleotides, which are immediately followed by a guanine nucleotide in the linear base sequence (in the $5' \rightarrow 3'$ direction). If this methylation affects a nucleotide in the regulatory region of a gene, then this may prevent the protein-coding sequence of the DNA from being expressed. The methylated cytosine nucleotides are thus accompanied by *silencing* of the gene. Gene expression is then switched off. Various other epigenetic mechanisms allow epigenetic modification of gene activity. These include a modification of the so-called histones as well as a control of gene activity via micro-ribonucleic acids, microRNAs for short (see Jones 2012).

The individual functioning of the basic psychoneural systems accordingly arises against the background of the individual genetic make-up and the respective experiences. Nowadays, however, these two factors are not assumed to have an additive effect. Instead, genes and environment interact in a variety of ways (Box: Gene-Environment Interactions).

Gene-Environment Interactions

... may vary:

1. Genes determine the importance of the environment. People with specific genetic variants react more sensitively to their environment than others. They are more likely to develop mental illnesses or behavioural disorders as a result of early experiences of stress than people with other gene variants.
2. The environment determines the importance of genes. Certain traits, such as intelligence or the way in which we react to stress, are influenced to a greater or lesser extent by genes, depending on the environment. For example, intelligence in children of educationally disadvantaged families is significantly influenced by genes, whereas in children affected by poverty it is almost exclusively shaped by the environment (Turkheimer et al. 2003).
3. The environment determines how active genes are. Experiences can epigenetically influence the expression of certain genes, such as the stress system, and thereby determine how humans react to their environment.

5.3.2.1 The Stress-Regulating System

People differ in how they deal with stress, i.e. whether they are good at coping with high demands, whether they "ramp up" quickly in the face of potentially threatening stimuli or whether they keep a clear head even in difficult situations. All of these are tasks of the stress regulating system. This system enables the organism to cope with physical and mental stress and challenges.

If we encounter such a situation, the sympathetic nervous system in the body, which is responsible for fight and flight reactions, is activated and drives up blood pressure and heart rate. In the brain, norepinephrine/noradrenalin is released in increased concentrations by the locus coeruleus in the brainstem. This increases attention and promotes emotional learning. In the hypothalamus and amygdala, there is production of the neuropeptide corticotropin-releasing factor (CRF), which in turn suppresses exploratory behavior, increases alertness, and in higher doses produces fear. Humans are on alert. If stress persists beyond a brief moment, this rapid reaction is followed by activation of the hypothalamic-pituitary-adrenal axis (HPA) axis (▶ Sect. 3.5.7.5), the end product of which is the hormone cortisol, which is released into the bloodstream by the adrenal cortices. The cortisol reaches the brain, initially promotes adequate processing of the stressful situation and somewhat later dampens the stress reaction via negative feedback.

The interaction of the genetic make-up and the respective experiences influences the individual function of the stress processing system. Prenatally, for example, stressful experiences of the expectant mother during pregnancy or earlier can have a negative effect on the fetus via the associated increased cortisol release in the body of mother and fetus. This **prenatal stress**, in turn, can affect the long-term expression of the gene for cortisol receptors in an epigenetic process. As this is involved in feedback inhibition of the stress response, this may modulate the level of activity of the stress processing system in the long term. If the ability to cope with stress is reduced due to genetic factors or prenatal stress experiences, the stress may be amplified by negative or traumatizing experiences, but attenuated by positive attachment experiences (for a detailed account, see Strüber 2019). According to this, in addition to the genetic make-up, it is above all the experiences of the first years of life that underlie the individual function of this system.

> **Prenatal Stress**
>
> Stressful experiences of the expectant mother can shape the functioning of the child's stress system in the long term via an epigenetic mechanism.

5

5.3.2.2 The Internal Calming System

As soon as a stressful situation has subsided, the organism should quickly regain its composure. In the body, this involves an activation of the parasympathetic nervous system, the counterpart of the sympathetic nervous system. In the brain, the return to rest mode is supported by various substances. Cortisol, which is released in the stressful situation, dampens the rapid stress reaction after a delay—as mentioned above—and serotonin can also have a self-soothing effect, primarily via activation of the 5-HT1A receptor.

Different genes of the serotonin system can be present in different variants, with the gene for the serotonin transporter and the gene for the 5-HT1A receptor being particularly well studied. These polymorphisms dictate how well serotonin can work. Numerous studies have found associations between the presence of certain polymorphisms of the serotonin system and certain character traits or predispositions to mental illnesses in which the ability to self-soothe is reduced, such as occurs in depression and anxiety disorders.

Animal models indicate that early experiences also influence the serotonin system in its later function. For example, adult rodents that grew up separated from their mothers or in social isolation during their development show lower serotonin concentrations in the hippocampus and the medial prefrontal cortex than animals that grew up under normal social conditions (Braun et al. 2000). Time and again, moreover, a gene-environment interaction is revealed—also in humans—and often in the form that the combination of certain polymorphisms with early stress experiences significantly increases the risk for later mental problems. Just like the stress processing system, the calming system also begins to stabilise its individual function in the first years of life.

5.3.2.3 The Evaluation and Reward System

Whenever we act, unconsciously or consciously, we follow the principle of striving for what seems pleasurable or beneficial and avoiding or terminating what is painful or detrimental. The brain derives its knowledge about the positive or negative consequences of our actions from the evaluation and reward system. It evaluates our personal experiences in terms of their consequences and develops reward and punishment expectations in similar situations (cf. ▶ Chap. 6).

Let us consider separately the two sub-functions of the evaluation and reward systems. One sub-system is the actual reward system. It produces a feeling of well-being when we experience something nice. Endogenous opioids, which act on their receptors in the shell region of the nucleus accumbens, in the ventral pallidum and in the amygdala, among other places, are particularly important for this hedonic experience and thereby unconsciously generate reward experiences. For a conscious experience of the feeling of pleasure and satisfaction associated with the receipt of rewards, cortical areas such as the orbitofrontal, ventromedial and insular cortex must also be activated.

Another subsystem is responsible for reward expectation. The substance dopamine plays the decisive role for this system. When we encounter potentially behaviorally relevant stimuli in everyday life, groups of dopaminergic neurons from the ventral tegmental area inform the brain about their reward value via a targeted and structured release of dopamine. They indicate the type, magnitude and probability of occurrence of an expected reward and thereby influence our motivation to act or not to act in a certain way.

The evaluation and reward system is also influenced by the respective genetic make-up as well as by experience. Various polymorphisms of the dopamine system, for example, are associated with characteristics such as sensation-seeking or the individual's propensity to take risks. But also early negative experiences and their interaction with the genetic make-up influence the function of this system in the long term. This can result in an increased urge for intense reward, drug addiction, but also in apathy and hopelessness.

5.3.2.4 The Impulse-Inhibition System

Another system is closely related to the reward system just presented. Often the environment requires postponement of the reward. The impulse to want a reward immediately must be inhibited. This is a task of the impulse inhibition system. However, it also comes into play when quick emotional reactions to a stimulus (for example, the aggressive reaction to a threatening speech) are not appropriate in the respective situation.

The serotonin system in particular is important for impulse inhibition. It acts as an antagonist of dopamine. If a stressful situation requires fast and impulsive action such as fight or flight, dopamine release is increased in structures such as the nucleus accumbens as well as in the orbitofrontal and ventromedial prefrontal cortex of the upper limbic level. However, if restraint is appropriate in a particular situation, for example because one cannot do anything about the stressor anyway, dopamine release is reduced and serotonin release is increased in many areas, especially in the orbitofrontal and ventromedial cortex. The serotonin seems to encourage doing nothing instead of reacting wrongly.

As we have already seen, the function of both dopamine and serotonin systems is influenced by genetic makeup and individual experience, with early experience being particularly relevant. Clearly, individual differences in function will also affect impulse inhibition. However, the characteristic

capacity for impulse inhibition often only gradually becomes apparent in later childhood, namely when the aforementioned cortical structures of the upper limbic level have sufficiently stable connections to the areas of the two lower limbic levels to exert their impulse-regulating influence on these structures under the influence of dopamine and serotonin.

5.3.2.5 The Attachment System

From the very beginning, the human being is a social being. The infant prefers to look at faces and much prefers to reach for a human thumb than for any other object. Already at an early age, he begins to smile purposefully at close people. He is prepared to form attachments. The neuropeptide oxytocin plays an essential role in this process. It is released during breastfeeding, touching, but also generally during trusting social contacts. It inhibits the stress system and increases the release of serotonin in a way that reduces feelings of anxiety. It increases the ability to recognize facial expressions and promotes motivation to form social bonds. Social emotions, trust and empathy towards pleasant social contacts are also enhanced, as is parental behaviour.

Various other substances are involved in the attachment system, such as endogenous opioids, which mediate the feeling of well-being during social contacts, and dopamine, via which the rewarding value of certain attachment figures is stored (for an overview, see Strüber 2016).

The individual function of the attachment system is also influenced by genetic polymorphisms—especially those of the oxytocin system. This produces, for example, differences in the tendency to act prosocially, empathically and trustingly (Kumsta and Heinrichs 2013). In terms of experience, it is primarily early experiences with attachment figures that influence the long-term functioning of the oxytocin system (e.g. Heim et al. 2009). Accordingly, the individual function of the system is relatively stable after the first years of life.

5.3.2.6 The System of Reality Sense and Risk Assessment

Even if we hardly notice it in everyday life, our brain constantly assesses the probability that a certain behaviour is associated with negative consequences. This is the task of the reality sense and risk assessment system. If the likelihood of negative consequences is ignored or undervalued relative to the likelihood of positive consequences, this manifests itself in the form of increased risk taking. In addition to sensory and cognitive functions and the reward-seeking dopamine system, the neuromodulatory substances norepinephrine/noradrenalin and acetylcholine play an important role in the ability to realistically assess risks. Norepinephrine increases general alertness and vigilance, while acetylcholine maintains neuronal activity in working memory and in the targeted recall of memory content, thereby helping to implement deliberate action. The activity of these neuromodulators is also shaped by genes and experience. However, the degree of maturity of the brain structures whose activity is influenced by the aforementioned neuromodulators in the course of risk assessment is also important for the development of a characteristic risk assessment. In addition to structures of the upper limbic level, this also includes the dorsolateral prefrontal cortex, which only gradually matures in the course of childhood and adolescence.

Six Basic Psychoneural Systems

- the stress-regulation system
- the internal calming system
- the evaluation and reward system
- the attachment system
- the impulse-inhibition system
- the system of realism and risk assessment

Maturation of Psychologically Relevant Brain Functions

Psychological development takes place at four levels of brain function, i.e. three limbic and one cognitive-linguistic level. Neuromodulatory molecules give rise to six basic psychoneural systems that affect the levels and thereby our psyche. The individual function of these systems is shaped by an interaction of genes and environment, and it has now frequently been shown that experience exerts its influence via epigenetic imprinting. The individual function of the psychoneural systems forms the basis for our personality, which will be discussed in ▶ Chap. 6.

References

Akers KG, Martinez-Canabal A, Restivo L, Yiu AP, De Cristofaro A, Hsiang HLL et al (2014) Hippocampal neurogenesis regulates forgetting during adulthood and infancy. Science 344(6184):598–602

Bale TL (2015) Epigenetic and transgenerational reprogramming of brain development. Nat Rev Neurosci 16(6):332

Braun K, Lange E, Metzger M, Poeggel G (2000) Maternal separation followed by early social deprivation affects the development of monoaminergic fiber systems in the medial prefrontal cortex of Octodon degus. Neuroscience 95:309–318

Casey BJ, Tottenham N, Liston C, Durston S (2005) Imaging the developing brain: what have we learned about cognitive development? Trends Cogn Sci 9(3):104–110

Changeux J-P, Danchin A (1976) Selective stabilisation of developing synapses as a mechanism for the specification of neuronal networks. Nature 264:705–712

Constantine-Paton M, Cline HT, Debski E (1990) Patterned activity, synaptic convergence, and the NMDA receptor in developing visual pathways. Annu Rev Neurosci 13:129–154

Croteau-Chonka EC, Dean DC III, Remer J, Dirks H, O'Muircheartaigh J, Deoni SC (2016) Examining the relationships between cortical maturation and white matter myelination throughout early childhood. NeuroImage 125:413–421

Gogolla NP, Luthi CA, Herry C (2009) Perineuronal nets protect fear memories from erasure. Science 325(5945):1258–1261

Greenough WT, Black JE, Wallace CS (1987) Experience and brain development. Child Dev 58:539–559

Greifzu F, Pielecka-Fortuna J, Kalogeraki E, Krempler K, Favaro PD, Schlüter OM, Löwel S (2014) Environmental enrichment extends ocular dominance plasticity into adulthood and protects from stroke-induced impairments of plasticity. Proc Natl Acad Sci 111(3):1150–1155

Gunnar MR, Van Dulmen MH (2007) Behavior problems in postinstitutionalized internationally adopted children. Dev Psychopathol 19(1):129–148

Harauzov A, Spolidoro M, DiCristo G, De Pasquale R, Cancedda L, Pizzorusso T et al (2010) Reducing intracortical inhibition in the adult visual cortex promotes ocular dominance plasticity. J Neurosci 30(1):361–371

Hebb DO (1949) The organization of behavior: a neuropsychological theory. Wiley, New York

Heim C, Young LJ, Newport DJ, Mletzko T, Miller AH, Nemeroff CB (2009) Lower CSF oxytocin concentrations in women with a history of childhood abuse. Mol Psychiatry 14(10):954–958

Hensch TK (2018) Critical periods in cortical development. In: Gibb R, Kolb B (eds) The neurobiology of brain and behavioral development. Academic, London, pp 133–151

Herculano-Houzel S (2009) The human brain in numbers: a linearly scaled-up primate brain. Front Hum Neurosci 3:31

Johnson MH (2005) Sensitive periods in functional brain development: problems and prospects. Dev Psychobiol 46(3):287–292

Jones PA (2012) Functions of DNA methylation: islands, start sites, gene bodies and beyond. Nat Rev Genet 13(7):484

Knudsen EI (2004) Sensitive periods in the development of the brain and behavior. J Cogn Neurosci 16(8):1412–1425

Li G, Lin W, Gilmore JH, Shen D (2015) Spatial patterns, longitudinal development, and hemispheric asymmetries of cortical thickness in infants from birth to 2 years of age. J Neurosci 35(24):9150–9162

Lyall AE, Shi F, Geng X, Woolson S, Li G, Wang L et al (2015) Dynamic development of regional cortical thickness and surface area in early childhood. Cereb Cortex 25(8):2204–2212

Maurer D (2017) Critical periods re-examined: evidence from children treated for dense cataracts. Cogn Dev 42:27–36

Ray RD, Zald DH (2012) Anatomical insights into the interaction of emotion and cognition in the prefrontal cortex. Neurosci Biobehav Rev 36:479–501

Roth G (2019) Warum es so schwierig ist, sich und andere zu ändern. Persönlichkeit, Entscheidung und Verhalten. Klett-Cotta, Stuttgart

Roth G, Strüber N (2018) Wie das Gehirn die Seele macht. Klett-Cotta, Stuttgart

Rutter M, Sonuga-Barke EJ, Beckett C, Castle J, Kreppner J, Kumsta R, et al (2010) Deprivation-specific psychological patterns: effects of institutional deprivation. Monographs of the Society for Research in Child Development, i–253

Shaw P, Kabani NJ, Lerch JP, Eckstrand K, Lenroot R, Gogtay N, Greenstein D et al (2008) Neurodevelopmental trajectories of the human cerebral cortex. J Neurosci 28:3586–3594

Silbereis JC, Pochareddy S, Zhu Y, Li M, Sestan N (2016) The cellular and molecular landscapes of the developing human central nervous system. Neuron 89(2):248–268

Singer BYW (1990) The formation of cooperative cell assemblies in the visual cortex. J Exp Biol 153:177–197

Sowell ER, Thompson PM, Leonard CM, Welcome SE, Kan E, Toga AW (2004) Longitudinal mapping of cortical thickness and brain growth in normal children. J Neurosci 24:8223–8231

Strüber N (2016) Die erste Bindung: wie Eltern die Entwicklung des kindlichen Gehirns prägen. Klett-Cotta, Stuttgart

Strüber N (2019) Risiko Kindheit. Die Entwicklung des Gehirns verstehen und Resilienz fördern. Klett-Cotta, Stuttgart

Strüber N, Roth G (2017) Infografik. So reift das Ich. Gehirn & Geist 7:12–19

Szyf M (2015) Nongenetic inheritance and transgenerational epigenetics. Trends Mol Med 21(2):134–144

Teicher MH, Samson JA, Anderson CM, Ohashi K (2016) The effects of childhood maltreatment on brain structure, function and connectivity. Nat Rev Neurosci 17(10):652

The St. Petersburg-USA Orphanage Research Team (2008) The effects of early social emotional and relationship experience on the development of young orphanage children. Monogr Soc Res Child Dev 73(3):1–297

Turkheimer E, Haley A, Waldron M, d'Onofrio B, Gottesman II (2003) Socioeconomic status modifies heritability of IQ in young children. Psychol Sci 14(6):623–628

Vetencourt JFM, Sale A, Viegi A, Baroncelli L, De Pasquale R, O'leary O et al (2008) The antidepressant fluoxetine restores plasticity in the adult visual cortex. Science 320(5874):385–388

Walhovd KB, Fjell AM, Giedd J, Dale AM, Brown TT (2017) Through thick and thin: a need to reconcile contradictory results on trajectories in human cortical development. Cereb Cortex 27(2):1472–1481

Watts DJ, Strogatz SH (1998) Collective dynamics of 'small-world' networks. Nature 393:440–442

Werker JF, Hensch TK (2015) Critical periods in speech perception: new directions. Annu Rev Psychol 66:173–196

Zeanah CH, Humphreys KL, Fox NA, Nelson CA (2017) Alternatives for abandoned children: insights from the Bucharest Early Intervention Project. Curr Opin Psychol 15:182–188

Emotion, Motivation, Personality and Their Neurobiological Foundations

Gerhard Roth and Nicole Strüber

Contents

© The Author(s), under exclusive license to Springer-Verlag GmbH, DE, part of Springer Nature 2023
G. Roth et al. (eds.), *Psychoneuroscience*, https://doi.org/10.1007/978-3-662-65774-4_6

In this chapter we will deal with three basic components of the human psyche: emotions or feelings, motivation and personality. In the spirit of bridging the gap between the psychosciences and the neurosciences, the first aim is to show what emotions and motivational states are from a psychological point of view, how they result in characteristics of a person's personality, and how these determine our state of mind and our behavior. Secondly, we want to ask to what extent these psychological phenomena can be linked to structures and functions of the brain. Particularly exciting is the question on which structural and functional level correlates can be found. We will see that the level of neuronal networks and processes at the synapses play a special role here, which in turn are influenced by genetic and epigenetic factors as well as prenatal and postnatal environmental influences and experiences.

Learning Objectives

After reading this chapter, the reader should be familiar with common psychological statements about emotions, motivation, and the development and structure of personality, as well as the neurobiological basis of these components of the human psyche.

6.1 Emotions

In psychology, the study of emotions or feelings experienced a first peak towards the end of the nineteenth century and in the early twentieth century with authors such as William James, Wilhelm Wundt and William McDougall. However, this development was hampered for decades by the dominance of **behaviorism**, for which "internal states" such as feelings, thoughts, or intentions had no explanatory value whatsoever. It was only in the 1980s that psychologists began to focus more and more on the question of what emotions are, how they arise and what influence they have on our thinking and behaviour. This then triggered the interest of neurobiologists, who were often also trained psychologists or psychiatrists. Milestones here were the works of Antonio Damasio, Jaak Panksepp and Joseph LeDoux—to name but a few.

Behaviorism

Behaviorism is a very influential psychological theory that developed since the beginning of the twentieth century, mainly in the USA, through researchers such as E. L. Thorndike, J. B. Watson and especially B. F. Skinner. F. Skinner and, in contrast to the continental European "psychology of understanding", advocated the view that psychology must be strictly limited to observable stimulus-response relationships when studying human and animal behaviour. Assumptions about "internal" mental-psychic states were useless. Moreover, it was held that animal and human behavior is more or less exclusively learned. For psychology and behavioral science, Skinner developed seminal "laws" of learning in the form of operant conditioning in addition to those of classical conditioning described by Pavlov. This also became the basis of widespread behavioral therapy.

6.1.1 What Are Emotions and Feelings?

In everyday psychology, the term "emotion" is equated with "Gefühle" in German. In the following, however, we want to define the term "emotion" more broadly in the sense of a state that "moves" us either unconsciously or consciously—according to its origin from the Latin word *movere*. "*Feelings*", on the other hand, we take to be a conscious state of experience and thus a *sub-form* of emotions that differs from cognitive states such as thinking, imagining and remembering. However, feeling states are usually associated with concrete perceptual and cognitive content: As we recognize, remember, or imagine certain things, we have certain feelings. However, some feelings, especially in the form of moods such as dejection, can also occur without content (one then does not even know why one feels so depressed), which is not true for perceptual and cognitive states.

Feelings can have different *intensities* and be *positive* or *negative*—the latter is called their **valence**. In addition, feelings, especially those of higher intensity, often also called **affects**, usually occur together with *physical forms of expression* such as facial *expressions*, gestures, posture, voice pitch as well as with *vegetative reactions* such as feelings of anxiety, sweating, trembling, rapid breathing and a higher pulse rate. Antonio Damasio (1994) refers to these as "somatic markers" that occur automatically along with perceptions and events. Finally, many feelings have a *motivational* effect, i.e. they drive us to do or seek out certain positive things (approach behaviour, *appetence*) or to avoid negative things (avoidance behaviour, *aversion*; ▶ Sect. 6.2).

In the psychology of emotions, **valence** refers to the positive or negative valence of a (conscious) feeling, depending on whether it is accompanied by joy, pleasure or even elation, or fear, anxiety, sadness and disgust. This then results in certain motives or action tendencies such as approach and avoidance.

While feelings are by the above definition conscious, there are also unconscious emotions. They may remain unconscious if the stimuli that trigger them are too brief or "masked" (▶ Sect. 6.1.3) or too weak to cross the threshold of consciousness; they may also be "repressed" from a psychoanalytic perspective. They may trigger vegetative reactions such as an increase in blood pressure or respiratory rate, behaviours such as avoidance behaviour, or musculo-skeletal tension, without us necessarily noticing.

6.1.2 How Many Emotions Are There and How Do They Arise?

There ist a long-standing debate of whether there is a basic emotional make-up with a certain number of independently existing affects or emotions that can also be found as separate "modules" in the brain, whether emotions or feelings have a continuum or whether all affects/emotions can be reduced to only *two basic polarities*—mostly "positive/desirable" vs. "negative/to be avoided" and low vs. strong arousal.

The psychologist Paul Ekman advocates a *modular* model of emotions, whereby he understands emotions to be short-term emotional states related to a specific stimulus (Ekman 1999, 2007). He assumes a total of 15 "*basic emotions*", namely *happiness/amusement, anger, contempt, contentment, disgust, embarrassment, excitement, fear, guilt, pride* in *achievement, relief, sadness/distress, satisfaction, sensory pleasure,* and *shame.* Other affective states such as *grief, jealousy, romantic* and *parental love* are for Ekman rather longer-term affective states or moods and therefore not necessarily to be considered as emotions. These 15 emotions are characterized for Ekman by a unique combination of external and internal physical features, e.g., a typical facial expression, a typical sound utterance (sounds of pain, sadness, joy, etc.), and a characteristic state of the autonomic-vegetative nervous system.

The Estonian-American neurobiologist Jaak Panksepp, who died in 2017, assumes that there are clearly definable *basic affective-emotional states that* are characterized by different neuronal "modules" in the brain.

In his opinion, these can be detected via targeted brain stimulation primarily in the periaqueductal gray (Panksepp 1998; Panksepp et al. 2017), although he bases this view primarily on animal experiments (rat). However, he arrives at a different classification than Ekman and distinguishes six *basic emotional systems*, namely *seeking/expectancy, rage/anger, lust/sexuality, care/nurturance, panic/separation*, and *play/joy*.

In contrast to this "modular" view of emotions, emotion researchers such as James Russell, David Watson and Auke Tellegen, but also the neurobiologist Edmund Rolls, argue that emotions are doubly "polar". According to this view, emotions differ on one axis by their valence (positive-pleasant vs. negative-unpleasant) and on the other axis by the strength of arousal (low vs. strong arousal; cf. Russell 2009).

Another classification of emotions is presented by the American emotion researchers Andrew Ortony, Gerald Clore and Allan Collins (cf. Clore and Ortony 2000), which is also followed by the Swedish researcher Arne Öhman (Öhman 1999). According to Ortony, Clore and Collins, emotions differ from affects in that they involve an evaluation of *goals, standards* and *attitudes*, which is not the case with affects and moods. According to these authors, emotions are always *intentional*, i.e. directed towards a goal.

While many authors regard affects and emotions on the one hand and cognitive performance on the other as independent, albeit interacting, mental states, Clore and Ortony are staunch advocates of a *cognitive theory of emotions*, as also advocated by the Swiss psychologist Klaus Scherer (Scherer 1999). For them, emotions are evaluative states (*appraisals*) and always have a cognitive component, in contrast to affects. They refer, whether consciously or unconsciously, to the grasping of the *meaning of* a situation or an object.

6.1.3 Unconscious Emotions and Conscious Feelings

There are numerous studies on the specific processing and effect of unconsciously perceived stimuli and stimulus situations on the limbic-emotional system (▶ Sect. 6.1.4). For example, subjects were studied who had a strong fear of snakes but not of other animate or inanimate objects. In them, a very brief (i.e., 30 ms long) and *masked* (i.e., flanked by two longer presented neutral pictures) presentation of snake pictures led to strong vegetative fear responses, although the fear-inducing pictures were not consciously perceived. This was not the case for these subjects with images that were not frightening to them. This indicates that the subjective threatening nature of the stimulus was *recognized unconsciously*. Similar results were obtained in experiments in which subjects were fear-conditioned to certain objects such as spiders, snakes or even neutral faces with the aid of a mild electric shock. When these objects were presented in a mask, the subjects showed marked vegetative responses. Similarly, in a visual search task, subjects who had been conditioned to spiders, snakes, or other objects as fearful objects recognized such objects more quickly when they were embedded in neutral or positive objects (flowers or mushrooms). The latter shows that in the domain of unconscious percepts, the *recognition of threatening stimuli* has priority over the recognition of neutral or positive stimuli. In this context, Öhman speaks of an *automated sensitivity to threats* (Öhman 1999).

6.1.4 The Neurobiological Basis of Emotions

Emotions are based on the activity of centers of the limbic system (◼ Fig. 6.1), which is described in detail in ▶ Chap. 2. Unconscious emotions are generated in sub-

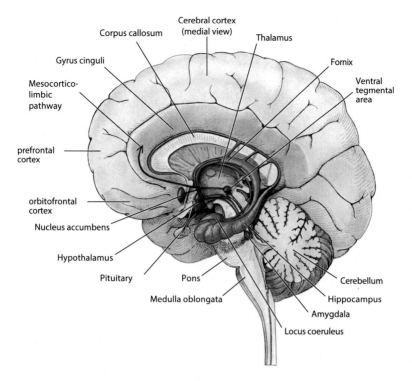

◘ Fig. 6.1 Median view of the human brain with the most important limbic centers (blue). These centers are sites of affect generation, of predominantly positive (nucleus accumbens, ventral tegmental area) and predominantly negative or strongly moving feelings (amygdala), of memory organization (hippocampus), of attention and consciousness control (basal forebrain, locus coeruleus, thalamus), and of control of vegetative functions (hypothalamus). (Modified after Gershon and Rieder 1992)

cortical centers of the limbic system; the conscious experience of these emotions as a subjective emotional state requires the additional activation of cortical areas.

The American neurobiologist Joseph LeDoux has dealt extensively with the relationship between unconscious and conscious processes in the context of his studies on fear conditioning, primarily in rodents (◘ Fig. 6.2). His starting point is the well-known fact that we usually react very quickly to a negative event with certain reactions (defence, flight, attack) before we have even recognised more precisely the event that triggered this reaction (LeDoux 1996). LeDoux has summarized this in a popularized model of two pathways of fear- and anxiety-related information, which assumes, on the one hand, a "fast", unconscious pathway that ends in the amygdala via the dorsal thalamus, and, on the other hand, a "slow" conscious pathway that also moves via the dorsal thalamus, but then to sensory and finally associative-limbic cortical areas (LeDoux 1996).

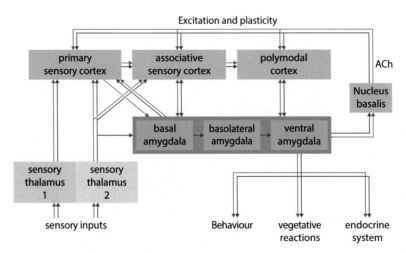

Fig. 6.2 LeDoux's model of two fear-conditioning paths and its correction. The connectivities proposed by LeDoux are shown with blue arrows, those of Pessoa and Adolphs with red arrows. According to LeDoux', auditory stimuli proceed from the inner ear via the brainstem to the thalamus and from there via the unconscious "fast" pathway directly to the amygdala (sensory pathway 2), where association with other stimuli (e.g., electric shock) occurs. Also from the thalamus, in the "slow" pathway (sensory pathway 1), signals go to the primary and secondary auditory cortex and to the associative cortex, where they are further processed until they become conscious. These conscious processes then have a feedback effect on the amygdala and other subcortical areas. According to Pessoa and Adolphs, in humans there is no direct thalamic pathway to the amygdala, rather it receives "rapid" descending inputs from primary sensory cortical areas. ACh = acetylcholine-mediated input from nucleus basalis to cortical areas. Further explanation in the text

---

Model of the Two Processing Paths

According to neurobiologist Joseph LeDoux's model, there is a "fast" and a "slow pathway" in the perception of threatening stimuli (**Fig. 6.2**). According to LeDoux's original model of auditory fear conditioning—based on experiments on rats—auditory stimuli travel from the inner ear via the brainstem to the thalamus and from there via the unconscious "fast" pathway directly to the amygdala, where the association with other stimuli (e.g. electric shock) takes place. Also from the thalamus, in the "slow" pathway, signals go to the primary and secondary auditory cortex and to the associative-limbic cortex, where they are further processed until they become conscious. These conscious processes then have a feedback effect on the amygdala and other subcortical areas. LeDoux has also extended this model to humans and to visual fear conditioning. However, the fast unconscious pathway seems to be designed differently in humans (as well as in other primates), where no direct sensory connections from the thalamus to the amygdala are found (cf. **Fig. 6.2**).

The neurobiologists Pessoa and Adolphs (2010) have pointed out that in humans, unlike the rat, LeDoux's object of study, there is no fast sensory thalamoamoamygdalar pathway. Rather, the "fast" pathway in humans and other primates basically runs from the thalamus first to the primary sensory cortex areas, which in turn have a direct connection to the basolateral amygdala in addition to pathways that draw to associative cortex areas (◘ Fig. 6.2). However, the notion of a fast and unconscious and a slow and finally conscious processing pathway still seems justified.

The unconscious emergence of emotions is predominantly a matter of the basolateral amygdala and the mesolimbic system. Appetitive behaviour, i.e. the pursuit of reward, is predominantly associated with activation of the nucleus accumbens, while the processing of aversive stimuli, i.e. the cessation or avoidance of pain and disappointment, mostly involves activation of the amygdala and the habenula.

Like all events in the brain, emotions thus basically arise *unconsciously in the* first instance (LeDoux 1996, 2017). The reason for this is that consciousness necessarily requires an unconscious run-up of 200–300 ms, within which certain "wake-up stimuli" from the reticular formation of the brain stem and unconscious excitations from the subcortical limbic system arrive in certain parts of the cerebral cortex and cause excitations there. They could, for example, cause the neurons that are excited there to become synchronously active (Dehaene 2014; see also ▶ Chap. 8).

The conscious experience of feelings arises in cortex areas belonging to the upper limbic level (see ▶ Sect. 6.3 on personality, see ▶ Chap. 5). On the one hand, these limbic cortex areas receive their own input from the subcortical and cortical sensory centers and associative areas. On the other hand, they receive massive inputs from the hypothalamus, the septum, the amygdala, and the mesolimbic system (VTA, nucleus accumbens). Through these ascending connections, the "unconscious" controls our conscious emotions (Roth and Strüber 2018). Moreover, the cortical limbic areas send massive outputs back there (Pessoa 2017). Through these retrograde connections, emotions can be consciously or unconsciously adapted to the situation at hand.

The *orbitofrontal cortex* (OFC), together with the *ventromedial prefrontal cortex* (vmPFC), is concerned with the detailed and contextual evaluation of our behavior and its consequences (i.e., prospect of reward or fear of loss) and, through this, with the control of our decisions, social behavior, morals and ethics, and, in general, with the regulation of our emotions. This also involves the ability to restrain strong emotions and impulses, which originate primarily in the hypothalamus, the amygdala, and the mesolimbic system, and not to let them overwhelm us (cf. ▶ Chap. 2).

A limbic area on the inner side of the cerebral cortex is the *anterior cingulate cortex* (ACC). Its dorsal part deals with internally directed (i.e., *top-down*) attention, error detection, assessing the risks of our behavior according to success and failure, while its ventral part deals with our own perception of pain and with feeling the suffering of others, i.e., *empathy* (Lavin et al. 2013; Misra and Coombes 2014).

At the transition between the frontal, parietal and temporal lobes lies—deeply sunken—the *insular* cortex. It has to do with taste, which, as is well known, together with smell, is very close to feelings, with pain perception and, in this context, just like the ACC, with empathy, namely the perception of pain in others (Decety and Michalska 2010), but also with one's own emotional pain in the case of humiliation, embarrassment and exclusion (Eisenberger et al. 2003; Singer et al. 2004, 2009; Eisenberger 2012).

The *hippocampus* plays an important role in the interaction of cognitive and emotional contents of the memory (◘ Fig. 6.1).

It is the *organizer of* the *declarative memory* that is capable of consciousness and can be formulated in language, i.e. it controls both the "reading in" and the "retrieval" of contents of long-term memory. On the one hand, it is under the influence of the entire associative cortex. On the other hand, the hippocampus receives inputs from the amygdala, the hypothalamus and the mesolimbic system. Via these inputs from the subcortical limbic system, unconscious emotional and motivational processes strongly influence the encoding and retrieval of memory content (Bocchio et al. 2017). Here, the interaction between hippocampal pyramidal cells and GABAergic interneurons, expressed in high-frequency discharges, seems to play an important role (Pastoll et al. 2013).

6.1.5 The Chemistry of Emotions

Feelings as consciously experienced emotions are always linked to the release of certain substances in the brain (Roth and Strüber 2018). *Positive* feelings such as contentment, happiness, joy, euphoria and ecstasy are caused by the release of a number of very different substances such as the neurotransmitter or modulator serotonin, which tends to calm and relax, "brain-own" (endogenous) drugs such as endorphins, enkephalins and endocannabinoids, which produce joy, pleasure and euphoria, and hormones or neuropeptides such as prolactin and oxytocin, which convey a sense of social well-being ("attachment"). Most of these substances also have *analgesic* and stress-reducing functions. These substances are produced in different nuclei of the hypothalamus and in the pituitary gland as well as in the raphe nuclei of the brain stem (serotonin) and are released in other limbic centres, especially in the amygdala, the mesolimbic system and the limbic cortex.

Negative emotional states are also triggered by very different neuropeptides and hormones. For example, the neuropeptide substance P generally mediates pain sensations and increases arousal, aggression and male sexual behavior. Vasopressin increases blood pressure and, in males, sexual appetence behavior and aggression, as does substance P. Cholecystokinin can trigger panic attacks, and corticotropin-releasing hormone (CRH) triggers feelings and reactions of stress and, in higher doses, fear and anxiety. The neurotransmitter adrenaline/norepinephrine increases general alertness, produces a general feeling of threat at higher doses and supports the consolidation of memory content (Valentino and van Bockstaele 2008).

These substances have a partly promoting, partly inhibiting effect on each other and occur in the most diverse combinations. They can be regarded as the "labels" of the results of limbic evaluation and conditioning processes and as reinforcers of their behavioral relevance. This is thought to occur primarily via the recursive interaction between the amygdala, nucleus accumbens, ventral pallidum, and hippocampus on the one hand, and the aforementioned limbic cortical areas on the other (Pessoa 2017).

Emotions and Brain

In contrast to cognitive functions, emotions have a close connection with physical states and with certain behavioral reactions (appetence and aversion). As *affects*, they are associated with basic biological needs such as food intake, defensive and sexual behavior, and brood care, and as *strong experiential states* such as anger and enthusiasm, they accompany the various attack, defense, and flight responses as well as sexual activities—i.e., our basic majority innate modes of response.

In addition, there are a number of "basic emotions" that seem to be common to all people and are associated with certain positive or negative experiences within the framework of classical conditioning. The number of such basic emotions is controversial—estimates range from five (e.g., joy, fear, surprise, disgust, and sadness) to 15. The extent to which they exist independently as fixed modules or combine with each other, or whether they can be arranged in a double polar fashion (low vs. high arousal and positive vs. negative), is equally controversial. Generally accepted, however, is the interpretation of emotions as the results of behavioural evaluation (*appraisal*), i.e. the assessment of whether something has gone well or badly from the organism's point of view, and a resulting orientation of future behaviour in the form of wishes, goals and expectations, or the ending or avoidance of negative events. In this sense, most emotions form the basis of *action tendencies*, i.e. motivation.

Emotions initially arise unconsciously as excitations of subcortical brain centers. Often, but not always, these unconscious excitations are accompanied by a conscious feeling, which requires activation of limbic cortical areas and is much more detailed.

In the generation of emotions, activities of the subcortical and cortical limbic areas in the brain are associated with the release of certain neuromodulators, neuropeptides, and neurohormones, e.g. brain-derived opioids and cannabinoids, serotonin and oxytocin, which induce calming or positive sensations, or substance P, vasopressin, or high doses of cortisol, which induce distressing, painful, or other unpleasant sensations such as disappointment (Roth and Strüber 2018).

6.2 Motivation

Motives are mental drive states for things that do *not* run automatically, but must overcome a certain threshold or certain resistances. The higher the resistance, the stronger the **motivation** for a certain action must be. But what is it that drives us?

> The term **motivation** describes the nature, direction and duration of a behavioural drive. Such behavioural drives are called "motives". A distinction is also often made between *unconscious* motives and *conscious* goals. They are usually accompanied by positive (appetitive) and negative (aversive) feelings.

The answer of motivational psychology is that people strive to bring about events that stimulate *positive* (appetitive) feeling states and to avoid those that lead to *negative* (aversive) feeling states (cf. Weiner 1986; Kuhl 2001; Neyer and Asendorpf 2018). This is called *affect optimization* in motivational psychology. It is meant to express that everyone strives to do maximally well under the given circumstances, i.e., to experience pleasure and joy, to have fun, to be in a good mood, to be optimistic about the future, etc. This usually means at the same time that he tries to avoid pain and negative emotional states (Puca and Langens 2005).

The striving for positive emotional states, mostly due to rewards of some kind, is of course not always equally strong, but depends on many factors, such as the type and attractiveness of the reward, its sustainability and expectability or the uncertainty of its achievement or occurrence, the effort that must be made, and many others. The same applies, of course, to aversive behaviour.

6.2.1 Psychological Motivation Models

In motivational psychology, a distinction is usually made between *biogenic* motives, which are part of our biological equipment, such as the satisfaction of needs in the form of hunger, thirst and sexuality, and *sociogenic* motives. Here, four motive domains in particular are mentioned, namely *attachment*, *intimacy*, *power*, and *achievement* (cf. Heckhausen and Heckhausen 2018; Neyer and Asendorpf 2018). However, this distinction is not particularly strict, as all sociogenic motives, in order to be effective, must ultimately be linked to biogenic motives.

Attachment is the pursuit of social closeness, i.e. security, friendship and affection. This motive can also have negative effects, because people who are dominated by it often feel a fear of losing connection, i.e. rejection and disregard or the end of close social relationships. This is often accompanied by the personality trait "neuroticism", i.e. increased anxiety and ego weakness, which in turn may have its roots in a deficient attachment experience (▸ Sect. 6.3). The *intimacy* motive, on the other hand, is found predominantly in extraverted, i.e. positive-minded individuals who themselves radiate trust, warmth and reciprocity. They are, for example, typical "listeners". Presumably this goes hand in hand with high oxytocin levels. Negative feelings can also occur in connection with the intimacy motive, for example as fear of distance and loneliness. The motive *power* is characterized by the striving for status, influence, control and dominance. Characteristic here is the connection with an increased level of testosterone—interestingly, this is more evident in women than in men. Testosterone levels are positively coupled with the release of dopamine ("do something!") and negatively coupled with serotonin ("it's good the way it is!"). The commonly suspected link between testosterone and aggression is only significant in violent offenders and appears to occur as a result of an interaction with the cortisol system (Roth and Strüber 2018). The pursuit of power is often accompanied by the fear of losing power and control (Neyer and Asendorpf 2018).

The *achievement* motive is complex and is expressed in the need to do things well or better, to surpass oneself and others, to master difficult tasks, to start something new, to conquer things, to overcome obstacles and to increase status. The achievement motive is coupled with curiosity. However, fear of failure also occurs with it (Heckhausen and Heckhausen 2018).

Types Motivation

A distinction is made between biogenic motives, i.e. the striving to fulfil basic biological needs, and sociogenic motives such as the striving for achievement, power, closeness and intimacy. Often these motives are coupled with fear of loss of achievement, power, rejection, and distance. The achievement motive is particularly well studied in motivational psychology. Here, a distinction is made between individuals who are confident of success and those who are fearful of failure, who differ significantly in the nature and attainability of their goals.

Many psychologists have dealt intensively with the achievement motive, e.g. the American psychologist J. W. Atkinson. In his "expectation-value model" Atkinson saw the *need for achievement* as a fundamental human motive and in this context dealt with the question of what goals a person strives for in order to succeed and at the same time avoid failure (Atkinson 1964). The product of this, according to Atkinson, then determines a person's performance behavior. However, today Atkinson's model is perceived as too simplistic (Myers 2014).

The American social psychologist B. Weiner, a student of Atkinson, has tried to combine various so-called attribution theories, i.e. attempts to explain motivational behaviour on the basis of certain preconceptions, with Atkison's expectation-value model. In this context, two personality types are distinguished. The first are the *success-assured*: they have a positive basic mood and generally set themselves realistic goals and moderately difficult tasks, i.e. those which they can achieve with some effort. They generally attribute successes to themselves. Persons exhibiting *fear of failure*, on the other hand, show a negative mood and usually choose either goals that are too high, which they do not believe they can achieve anyway, or goals that are too low, the achievement of which does not give them a real feeling of reward. They fear failure rather than looking forward to success (Weiner 1986).

6.2.2 Congruence and Incongruence of Motives and Goals

In motivational psychology, a distinction is often made between motives and goals (cf. Puca and Langens 2005). According to this, motives are *unconscious*, whereas goals are *conscious* drives for action. If we follow this distinction, we can say that motives are determined by both phylogenetic and attachment-related drives to act acquired in early childhood, whereas goals are determined by drives that arise in later childhood, adolescence, and adulthood as a result of conscious experiences. Goals are shaped in particular by *ideas* about states to be achieved.

While motives are deeply and unconsciously rooted in the personality, goals are consciously processed. At the level of consciousness, we speak of *extrinsic* and *intrinsic* goals. Extrinsic goals are those that consist of material, e.g. financial, incentives or social incentives such as recognition, influence and power. Intrinsic goals, on the other hand, are those goals that correspond to personality development and are correspondingly self-rewarding. According to the well-known motivation theory of Deci and Ryan (1985), the main characteristics of intrinsic motivation are striving for competence, for inclusiveness, and self-determination/autonomy. Other authors give as examples of intrinsic rewards an increase in self-efficacy, the feeling of being better than others, or contributing to an important cause. According to Di Domenico and Ryan (2017), intrinsic motivation best predicts the traits of achievement, competence, and autonomy, and thus social and career success.

Conflicts or **"incongruities"** can arise between motives and goals. This can already occur at the level of unconscious motives (lower and middle limbic level, ▶ Sect. 6.3), for example between the striving for attachment and the striving for independence, but also between motives and goals (lower and middle limbic level vs. upper limbic and cognitive level) such as the longing for attachment and a professional striving for success. A person may be dissatisfied with a career chosen for external, e.g., material, reasons and may mourn his or her childhood dream of becoming an actor. This incongruence can manifest itself in increased psychological stress (Grawe 2007).

Congruence and Incongruence of Motives and Goals

Already at the lower unconscious level of the personality, conflicts can arise between different drives such as the search for attachment and the striving for autonomy, but also between unconscious motives (such as the striving for closeness) and conscious goals (such as the will to make a career). Such incongruities often lead to psychological conflicts and even serious mental illness. It is one of the main goals of psychotherapy and coaching to eliminate such incongruencies.

Congruence of motives and goals is the prerequisite for what the Canadian-American psychologist Albert Bandura (born 1925) called *self-efficacy*, namely the subjective assessment that the achievement of goals can be influenced by one's own behaviour (Bandura 1997). Self-efficacious people show *persistence*, i.e. a tenacity in the pursuit of goals. *Avoiders* are the opposite: they see obstacles not as a challenge but as a threat and a danger of failure. Persistence is not the only prerequisite for self-efficacy, however; the other is *reality orientation*. Indeed, one can be very persistent in pursuing a particular goal without realizing that one will never achieve that goal or that this goal is not at all as rewarding as it looked. Reality orientation means being able to assess which effort is worthwhile for which goal (Neyer and Asendorpf 2018).

6.2.3 The Neurobiological Basis of Motives and Goals

Some motives, especially those for securing our biological existence, are genetically determined, but the majority are based on learning processes that are also part of the development of the individual personality (▶ Sect. 6.3). In connection with the already mentioned unconscious and conscious *evaluation*, the brain determines whether and in what way certain events or one's own actions have positive or negative consequences. This is then stored in the memory of experience and forms the basis for the orientation of future motives.

At the unconscious middle limbic level (▶ Sect. 6.3), numerous centers are involved in these processes, including the basolateral amygdala, lateral habenula, ventral tegmental area (VTA), nucleus accumbens, ventral pallidum, and dorsal raphe nucleus. The result of this cooperation is transmitted to the dorsal striato-pallidum as a subcortical coordination center for actions. In the above-mentioned centers, especially in the VTA and in the nucleus accumbens/ventral striatum, there is a variety of neurons that process quite different portions of unconscious action planning and, in their action on cortical areas, also become the basis of conscious action planning.

Central aspects concern, on the one hand, the distinction between the subjective positive or negative state of experience, i.e. pleasure (*liking*) and displeasure, on the one hand, and the striving to attain or avoid such states of experience (appetence and aversion) on the other (Berridge and Kringelbach 2015). Both functions are based on different neuronal systems, which usually interact with each other, but can also act independently of each other.

The occurrence of pleasure states in reward situations is primarily linked to the release of endogenous opioids (endorphins and enkephalins) and cannabinoids, which evoke corresponding feelings by binding to different receptors (mostly mu and kappa receptors or CB1 receptors). This occurs in so-called *hedonic hotspots*, i.e., small areas in different limbic centers such as the nucleus accumbens, amygdala, ventral pallidum, VTA, and cortical limbic areas (e.g., OFC, insular cortex; Berridge and Kringelbach 2015; Wenzel and Cheer 2018). These *hedonic hotspots* are spatially separated from *coldspots* that counteract positive experiential states, for example, in the nucleus accumbens, VTA, and ventral pallidum.

The second effect of endogenous opioids and cannabinoids is the inhibition of GABAergic interneurons in the VTA, which in turn inhibit dopaminergic neurons. By *inhibiting* these inhibitory neurons, the dopaminergic neurons in the VTA are activated and can influence the nucleus accumbens, ventral pallidum, and other behaviorally relevant limbic areas via their efferents. Via this effect, endogenous opioids and cannabinoids released in reward situations can trigger appetitive behavior. Aversive stimuli, on the other hand, have a reinforcing effect on inhibitory GABAergic interneurons and decrease or block the activity of dopaminergic neurons. The lateral habenula, which is under the influence of both the amygdala and limbic cortical areas (e.g., OFC, mPFC), plays an important role in mediating aversive stimuli with its projection to caudal parts of the VTA (Baker and Mizumori 2017).

According to a model developed by W. Schultz and colleagues, dopaminergic neurons signal two different stimulus classes with their activity (Stauffer et al. 2016). A first and fast response occurs to any kind of *salient* stimuli regardless of their reward character (*saliency response*), which directs the brain's attention to these stimuli. Only a second, slower response is reward-specific and signals whether a particular reward expectation will be fulfilled or is stronger or weaker than expected—somewhat confusingly called "*prediction error*" (Schultz 2016), a better term would be "deviation from expectation". Normally, dopaminergic cells are uniformly active at low frequency (*tonic* activity). The unexpected receipt of a reward (i.e., a high positive deviation from expectation) leads to an additional volley of action potentials, i.e., a *phasic* response. If it

is learned that a particular stimulus always precedes a reward and thus predicts it, then the phasic dopamine response occurs shortly after the announcing stimulus, but not at the time of the reward. This signals *reward anticipation*. However, if the reward fails to occur despite being announced, then tonic dopamine activity falls *below* the normal tonic level at the time of the missing reward (Schultz 2007).

In addition to the occurrence and eventual degree of reward, certain dopaminergic cells signal the degree of *uncertainty of* a reward (Fiorillo et al. 2003). This is encoded by a slow and moderate activation that occurs between the first cue of a reward and the time of its occurrence and is higher the greater the uncertainty about whether the cue stimulus and the accompanying phasic dopamine response actually herald a reward. Thus, the target cell receives another signal: a fast and high dopamine release informs that a reward is expected, a slow, moderate dopamine release signals *uncertainty* about whether the reward will actually occur.

However, *risk awareness* also plays a role here: If the reward value is high and the uncertainty about the occurrence of the reward is also high, then cautious individuals with a high risk awareness feel no motivation to act, whereas particularly risk-seeking individuals react to this pattern with a great willingness to behave. For them, the slow dopamine signal of uncertainty itself has a rewarding effect and reinforces risky behavior. This explains, for example, why some individuals are willing to wager large amounts in gambling even though the uncertainty about a possible win is extremely high (Fiorillo et al. 2003).

Some dopaminergic cells also signal *aversive* events such as punishment or reward deprivation via a slow, sustained reduction in spontaneous activity due to the influence of the aforementioned inhibitory (GABAergic) neurons in the VTA (Luo et al. 2011). Such signals reinforce withdrawal behavior and lead to avoidance of a particular stimulus.

However, there is increasing evidence that there are dopaminergic neurons in the VTA that are *directly* excited by aversive stimuli (Holly and Miczek 2016). These influence neurons in medial areas of the nucleus accumbens via their terminals, which then mediate the occurence of aversive stimuli (De Jong et al. 2019).

With regard to the expectation of rewards, it has also been shown that serotonergic neurons in the dorsal raphe nucleus are significantly involved in the action of dopaminergic neurons (Fischer and Ullsperger 2017). For its part, the dorsal raphe nucleus is under cortical (medial PFC) and subcortical control (lateral habenula), among others, and projects massively to the VTA, among many other brain regions. The serotonin release that takes place leads to increased activity of the dopaminergic neurons there. It is possible that they are particularly involved in the detection of surprising stimuli (Fischer and Ullsperger 2017).

Parallel to subcortical processing, information is transmitted via the mesocortical pathway system to the upper limbic and cognitive levels, where conscious desires, goals and concrete intentions arise. This primarily involves the orbitofrontal, anterior cingulate, ventromedial, and ventrolateral prefrontal cortex, which are responsible for conscious intentions to act, the dorsolateral prefrontal cortex, where mental action planning occurs, and the posterior parietal cortex, which is responsible for the spatial embedding of behaviors. Of particular importance are the functions of the orbitofrontal and ventromedial cortex, which contain neurons that also encode the social desirability or nondesirability of desires and intentions. This occurs in interaction with the insular cortex, which, like the nucleus accumbens, signals physical pleasure and physical pain via hotspots and coldspots (Berridge and Kringelbach 2015).

Dopaminergic Reward System

The limbic evaluation system classifies everything we experience or do as positive or negative. This results in the tendency to repeat what evoked positive states or feelings (appetence) and to avoid negative things (aversion). The interaction between the amygdala, lateral habenula, nucleus accumbens and VTA plays the decisive role here, whereby the amygdala in humans via the lateral habenula is more "responsible" for the negative, surprising or strongly emotionalising, while the nucleus accumbens and VTA, with the involvement of the serotonergic dorsal raphe nucleus, are more "responsible" for the positive, rewarding. Positive experiences result in reward expectations, which are represented in dopamine signals in the VTA and nucleus accumbens. These encode different aspects of reward occurrence and expected reward such as type, strength, probability of occurrence, effort, risk and uncertainty. However, there are also dopaminergic neurons that directly encode aversive stimuli.

Thus, a complex network exists in the brain that registers the occurrence of positive and negative stimuli in a finely graduated manner and becomes the basis of motivation in the form of appetitive or aversive behavior. In this process, all conceivable aspects of possible action goals are taken into account, such as the strength and the temporal and spatial attainability of a goal, its sustainability, the effort to be expended, and the certainty or uncertainty of its occurrence. Schultz and colleagues were able to show that, based on the activity of this network, a behavior is generated that can be interpreted as "purposive" (Pastor-Berniera et al. 2017).

6.2.4 How Is Motivation Translated into Behaviour?

Motivation, we have heard, is the formation of unconscious motives and conscious goals, and thus of behavioral tendencies. But how does the actual conversion of such tendencies into behavior take place in the brain?

It was long believed that the dorsolateral prefrontal cortex (dlPFC), located in the upper lateral frontal brain, as the seat of logical operations, was the "supreme decision-making center", but it turned out that behaviorally relevant decisions are made by a complex network of brain centers. The dlPFC maintains goals in the process and is something of a "rational advisor" to which, figuratively speaking, the actual action-controlling cortical and subcortical brain centers can, but need not, listen. The dlPFC might not exert any direct control function at all, because it has only sparse connections to the decision centers, while they exert a strong influence on it (Ray and Zald 2012). This explains why reason and rationality often have little effect, while stronger and seemingly irrational emotions can carry us along.

As illustrated, there are unconscious- or conscious-emotional and conscious-rational operating decision-making instances. The unconsciously operating instance includes the hypothalamus, the septum, the PAG, the amygdala and the nucleus accumbens at the lower and middle limbic level; they comprise partly innate and partly early acquired motives. If the activity of these centers remains, we act *without knowing why*. The conscious instances are represented by cortex areas on the upper limbic level. These include the intuitive behavioral tendencies, often called "gut feelings." The third instance

is located at the cognitive-linguistic level, primarily in the dlPFC, and includes the rational-thought evaluation level. All three instances mediate motives and the goals, between which there are often multiple "battles" in the sense of a clearing of neuronal excitations until the dominance of one motive or goal is established. As mentioned above, the rational level has the weakest voice—unless the rational arguments are supported by emotional states—and the unconscious motivational level the strongest voice—unless the conscious level can offer strong counter-arguments, for example in the form of imagined losses or unpleasant social consequences (Ray and Zald 2012).

The convergence center of this "power poker" is the dorsal striatum, which is our action memory. This is where all our actions are stored that were once successful. All unconscious and conscious intentions to act must be aligned with this memory. The dorsal striatum is connected by many recursive pathways to both the cortical and subcortical decision centers. It is in these circuits that the sometimes short, sometimes long process takes place in which intentions and desires become a concrete willingness to act, resulting in either a volitional decision or a pressure to act (or both) (Ashby et al. 2010). These findings of neuroscience substantially contradict the still highly regarded "Rubicon model" of decision-making as developed by Heckhausen and Gollwitzer (1987) decades ago, in which *unconscious* deliberation and decision-making processes do not occur at all.

Motivation
Motivation is understood to be psychological drive states that are directed towards the fulfilment of biological, individual and social needs. The former result from genetic drives, the latter from unconscious or conscious experiences of a posi-

6

tive and negative nature that guide our further actions in such a way as to seek the repetition and more detailed exploration of positive states (appetence) or the termination or avoidance of negative states (aversion). In the brain, this is controlled in an unconscious way by subcortical limbic centers such as the amygdala, nucleus accumbens, and VTA, and in a conscious way by limbic cortex areas such as the orbitofrontal, ventromedial, and insular cortex.

6.3 Personality

Our experience teaches us that no two people are alike, that everyone is somehow different from others in terms of how they look, think, feel and act. At the same time, we also observe that despite all variability, there are *basic patterns of feeling*, *thinking* and *acting* that we can often use to describe people quite effectively.

However, these are not regularities in the strict sense, but *dispositions* which we expect a person to have with varying degrees of probability on the basis of certain prior knowledge and prior experience. A sufficiently high degree of expectability of the human personality and the resulting action is the basis of social coexistence, whereas too much rigidity of personality as well as too much variability would make social life impossible.

6.3.1 How Do We Capture a Person's Personality?

Even in antiquity, people thought about how best to assess the personality and psyche of a person. The best known of these is the "doctrine of the temperaments", which goes back to Hippocrates and Galenos and divides people into four per-

sonality types: choleric, melancholic, phlegmatic and sanguine. Modern personality psychology, on the other hand, does not look for rigid types, but for the existence of individual, statistically well-definable (if possible, non-overlapping) personality traits that are found in stronger or weaker forms in all people. Accordingly, the individual personality of a person consists of a *unique combination of* such characteristics (for an overview, see Stemmler et al. 2016; Neyer and Asendorpf 2018).

The approach commonly used in personality psychology is usually based on the so-called lexical method, which was first developed in the 1930s by the psychologists Allport and Odbert (Allport and Odbert 1936). In this process, starting from everyday psychology, one takes from common lexicons all conceivable vocabulary describing human characteristics. There are many thousands (in English almost 18,000) of such words, which, however, overlap considerably in their meaning. By repeatedly combining overlapping characteristics, usually with the help of so-called **factor analysis**, we arrive at fewer and fewer overlapping personality attributes, until finally a few basic characteristics emerge. These should be maximally free of overlap (mutually "orthogonal").

Factor Analysis

is a statistical method that is used to derive a few basic factors from a set of observable characteristics that exist as independently as possible and are as free of overlap ("orthogonal") as possible. This procedure is used for data and dimension reduction. In psychology, for example, it is used to identify the basic characteristics of a person's personality.

The personality tests in use today are usually based on three to six basic factors (cf. Neyer and Asendorpf 2018). The well-known "Big Five" personality test was developed by the psychologists Costa and McCrae in the 1980s and 1990s, building on preliminary work by the German-British psychologist Hans-Jürgen Eysenck (Costa and McCrae 1989, 1992). In the meantime, a revised version (NEO-PI-R) is available. A German version of this version was published by Ostendorf and Angleitner (2004). In this test, the basic factors are extraversion, neuroticism, agreeableness, conscientiousness and openness/intellect.

The "NEO-PI-R" is the revised version of the Five-Factor Personality Test by Costa and McCrae. A German version was presented by Ostendorf and Angleitner in 2004. With 240 items, this test attempts to capture the basic personality traits of human beings. A more differentiated view of the five main factors is to be achieved by a total of 30 facets.

Let us look at the five basic factors. Each of these basic factors can be present in different degrees of expression—usually this is indicated in the form of a five-point *Likert scale* from "strongly pronounced" to "weakly pronounced" or "not at all" with three intermediate levels. Let's take a look at these "Big Five":

- *Openness-Intellect* denotes, in its strongest form, the characteristics broadly interested, imaginative, intelligent, original, inquisitive, intellectual, artistic, clever, inventive, witty, and wise, and, in its weakest form, the characteristics ordinary, one-sidedly interested, simple, without depth, and unintelligent.
- The factor *Conscientiousness* includes the traits organized, careful, planning, effective, responsible, reliable, accurate, practical, cautious, deliberate and conscientious in strong expression and the traits careless, untidy, reckless, irresponsible, unreliable and forgetful in weak expression.

- The factor (Extraversion) *Extraversion* in its strong expression includes the traits talkative, determined, active, energetic, open, dominant, enthusiastic, social and adventurous and in its weak expression the traits quiet, reserved, shy and withdrawn.
- The factor *Agreeableness* denotes the traits compassionate, kind, admiring, cordial, soft-hearted, warm, generous, trusting, helpful, indulgent, friendly, cooperative, and sensitive when strongly expressed, and the traits cold, unfriendly, quarrelsome, hard-hearted, cruel, ungrateful, and stingy when weakly expressed.
- The factor (Neuroticism) *Neuroticism* refers in strong expression to the traits tense, anxious nervous, moody, worried, sensitive, irritable, fearful, self-pitying, unstable, despondent and despondent and in weak expression to the traits stable, calm and content. It should be noted that this factor has a negative-positive polarity, while the others show a positive-negative polarity.

In English, the best way to remember the "Big Five" is by the acronym OCEAN.

Personality tests of the Big Five type are commonly used to determine a person's personality in relation to his or her suitability for a particular job, whether it be a managerial position in business or government or in politics. It attempts to determine the degree to which a person is "extraverted", "neuroticistic", or "conscientious", etc. This results in a personality profile of the person in question.

6.3.2 Criticism of the "Big Five", Additions and Alternatives

Within personality psychology, the Big Five approach is not without controversy (cf. Neyer and Asendorpf 2018). One fundamental criticism concerns the fact that the Big Five are essentially taken from everyday psychology and have no further explanatory value. Likewise, it is criticized that personality tests of the "Big Five" type usually involve self-reporting by the persons tested, which as "questionnaire psychology" is regarded by experts as not reliable, since people are generally not good at assessing their own personality, not to mention pretence. As a rule, people tend to "whitewash" their own abilities and achievements; however, neuroticistic individuals show a clear tendency to "blackwash" (cf. Myers 2014; Fletcher and Schurer 2017). It is also apparent that the five basic factors are sometimes significantly correlated with each other, which affects their discriminatory power. Indeed, Neuroticism and Conscientiousness have considerable proximity to each other, as do Extraversion, Agreeableness, and Openness/Intellect. Finally, the Big Five have not been shown to apply well to other populations and cultures, leading in part to an expansion and in part to a reduction in the number of basic factors or subfactors (*facets*) (Neyer and Asendorpf 2018).

The efforts of the British psychologist and personality researcher Jeffrey Gray were also aimed at reducing the five basic factors. Gray assumed three basic personality-related behavioral patterns, namely a "*behavioral approach system*" (BAS), at the center of which is reward orientation; a "*behavioral inhibition system*" (BIS), which is essentially characterized by passive avoidance behavior; and a "*fight, flight, and freeze system*" (FFFS) *fight-flight-freezing system*, which involves rapid, active avoidance behavior (Gray 1990). The BAS shows great similarity to "extraversion" in that it includes strong reward orientation, impulsivity, sensation seeking, as well as sociability and generally positive feelings. The BIS, in turn, has great similarity to "Neuroticism" in that it includes increased attention to negative things, rumination, anxiety, and depression. The FFFS, on the other hand, has no equivalent in the Big Five.

Definition

Proximity system	(*behavioral approach system*, BAS) refers to reward orientation.
Inhibition system	(*behavioral inhibition system*, BIS) refers to passive avoidance behavior.
Fight, Escape and Freezing System	(*fight-flight-freezing system*, FFFS) refers to fast, active avoidance behaviour.

Current efforts by personality psychologists are also aimed at identifying "super traits" along the lines of Gray's BAS and BIS. According to the American psychologist Colin DeYoung and his colleagues, these are *Stability* and *Plasticity* (DeYoung 2006; DeYoung et al. 2013, 2016). The Stability super trait encompasses the three Big Five traits of Neuroticism, Agreeableness and Conscientiousness, which have the core trait of risk avoidance, playing it safe to the point of absolute passivity and complete withdrawal into depression. The super trait Plasticity comprises the two Big Five traits Extraversion and Openness/Intellect, which revolve around a desire for novelty and adventure up to high-risk behavior and sensationalism.

Asendorpf and Neyer assume that people can be grouped into three *main types*, namely the *resilient*, the *over-controlled* and the *under-controlled* person (cf. Neyer and Asendorpf 2018). In this context, the *resilient* person turns out to be attentive, proficient, skilled, self-confident, fully engaged and curious. However, she may also have marked mood swings, also exhibits immature behavior under stress, loses control easily, is quick to snap, and starts crying easily. The *over-controlled* person is agreeable, considerate, helpful, obedient, compliant, understanding-reasonable, has self-confidence, is self-assured, but is also aggressive and annoys others. Finally, the *under-controlled* person is lively, fidgety, does not keep to boundaries, has negative feelings, blames others, is fearful-anxious, gives in to conflicts, makes high demands on himself, is inhibited and tends to brood.

Experts agree that there are important personality traits that are not accurately captured by the Big Five. These include the trait impulsivity, which, however, has to do with very different and poorly connected sub-traits such as high plasticity and low *stability*, *urgency*, *lack of stamina*, *lack of foresight* and *low tolerance of reward deferral* (cf. Heinz and Rothenberg 1998; Heinz et al. 2011). Other authors cite traits that are not well captured by or are "cross-cutting" to the Big Five approach as: *Distress tolerance* (Chowdhury et al. 2018), *sensation seeking* or *hunger for experience* (Mann et al. 2017), *psychological flexibility* (Steenhaut et al. 2018), and *grit* (determination, commitment), by which is meant especially the persistent pursuit of long-term goals (Tucker-Drob et al. 2016; Wang et al. 2017), and *self-control* (Myers 2014). Both of the latter traits predict academic and career success better than the Big Five.

An important criticism of the Big Five model concerns the fact that it is not based on any statements about the *development of individual personality*. Indeed, certain basic personality traits are already visible at birth or shortly thereafter and are referred to as **temperament** (Thomas and Chess 1977; Buss and Plomin 1984; Blatný et al. 2015; ► Chap. 5). Thus, one baby or toddler is relatively calm, the other more "whiny" or even a "cry baby"; one child is open, friendly, the other more closed, difficult to approach, and so on, and these characteristics do not change significantly throughout life. Many psychologists therefore believe that temperament is essentially genetic and thus subject to the "lottery of genes or gene alleles".

However, there is much evidence of a lasting influence by the prenatal environment, namely via the mother's body and brain (see ► Chap. 5). For this reason, the term "congenital" may only be understood as "already present at birth" and not necessarily as "genetically determined". Irrespective of this, it can be assumed that the temperament of a newborn or young child has an important function in setting the course for the further development of the personality. Thus, the caring behaviour of the primary caregivers is often unintentionally very different in the case of a calm or difficult temperament, and this can significantly influence the child's attachment experience. This, in turn, can provide the basis for the child's later attachment model, unless psychologically powerful events occur later in life (Fletcher and Schurer 2017).

Temperament

Temperament refers to basic emotional and motor characteristics of a person that appear very early, often shortly after birth, and remain relatively constant over the lifespan. They mainly concern a person's degree of sensory and emotional excitability, his or her readiness and strength to react, the degree of openness or closedness to other people and new things, the tendency to be calm or active, etc. Temperament can be determined genetically as well as prenatally epigenetically or early postnatally.

Overview

Current personality psychology is concerned with determining basic characteristics from the multitude of personality traits with the aid of statistical procedures (e.g. factor analysis) that are relatively persistent over time and as free of overlap as possible. Most models are based on a few, usually three to six basic personality factors, which are present in varying degrees of intensity. They are mainly based on everyday psychology, and their lack of overlap is controversial. They are also based on self-report, which many experts believe is an unreliable tool. Beyond the Big Five, many authors see important personality traits such as impulsivity and attachment as not being included, while other authors try to reduce them to the two "super traits" of approach and avoidance or stability and plasticity.

6.3.3 Genetic Foundations, Stability and Changeability of Personality Traits

The question of the genetic determinacy of basic personality traits is widely disputed in personality psychology. On the basis of classical twin research, heritability values of the Big Five traits of 40–60% have been arrived at so far (Bouchard Jr and McGue 2003). Due to methodological inadequacies of twin research and more recent findings on the role of genes in the development of psychological traits, genetic studies based on so-called single nucleotide polymorphisms (SNPs) have been carried out in recent years. These showed, on the one hand, that hundreds to thousands of different genes are involved in basic personality traits and, on the other hand, that there is a much lower heritability rate of the Big Five traits.

For example, a heritability rate of only 15% was found for neuroticism, a rate of 21% for openness to experience, and no meaningful values at all for extraversion, conscientiousness and agreeableness (Power and Pluess 2015). However, these findings by no means imply that the Big Five traits or the other traits mentioned have weak genetic underpinnings, but merely that they are undetectable with today's standard genetic screening based on SNPs. In particular, the much more significant epigenetic factors have hardly been studied to date (▶ Chaps. 5 and 7).

In the popular psychological literature, opinions about the stability of personality or personality traits over the life span vary widely. While some assume a high degree of stability from childhood to old age, many popular authors assume a constant lifelong changeability, whether due to changing life circumstances or of one's own volition. However, serious research comes to a different conclusion (cf. Neyer and Asendorpf 2018). Different personality traits vary in stability: intelligence (measured by IQ) is the most stable, ranging from 11 to 69 years, while the Big Five personality traits, understood as personality profiles, have an average stability of up to 0.65 (extraversion and neuroticism). In general, the variability of personality traits is greater in childhood and adolescence, where environmental influences have a greater impact and traits experience a temporary destabilization at puberty. In early adulthood up to the age of 60–70, the traits stabilize significantly (up to 0.8), but become more variable again towards older age, mostly due to neurological degradation processes. The whole can be understood as a product of the interaction of "disposition" and "environment", with a clear tendency towards self-stabilization. As Asendorpf and Wilpers (1998) note, the ability to either influence one's own environment or to seek out the environment that suits one's own personality increases greatly with increasing age.

Certain genetic, epigenetic and early-childhood factors, especially negative ones, can *in combination* strongly influence personality development at a very early age, as long-term cohort studies, such as the well-known Dunedin study, have shown (Moffit and Caspi 2001), and can be linked to neurobiological factors. With regard to certain psychiatric disorders and antisocial behaviour, a very negative channelling effect may occur here that is then difficult to interrupt (see ▶ Chap. 7).

The stability of personality traits mentioned here by no means implies that people behave in a certain way across different situations. Rather, it is part of the nature of a healthy person's personality to behave, sometimes very differently, in different social contexts (Mischel et al. 1989; cf. Myers 2014). Constancy here refers to the pattern of difference in contextual behavior. Unfortunately, there is little robust empirical research on this.

6.3.4 The Neurobiological Foundations of Personality

The psychological personality typologies presented are predominantly purely descriptive and usually do not provide any deeper reasoning as to why it is precisely these basic factors that best describe a person's personality. Nor do they answer the question of *why* one person is more extraverted and another more neuroticistic. In recent years, a number of personality psychologists and neurobiologists have sought to provide a neurobiological rationale, although the results have so far been unsatisfactory (cf. DeYoung and Gray 2009; Corr et al. 2013; Di Domenico and Ryan 2017). Currently, common methods include measurements of properties of the so-called *default-mode* network (DMN), which is active when no cognitively demanding tasks are currently being processed. It is assumed here that different properties of the *default-mode* network are recognizable in different personality types. This is measured using various imaging methods, primarily fMRI and EEG. In research of this type by Toschi et al. 2018, only the Big Five trait Conscientiousness was found to be significantly related to structural and functional connectivity in the left fronto-parietal network, i.e., this trait was more strongly present the more pronounced the trait Conscientiousness was. The authors interpret this finding as an indication of increased cognitive control and behavioral flexibility. With regard to the other Big Five traits, there were no significant correlations with states of the DMN.

Another approach to linking characteristics of this network with personality traits is the determination of low-frequency oscillations in a frequency range of 0.01–0.25 Hz in total, which is usually divided into five frequency bands. Here, for the trait Extraversion, there was a significant correlation of resting activity in all five frequency bands and for Conscientiousness in frequency band 2 (0.138–0.25). The other three Big Five traits showed no significant correlation (Ikeda et al. 2017).

The application of further neurobiological methods such as the determination of brain surface morphology revealed correlations with extraversion and agreeableness and neuroanatomical properties such as surface area or thickness of certain cortical areas, which did not correspond well with previous research results and are also difficult to interpret (cf. Li et al. 2016).

Overall, it appears that the described measurements of the brain's resting activity have so far not yielded any meaningful results about the neurobiological foundations of personality traits. It can be assumed that such measurements have so far been far too crude to capture the neurobiological bases of complex personality traits. Currently, studies of the relationship between psychologically well-recorded mental states and behavioral performance, functional anatomical conditions, and neurophysiological-pharmacological processes are far more informative.

In the following, we will explain the emergence of basic personality traits on the basis of such correlations, using the four-level model of personality already presented in ► Chap. 5 and the basic psychoneural systems also presented there.

6.3.4.1 The Four-Level Model of Personality

Roth and Cierpka's four-level model of personality (cf. Roth and Strüber 2018) assumes the presence of four anatomical and functional brain levels, namely three limbic levels and one cognitive level (◨ Fig. 6.3), based on a large body of neuroscientific research (cf. ► Chap. 2).

— The *lower limbic level* contains, through the activity of centres such as the hypothalamus-pituitary, septum, central amygdala, PAG and centres of the pons and medulla oblongata, mechanisms which serve to sustain life and fulfil primary bodily needs; but it also contains those characteristics which are considered to be part of temperament (see ► Sect. 6.3.2). The processes taking place at the lower limbic level are and remain unconscious; they belong to the *primary unconscious* and are difficult to change from the outside.

— The *middle limbic level*, primarily represented by the activity of the mesolimbic system (nucleus accumbens, VTA) and the basolateral amygdala, is significantly shaped by the experiences of the infant and toddler over the course of the first 3 years, with the experiences of interaction with the primary caregiver, often the mother, being particularly important. These experiences become deeply imprinted, and their influence is difficult to change, and only through targeted action. Infants and toddlers are at least partially conscious of these experiences. However, these experiences cannot be stored in the long term, because in the first years of life there is no long-term memory capable of remembering. Since Sigmund Freud, this phase has been called "infantile amnesia". It belongs to the *secondary unconscious* because of its fundamental nonrememberability.

— At the *upper limbic level*, represented by activities of the limbic cortex (orbitofrontal, ventromedial, anterior cingulate and insular cortex), those processes take place which are suitable for bringing our primary personality into harmony with the requirements of social coexistence, from the family through kindergarten and school to adulthood. Here it is a

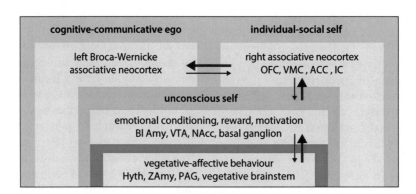

◨ **Fig. 6.3** Roth and Cierpka's four-level model of personality. The lower limbic level of vegetative-affective behaviour and the middle limbic level of emotional conditioning, evaluation and motivation together form the "unconscious self". At the conscious level, the upper limbic level forms the "individual-social self", which is contrasted with the "cognitive-communicative self". *ACC* anterior cingulate cortex, *basal gang.* basal ganglia, *Bl Amy* basolateral amygdala, *Hyth* hypothalamus, *IC* insular cortex, *NAcc* nucleus accumbens, *PAG* periaqueductal gray, *OFC* orbitofrontal cortex, *VMC* ventromedial prefrontal cortex, *VTA* ventral tegmental area, *ZAmy* central amygdala. (From Roth and Strüber 2018)

matter of the formation of cooperative-ness, consideration, patience, the ability to compromise, empathy, but also of determination, the will to assert oneself, self-efficacy, self-realisation, etc.

- At the *cognitive-linguistic level*, mediated by the activities of the frontal, temporal and parietal associative cortex, the acquisition of experience and knowledge as well as linguistic communication take place as the basis of factual-logical thinking, ideas and action planning. The emotional components of such events are added by the instances of the upper limbic level. The cognitive-linguistic level can be strongly influenced by the limbic levels, but itself only has an influence on our behavioural decisions by "address-ing" emotional contents of the limbic levels.

6.3.4.2 The Six Basic Psychoneural Systems as Determinants of Personality

Personality and psyche develop on the three limbic levels mentioned above and with the involvement of the cognitive level. This occurs within the framework of the func-tions of six "basic psycho-neural systems", namely stress processing, emotional control and self-soothing, reward and reward expec-tancy/motivation, attachment behaviour/ empathy, impulse control and reality sense-risk perception (see ▶ Chap. 5). The systems influence each other in both positive and negative ways (cf. Roth and Strüber 2018) and form a tight network of interactions. Their respective activity is associated with various personality traits.

- **Stress Management**
The way a person deals with physical stresses such as illness and pain, as well as psychological stresses such as threats, chal-lenges, disappointments and defeats, shame and exclusion, forms a basic feature of his

or her personality. This trait forms very early and is essentially related to the prena-tal and postnatal development of the corti-sol system. Here, early aversive experiences have a clear negative effect (Fletcher and Schurer 2017). This is already evident in the very normal diurnal pattern of "basal" cor-tisol release. For example, individuals who are attributed with a high degree of emo-tional instability in the sense of "neuroti-cism" of the Big Five often react to waking up in the morning with a high release of cortisol. In a recent study, individual corti-sol parameters were found to be less related to neuroticism and negative affect, and instead closely related to extraversion and positive affect. Indeed, the latter traits are associated with only a low morning cortisol release (Miller et al. 2016).

The actual stress-related releases of cor-tisol sit as "pulses" on top of the normal cortisol diurnal cycle. Apparently, the high resting cortisol release interferes with an adequate stress-related cortisol response, because neuroticistic individuals respond to a stressful situation with an attenuated corti-sol response (Oswald et al. 2006).

- **Emotional Control and Self-Soothing**
The psychoneural self-soothing system is closely linked to the serotonin system (primar-ily the 5HT1A-receptors). Similar to the stress processing system, it partially develops prena-tally. Sufficient serotonin levels are important for the perception of emotional states, i.e. they promote emotion control, goal-directed behaviour and inhibit hasty reactions to pos-sible dangers (see below). A *deficiency* of sero-tonin is observed in a context of continuous preoccupation with stressful stimuli. This can manifest itself in inner restlessness and, mainly in men, in impulsivity and reactive aggression (Cleare and Bond 1995).

The serotonin system is involved in almost all Big Five traits as well as nega-tively in the trait impulsivity. A low activity of the self-soothing system leads to the predominance of neuroticistic traits such as an increased sense of threat, low frustration and loss tolerance, brooding, anxiety and depression up to complete apathy.

- ■ **Reward and Reward Expectation (Motivation)**

The system of reward and reward expectancy as the basis of motivation is closely related to the dopamine system described in ▶ Sect. 6.2. This system develops postnatally from the first years of life until well into adulthood. It is usually associated with the trait *Extraversion* and with Gray's *behavioral activation system* mentioned above—the latter describes the individual's *reward sensitivity*.

Depue and Collins (1999) distinguish "attachment-oriented" extraversion (*affiliative extraversion*), i.e. increased sociability, in contrast to "action-oriented" extraversion (*agentic extraversion*), which is associated with the characteristics "energetic" and "success- and reward-oriented".

An increase in attachment-oriented extraversion leads to a strong need for sociability and social integration. A stronger expression of action-related extraversion, on the other hand, leads to ambition, dominance, a desire for power, and a desire for adventure and sensation. Action-related extraversion is influenced by other substance systems in addition to dopamine. In some studies, a high level of extraversion is associated with low serotonin and high norepinephrine (noradrenalin) levels (Cloninger 1987, 2000). In addition to extraversion, according to a number of studies, personality traits such as curiosity, sensation seeking, and creativity are also associated with increased release of dopamine. Highly creative individuals, for example, have a low density of inhibitory D2 receptors in thalamic nuclei that project to the prefrontal cortex (De Manzano et al. 2010), and therefore may be able to live out their ingenuity.

- ■ **Attachment Behaviour and Empathy**

A person's attachment behavior is also a central personality trait that is "transverse" to the Big Five, as it influences the expression of several Big Five traits. It correlates positively with traits of extraversion, agreeableness and openness and negatively with traits of neuroticism, namely anxiety and withdrawal. From a neurobiological perspective, attachment orientation is equally determined by oxytocin, endogenous opioids, and dopamine as the basis of *attachment-oriented* extraversion. Individuals with a highly active oxytocin system are often characterized by a marked sensitivity to others (Meyer-Lindenberg et al. 2011; Carter 2014). Several studies have shown that a single administration of oxytocin via nasal spray can transiently affect numerous traits. These include, for example, trust and generosity, but also negative traits such as schadenfreude. The effect of oxytocin is also dependent on the expression of the personality trait *extraversion*. In individuals with a low expression of this trait, the administration of oxytocin is associated with increased prosocial behavior and increased trust in an interaction partner. In individuals with a high expression of this trait, no comparable effect of oxytocin administration could be demonstrated (Human et al. 2016).

- ■ **Impulse Control**

Impulse control, like attachment behavior and sensation seeking, is "across" the Big Five and related to components of neuroticism and conscientiousness (Mann et al. 2017). Serotonin plays an important role here by contributing to emotional control and thus behavioral inhibition (Daw et al. 2002). With impulsivity, according to DeYoung and Gray (2009), a distinction must be made between active and reactive impulsivity. *Active* (or "agentic") impulsivity is associated with high scores on the trait extraversion and with seeking immediate rewards, dominance, power-seeking, sensation-seeking, and lack of risk perception. At the same time, active impulsivity is related to high levels of dopamine and testosterone. Similarly, actively impulsive individuals have low levels of neuroticism, agreeableness, and conscientiousness.

Different from this is *reactive* impulsivity, which is associated with low serotonin levels and high cortisol and noradrenaline levels and is based, among other things, on a reduced ability to distinguish threatening stimuli from non-threatening stimuli. Similarly, a diminished capacity to regulate one's emotions occurs. This in turn brings with it a high level of insecurity and a general negative emotionality (see Depue 1995). Reactively impulsive persons are not impulsive all the time, but only in situations that appear threatening, in which they defend themselves because they see no other possibilities for action.

■ **Sense of Reality and Risk Perception**

A balanced personality includes the ability to perceive the situation in which one finds oneself appropriately and to realistically assess its relevance for one's own behaviour. In addition, there is the ability to evaluate the short- and long-term consequences of one's own actions, not to overestimate or underestimate one's own strengths, to correctly grasp the intentions of others, to recognize opportunities and risks and to take them into account in one's own actions.

This important personality trait is also not centrally contained in the Big Five, but is distributed across almost all five basic traits. A good perception of reality and risk includes, on the one hand, a balanced relationship between extraversion, i.e. positive thinking and risk-taking, and neuroticism, i.e. critical thinking and risk aversion, and, on the other hand, a balance between conscientiousness and openness/intellect. From a neurobiological perspective, this implies a balance between the serotonergic and dopaminergic systems, but at the same time also a

high activity of the cholinergic system, which forms the basis of attention, willingness to learn, the rapid grasp and classification of reward and punishment stimuli, as well as low distractibility and high goal focus (Hasselmo and Sarter 2011).

We see that there is no "one-to-one" relationship whatsoever between basic personality traits, as treated in the Big Five or modified variants, and the six psychoneural systems listed, and certainly not—as originally assumed—between the Big Five personality traits presented and the amount of neurotransmitters, peptides and hormones released. Rather, the psychoneural systems and their active substances are involved in the various traits in a complex but empirically ascertainable manner.

There is a complex positive (*agonistic*) and/or negative (*antagonistic*) correlation of effects between the six basic systems mentioned (details in Roth and Strüber 2018). Thus, the stress processing system and the self-soothing system must work closely together to achieve a level of activation appropriate to the problem on the one hand and to bring the organism back to calm after the stress has ended on the other. Severe stress, on the other hand, suppresses the serotonergic system in its calming function. A strong link exists between the self-soothing system and the attachment system in that the release of oxytocin causes an increase in serotonin levels as well as a release of brain-derived opioids. Oxytocin, like serotonin, can also reduce stress levels.

Impulse control and sense of reality/ perception of risk are negatively coupled. A low level of impulse control can override the sense of reality and risk perception; a high level of sense of reality and risk perception contributes significantly to impulse control. Finally, the motivational system can connect with the other basic systems in almost any way, evaluating their respective states as pleasurable or desirable or as painful and to be avoided, depending on the personality. One person loves excitement and the thrill, another the quiet; one person is happy only in company, another wants to be by himself, etc.

The four levels and six basic psychoneural systems determine temperament and personality, and thus the psyche of a person, in their respective manifestations, within the framework of the interaction of the factors genes, epigenetic factors, prenatal and postnatal influences and experiences.

6.3.4.3 A Neuroscientifically Based Personality Typology

How can the personality systems just described be reconciled with the findings of psychological and neuroscientific personality research? In the following, we will show that this can be achieved particularly well if we consider the two basic characteristics of plasticity/dynamics and stability from the newer approaches presented in ▶ Sect. 6.3.2. As

shown schematically in ◘ Fig. 6.4, these two traits form the basis for two different personality types, whereby we want to assign a particularly change-ready *dynamic* personality to the trait plasticity and a *stable* personality to the trait stability. The main reason for this dichotomy is probably the different dominance of the dopaminergic reward expectancy system on the one hand and the risk avoidance and impulse inhibition systems on the other. The two basic types that can be described in this way can in turn be subdivided into two subtypes—possibly on the basis of a characteristic activity of the oxytocin-attachment system in each case. Under certain conditions, and especially when individual genetic predisposition and/or significant early or even later stressful experiences produce an over- or underfunctioning of the stress system, the four subtypes may each develop a characteristic

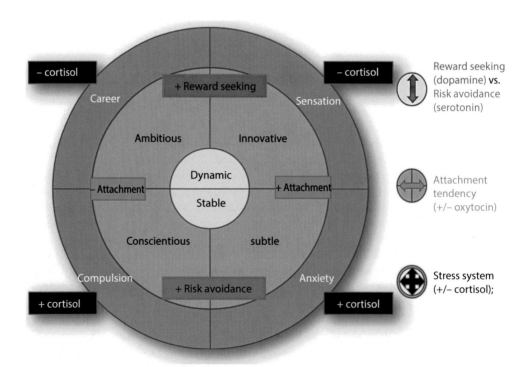

◘ **Fig. 6.4** Personality and neuromodulators, Details in text

psychopathology. Some of this is still somewhat speculative, but quite compatible with current findings.

The first of the two basic types, the *Dynamic*, shows a high degree of enterprise, daring, openness to other people and a greater willingness to change. The opposite type of the Dynamic is the *Stable*. He loves order, ensures that private matters and professional business run smoothly, carefully weighs opportunities and risks, leaves opportunities unused if they are associated with higher uncertainty, values sincerity, punctuality and reliability.

Underlying these two basic personality types, "Dynamic" and "Stable", appears to be a characteristic expression of reward sensitivity, risk assessment and impulse inhibition in each case. The dynamic person shows a high reward orientation, combined with a lower risk avoidance and impulse inhibition. In the Stable, on the other hand, reward orientation takes a back seat in favor of pronounced risk avoidance and high impulse inhibition. In the case of the Dynamic, two subtypes can be distinguished, the *ambitious* and the *innovative*, each of which forms the basis for a characteristic psychopathology. The individual development of the attachment system then determines which of the two subtypes is formed—although there is still a lot of research to be done here. The ambitious person has a strong will to succeed. He sees himself as strong and independent and attaches great importance to material possessions and recognition. These traits are also characteristic of individuals with an insecure-distant inner model of attachment, that is, individuals who have dialed down their own emotional and attachment needs. Under certain conditions, such as when there is an under-functioning of the stress system following early childhood traumatic experiences or later stress, the Ambitious subtype may develop into the *Careerist*. This person is strongly reward-oriented, likes to announce big goals and shows a high ruthlessness in achieving his

goals. He often takes high risks and has clear empathy deficits. At the same time, these individuals experience a constant inner restlessness and "emptiness", which is only briefly eliminated by experiences of success and happiness, which provide the "kick". This type is found in people with antisocial personality disorder ("psychopathy") or malignant narcissism.

The other subtype of the dynamic, the *innovative*, can be enthusiastic about creative innovations. He usually enjoys being with people and is open to change. In this case, the attachment system is probably more active than in the ambitious person—after all, one of the properties of the oxytocin that is central to this system is to promote creativity and openness. The innovative person can also develop problematic characteristics under certain conditions, such as a strong sense of success. Thus the innovative *person* can become a *sensation* seeker. This person loves change for change's sake, likes the excitement that comes with change, always has new plans before he has finished previous projects, shows an increased carelessness and unreliability. He is willing to sacrifice goal-oriented action for the sake of the "thrill" that comes with change. As with the ambitious dynamic, a characteristic under-functioning of the stress system is likely to be involved in this elevation of extraverted, creative, and change-ready behavior into the pathological.

In the *Stable*, whose characteristic trait is risk avoidance, it is apparently the individual level of activity of the attachment system that determines whether he is more conscientious or sensitive. The *conscientious person* is driven by a desire for order and correct procedures, and his motives and goals are often antithetical to those of the innovative dynamist; he appears to have a lowered need for social relationships. The *conscientious person* can develop into an *obsessive-dogmatic person*. This person is exclusively concerned with correctness ("continuing to act as before"), rigidly adheres to his views and the usual procedures, even if this brings disadvantages, and insists on principles. Increasing these tendencies, he shows a clear aversion to change and a deeply rooted conservatism; new things are fundamentally rejected because they are perceived as threatening. This type is prone to obsessive-compulsive disorders. It can be assumed that genetic predispositions, early experiences of stress and/or the experience of later chronic stress have led to an over-functioning of the stress system.

This is different with the *sensitive person*, who reacts socially and empathically despite his pronounced risk avoidance. In contrast to the conscientious stable person, the attachment system is obviously very active. In its negative manifestation, the sensitive person can easily become the *fearfully insecure person* who immediately thinks of everything that could happen in his environment and also to him. He is easily thrown off balance. He perceives the world more negatively than the average, and is therefore often worried, ashamed, insecure, embarrassed, nervous or sad. In him, as in the Conscientious Stable, genetic predisposition and/or prenatal, early childhood, or later chronic stress experiences are likely to have caused his stress system to tend to be overactive and his ability to self-soothe to be deficient. He overreacts to high demands and has difficulty ending his internal state of alarm—he cannot calm down and continues to ruminate on his difficulties for a long time following the high demands. A deficiency of endogenous opioids may also diminish his ability to feel pleasure and joy. Psychopathologically, he may develop towards anxiety disorders or depression.

So we see that from the four "normal types" of personality, namely in the area of the dynamic the ambitious and the innovative and in the area of the stable the sensitive and the conscientious, four "deviant types" can be derived relatively easily, namely the careerist, the change addict, the obsessive-dogmatic and the anxious-insecure. In these deviant types, a transition to full-blown mental disorders is possible in the form of antisocial personality disorder, sensation-seeking, obsessive-compulsive disorder, depression, and anxiety.

6.4 Summary: Brain and Personality

Contemporary personality psychology assumes that a person's personality is a highly individual combination of characteristics. With regard to the development of personality and aptitude tests, it attempts to reduce the large number of such characteristics to a few and more or less selective basic characteristics or basic factors, which then in turn have certain sub-factors. The basic factors include the "Big Five", namely extraversion, neuroticism, agreeableness, conscientiousness and openness/intellect. However, it is disputed whether these basic factors are actually all separable—some authors propose a grouping of the Big Five into *stability* (neuroticism, agreeableness, conscientiousness) and *plasticity* (extraversion, openness). Other authors assume a basic polarity between *extraversion-approach* and *neuroticism-avoidance*. Finally, there are a number of personality traits such as *impulsivity, attachment ability, flexibility*, and *stress tolerance* that experts believe are "at cross-purposes" with the Big Five.

A neurobiological foundation of the most important personality traits is achieved if we start from four levels of personality and the six basic psychoneural systems located on them. We then understand more precisely how the stress processing system, the self-soothing system or the system of emotional control, the reward and reward expectation system, the attachment system, the impulse control system and the reality perception system build on each other and interact positively and negatively with each other. The stress processing system and the emotional control and self-soothing system play the most important role here, as they develop first and influence the formation of the other systems. Favorable development of these two systems, some of which occurs before birth, is the most important prerequisite for the development of a balanced, introspective personality with moderate to high scores on the traits extraversion, agreeableness, conscience, and openness and low scores on the trait neuroticism. It also forms a robust resistance, **resilience**, to subsequent negative developments.

Resilience

In the context of personality development, **resilience** refers to resistance to psychological stress such as stress, abuse, etc., whereas vulnerability refers to susceptibility to such negative experiences.

If, on the other hand, prenatal disorders occur in the formation of these systems, this can lead to a **vulnerability** to the effects of negative experiences. This is accompanied by low values in the traits extraversion, agreeableness and openness, often high values in the trait conscientiousness and equally high values in the trait neuroticism. An anxious, insecure, withdrawn and sensitive personality develops.

Summary

In this chapter we have dealt with the development and structure of the personality, the effect of emotions and the emergence of motives and goals from a psychological point of view. This is about the answer to the question, which is still central today, why people are the way they are and why they do what they do. No single science can provide a satisfactory answer, not even psychology or neurobiology. It is important to identify precisely the manifestations of personality, emotion and motivation from a psychological point of view, but such findings "hang in

the air" if one does not ask about their neurobiological foundations.

Everything we perceive, feel, think and do is closely related to brain processes, which in turn are conditioned by genetic and epigenetic factors as well as prenatal and postnatal influences. Recognizing these connections enables us to better understand phenomena of human action, for example the fact that rational arguments and mere insight often have no effect on our behavior, that strong emotions (affects) can lead us to act completely "irrationally". Or why a good resolution is often not put into practice. Psychological explanations, which always have counterarguments, can only be made plausible in all these cases by the findings of brain research. This is a great step forward. However, no serious brain researcher will deduce from this that brain research can ever replace psychology, because it cannot do without precise knowledge of psychological processes.

References

Allport GW, Odbert HS (1936) Trait-names: a psycholexical study. Psychol Monogr 47(1)

Asendorpf JB, Wilpers S (1998) Personality effects on social relationships. J Pers Soc Psychol 74:1531–1544

Ashby FG, Turner BO, Horvitz C (2010) Cortical and basal ganglia contributions to habit learning and automaticity. Trends Cogn Sci 14:208–215

Atkinson JW (1964) An introduction to motivation. Van Nostrand, Princeton

Baker PM, Mizumori JY (2017) Control of behavioral flexibility by the lateral habenula. Pharmacol Biochem Behav 162:62–68. https://doi.org/10.1016/j.pbb.2017.07.012

Bandura A (1997) Self-efficacy: the exercise of control. Freeman, New York

Berridge JC, Kringelbach ML (2015) Pleasure systems in the brain. Neuron 86:646–664. https://doi.org/10.1016/j.neuron.2015.02.018

Blatný M, Millová K, Jelínek M, Osecká T (2015) Personality predictors of successful development: toddler temperament and adolescent personality traits predict well-being and career stability in middle adulthood. PLoS One 1:1–21. https://doi.org/10.1371/journal.pone.0126032

Bocchio M, Nabavi S, Capogna M (2017) Synaptic Plasticity, engrams, and network oscillations in amygdala circuits for storage and retrieval of emotional memories. Neuron 94:731–743

Bouchard TJ Jr, McGue M (2003) Genetic and environmeltal influences on human psychological differences. J Neurobiol 54:4–45

Buss AH, Plomin R (1984) Temperament: early developing personality traits. Erlbaum, London

Carter CS (2014) Oxytocin pathways and the evolution of human behavior. Annu Rev Psychol 65:1–23

Chowdhury N, Kevorkian S, Hawn SE, Amstadter AB, Dick D, Kendler KE, Berenz EC (2018) Associations between personality and distress tolerance among trauma-exposed young adults. Pers Individ Dif 120:166–170. https://doi.org/10.1016/j.paid.2017.08.041

Cleare AJ, Bond AJ (1995) The effect of tryptophan depletion and enhancement on subjective and behavioural aggression in normal male subjects. Psychopharmacology 118:72–81

Cloninger CR (1987) A systematic method for clinical description and classification of personality variants. Arch Gen Psychiatry 44:573–588

Cloninger CR (2000) Biology of personality dimensions. Curr Opin Psychiatry 13:611–616

Clore CL, Ortony A (2000) Cognitive neuroscience of emotion. In: Lane DR, Nadel L, Ahern GL, Allen J, Kaszniak AW (eds) Cognitive neuroscience of emotion. Oxford University Press, Oxford, pp 24–61

Corr PT, DeYoung CG, McNaughton N (2013) Motivation and personality: a neuropsychological perspective. Soc Personal Psychol Compass 7:158–175

Costa PT, McCrae RR (1989) The NEO-PI/NEO-FFl manual supplement. Psychological Assessment Resources, Odessa

Costa PT Jr, McCrae RR (1992) Normal personality assessment in clinical practice: the NEO personality inventory. Psychol Assess 4:5–13

Damasio AR (1994) Descartes' Irrtum. List, München, Fühlen, Denken und das menschliche Gehirn

Daw ND, Kakade S, Dayan P (2002) Opponent interactions between serotonin and dopamine. Neural Netw 15:603–616

Decety J, Michalska DJ (2010) Neurodevelopmental changes in the circuits underlying empathy and sympathy from childhood to adulthood. Dev Sci 13:886–899. https://doi.org/10.1111/j.1467-7687.2009.00940.x

Deci EL, Ryan RM (1985) Intrinsic motivation and self-determinationin human behavior. Plenum, New York

Dehaene S (2014) Consciousness and the Brain. Viking, New York

De Jong JW, Afjei SA, Pollak Dorocic I, Peck JR, Liu C, Kim CK, Tian L, Deisseroth K, Lammel S (2019) A neural circuit mechanism for encoding aversive stimuli in the mesolimbic dopamine system. Neuron 101:133–151.e7. https://doi.org/10.1016/j.neuron.2018.11.005

De Manzano O, Cervenka S, Karabanov A, Farde L, Ullén F (2010) Thinking outside a less intact box: thalamic dopamine D2 receptor densities are negatively related to psychometric creativity in healthy individuals. PLoS One 5:e10670

Depue RA (1995) Neurobiological factors in personality and depression. Eur J Personal 9:413–439

Depue RA, Collins PF (1999) Neurobiology of the structure of personality: dopamine, facilitation of incentive motivation, and extraversion. Behav Brain Sci 22:491–517

DeYoung CG (2006) Higher-order factors of the Big Five in a multi-informant sample. J Pers Soc Psychol 91:1138–1151

DeYoung CG, Gray JR (2009) Personality neuroscience: explaining individual differences in affect, behavior, and cognition. In: Corr PJ, Matthews G (eds) Cambridge handbook of personality psychology. Cambridge University Press, New York, pp 323–346

DeYoung CG, Weisberg YJ, Quilty LC, Peterson JB (2013) Unifying the aspects of the Big Five, the interpersonal circumplex, and trait affiliation. J Pers 81:465–475

DeYoung CG, Carey BE, Krueger RF, Ross SR (2016) 10 aspects of the Big Five in the personality inventory for DSM-5. Pers Disord 7:113–123

Di Domenico SI, Ryan RM (2017) Commentary: primary emotional systems and personality: an evolutionary perspective. Front Psychol. https://doi.org/10.3389/fpsyg2017.01414

Eisenberger N (2012) The pain of social disconnection: examining the shared neural underpinnings of physical and social pain. Nat Rev Neurosci 13:421–434

Eisenberger NI, Lieberman MD, Kipling D, Williams KD (2003) Does rejection hurt? An fMRI study of social exclusion. Science 302:290–292

Ekman P (1999) Facial expressions. In: Dagleish T, Power MJ (eds) Handbook of cognition and emotion. Wiley, Chichester, pp 301–320

Ekman P (2007) Emotions Revealed. Recognizing Faces and Feelings to Improve Communication and Emotional Life. Henry Hot & Company, New York

Fiorillo CD, Tobler PN, Schultz W (2003) Discrete coding of reward probability and uncertainty by dopamine neurons. Science 299:1898–1902

Fischer AG, Ullsperger M (2017) An update of the role of serotonin and its interplay with domapine in the role of reward. Front Neurosci 11:1–10

Fletcher JM, Schurer S (2017) Origins of adulthood personality: the role of adverse childhood experiences. BEJ Econom Anal Policy 17:1–29

Gershon ES, Rieder RO (1992) Major disorders of mind and brain. Sci Am 267:126–133

Grawe K (2007) Neuropsychotherapy. How the Neurosciences Inform Effective Psychotherapy. Routledge, London

Gray JA (1990) Brain systems that mediate both emotion and cognition. Cogn Emot 4:269–288

Hasselmo ME, Sarter M (2011) Modes and models of forebrain cholinergic neuromodulation of cognition. Neuropsychopharmacology 36:52–73

Heckhausen H, Gollwitzer PM (1987) Thought contents and cognitive functioning in motivational versus volitional states of mind. Motiv Emot 11(2):101–120. https://doi.org/10.1007/bf00992338

Heckhausen J, Heckhausen H (2018) Motivation and action. Springer, Berlin

Heinz A, Rothenberg J (1998) Meddling with monkey metaphors—capitalism and the threat of impulsive desires. Soc Justice 25:44–64

Heinz AJ, Beck A, Meyer-Lindenberg A, Heinz A (2011) Cognitive and neurobiological mechanisms of alcohol-related aggression. Nat Rev Neurosci 12:400–413

Holly EN, Miczek KA (2016) Ventral tegmental area dopamine revisited: effects of acute and repeated stress. Psychopharmacology 233:163–186. https://doi.org/10.1007/s00213-015-4151-3

Human LJ, Thorson KR, Mendes WB (2016) Interactive effects between extraversion and oxytocin administration: implications for positive social processes. Soc Psychol Personal Sci 7:735–744

Ikeda S, Takeuchi H et al (2017) A comprehensive analysis of the correlations between resting state oscillations in multiple-frequency bands and big five traits. Front Hum Neurosci 11:321. https://doi.org/10.3389/fnhum.2017.00321

Kuhl J (2001) Motivation und Persönlichkeit. Hogrefe, Göttingen

Lavin C, Melis C, Mikulan E, Gelormini C, Huepe D, Ibañez A (2013) The anterior cingulate cortex: an integrative hub for human socially-driven interactions. Front Neurosci 7:64

LeDoux J (1996) The emotional brain: the mysterious underpinnings of emotional life. Simon and Schuster, New York

LeDoux JE (2017) Semantics, surplus meaning, and the science of fear. Trends Cogn Sci 21:303–306

Li Y, Vanni-Mercier G, Isnard J, Mauguière F, Dreher J-C (2016) The neural dynamics of reward value and risk coding in the human orbitofrontal cortex. Brain 139:1295–1309. https://doi.org/10.1093/brain/awv409

Luo AH, Tahsili-Fahadan P, Wise RA, Lupica SR, Aston-Jones G (2011) Linking context with reward. Science 333:353–357

Mann FD, Briley DA, Tucker-Drob EM, Harden KP (2017) A behavioral genetic analysis of callous-unemotional traits in big five personality in adolescence. J Abnorm Psychol 124:982–993

Meyer-Lindenberg A, Domes G, Kirsch P, Heinrichs M (2011) Oxytocin and vasopressin in the human brain: social neuropeptides for translational medicine. Nat Rev Neurosci 12:524–538

Miller KG, Wright AG, Peterson LM, Kamarck TW, Anderson BA, Kirschbaum C et al (2016) Trait positive and negative emotionality differentially associatewith diurnal cortisol activity. Psychoneuro 68:177–185

Mischel W, Shoda Y, Rodriguez ML (1989) Delay of gratification in children. Science 244:933–938

Misra G, Coombes S (2014) Neuroimaging of motor control and pain processing in the human midcingulate cortex. Cereb Cortex 25:1906–1919

Moffit TE, Caspi A (2001) Childhood predictors differentiate life-course persistent and adolescence-limited antisocial pathways among males and females. Dev Psychopathol 13:355–375

Myers D (2014) Neuropsychologie. Springer, Berlin

Neyer FJ, Asendorpf J (2018) Psychologie der Persönlichkeit. Springer, Berlin

Öhman A (1999) Distinguishing unconscious from conscious emotional processes: methodological considerations and theoretical implications. In: Dagleish T, Power MJ (eds) Handbook of cognition and emotion. Wiley, Chichester, pp 321–352

Ostendorf F, Angleitner A (2004) NEO-PI-R—NEO Persönlichkeitsinventar nach Costa & McCrae—Revidierte Fassung (PSYNDEX Tests Review). Hogrefe, Göttingen

Oswald LM, Zandi P, Nestadt G, Potash JB, Kalaydjian AE, Wan GS (2006) Relationship between cortisol responses to stress and Personality. Neuropsychopharmacology 31:583–1591

Panksepp J (1998) Affective neuroscience. The foundations of human and animal emotions. New York, Oxford University Press

Panksepp J, Lane RD, Solms M (2017) Reconciling cognitive and affective neuroscience perspectives on the brain basis of emotional experience. Neurosci Biobehav Rev 76:187–215

Pastoll H, Solanka L, van Rossum MC, Nolan MF (2013) Feedback inhibition enables θ-nested γ oscillations and grid firing fields. Neuron 77:141–154

Pastor-Berniera A, Plott CR, Schultz W (2017) Monkeys choose as if maximizing utility compatible with basic principles of revealed preference theory. Proc Natl Acad Sci U S A 114(10):E1766–E1775. https://doi.org/10.1073/pnas.1612010114

Pessoa L (2017) A network model of the emotional brain. Trends Cogn Sci 21:357–371

Pessoa L, Adolphs R (2010) Emotion processing and the amygdala: from a 'low road' to 'many roads' of evaluating biological significance. Nat Rev Neurosci 11:773–782

Power RA, Pluess M (2015) Heritability estimates of the Big Five personality traits based on common genetic variants. Transl Psychiatry 5:e604. https://doi.org/10.1038/tp.2015.96

Puca RM, Langens TA (2005) Motivation. In: Müsseler J, Prinz W (eds) Allgemeine Psychologie. Spektrum Akademischer, Heidelberg, pp 225–269

Ray RD, Zald DH (2012) Anatomical insights into the interaction of emotion and cognition in the prefrontal cortex. Neurosci Biobehav Rev 36:479–501

Roth G, Strüber N (2018) Wie das Gehirn die Seele macht. Klett-Cotta, Stuttgart

Russell JA (2009) Emotion, core affect, and psychological construction. Cogn Emot 23:1259–1283

Scherer KR (1999) Appraisal theory. In: Dagleish T, Power MJ (eds) Handbook of cognition and emotion. Wiley, Chichester, pp 637–663

Schultz W (2007) Behavioral dopamine signals. Trends Neurosci 30:203–210

Schultz W (2016) Reward functions of the basal ganglia. J Neural Transm 123:679–693. https://doi.org/10.1007/s00702-016-1510-0

Singer T, Seymour B, O'Doherty J, Kaube H, Dolan RJ, Frith CD (2004) Empathy for pain involves the affective but not sensory components of pain. Science 303:1157–1162

Singer T, Critchley HD, Preuschoff K (2009) A common role of insula in feelings, empathy and uncertainty. Trends Cogn Sci 13:334–340

Stauffer WR, Lak A, Kobayashi S, Schultz W (2016) Components and characteristics of the dopamine reward utility signal. J Comp Neurol 524:1699–1711. https://doi.org/10.1002/cne.23880

Steenhaut P, Rossi G, Demeyer I, De Raedt R (2018) How is personality related to well-being in older and younger adults? The role of psychological flexibility. Int Psychogeriatr:1–11. https://doi.org/10.1017/s1041610218001904

Stemmler G, Hagemann D, Amelang M, Spinath F (2016) Differentielle Psychologie und Persönlichkeitsforschung. Kohlhammer, Berlin

Thomas A, Chess S (1977) Temperament and development. Brunner/Mazel, New York

Toschi N, Riccelli R, Indovina I, Terracciano A, Passamonti L (2018) Functional connectome of the five-factor model of personality. Personal Neurosci. https://doi.org/10.1017/pen.2017.2

Tucker-Drob EM, Briley DA, Engelhardt LE, Mann FD, Harden KP (2016) Genetically-mediated associations between measures of childhood character and academic achievement. J Pers Soc Psychol 111:790–815. https://doi.org/10.1037/pspp0000098

Valentino RJ, van Bockstaele E (2008) Convergent regulation of locus coeruleusactivity as an adaptive response to stress. Eur J Pharmacol 583:194–203

Wang S, Zhou M, Chen T, Yang X, Chen G, Wang M, Gong Q (2017) Grit and the brain: spontaneous activity of the dorsomedial prefrontal cortex mediates the relationship between the trait grit and academic performance. Soc Cogn Affect Neurosci:452–460. https://doi.org/10.1093/scan/nsw145

Weiner B (1986) An attributional theory of motivation and emotion. Springer, New York

Wenzel JM, Cheer JF (2018) Endocannabinoid regulation of reward and reinforcement through interaction with dopamine and endogenous opioid signaling. Neuropsychopharmacol Rev 43:103–115

Neurobiological Consequences of Early Life Stress

Andrea J. J. Knop, Nora K. Moog, and Christine Heim

Contents

Andrea J.J. Knop and Nora K. Moog contributed equally to this work.

English Translation for:
Roth, Heinz, Walter (Eds.). Psychoneurowissenschaften. Springer-Verlag.

Trailer

The foundation for the development of a broad spectrum of stress-related disorders is laid early in development. Early-life stress, including prenatal stress exposure, increases individual susceptibility for the development of mental disorders and physical diseases across the lifespan. Persistent alterations in the brain as well as in endocrine, immune, and metabolic systems underlie this developmental programming of disease susceptibility. Thus, stress exposure in early developmental stages results in neurobiological traces or "scars" in the central nervous system that render individuals susceptible to developing a broad range of diseases throughout the lifespan. Recent evidence suggests that this risk can be passed on to subsequent generations. Genetic factors and the developmental timing of adverse exposures moderate the clinical and biological consequences of early-life stress as well as individual vulnerability to disease and course of disease. A better understanding of the neurobiological mechanisms that link exposure to early life stress with disease risk will allow for the identification of measurable parameters to help identify individuals at risk of disease and susceptibility to a specific intervention. A precise understanding of the processes of biological embedding of early life stress will further enable the development of mechanism-derived targets and time windows for interventions and prevention strategies.

Learning Objectives

This chapter summarizes current findings from human clinical studies investigating the mechanisms by which early life stress affects neurobiological systems, as well as regulatory outflow systems of the brain, and influences susceptibility for psychiatric disorders and a wide range or physical diseases. The reader will be introduced to the concept of developmental programming of disease vulnerability and neurobiological changes resulting from early-life stress.

7.1 Early-Life Stress

The foundation for health and disease is laid early in development. Early life stress (ELS) is one of the most potent and pervasive risk factors for the development of psychiatric disorders and predicts a broad spectrum of physical diseases and increased mortality across the lifespan (Gilbert et al. 2009; Grummitt et al. 2021). Early life stress involves adverse experiences during childhood, such as exposure to various forms of severe stressors, including parental loss, unstable family situations, inadequate parental care due to mental or physical illness, and poverty. The most salient form of ELS may arguably be maltreatment, which encompasses neglect of care or supervision (emotional and physical neglect) and emotional, physical, and sexual abuse in childhood.

Exposure to ELS is alarmingly common in our society. Globally, prevalence estimates for maltreatment range from 13% for sexual abuse to 36% for emotional abuse (Stoltenborgh et al. 2015).When other forms of ELS are considered, prevalence estimates rise to nearly 50% of children affected and various types of ELS often coexist (for an overview, see Heim et al. 2019).

Childhood Maltreatment

The Centers for Disease Control and Prevention (CDC) define childhood maltreatment as "any act or series of acts of commission (i.e., abuse) or omission (i.e., neglect) by a parent or other caregiver that results in harm, potential for harm, or threat of harm to a child". In this definition, harm to a child may not be the intended consequence of these acts, however, the act itself has to be deliberate and intentional (Leeb et al. 2008).

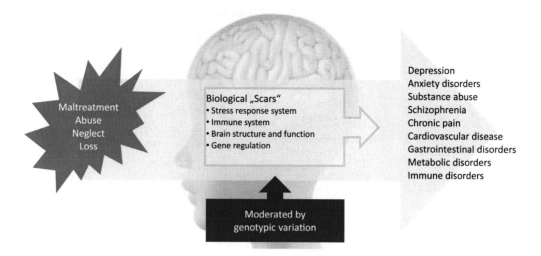

Fig. 7.1 ELS, biological effects and disease risk. (Reprinted from Heim et al. 2019)

Research demonstrates that ELS exerts pronounced effects on neural systems, as well endocrine, immune, and metabolic regulatory systems, that are fundamental to the organism's adaptation to stress. Changes in these systems that occur as a function of ELS may mediate increased risk for stress-related diseases (■ Fig. 7.1). The developing brain is particularly susceptible to the organizing effects of experiences. According to the concept of developmental programming, neural plasticity is particularly pronounced during early developmental periods (Lupien et al. 2009): During such times of heightened plasticity, positive and nurturing social-emotional experiences may be required for an optimal development of neural circuits that mediate adaptation to stress and emotional regulation, whereas any type of ELS occurring within such developmental periods may promote disruptions in the development of these circuits, leading to long-term "scars", which may result in maladaptive regulation upon further stress exposure an increased susceptibility for stress-related diseases across the lifespan.

> **Disease Vulnerability**
>
> An individual's susceptibility for developing disease across the lifespan. The extent of disease vulnerability may be dependent on whether the exposure occurs during critical periods of development during which ELS has a particularly strong and specific effect on the brain and its regulatory outflow systems. Vulnerability is dependent on interactions of ELS and genetic factors and such gene-environment interactions may be mediated by epigenetic programming (see Heim and Binder 2012).

The well-established link between ELS and increased risk for disease across the lifespan raises several questions: How does ELS exposure get "under the skin"? Which mechanisms mediate the increased long-term susceptibility to various stress-related diseases? Will each individual with a history of ELS develop some form of stress-related disease over the life course or do some individuals demonstrate resilience to the lasting consequences of ELS? Advances from neuroscience and molecular biology research have provided compelling answers to these questions, which are summarized in the following sections.

7.2   Clinical Consequences of ELS

Clinical and epidemiological research demonstrates a robust and substantial increase in the risk for both psychiatric disorders and physical diseases following ELS exposure (for an overview see Heim and Binder 2012). A particularly strong association can be found for affective and anxiety disorders, including post-traumatic stress disorder (PTSD). Further, there are established dose-response relationships between childhood adversity and psychiatric disorders in adulthood (Edwards et al. 2003). Early life stress is not only associated with increased prevalence rates of these disorders, but also predicts earlier onset, chronic course, and greater severity of disease as well as poor treatment response (see Heim and Nemeroff 2001; Nanni et al. 2012). Moreover, ELS is a consistent risk factor for suicidality across disorders (see Heim and Binder 2012). In addition to its adverse effects on mental health, ELS is associated with markedly increased risk for chronic physical diseases, including cardiovascular, immune-related and respiratory diseases, diabetes and obesity, chronic pain, and reduced longevity (Felitti et al. 1998; Norman et al. 2012; Shonkoff et al. 2012). Individuals with a history of ELS often exhibit multiple comorbid psychiatric and physical disorders, suggesting the existence of an ELS-related core "lesion" across neural and peripheral regulatory systems that promotes maladaptation and disease.

7.3   Long-Term Biological Consequences of ELS

The precise mechanisms that mediate the detrimental effects of ELS on disease risk have been subject to basic and clinical investigation over the past decades. Lasting effects of ELS on the brain and its regulatory outflow systems, including the autonomic, endocrine, immune, and metabolic systems, may lead to increased sensitivity to stress and risk for a range of psychiatric and physical diseases. Studies in animal models involving maternal separation or natural variation of the quality of maternal care provide causal evidence that ELS leads to structural and functional changes in neural circuits that are involved in the mediation of stress responses, autonomic and neuroendocrine control, emotion regulation, and fear conditioning. These neurobiological changes promote exaggerated behavioral and physiological reactivity to stressors later in life (stress sensitization). For instance, adult rodents exposed to maternal separation or naturally occurring low maternal care in early life exhibit hyperactivity of the central stress-mediating neuropeptide corticotrophin releasing hormone (CRH) system and sensitization of the hypothalamic-pituitary-adrenal (HPA) axis as well as behavioral responses reminiscent of symptoms of depression and anxiety (see Heim and Binder 2012; Heim et al. 2019).

In accordance with findings from animal models, adult women with a history of ELS exhibit markedly increased pituitary-adrenal and autonomic responses to psychosocial laboratory stress, induced by the Trier Social Stress Test (TSST). This effect was particularly pronounced in abused women with concurrent major depression. Neuroendocrine alterations following ELS were demonstrated at multiple levels of stress regulation, including reduced adrenal capacity and dysregulated negative feedback of the HPA axis due to relative glucocorticoid receptor resistance as measured with the dexamethasone/CRH challenge test. Furthermore, increased cerebrospinal fluid (CSF) concentrations of CRH and decreased CSF levels of the neuropeptide oxytocin were reported as a function of severity of ELS, suggesting that stress-mediating systems are upregulated as a consequence of ELS whereas stress-buffering neuropeptide systems are downregulated. Taken together, these findings suggest a sensitization of the endocrine and autonomic stress responses and a disturbed balance between stress-mediating and stress-protective neural systems after ELS exposure, converging into increased stress vulnerability (see Heim et al. 2008).

The neuroendocrine system is tightly linked to the immune system and systemic inflammation is one of the most replicated biological correlates of ELS. Adults exposed to ELS exhibit significantly increased plasma levels of interleukin-6 (IL-6) and C-reactive protein (CRP), particularly those with depression (Baumeister et al. 2016). Notably, studies in 12 year-old and 3- to 5-year-old children suggest that the effect of ELS on inflammation emerges already in childhood and in the immediate aftermath of exposure (Danese et al. 2011; Entringer et al. 2020). One potential pathway through which ELS can induce an increased release of these inflammatory mediators may involve the above-described dysfunction of the glucocorticoid receptor (GR), a key regulator of the immune response (see Raison et al. 2006). In addition, ELS is associated with metabolic dysregulation. Elevated inflammatory levels and cortisol secretion following ELS may lead to decreased sensitivity to insulin contributing to the development of metabolic disorders, such as type 2 diabetes or metabolic syndrome. Inflammatory processes and metabolic abnormalities may also promote atherosclerosis progression contributing to the development of cardiovascular disease (see Danese and McEwen 2012).

At the central nervous system level, exaggerated concentrations of cortisol or inflammatory cytokines may exert neurotoxic effects on brain structures that are implicated in stress and emotion regulation. During early developmental periods of pronounced plasticity, ELS may shape the development of these brain regions (◘ Table 7.1). Volumetric alterations as a function of ELS have been shown specifically in cortical and subcortical regions that are particularly sensitive to glucocorticoid exposure and hence vulnerable the detrimental effects of stress, including the hippocampus (see Teicher et al. 2016). The hippocampus is critically involved in contextual aspects of fear conditioning and one of the most plastic central regions, exhibiting a high degree of synaptic reorganization and neurogenesis across the lifespan. With a high density of GRs, the hippocampus exerts an inhibitory control of hypothalamic CRH neurons. Several magnetic resonance imaging (MRI) studies in adults have demonstrated a small hippocampal volume in

◘ **Table 7.1** Brain changes in relation to early life stress experience

Structural Changes
- Smaller hippocampus
- Altered amygdala volume
- Smaller prefrontal cortex and anterior cingulate cortex
- Smaller cerebellum
- Reduced cortical thickness or volume in sensory processing areas (visual cortex, auditory cortex, somatosensory cortex)
- Reduced structural connectivity (corpus callosum, cingulum, fornix, fasciculus arcuatus, fasciculus uncinatus, fasciculus longitudinalis superior)

Functional Changes
- Increased reactivity of the amygdala to emotional stimuli (especially threat)
- Reduced activation in the striatum during the expectation of reward
- Reduced functional frontal-limbic connectivity

association with ELS (hippocampal atrophy), suggesting a dysfunctional inhibition of the stress response. Region-specific investigations of the hippocampus with high-resolution imaging demonstrate a pronounced decrease in volume in the CA3 region, the dentate gyrus and the left subiculum in ELS-exposed individuals (Teicher et al. 2012). Furthermore, structural and functional changes in cortico-limbic circuits have been reported as a function of ELS. The prefrontal cortex (PFC) is critically involved in executive functioning, regulation of goal-directed behavior, and impulse inhibition. The medial PFC is particularly relevant for emotion regulation via structural connections to the anterior cingulate cortex (ACC) and the amygdala. ELS has consistently been associated with volume loss in the PFC, including the medial PFC and ACC (see Heim et al. 2019). In addition, neuroimaging studies suggest structural and functional changes of the amygdala following ELS. The amygdala plays a key role in fear conditioning, emotion processing, as well as evaluating potentially threatening stimuli and eliciting an appropriate stress response. Whereas findings on volumetric alterations of the amygdala are inconsistent, functional

MRI studies consistently demonstrate a sustained hyperactivity of the amygdala in response to emotionally threatening stimuli following ELS (Dannlowski et al. 2013). With the PFC exerting inhibitory and the amygdala exerting excitatory regulation of hypothalamic CRH neurons (Ulrich-Lai and Herman 2009), these findings may reflect a dysfunctional "top down" control of emotion regulation, fear conditioning, and stress responses that promotes disease risk.

Several studies suggest a specific impact of ELS on sensory representation areas implicated in the perception of the very nature of the abusive experience. Using whole mantle cortical thickness analysis in adults, we observed pronounced cortical thinning of the somatosensory genital field as a function of childhood sexual abuse. Emotional abuse was specifically associated with cortical thinning in the precuneus, a region that is relevant for self-awareness and self-evaluation, as well as thinning in the anterior cingulate cortex, which is relevant for emotional regulation (Heim et al. 2013; see ◘ Fig. 7.2). These findings suggest that experience-dependent plasticity leads to effects of ELS on sensory and associative

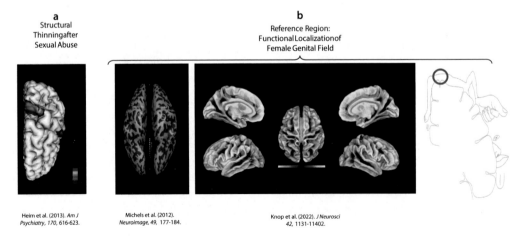

a
Structural
Thinning after
Sexual Abuse

b
Reference Region:
Functional Localization of
Female Genital Field

Heim et al. (2013). *Am J Psychiatry, 170*, 616-623.

Michels et al. (2012). *Neuroimage, 49*, 177-184.

Knop et al. (2022). *J Neurosci 42*, 1131-11402.

◘ **Fig. 7.2** Specific thinning of somatosensory genital field in adult women with childhood sexual abuse (**a**; Heim et al. 2013) and reference regions from sensory tactile functional imaging studies to localize the genital representation field (**b**; Michels et al. 2010; Knop et al. 2022)

processing areas in a highly region-specific manner. This specific cortical thinning in sensory processing areas may represent the most adaptive and protective response of the developing brain that may "shield" the child living under these conditions from the abusive experience, similar to sensory gating. In later life, these neurostructural changes may represent a direct biological substrate for behavioral disorders, such as sexual dysfunction. Of note, Teicher and colleagues report similar findings for other sensory modalities, including thinning of the visual cortex after witnessing domestic violence and thinning of the auditory cortex after verbal abuse (see Teicher et al. 2016).

7.4 Molecular Consequences of ELS: Epigenetic Programming and Telomere Biology

A major research question in the field of ELS research concerns the molecular effects by which early environmental exposures interact with genetic factors to produce a neurobiological phenotype with elevated vulnerability for stress and disease. In this context, epigenetic alterations as a mechanism underlying the biological embedding of the above-described stable physiologic and behavioral changes that result from ELS exposure have received much attention (Box: Epigenetic Programming, see Anacker et al. 2014).

Epigenetic Programming

The process by which the environment regulates DNA transcriptional activity without altering DNA sequence. Epigenetic changes are produced by DNA (de-) methylation, histone modifications, and non-coding RNAs (for an overview, see Jaenisch and Bird 2003).

Over the past decade, numerous studies have investigated whether exposure to ELS can leave persistent epigenetic marks in the genome, thereby altering gene expression and ultimately neurobiological substrates (Parade et al. 2021; Provençal and Binder 2015). Initial studies have typically employed a candidate gene approach, focusing on epigenetic variation in specific stress-relevant genes, including the GR gene, *NR3C1*, and *FKBP5*, a key modulator of glucocorticoid signaling. For instance, ELS has been associated with increased DNA methylation of the promoter region of *NR3C1* in the human postmortem hippocampus. This hypermethylation was associated with decreased NGFI-A transcription factor binding and reduced gene transcription (McGowan et al. 2009; Perroud et al. 2011). ELS has also been linked with DNA demethylation in functional glucocorticoid response elements of *FKBP5* in carriers of the "risk"-allele of a functional polymorphism within the *FKBP5* gene, resulting in an increased risk of developing affective disorders in adulthood (Klengel et al. 2013). More recently, research interest has shifted away from candidate gene approaches toward investigating epigenetic variation across the entire epigenome in a hypothesis-free design (Epigenome-Wide Association Studies, EWAS). EWAS have consistently demonstrated presence of methylation changes in individuals exposed to ELS (e.g., Cicchetti et al. 2016; O'Donnell et al. 2018), however, to date no clear patterns of functional enrichment within the epigenome-wide methylation profiles associated with ELS have emerged (Cecil et al. 2020; Parade et al. 2021). A recent study from our group suggests that epigenetic modifications associated with ELS may, in part, also reflect co-occurring concurrent adversities and prenatal exposures (Martins et al. 2021). Epigenetic marks can also be used to estimate a form of biological ageing that takes into account time-dependent changes in DNA methylation at specific sites in the genome. The difference between this epigenetic age and the chronological age can be interpreted as accelerated or decelerated biological ageing. Our group recently observed accelerated epigenetic ageing in children with internalizing problems, but only if they had been exposed to ELS, whereas no acceleration of ageing was observed in children with internalizing disorder who had no history of ELS, suggesting that these molecular alterations may already be present shortly after the exposure and are associated with health status (Dammering et al. 2021).

Cellular ageing is another molecular process and form of biological ageing that has the potential to mediate the ELS-associated increased risk of developing both physical and mental health problems across the lifespan. Cellular ageing is a function of the integrity of telomeres, DNA-protein complexes at the end of chromosomes promoting chromosomal stability (Box: Telomere Biology). Telomeres shorten with each cell replication cycle until a critical limit is reached and the cell enters a state of senescence or undergoes apoptosis. Telomere length can be maintained by the enzyme telomerase, however, in most somatic tissues, telomerase levels and activity are very low so that telomeres cannot be maintained indefinitely. Thus, telomere length can serve as a biomarker for biological ageing and time until senescence and research suggests that it may also be a risk marker for a variety of age-related diseases (e.g., cancer, hypertension) as well as shorter lifespan (Price et al. 2013).

Telomere Biology

A system that plays a central role in maintaining the integrity of the genome and the cell. Telomere biology refers to the structure and function of two related entities: telomeres, a complex of non-coding double-stranded repeats of guanine-rich DNA sequences and the shelterin protein structures that serve to protect the ends of chromosomes, as well as the enzyme telomerase, a ribonucleoprotein that adds telomeric DNA to the ends of chromosomes and thus elongates and maintains telomeres (for an overview, see Entringer et al. 2018).

In addition to chronological age, telomere length is also influenced by a wide range of environmental and behavioral factors, including stress. Chronic psychological stress, such as ELS, may shape the biochemical milieu of the cellular environment, promoting conditions of inflammation and oxidative stress which can lead to telomere damage (Barnes et al. 2019). The telomere system has been shown to play an important role in the development of depression and other psychiatric disorders (Ridout et al. 2016). As such, reduced telomere length and/or telomerase activity may be underlying, in part, the association between ELS and adverse mental and physical health outcomes (see Entringer et al. 2018).

In the last decade, numerous studies have provided evidence linking ELS with reduced telomere length (see meta-analysis by Ridout et al. 2018). The developmental timing of ELS exposure appears to influence the size of the effect, with adversity earlier in development showing greater negative associations with telomere length (Ridout et al. 2018). These findings are consistent with the neurobiological findings in that they suggest that early childhood seems to be a particularly sensitive period in which stress exposure can exert long-lasting effects on various systems, including the telomere system.

7.5 Sensitive Periods for the Effects of ELS

The above-referenced findings give rise to the question as to whether or not there are sensitive periods during human childhood, where the brain is particularly sensitive to the environment, including the effects of ELS. Of note, one important factor that max contribute to variability of the outcomes of ELS between individuals may depend on the developmental timing of the ELS exposure within childhood. In other words, there may exist discrete sensitive periods during development, during which adverse exposures may be specifically detrimental, leading to longterm change. Little is known to date about such circumscribed time windows for the effects of ELS. Teicher and colleagues have conducted statistical analyses to identify sensitive periods for the effects of ELS on brain regional development. They report that the amygdala is particularly sensitive to the effects of abuse at the age of around 10 years, whereas the hippocampus has heightened sensitivity at an earlier age and the prefrontal cortex seems to be particularly amenable around puberty (see Teicher et al. 2016). Whether or not such temporally differential effects of ELS on brain regions are associated with specific symptom constellations remains poorly understood. A precise identification of sensitive periods for the effects of ELS may enable the development of timing-specific interventions that make use of sensitive periods to induce positive change with lasting effects. Further, a precise understanding of the mechanisms that determine the opening and closing of sensitive time windows of developmental plasticity may have therapeutic benefit, enabling the development of novel pharmacological targets for augmenting effects of psychotherapy by increasing plasticity. Of note, rodent models of maternal separation or naturally occurring low care typically focus on the first 2 weeks of life, which developmentally corresponds to fetal life in humans. Hence, it is conceivable that prenatal stress has profound impact on adult health and adaptation in humans, as discussed next.

7.6 Fetal Programming of Health and Disease

While most research has been conducted on ELS occurring postnatally, the intrauterine period of life represents another sensitive developmental window during which stress exposure can have long-term or even permanent consequences for health and disease susceptibility (Box: Fetal Programming; for an overview see Entringer et al. 2015).

Fetal Programming

The concept of fetal programming describes the process by which the embryo/fetus seeks, receives, and responds to signals from the gestational environment to incorporate this information into its development. The concept is based on the assumption that the rapid and foundational nature of developmental processes occurring during intrauterine life render them particularly vulnerable to environmental perturbations with lifelong consequences for disease susceptibility (see Entringer et al. 2015; Gluckman and Hanson 2004). The research field has its origins in a set of epidemiological studies, demonstrating that a person's birth weight is associated with the risk for cardiovascular disease, as well as a variety of other conditions, including depression, obesity, and diabetes, later in life (see Wadhwa et al. 2009). In this context, birth weight is assumed to reflect the developmental environment in the womb, which, in interaction with the genetic make-up, elicits context-dependent and long-term adaptations in cells, tissues, organ systems and homeostatic set points of the developing embryo/fetus.

In recent years, an impressive body of research has collected evidence suggesting that maternal psychosocial stress and emotional state during pregnancy may affect child neurodevelopment as well as social-emotional and cognitive development and thus may increase risk for a variety of adverse physical and mental health outcomes, including mood disorders, attention problems, asthma, and obesity (Entringer et al. 2015; Lautarescu et al. 2020; Madigan et al. 2018). Since there are no direct vascular or neural connections between the maternal and fetal compartments all exchange of signals and communication is mediated by biological processes via the placenta. Stress-related biological processes appear to play a role as key sensors, transducers and effectors of maternal stress on the developing fetus. It is important to note that these stress-related biological mediators participate directly or indirectly in the process of phenotypic specification of the brain and other organ systems and should not be considered as developmental disruptors.

Several studies characterize alterations in brain structure and function in association with exposure to maternal stress-related biological mediators. For instance, elevated maternal cortisol concentrations have been associated with amygdala volume, microstructure and connectivity in newborns and children, with consequences for internalizing and affective problems (Buss et al. 2012; Graham et al. 2019). The pro-inflammatory cytokine interleukin-6 (IL-6) is another potential stress-related biological mediator of environmental conditions with an important role in fetal brain development. Several studies support a link between maternal IL-6 concentrations and offspring amygdala volume, structural and functional connectivity as well as connectivity within and between networks involved in sensory processing and higher order cognition (see Heim et al. 2019).

Maternal cortisol concentrations during pregnancy have also been repeatedly associated with offspring body composition and adiposity (Entringer et al. 2017; Van Dijk et al. 2012), suggesting that cortisol may also be involved in the prenatal programming of metabolism. Alterations in metabolic function appear to be present very early in postnatal life as demonstrated by one study showing that cortisol production particularly during the third trimester of pregnancy was associated with a greater change in infant percent body fat from 1 to 6 months assessed with Dual-energy X-ray absorptiometry imaging (Entringer et al. 2017).

Furthermore, prenatal stress exposure seems to have an effect on the child's telomere length. Maternal psychosocial stress during pregnancy has been associated with shortened offspring telomeres in young adulthood and in the newborn period, whereas maternal psychological resiliency during pregnancy has been linked to increased newborn telomere length (for an overview, see Heim et al. 2019; Verner et al. 2021). The effects of maternal stress on fetal telomere biology may be mediated by alterations in gestational biology, including increased cortisol concentrations and a higher pro-inflammatory milieu (Bosquet Enlow et al. 2019; Lazarides et al. 2019). The initial telomere length at birth may have lifelong implications for telomere biology and health, the precise relationship of which remains to be explored.

Taken together, this evidence suggests that in utero exposure to maternal psychosocial stress may confer increased long-term risk of a range of negative health outcomes mediated by adaptations in organ systems during intrauterine development in response to stress-related biological signals. The presence of alterations in brain structure and function and in body composition in association with maternal biological mediators of stress so close to birth provides compelling evidence for a programming effect of the gestational environment that is independent of postnatal environmental influences.

7.7 Intergenerational Transmission of the Effects of Early Life Stress

Over the last decade, the notion that the deleterious consequences of ELS may be transmitted across generations has received increasing attention. A steadily growing body of studies have explored the effects of ELS exposure in the parental generation on neurodevelopmental outcomes in the offspring generation and observed many of the same sequelae that are well-established consequences of ELS in exposed individuals (Box: Intergenerational Transmission, Buss et al. 2017; Moog et al. 2022).

Empirical evidence points to an increased risk for a range of behavioral and emotional problems, including internalizing and externalizing problems, conduct disorder, self-regulation difficulties, and anxiety disorders as well as neurodevelopmental disorders, including autism and attention-deficit hyperactivity disorder in children of mothers who experienced ELS. In addition, maternal ELS is associated with an increased offspring risk of developing physical health problems and risk factors, including asthma, allergy and obesity (see Moog et al. 2022). As in the directly exposed generation, structural and functional alterations in the brain as well as dysregulations in autonomic, endocrine and immune system may be underlying these mental and physical health problems. We showed that infants of mothers with experiences of ELS had less cortical gray matter, which contributed to an overall lower brain volume compared to infants of mothers without ELS (Moog et al. 2018).

Intergenerational Transmission

The contribution of parental experiences and exposures (e.g., of early life stress) in shaping the development and phenotype of the offspring.

The mechanisms underlying the intergenerational transmission of ELS effects have not been fully clarified. One potential mechanism that has been investigated mainly in animal models is epigenetic inheritance. The term epigenetic inheritance refers to germline transmission of epigenetic information between generations independent of the DNA sequence, either through direct transfer of epigenetic marks or through reconstruction and reestablishment of germline epigenetic alterations in the zygote. As reviewed in ▶ Sect. 7.4, ELS exposure has been associated with persistent epigenetic alterations in certain tissues, including the germline in humans (Provençal and Binder 2015; Roberts et al. 2018). However, while epigenetic transmission of paternal ELS has been demonstrated in animal models (Gapp et al. 2020), conclusive evidence

for the existence of epigenetic inheritance in humans is still lacking. Another pathway that has been debated is based on the observation that ELS-related physiological dysregulations in the stress, immune and metabolic systems may be carried forward into pregnancy and thus affect child health outcomes via a fetal programming mechanism (see ▶ Sect. 7.6; for an overview see Moog et al. 2022). Lastly, a caregiving environment characterized by maternal psychopathology or difficulty providing high-quality parenting due to ELS-related personal, social and socio-economic constraints is another important potential mediator of the intergenerational transmission of ELS (Plant et al. 2018).

7.8 Gene-Environment (GxE) Interactions

While the increase in disease risk in association with ELS is substantial and alarming, not all children exposed to ELS go on to develop stress-related disorders, even if additional stressors occur later in life. Genetic factors likely play a moderating role in the extent to which environmental conditions, such as ELS exposure, may program neurobiological structures and functions as well as in the individual susceptibility versus resilience to developing stress-related disorders. A variety of studies have identified candidate genes in stress-regulatory systems that moderate the link between ELS and risk for depression and other disorders, including the CRH receptor 1 (*CRHR1*), GR (*NR3C1*), the GR-regulating FK506 binding protein 5 (*FKBP5*), serotonin transporter (*SLC6A4*), brain-derived neurotrophic factor (*BDNF*), and oxytocin receptor (*OXTR*) genes (for an overview, see Heim and Binder 2012). Notably, most functional polymorphisms in these genes that confer risk for depression in combination with ELS are associated with GR resistance or enhanced stress hormone system activity. However, there has been increasing awareness that multiple gene variants likely work together to shape disease risk, such that in recent years the field has moved towards employing polygenic approaches. Polygenic risk scores incorporate the contributions of many common genetic variants and are derived from genome-wide association studies (for an overview, see Halldorsdottir and Binder 2017). For instance, a polygenic risk factor for major depressive disorder has been demonstrated to moderate the association between ELS and depression (Peyrot et al. 2014). In the future, a deeper and comprehensive understanding of GxE interactions is critically important to identify cases that are vulnerable to the pathogenic effects of ELS and require preventive intervention. Genetic markers may also be used to develop more targeted treatments, which will be discussed in the following section.

7.9 Implications for Intervention

Research findings on ELS as a risk factor for a wide range of diseases, as reviewed in this chapter, have the potential to inform the development of novel intervention strategies that target different aspects of biological embedding, disease manifestation and transmission of disease risk across generations (see Heim et al. 2019). Results from our studies demonstrating neurobiological differences and differential responses to drug or psychotherapy as a function of ELS suggest that developmental factors should be included in individual treatment decisions for patients with affective disorders. The development of algorithms based on biomarkers, genetic factors and symptom constellations could lead to personalized interventions as a form of precision medicine in the field of psychiatry. However, we propose that it is even more efficient to take advantage of the high levels of plasticity

during early development and intervene before the clinical manifestation of disease to counteract, reverse or compensate biological "scars" of ELS. These interventions target the underlying mechanisms rather than symptoms and may involve "top down" compensatory regulation of the altered neural and physiological systems (e.g., via psychotherapeutic interventions starting as soon as possible after the exposure) or "bottom up" approaches directly counteracting biological embedding effects of ELS (e.g., FKBP51 antagonists). Furthermore, a better understanding of the molecular mechanisms that determine sensitive periods of brain development may enable the development of entirely novel treatment approaches that restore such a state of increased plasticity in order to reverse the programming effects of ELS (Bavelier et al. 2010). It is important to note, however, that biological alterations may not be detrimental per se but can represent adaptations that confer short-term benefits to help the system function under conditions characterized by ELS. Our research on the intergenerational transmission of ELS-related risk suggest that interventions during pregnancy that target gestational physiology, stress reduction, trauma coping and different forms of social and personal constraints that can be consequences of ELS-exposure could minimize the intergenerational effects on the offspring.

7.10 Conclusion

In sum, ELS is a profound and nonspecific risk factor for a wide range of diseases. Exposure to ELS during sensitive developmental periods, including the prenatal period, appears to lead to immediate processes of biological embedding. This biological embedding of ELS may involve epigenetic modifications in stress-regulatory genes, with subsequent dysregulation of endocrine and immune stress response systems, metabolic dysregulation, structural and functional changes in brain regions regulating stress and emotion, as well as accelerated biological ageing. These physiological alterations may lead to manifestation of adverse mental and physical health outcomes, depending also on the presence of additional stressors later in life as well as genetic factors. The phenotypic consequences of ELS may be transmitted into the next generation, thereby multiplying the number of affected individuals. Existing and future research will inform novel approaches that make use of developmental plasticity in order to promote optimal development, health, and longevity in all children.

Acknowledgement This work was funded by the Deutsche Forschungsgemeinschaft (DFG, German Research Foundation) under Germany's Excellence Strategy—EXC-2049—390688087 (to CH) and DFG-441735381 (to NM) and a scholarship from the Einstein Center for Neuroscience Berlin (to AJJK).

References

Anacker C, O'Donnell KJ, Meaney MJ (2014) Early life adversity and the epigenetic programming of hypothalamic-pituitary-adrenal function. Dialogues Clin Neurosci 16(3):321–333

Barnes RP, Fouquerel E, Opresko PL (2019) The impact of oxidative DNA damage and stress on telomere homeostasis. Mech Ageing Dev 177:37–45. https://doi.org/10.1016/j.mad.2018.03.013

Baumeister D, Akhtar R, Ciufolini S, Pariante CM, Mondelli V (2016) Childhood trauma and adulthood inflammation: a meta-analysis of peripheral C-reactive protein, interleukin-6 and tumour necrosis factor-α. Mol Psychiatry 21(5):Article 5. https://doi.org/10.1038/mp.2015.67

Bavelier D, Levi DM, Li RW, Dan Y, Hensch TK (2010) Removing brakes on adult brain plasticity: from molecular to behavioral interventions. J Neurosci 30(45):14964–14971. https://doi.org/10.1523/JNEUROSCI.4812-10.2010

Bosquet Enlow M, Sideridis G, Bollati V, Hoxha M, Hacker MR, Wright RJ (2019) Maternal cortisol output in pregnancy and newborn telomere length: evidence for sex-specific effects. Psychoneuroendocrinology 102:225–235. https://doi.org/10.1016/j.psyneuen.2018.12.222

Buss C, Poggi Davis E, Shahbaba B, Pruessner JC, Head K, Sandman CA (2012) Maternal cortisol over the course of pregnancy and subsequent child amygdala and hippocampus volumes and affective problems. Proc Natl Acad Sci U S A 109:E1312–E1319. https://doi.org/10.1073/pnas.1201295109

Buss C, Entringer S, Moog NK, Toepfer P, Fair DA, Simhan HN, Heim CM, Wadhwa PD (2017) Intergenerational transmission of maternal childhood maltreatment exposure: implications for fetal brain development. J Am Acad Child Adolesc Psychiatry 56(5):373–382. https://doi.org/10.1016/j.jaac.2017.03.001

Cecil CAM, Zhang Y, Nolte T (2020) Childhood maltreatment and DNA methylation: a systematic review. Neurosci Biobehav Rev 112:392–409. https://doi.org/10.1016/j.neubiorev.2020.02.019

Cicchetti D, Hetzel S, Rogosch FA, Handley ED, Toth SL (2016) An investigation of child maltreatment and epigenetic mechanisms of mental and physical health risk. Dev Psychopathol 28(4pt2):1305–1317. https://doi.org/10.1017/S0954579416000869

Dammering F, Martins J, Dittrich K, Czamara D, Rex-Haffner M, Overfeld J, de Punder K, Buss C, Entringer S, Winter SM, Binder EB, Heim C (2021) The pediatric buccal epigenetic clock identifies significant ageing acceleration in children with internalizing disorder and maltreatment exposure. Neurobiol Stress 15:100394. https://doi.org/10.1016/j.ynstr.2021.100394

Danese A, McEwen BS (2012) Adverse childhood experiences, allostasis, allostatic load, and age-related disease. Physiol Behav 106(1):29–39. https://doi.org/10.1016/j.physbeh.2011.08.019

Danese A, Ouellet-Morin I, Arseneault L (2011) Elevated salivary C-reactive protein in maltreated children. Brain Behav Immun 25:S210. https://doi.org/10.1016/j.bbi.2011.07.110

Dannlowski U, Kugel H, Huber F, Stuhrmann A, Redlich R, Grotegerd D, Dohm K, Sehlmeyer C, Konrad C, Baune BT, Arolt V, Heindel W, Zwitserlood P, Suslow T (2013) Childhood maltreatment is associated with an automatic negative emotion processing bias in the amygdala. Hum Brain Mapp 34(11):2899–2909. https://doi.org/10.1002/hbm.22112

Edwards VJ, Holden GW, Felitti VJ, Anda RF (2003) Relationship between multiple forms of childhood maltreatment and adult mental health in community respondents: results from the adverse childhood experiences study. Am J Psychiatr 160(8):1453–1460. https://doi.org/10.1176/appi.ajp.160.8.1453

Entringer S, Buss C, Rasmussen JM, Lindsay K, Gillen DL, Cooper DM, Wadhwa PD (2017) Maternal cortisol during pregnancy and infant adiposity: a prospective investigation. J Clin Endocrinol Metabol 102(4):1366–1374. https://doi.org/10.1210/jc.2016-3025

Entringer S, Buss C, Wadhwa PD (2015) Prenatal stress, development, health and disease risk: a psychobiological perspective-2015 Curt Richter Award Paper. Psychoneuroendocrinology 62:366–375.   https://doi.org/10.1016/j.psyneuen.2015.08.019

Entringer S, de Punder K, Buss C, Wadhwa PD (2018) The fetal programming of telomere biology hypothesis: an update. Philos Trans R Soc Lond Ser B Biol Sci 373(1741):20170151. https://doi.org/10.1098/rstb.2017.0151

Entringer S, de Punder K, Overfeld J, Karaboycheva G, Dittrich K, Buss C, Winter SM, Binder EB, Heim C (2020) Immediate and longitudinal effects of maltreatment on systemic inflammation in young children. Dev Psychopathol 32(5):1725–1731. https://doi.org/10.1017/S0954579420001686

Felitti VJ, Anda RF, Nordenberg D, Williamson DF, Spitz AM, Edwards V, Koss MP, Marks JS (1998) Relationship of childhood abuse and household dysfunction to many of the leading causes of death in adults: the Adverse Childhood Experiences (ACE) study. Am J Prev Med 14(4):245–258. https://doi.org/10.1016/S0749-3797(98)00017-8

Gapp K, van Steenwyk G, Germain PL, Matsushima W, Rudolph KLM, Manuella F, Roszkowski M, Vernaz G, Ghosh T, Pelczar P, Mansuy IM, Miska EA (2020) Alterations in sperm long RNA contribute to the epigenetic inheritance of the effects of postnatal trauma. Mol Psychiatry 25(9):Article 9. https://doi.org/10.1038/s41380-018-0271-6

Gilbert R, Widom CS, Browne K, Fergusson D, Webb E, Janson S (2009) Burden and consequences of child maltreatment in high-income countries. Lancet 373(9657):68–81. https://doi.org/10.1016/S0140-6736(08)61706-7

Gluckman PD, Hanson MA (2004) Developmental origins of disease paradigm: a mechanistic and evolutionary perspective. Pediatr Res 56(3):3. https://doi.org/10.1203/01.PDR.0000135998.08025.FB

Graham AM, Rasmussen JM, Entringer S, Ward EB, Rudolph MD, Gilmore JH, Styner M, Wadhwa PD, Fair DA, Buss C (2019) Maternal cortisol concentrations during pregnancy and sex-specific associations with neonatal amygdala connectivity and emerging internalizing behaviors. Biol Psychiatry   85(2):172–181.   https://doi.org/10.1016/j.biopsych.2018.06.023

Grummitt LR, Kreski NT, Kim SG, Platt J, Keyes KM, McLaughlin KA (2021) Association of childhood adversity with morbidity and mortality in US adults: a systematic review. JAMA Pediatr 175(12):1269–1278. https://doi.org/10.1001/jamapediatrics.2021.2320

Halldorsdottir T, Binder EB (2017) Gene × environment interactions: from molecular mechanisms to behavior. Annu Rev Psychol 68(1):215–241. https://doi.org/10.1146/annurev-psych-010416-044053

Heim C, Binder EB (2012) Current research trends in early life stress and depression: review of human studies on sensitive periods, gene–environment interactions, and epigenetics. Exp Neurol 233(1):102–111. https://doi.org/10.1016/j.expneurol.2011.10.032

Heim CM, Entringer S, Buss C (2019) Translating basic research knowledge on the biological embedding of early-life stress into novel approaches for the developmental programming of lifelong health. Psychoneuroendocrinology 105:123–137. https://doi.org/10.1016/j.psyneuen.2018.12.011

Heim CM, Mayberg HS, Mletzko T, Nemeroff CB, Pruessner JC (2013) Decreased cortical representation of genital somatosensory field after childhood sexual abuse. Am J Psychiatry 170(6):616–623. https://doi.org/10.1176/appi.ajp.2013.12070950

Heim C, Nemeroff CB (2001) The role of childhood trauma in the neurobiology of mood and anxiety disorders: preclinical and clinical studies. Biol Psychiatry   49(12):1023–1039.   https://doi.org/10.1016/s0006-3223(01)01157-x

Heim C, Newport DJ, Mletzko T, Miller AH, Nemeroff CB (2008) The link between childhood trauma and depression: insights from HPA axis studies in humans. Psychoneuroendocrinology 33(6):693–710. https://doi.org/10.1016/j.psyneuen.2008.03.008

Jaenisch R, Bird A (2003) Epigenetic regulation of gene expression: how the genome integrates intrinsic and environmental signals. Nat Genet 33(Suppl):245–254. https://doi.org/10.1038/ng1089

Klengel T, Mehta D, Anacker C, Rex-Haffner M, Pruessner JC, Pariante CM, Pace TW, Mercer KB, Mayberg HS, Bradley B, Nemeroff CB, Holsboer F, Heim CM, Ressler KJ, Rein T, Binder EB (2013) Allele-specific FKBP5 DNA demethylation mediates gene-childhood trauma interactions. Nat Neurosci 16(1):33–41. https://doi.org/10.1038/nn.3275

Knop A, Spengler S, Bogler C, Forster C, Brecht M, Haynes JD, Heim C (2022) Sensory-tactile functional mapping and use-associated structural variation of the human female genital representation field. J Neurosci 42:1131–11402. https://doi.org/10.1523/JNEUROSCI.1081-21.2021

Lautarescu A, Craig MC, Glover V (2020) Prenatal stress: effects on fetal and child brain development. In: International Review of Neurobiology, vol 150. Elsevier, pp 17–40. https://doi.org/10.1016/bs.irn.2019.11.002

Lazarides C, Epel ES, Lin J, Blackburn EH, Voelkle MC, Buss C, Simhan HN, Wadhwa PD, Entringer S (2019) Maternal pro-inflammatory state during pregnancy and newborn leukocyte telomere length: a prospective investigation. Brain Behav Immun 80:419–426. https://doi.org/10.1016/j.bbi.2019.04.021

Leeb, R., Paulozzi, L., Melanson, C., Simon, T., & Arias, I. (2008). Child maltreatment surveillance: uniform definitions for public health and recommended data elements. Centers for Disease Control and Prevention (CDC)

Lupien SJ, McEwen BS, Gunnar MR, Heim C (2009) Effects of stress throughout the lifespan on the brain, behaviour and cognition. Nat Rev Neurosci 10(6):6. https://doi.org/10.1038/nrn2639

Madigan S, Oatley H, Racine N, Fearon RMP, Schumacher L, Akbari E, Cooke JE, Tarabulsy GM (2018) A meta-analysis of maternal prenatal depression and anxiety on child socioemotional development. J Am Acad Child Adolesc Psychiatry 57(9):645–657.e8. https://doi.org/10.1016/j.jaac.2018.06.012

Martins J, Czamara D, Sauer S, Rex-Haffner M, Dittrich K, Dörr P, de Punder K, Overfeld J, Knop A, Dammering F, Entringer S, Winter SM, Buss C, Heim C, Binder EB (2021) Childhood adversity correlates with stable changes in DNA methylation trajectories in children and converges with epigenetic signatures of prenatal stress. Neurobiol Stress 15:100336. https://doi.org/10.1016/j.ynstr.2021.100336

McGowan PO, Sasaki A, D'Alessio AC, Dymov S, Labonté B, Szyf M, Turecki G, Meaney MJ (2009) Epigenetic regulation of the glucocorticoid receptor in human brain associates with childhood abuse. Nat Neurosci 12(3):342–348. https://doi.org/10.1038/nn.2270

Michels L, Mehnert U, Boy S, Schurch B, Kollias S (2010) The somatosensory representation of the human clitoris: an fMRI study. NeuroImage 49:177–184. https://doi.org/10.1016/j.neuroimage.2009.07.024

Moog NK, Entringer S, Rasmussen JM, Styner M, Gilmore JH, Kathmann N, Heim CM, Wadhwa PD, Buss C (2018) Intergenerational effect of maternal exposure to childhood maltreatment on newborn brain anatomy. Biol Psychiatry 83(2):120–127. https://doi.org/10.1016/j.biopsych.2017.07.009

Moog NK, Heim CM, Entringer S, Simhan HN, Wadhwa PD, Buss C (2022) Transmission of the adverse consequences of childhood maltreatment across generations: focus on gestational biology. Pharmacol Biochem Behav 215:173372. https://doi.org/10.1016/j.pbb.2022.173372

Nanni V, Uher R, Danese A (2012) Childhood maltreatment predicts unfavorable course of illness and treatment outcome in depression: a meta-analysis. Am J Psychiatr 169(2):141–151. https://doi.org/10.1176/appi.ajp.2011.11020335

Norman RE, Byambaa M, De R, Butchart A, Scott J, Vos T (2012) The long-term health consequences of child physical abuse, emotional abuse, and neglect: a systematic review and meta-analysis. PLoS Med 9(11):e1001349. https://doi.org/10.1371/journal.pmed.1001349

O'Donnell KJ, Chen L, MacIsaac JL, McEwen LM, Nguyen T, Beckmann K, Zhu Y, Chen LM, Brooks-Gunn J, Goldman D, Grigorenko EL, Leckman JF, Diorio J, Karnani N, Olds DL, Holbrook JD, Kobor MS, Meaney MJ (2018) DNA methylome variation in a perinatal nurse-visitation program that reduces child maltreatment: a 27-year follow-up. Transl Psychiatry 8(1):15. https://doi.org/10.1038/s41398-017-0063-9

Parade SH, Huffhines L, Daniels TE, Stroud LR, Nugent NR, Tyrka AR (2021) A systematic review of childhood maltreatment and DNA methylation: candidate gene and epigenome-wide approaches. Transl Psychiatry 11(1):1–33. https://doi.org/10.1038/s41398-021-01207-y

Perroud N, Paoloni-Giacobino A, Prada P, Olié E, Salzmann A, Nicastro R, Guillaume S, Mouthon D, Stouder C, Dieben K, Huguelet P, Courtet P, Malafosse A (2011) Increased methylation of glucocorticoid receptor gene (NR3C1) in adults with a history of childhood maltreatment: a link with the severity and type of trauma. Translational. Psychiatry 1(12):12. https://doi.org/10.1038/tp.2011.60

Peyrot WJ, Milaneschi Y, Abdellaoui A, Sullivan PF, Hottenga JJ, Boomsma DI, Penninx BWJH (2014) Effect of polygenic risk scores on depression in childhood trauma. Br J Psychiatry J Ment Sci 205(2):113–119. https://doi.org/10.1192/bjp.bp.113.143081

Plant DT, Pawlby S, Pariante CM, Jones FW (2018) When one childhood meets another—maternal childhood trauma and offspring child psychopathology: a systematic review. Clin Child Psychol Psychiatry 23(3):483–500. https://doi.org/10.1177/1359104517742186

Price LH, Kao H-T, Burgers DE, Carpenter LL, Tyrka AR (2013) Telomeres and early-life stress: an overview. Biol Psychiatry 73(1):15–23. https://doi.org/10.1016/j.biopsych.2012.06.025

Provençal N, Binder EB (2015) The effects of early life stress on the epigenome: from the womb to adulthood and even before. Exp Neurol 268:10–20. https://doi.org/10.1016/j.expneurol.2014.09.001

Raison CL, Capuron L, Miller AH (2006) Cytokines sing the blues: inflammation and the pathogenesis of depression. Trends Immunol 27(1):24–31. https://doi.org/10.1016/j.it.2005.11.006

Ridout KK, Levandowski M, Ridout SJ, Gantz L, Goonan K, Palermo D, Price LH, Tyrka AR (2018) Early life adversity and telomere length: a meta-analysis. Mol Psychiatry 23(4):Article 4. https://doi.org/10.1038/mp.2017.26

Ridout KK, Ridout SJ, Price LH, Sen S, Tyrka AR (2016) Depression and telomere length: a meta-analysis. J Affect Disord 191:237–247. https://doi.org/10.1016/j.jad.2015.11.052

Roberts AL, Gladish N, Gatev E, Jones MJ, Chen Y, MacIsaac JL, Tworoger SS, Austin SB, Tanrikut C, Chavarro JE, Baccarelli AA, Kobor MS (2018) Exposure to childhood abuse is associated with human sperm DNA methylation. Translational. Psychiatry 8(1):1. https://doi.org/10.1038/s41398-018-0252-1

Shonkoff JP, Garner AS, The Committee on Psychosocial Aspects of Child and Family Health, C. on E. C, Siegel BS, Dobbins MI, Earls MF, Garner AS, McGuinn L, Pascoe J, Wood DL (2012) The lifelong effects of early childhood adversity and toxic stress. Pediatrics 129(1):e232–e246. https://doi.org/10.1542/peds.2011-2663

Stoltenborgh M, Bakermans-Kranenburg MJ, Alink LRA, van Ijzendoorn MH (2015) The prevalence of child maltreatment across the globe: review of a series of meta-analyses. Child Abuse Rev 24(1):37–50. https://doi.org/10.1002/car.2353

Teicher MH, Anderson CM, Polcari A (2012) Childhood maltreatment is associated with reduced volume in the hippocampal subfields CA3, dentate gyrus, and subiculum. Proc Natl Acad Sci 109(9):E563–E572. https://doi.org/10.1073/pnas.1115396109

Teicher MH, Samson JA, Anderson CM, Ohashi K (2016) The effects of childhood maltreatment on brain structure, function and connectivity. Nat Rev Neurosci 17(10):652–666. https://doi.org/10.1038/nrn.2016.111

Ulrich-Lai YM, Herman JP (2009) Neural regulation of endocrine and autonomic stress responses. Nat Rev Neurosci 10(6):397–409. https://doi.org/10.1038/nrn2647

Van Dijk AE, Van Eijsden M, Stronks K, Gemke RJBJ, Vrijkotte TGM (2012) The relation of maternal job strain and cortisol levels during early pregnancy with body composition later in the 5-year-old child: the ABCD study. Early Hum Dev 88(6):351–356. https://doi.org/10.1016/j.earlhumdev.2011.09.009

Verner G, Epel E, Lahti-Pulkkinen M, Kajantie E, Buss C, Lin J, Blackburn E, Räikkönen K, Wadhwa PD, Entringer S (2021) Maternal psychological resilience during pregnancy and newborn telomere length: a prospective study. Am J Psychiatr 178(2):183–192. https://doi.org/10.1176/appi.ajp.2020.19101003

Wadhwa PD, Buss C, Entringer S, Swanson JM (2009) Developmental origins of health and disease: brief history of the approach and current focus on epigenetic mechanisms. Semin Reprod Med 27(5):358–368. https://doi.org/10.1055/s-0029-1237424

7

Psychological and Neurobiological Foundations of Consciousness

John-Dylan Haynes

Contents

G. Roth et al. (eds.), *Psychoneuroscience*, https://doi.org/10.1007/978-3-662-65774-4_8

This chapter deals with a highly popular and at the same time fiercely controversial psycho-neurobiological topic, namely consciousness and its possible neural foundations. In the following, we will set aside the philosophical dimensions of this topic, even though we appreciate their importance, and concentrate on approaches that are based on empirical psychological and neurobiological studies. Research on the neural correlates of consciousness seeks to determine what happens differently in the brain when a stimulus is consciously recognized. What differences characterize the neural processing of consciously perceived stimuli compared to stimuli that do not reach consciousness? Is there evidence that stimuli do not reach consciousness but are still processed by the brain? Can the brain decide what to be consciously aware of and what not? This chapter will deal with these and similar questions.

Learning Objectives

After reading this chapter, you should have gained a deeper insight into the findings of empirical consciousness research. This includes knowledge of the applied psychological and neuroscientific methods, as well as the experimental distinction between unconscious and conscious perceptual and cognitive performance. Likewise, you will become familiar with core questions in the neuroscience of consciousness.

8.1 Methodology of Consciousness Research

The neuroscience of consciousness aims to identify differences in how conscious and unconscious stimuli are processed in the brain. A simple first approach is to compare processing of two types of stimuli: First, weak or masked stimuli that don't cross the threshold to consciousness (i.e., **subliminal or subthreshold stimuli**, from *limes*, Latin for the boundary); second, stimuli that are unmasked or sufficiently strong to cross the threshold to consciousness (i.e., **suprathreshold or supraliminal stimuli**). Comparing these two cases should allow us to determine which brain regions are more involved in conscious stimulus processing than in subthreshold stimulus processing (e.g., Dehaene et al. 2001; Haynes and Rees 2005a).

This approach, however, is rather crude in providing only two conditions on the continuum between invisibility and visibility. A more fine-grained approach would be to cross the **perceptual threshold in more fine-grained steps**. This seems to be simple at first glance. One starts with a simple, faint and brief stimulus and then makes it either longer or more intense in order to determine the presentation duration or the intensity at which a person just begins to perceive a stimulus. One important finding of such studies is that the threshold of perception is not an abrupt, discontinuous transition from intensities at which the stimulus is never seen to intensities at which the stimulus is always seen (see Gescheider 1997). Around the threshold, there are intensities where perception of the stimulus is not fully determined and it is seen only with some probability. This results in an S-shaped rather than a step-shaped threshold function.

The continuous and gradual nature of the perceptual threshold raises the question how it can be defined technically. Typically, a threshold is defined as the stimulus intensity (duration) at which a certain percentage of responses are correct (e.g., 75%). Note that with such a definition performance is already above chance at the threshold. So why not define the threshold as the intensity at which the visibility first starts deviating from baseline? This has statistical reasons, because it is difficult to define the threshold as the first intensity or duration where a stimulus begins to be seen barely above chance. Take an accuracy of 51%. A lot of trials would be needed to tell if a measured accuracy of 51% really is above chance, or if this just reflects random fluctuations in people's behavior. In order to investigate truly unconscious processing and ensure that the subject has not seen a stimulus even on a few trials, the recognition of the stimulus must be at the chance level (e.g. 50% for two equally likely stimuli). For this, Bayesian statistics can be useful because they provide a framework for quanitifying evidence for the absence of an effect.

8.1.1 Criteria for Conscious Perception I: Subjective Threshold

One problem with determining the perceptual threshold is that there are different criteria for whether a stimulus has been perceived. At first glance, one could simply ask the subject whether they saw a stimulus or not and base the threshold on their judgment. There are numerous examples of the use of such *subjective judgments* in research on **unconscious stimulus processing**. One example is a study by Berti and Rizzolatti (1992) on neglect patients. Patients with right parietal lesions often show an attentional deficit for stimuli in the left visual field, especially when these stimuli are in competition for attention with stimuli in the right visual field. This visual hemineglect is not due to perceptual deficits, as single isolated stimuli in the left visual field can be readily detected. Berti and Rizzolatti (1992) presented prime stimuli to the neglected visual field and target stimuli to the intact visual field and investigated whether the invisible primes had an effect on target perception. They inferred the invisibility of the primes from the subjective reports of the patients, who reported seeing only the stimulus in the right visual field. From this it was possible to conclude that unconscious primes have an influence on the processing of conscious targets (see Box: Confidence Judgments).

A study by Moutoussis and Zeki (2002) on the processing of unconscious object stimuli was also based on subjective judgments about whether subjects were consciously aware of stimuli. Face and house stimuli were presented in two different visibility conditions. Stimuli had either the same color or opposite colors in both eyes (green on a red background in one eye and red on a green background in the other eye). As a measure of visibility, subjects were asked to give one of three possible responses depending on whether they thought they saw "house," "face," or "neither." In the countercolored condition, the perceptibility of objects was greatly reduced. Subjects reported seeing no picture in most cases, so according to their subjective judgment they were unaware of the stimuli. However, an objective discrimination test revealed that some subjects were disproportionately good at guessing which picture had been shown. This typical dissociation between subjective reports and objective discrimination measures is discussed in more detail below in ▶ Sect. 8.2.

Experiments on unconscious stimulus processing were already conducted in the early days of experimental psychology, even with far more detailed distinctions between "conscious" and "unconscious" (Peirce and Jastrow 1884). As early as 1884, Peirce and Jastrow presented weight stimuli when investigating *tactile* perception and asked subjects to discriminate between positive and negative changes in weights. To investigate the role of consciousness in this, *confidence* judgments were collected, ranging from 0 ("No preference for one judgment over the other, the question seemed meaningless") to 3 ("High confidence of having answered correctly"). Subjects were disproportionately correct even at the lowest confidence level. Later experiments also showed that even at minimal confidence levels, discrimination performance can be disproportionately good (for an early review, see Adams 1957). Confidence judgments are also occasionally collected in experiments on signal detection theory (Green and Swets 1966). As a rule, subjects are overconfidently good even at the lowest confidence ratings (i.e., above the identity in the hits-versus-false alarms diagram).

8.1.2 Criteria for Conscious Perception II: Objective Threshold

However, it was already criticized in the 1960s that subjective methods could be influenced by possible conservative response tendencies (Eriksen 1954, 1960; Kunimoto et al. 2001). This is understood to mean that a subject may not be entirely sure that he or she has seen the stimulus when the perception is faint and unclear, but may prefer to answer "no" once too often rather than "yes" too often, i.e., have a tendency toward false negative rather than false positive judgments. This problem is particularly glaring in the field of *perceptual defense* research (e.g., McGinnies 1949), in which the perceptual threshold for taboo words (*dirty words*; Eriksen 1954) is reported to be elevated. Rather than a higher perceptual threshold for taboo words, it is also possible that subjects just have a reluctance to report these words (Eriksen 1954).

The problem of **partial information** also plays a major role here. Even if a subject did not *fully* recognize a stimulus, any partial or fragmentary information might be sufficient for the required discrimination (Kunimoto et al. 2001; Kouider and Dupoux 2004). For example, in an experiment by Sidis, subjects were asked to recognize numbers and letters on cards (Sidis 1898). The cards were presented at such a distance that subjects reported perceiving only a blurred dot. This was taken as evidence that the stimuli were not consciously perceived. Nevertheless, subjects were able to discriminate between letters and numbers disproportionately well. However, since the subjects had weak, albeit undifferentiated, perception, it may be that even fragmentary conscious perception allowed them to distinguish between numbers and letters (e.g., numbers tend to have more curves than letters). The problem of partial information is compounded by the fact that a dichotomous judgment of per-

ception as "conscious" or "unconscious" forces subjects to split a possible continuous graduation of awareness between two categories (Kunimoto et al. 2001). Frame and anchor effects may occur, so that the separation between "conscious" and "unconscious" responses may simply be oriented towards a median of visibility, with the consequence that even partially conscious stimuli are classified as "unconscious".

8.1.3 Experimental Implementation

Even once a decision has been made in favor of subjective or objective threshold measurements, the determination of a perceptual threshold is fraught with numerous further difficulties. For example, as already pointed out by Fechner (1860), the result depends on the temporal order used in which stimuli of different intensities are presented. In the *method of adjustment* the subject themself is allowed to set the intensity at which a stimulus is just perceived. In the *method of limits*, stimuli of increasing intensity are presented and the time at which the stimulus is perceived is noted. To avoid hysteresis effects, one then switches from increasing to decreasing intensity and notes at what point the stimulus is no longer perceived. This is repeated a few times, and the threshold is obtained as the mean value of the measured values of ascending and descending series of measurements. The most reliable method is the *method of constant stimuli*, in which different intensities are presented randomly in the threshold range, which allows sequence effects and expectations to be excluded. Likewise, it should be taken into account that often during a longer experiment stimuli that cannot be detected at first are seen later after some experience, an effect that can be attributed to perceptual learning (Kahnt et al. 2011; Watanabe et al. 2001).

Background Information

Perceptual framing effects refer, for example, to the fact that a stimulus is judged differently depending on the composition of the set of other stimuli it is presented with in an experiment (anchor effects, see e.g. Cannon 1984; Gescheider 1988). When determining thresholds, one usually cannot isolate individual variables but must consider the entire physical physical properties of the stimulus. For example, sometimes people report a "flicker fusion frequency", above which the flickering of a light source is no longer perceived and it appears to be continuously illuminated. However, this frequency depends (among other factors) on stimulus intensity, so it is not possible to state a single, universal flicker fusion frequency (Watson 1986). Not only perceptual, but also cognitive context effects must be taken into account. Consider, for example, a priming experiment in which the effect of a subthreshold priming stimulus on the judgment of a suprathreshold target stimulus is to be examined. For example, one could work with pattern masking and present the following stimulus sequence: Mask (100 ms)—Prime (16 ms)—Mask (100 ms)—Target stimulus (200 ms). In order to ensure that the prime was truly invisible one could randomly test on some proportion of trials whether the participant can see the prime. In the remaining trials one can then test for a priming effect on the perception of the target. However, the difference between these two cognitive tasks (judging the prime versus the target) will have an effect on the visibility of the prime, presumably leading one to over-estimate prime visibility. An alternative practice is to test the visibility before or after the series of measurements. But again, with this different task prime visibility could be overestimated, and perceptual learning throughout the experiment might also influence the visibility.

8.1.4 Criteria for Subliminal Processing

To investigate subthreshold processing, it is necessary, on the one hand, to measure the threshold of consciousness, as explained in ▶ Sect. 8.1.3. On the other hand, some behavioral or neuroscientific evidence is also required that indicates whether the subliminal stimulus is still processed by the brain (◘ Fig. 8.1). For example, one could test whether the subliminal—despite being unseen—has an effect on word stem comple-

tion or on some signature of brain activity. Measuring the conscious perceptibility of a stimulus is called a *direct measure of* processing (Reingold and Merikle 1988)—*direct* because the task directly and explicitly refers to seeing the stimulus. The measure of hypothetical subliminal processing of the stimulus is called *indirect* because it does not refer directly to the stimulus but allows the indirect effects of a stimulus on a task to be measured.

One possibility is to look for **qualitative differences** between conscious and unconscious processing (e.g., Merikle et al. 1995), in which case implicit effects cannot simply reflect a weak form of perception because it has very different properties. Another method is the **dissociation between confidence judgments and discrimination performance** (Kunimoto et al. 2001), in which a subject's discrimination performance is assumed to be unconscious when he is unable to say in which trials his performance is good or bad. The process dissociation method is also applied to subliminal perception (Debner and Jacoby 1994). The logic of this procedure is that it should be impossible to rule out the effect of a stimulus that is presented unconsciously, so that it nevertheless has a nonconscious and therefore uncontrollable influence on a word stem completion. This connection between consciousness and intentional controllability has been pointed out by several other researchers (Marcel 1983; Holender 1986). But even these newer approaches are not without controversy, so that presumably only the development of explicit mathematical models will bring about a clarification of the measurement problems (Schmidt and Vorberg 2006).

As mentioned above, it is also possible to take a completely different approach and use brain processes triggered by unconscious stimuli as evidence for implicit processing

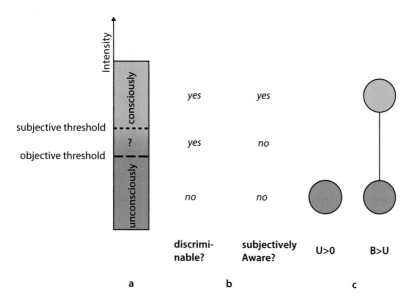

◘ Fig. 8.1 Thresholds and experimental contrasts. **a, b** Differential awareness of stimuli presented with increasing intensity. Two ranges are uncontroversial: Below the objective threshold, a stimulus is no longer discriminable and possible residual processing is unconscious. Above the subjective threshold, the stimulus is consciously recognized and can be discriminated. There are different interpretations of the range between the subjective and objective thresholds in which a stimulus was not considered conscious by the subject, but he was still able to discriminate it correctly. According to one view, the stimulus is unconscious and the residual discriminability is due to unconscious processing. According to another inter-

pretation, the stimulus in this domain is conscious, except that subjects are very conservative in their response behavior. **c** Two important statistical contrasts (comparisons) that examine different aspects of conscious and unconscious information processing and are commonly used in neuroimaging experiments. Contrast U > 0 tests whether significant activation (greater than zero) occurs with unconscious stimuli. This can be used to identify regions where unconscious stimuli are processed. The contrast B > U tests for regions where conscious stimuli elicit greater activity than unconscious stimuli. This allows us to determine what cortical activity is "added" during conscious processing

instead of implicit measures of behavior (e.g., Haynes and Rees 2005a). This will be illustrated in the following.

8.2 Neuronal Correlates of Conscious and Unconscious Stimulus Processing

Assuming that one has managed to avoid all the pitfalls and that one has experimentally generated a clean distinction between subliminal and suprathreshold stimuli. Then one can set about the task of comparing the neural processing of consciously perceived and subliminal stimuli in order to determine

something about the neural correlates of consciousness. First, one can examine what additional neural processing occurs during conscious compared to unconscious stimulus processing (◘ Fig. 8.1c). On the other hand, one can investigate the depth to which unconscious stimuli are processed (◘ Fig. 8.1b).

8.2.1 Conscious Neural Processing

A number of studies have examined the effects of awareness of visual information on brain activity in humans and in non-human primates. Most notably, conscious stimuli have been found to undergo more extensive cortical processing compared to unconscious stimuli. In one study, Dehaene et al. (2001) used functional magnetic resonance imaging (fMRI) to examine the cortical processing of masked words that were either visible or invisible. The subthreshold invisible word stimuli activated the brain, but the activation remained restricted to visual brain regions. In contrast, the suprathreshold stimuli were characterized by widespread activation in parietal and prefrontal brain regions as well. This was interpreted by Dehaene et al. as an indication of a *distribution of* sensory information in the brain during conscious perception and was interpreted in terms of the global workspace theory (▶ Sect. 8.5).

Suprathreshold and subthreshold stimuli are physically different stimuli. As a rule, one has to increase the stimulus intensity or attenuate a masking in order to cross the threshold of consciousness. However, this complicates the interpretation of the measured differences: a brain region that responds more strongly to a consciously perceived stimulus than to a subliminal one could possibly be responding only to the fact that the latter has a higher stimulus intensity, that it entails a stronger exogenous allocation of attention, or that the subject prepares and executes a response to the stimulus (Dehaene et al. 2001).

Some of these problems can be avoided by a more fine-grained examination of the perception threshold. As explained in ▶ Sect. 8.1, the perception threshold does not jump abruptly from subthreshold to suprathreshold with increasing intensity, but there is a range in which perception is only partially determined and the subject consciously recognizes even a constant stimulus only in a portion of the runs. The S-shaped transition at threshold can be approximated mathematically as a cumulative Gaussian normal distribution (Gescheider 1997). This makes it possible to correlate the individual course of a threshold for a particular stimulus and a particular person with the neuronal fMRI responses in different brain regions in order to search for areas in which the activity level reflects the threshold course.

It was shown that in the perception of masked objects, the profile of the perceptual threshold correlated with the level of activity in the lateral occipital cortex (LOC) but not in the primary visual cortex (V1) (Grill-Spector et al. 2000). Similarly, masking of simple brightness stimuli showed that the shape of the masking function was reflected in the connectivity between early (V1) and later visual areas (in the fusiform gyrus) (Haynes et al. 2005a, b). This direct comparison between *psychometric* threshold functions and neural or *neurometric* response curves can also be applied to the response behavior of single cells and cell populations in sensory brain regions (Parker and Newsome 1998).

Fluctuations in perceptual judgments of the same physical stimulus have been shown to be accompanied by fluctuations in neural activity already in early visual cortex. For example, in trials in which the stimulus is seen, activity in primary visual cortex V1 is higher than in trials in which the stimulus is not seen (Ress and Heeger 2003).

Fluctuations in perceptual judgments not only occur between individual stimulus presentations, but there are also other, slower stochastic fluctuations in perception. Tononi et al. (1998) used MEG to investigate fluctuations in awareness of rival stimuli. Conscious stimuli were found to produce more extensive activation and coherence in regions beyond the visual system. In a study of *change blindness*, Beck et al. (2001) showed that conscious awareness of a change in a stimulus display leads to increased activity in frontoparietal networks. Similarly, Vuilleumier et al. (2001) found the same in a study of perception in a neglect patient. In runs in which the patient consciously recognized a stimulus in the neglected visual field, brain activity was also increased in parietal regions.

Moreover, there was a widespread increase in effective connectivity between visual, parietal, and prefrontal regions during conscious perception. Similar effects of conscious stimulus perception on functional connectivity measures were also found in other studies (e.g., Lumer and Rees 1999; Dehaene et al. 2001; Haynes et al. 2005a, b).

8.2.2 Unconscious Neural Processing

Research on the neural processing of unconscious stimuli shows that they are processed very deeply. On the one hand, stimulus properties that are not consciously recognized are already represented at the level of the primary visual cortex in the activity levels of individual neurons as well as in fMRI measurements of neuronal populations (Gur and Snodderly 1997; Haynes and Rees 2005a), which had already been indicated previously in behavioral studies (He et al. 1996). However, processing of unconscious stimuli is not restricted to early stages of the visual system, but also reaches higher, content-specific processing stages (Moutoussis and Zeki 2002).

Moutoussis and Zeki (2002) were able to show that invisible images of houses and faces are processed in the corresponding regions parahippocampal place area (PPA) and fusiform face area (FFA, Kanwisher et al. 1997), respectively. Activation for invisible stimuli was significant, but was significantly weaker than for visible stimuli. Studies in neurological patients have shown that even in neglect patients, stimuli that are detected in isolation but do not reach consciousness due to competition with stimuli in other regions of the visual field nevertheless selectively activate content-specific brain regions (Rees et al. 2002). Fang and He (2005) used interocular suppression stimuli to show that the *dorsal* visual pathway is also activated by invisible objects. Interestingly, activation was almost as high for invisibly presented images of tools as for visible images, which fits with the theory that the dorsal pathway is relevant to action control (Goodale and Milner 1992). Moreover, emotional aspects of invisible pictures can activate emotion networks in amygdala and orbitofrontal cortex (Vuilleumier et al. 2002). The unconscious neural processing of stimuli may even lead to conditioning processes in the basal ganglia (Pessiglione et al. 2008), which may have direct effects on choice behavior.

8.2.3 The Phenomenon of Blindsight

An interesting phenomenon is *blindsight* (Weiskrantz et al. 1974; Weiskrantz 2004). After V1 lesions, patients have perceptual deficits (scotomata) in subregions of the visual field corresponding to the deficient V1 region. They claim not to notice and not to be able to identify stimuli presented at these locations in the visual field. However, if they are allowed to guess the stimulus, their recognition performance may be above chance, even if they subjectively appear to be only guessing. One explanation for this perceptual performance is that visual information from the visual thalamus can reach the rest of the cortex not only via V1, but also via direct projection to extrastriate cortex areas (see, e.g., Bullier and Kennedy 1983). Consistent with this, stimuli in blind regions of the visual field can induce extensive activation of the extrastriate visual cortex (MT+, V4/V8, LOC) in blindsight patients (Goebel et al. 2001).

8.3 Contents of Consciousness

Conscious perception cannot only be characterized at the level of a dichotomous distinction into conscious-unconscious, but also involves the representation of **conscious content**. This calls for the study of the representation of conscious content in the brain, all the more so because various theories of neural correlates make specific predictions about the effect of consciousness on representation of content (Crick and Koch 1998; Tononi and Edelman 1998; Dehaene and Naccache 2001).

Three different "coding levels" must be distinguished in the representation of consciousness contents.

8.3.1 Coding Levels

At the top level of encoding are contents of different **sensory modalities**. In PET and fMRI, the cortical processing of the different modalities is clearly separable (Binder et al. 1994; Tootell et al. 1996). At the middle level, the **representation of the submodalities** occurs. In the visual system, these would be, for example, color, brightness, motion, or object perception. They can usually be traced back to separate regions within the individual processing pathways.

Background Information
An early example of the assignment of visual submodalities to specific brain regions can be found in a PET study by Zeki et al. (1991). Using positron emission tomography, brain activity was measured in healthy subjects while they viewed visual presentations with different content.

In addition, a closed-eye condition was measured, which served as a "baseline" to which the activities in the other conditions were compared. A brain region responsible for recognizing *motion* should show stronger activity for the moving stimuli than for the static stimuli. A region responsible for recognizing *color* should show stronger activity for viewing colored stimuli than for gray stimuli. This allowed the researchers to identify an area of movement they called V5 (now widely known as MT+). Similarly, at the ventral border of occipital and temporal cortex, they found a region that was more responsive to color stimuli, which they named V4. Although the designation of this region as "V4" is controversial (Hadjikhani et al. 1998), there is no doubt that a color-selective region is located in the ventral temporal cortex. This was the first evidence in humans of a specialization of the visual cortex for different submodalities.

Below the submodalities there is the **level of the specific contents**. Thus, in color perception, the specific color qualities (hue, saturation, brightness) that are present at a particular location in the visual field must be distinguished. In motion perception, the different directions and speeds of movement must be encoded, and in object perception, different shapes as well as exemplars and categories of objects must be encoded.

Another and more flexible way of exploring the contents of consciousness is offered by the class of "multistable stimuli", which allow different interpretations in perception. Most famous are the "Necker cube" (Necker 1832) and "My wife and mother-in-law" (Boring 1930). If one looks at the stimulus for a long time, the perception suddenly flips and one sees another possible interpretation. Since the two interpretations are based on the same physical stimulus, this allows changes in conscious perception to be studied without changes in physical stimulation. A variant of this multistable perception is binocular rivalry (Leopold and Logothetis 1996, 1999), in which conflicting images are presented to both eyes, with the effect that perception does not fuse these images but spontaneously switches back and forth between the two images. Single-derivation studies in awake monkeys (Leopold and Logothetis 1996) and fMRI studies in humans have shown that higher, content-specific cortical regions of the visual system change their activity in concert with conscious perception. Later, further evidence was added that even early stages of visual processing (V1, LGN) are involved in binocular rivalry (Tong and Engel 2001; Haynes et al. 2005a, b; Wunderlich et al. 2005), presumably due to altered representation of simple features of conscious perception such as edges and brightnesses. However, the role of attention in these early correlates has not been definitively established (Watanabe et al. 2011).

> **Binocular Rivalry**
>
> A type of multistable perception in which both eyes are presented with images that do not merge into a unified image, but rather the perception spontaneously switches back and forth between the two images.

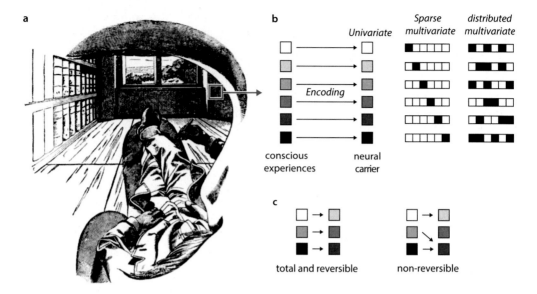

Fig. 8.2 Coding principles for perceptual content. **a** A drawing by Ernst Mach (1886) in which he attempts to represent his visual experiences pictorially. **b** left: In a univariate code, ordinally scalable contents of consciousness (such as brightness) are represented in a continuous neural parameter (e.g., firing rate). Middle: In a *sparse* multivariate code, a separate, specialized neuron is active for each content (so-called grandmother or cardinal cell code). Right: in a distributed multivariate code, all neurons are involved in encoding all content. **c** The connection between consciousness contents and neuronal representations must meet certain mapping criteria. Totality means that a neural correlate can be identified for each content of consciousness. More important is reversibility, which means that no information is lost because each content of consciousness is assigned its own neural correlate and thus the content can be decoded again from the neural representation at any time. (Illustration from Haynes 2009)

8.3.2 Coding of Consciousness Contents

The more recent approach of multivariate decoding (▶ Sect. 8.3.3) makes it possible to investigate directly how certain contents of consciousness (colours, brightness, edges, movement, objects) are realised in the brain and how these representations change when people become conscious (Haynes 2009). For this purpose, the nervous system can be thought of as a kind of *carrier* into which the contents of consciousness are encoded. Encoding here does not mean that there is a homunculus that has to read out an encrypted code, but that there is a stable mapping relationship between states of consciousness (such as different sensations of brightness) and the states of the neuronal carrier. So you have to look for a neural carrier in the brain that has a stable association with some content, so that every time the content is in consciousness, the same state of the carrier goes with it (Haynes 2009).

> ▶ **Example**
> Let us take as an example a drawing by Ernst Mach (1886). ▣ Figure 8.2a shows Mach's visual perceptual space as he sits on a sofa and looks out of his left eye into his study. The black and white drawing shows mainly edges and brightness. The brightness sensation in a region in the upper right quadrant of his visual field was presumably encoded in some parameter of neural activity of a particular brain area. One possibility would be that each brightness was encoded in the *activity rate of* individual cells in a region (say

V1 or V4) (see, e.g., Haynes et al. 2004), so that greater brightnesses would be associated with higher firing rates. This "univariate" code could be found by a simple correlation between brain activity level and perceptual intensity (Haynes et al. 2004). However, other coding formats are conceivable (◘ Fig. 8.2b). In a "multivariate" representation, multiple cells or cell populations are involved in the encoding, so that the content can no longer be explained by the activity of individual cells. In a "*sparse*" code, each perceptual content is associated with a dedicated cell, so that this cell (and only this cell) is always active when a person has a particular brightness experience. This representational format is also known as "grandmother cell" or "cardinal cell" code. In contrast, with a *distributed* multivariate representation, the assignment of experiences to individual cells is no longer possible at all; instead, a dedicated overall population activity state encodes each individual consciousness content. This code is called a *distributed* code. Importantly, sparse and distributed codes can no longer be explained by a simple correlation between activity in a region and perceptual content. ◄

Many studies have adopted a priori the **univariate encoding model** and searched for the neural "correlates" of specific perceptual content (e.g., Tong et al. 1998; Ress and Heeger 2003; Haynes et al. 2004). However, to identify any sparsely or distributedly encoded neural carriers, one must use dedicated analysis techniques, such as multivariate correlation, regression, and classification. The latter is also referred to as multivariate decoding and is presented in ► Sect. 8.3.3.

8.3.3 Multivariate Decoding

A general procedure for identifying neuronal populations that encode specific perceptual content is multivariate decoding (◘ Fig. 8.3). This involves determining how well a perceptual content can be reconstructed from a neuronal population response. If this works well, then the population has a lot of information about the content and there is a stable mapping relationship between brain states and a particular category of conscious content. Multivariate decoding can be driven by different types of population signals, i.e., multiple leads from single cells (Quiroga et al. 2005), fMRI voxel sets (Haynes and Rees 2006), or multiple EEG electrodes (Blankertz et al. 2003).

Multivariate Decoding

A procedure for determining how well a perceptual content can be reconstructed from a neural population response.

Voxel

Data point in a three-dimensional representation space, corresponds to a pixel in a two-dimensional representation space.

► Example

◘ Figure 8.3 shows the procedure using the example of an fMRI measurement. The aim is to determine in which cortical regions the content of consciousness is encoded when looking at one of the two vehicles shown on the top right (◘ Fig. 8.3a; Cichy et al. 2011a, b). First, one can start the analysis at a location in the visual system. One extracts there—averaged over a short measurement run—the activity level in the spatial neighborhood of this starting point while the subject is looking at the first car. Then one repeats this measurement a few times, while the subject sometimes sees the first car and sometimes the second car. Thus, one measures several "samples" of the local brain activity patterns for both images.

The individual measurements can be thought of as points in a high-dimensional coordinate system with as many dimensions

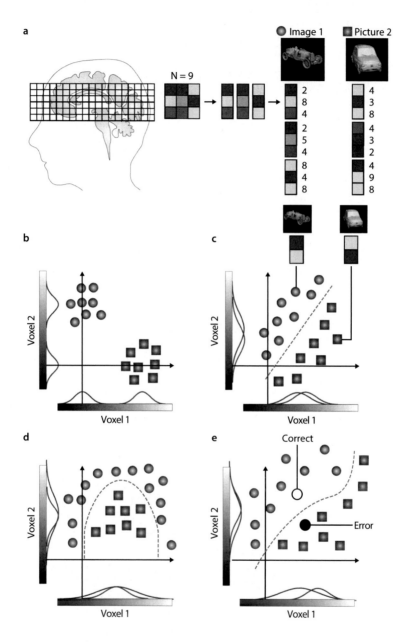

◧ Fig. 8.3 Decoding of consciousness content using multivariate pattern recognition (see text). (Modified after Haynes and Rees 2006)

as voxels (so nine in this example). Since people find it difficult to imagine a 9-dimensional space, the analysis can be visualized for two dimensions. The measurements in voxel 1 can be plotted as *x-values* and the measurements in voxel 2 as *y-values* in a coordinate system. This would be a point (x, y) for each measurement resulting from the first two dimensions. Several possible coding examples are shown in the figure. ◘ Figure 8.3b shows a sparse code. High values in voxel 1 (x) occur only when image 2 is perceived, and high values in voxel 2 (y) occur only when image 1 is perceived. The perception of a content can always be attributed to a dedicated voxel, and the classification of whether the subject was looking at car 1 or car 2 during a given measurement would even be possible based on one voxel alone (if there were more than two objects, a sparse code in two dimensions would no longer be possible). Classification becomes more difficult in the case of ◘ Fig. 8.3c. Here, the assignment of consciousness contents to individual voxels is no longer possible. In this case the classification can also be done, but only if the values of both voxels are known at the same time. In this example, the decision can be made based on a linear separation between the two groups, but there may also be situations that have nonlinear decision boundaries and require the use of specialized nonlinear classification algorithms (◘ Fig. 8.3d).

To test whether the classification of consciousness content from a voxel population is possible, one uses a two-step procedure (◘ Fig. 8.3e). In the first step, one uses only a portion of the data, the **training dataset**, to train a classification algorithm to optimally separate the contents. The decision boundary can be learned via various algorithms, such as linear discriminant analysis or so-called support vector classification (Haynes and Rees 2006). In a second step, one then takes a statistically independent, new part of the data (*test data set*) and classifies it using the previously trained decoder. If the measurement points in the test data fall on the correct side of the decision boundary (◘ Fig. 8.3e, "cor-

rect") and are thus correctly assigned, it is a hit. If they fall on the wrong side of the decision boundary (◘ Fig. 8.3e, "error"), it is a miss. The goodness with which the test data can be classified overrandomly well provides information about how much information about the perceptual content was contained in the voxel population. ◄

There are several ways to define the region used for classification. The **region** *of interest* (ROI) **method** uses an a priori defined region, such as the primary visual cortex, which can be defined on the basis of independent retinotopic mapping (Tootell et al. 1996). Alternatively, separate *localiser* runs can be considered, where a region is defined functionally (such as the fusiform face area, which is defined as the region that shows stronger responses to faces than to objects). The *searchlight* **method** starts with one point and notes the information content in the local environment around the starting point. The procedure is then repeated at many locations in the brain, creating a three-dimensional map showing the local information content at all locations in the brain (Haynes and Rees 2006). In the *whole brain* **method**, the voxels of the entire brain are used for classification. This last method is more commonly used in brain-computer interfaces that need to extract maximum information from brain signals for technical optimization. However, since the whole brain method is limited for clarifying regional hypotheses, it is less commonly used in cognitive neuroscience.

Multivariate decoding has been used in numerous studies to investigate the encoding of consciousness content (see Haynes 2009 for a review). In one study (Haynes and Rees 2005a), it was shown that subliminally presented masked orienting stimuli are nevertheless encoded in V1 in a feature-specific manner, confirming earlier results from psychophysics (He et al. 1996). Moreover, the encoding of specific consciousness cues can be clearly assigned to individual brain regions. Thus, different pictorial visual ideas

☐ Fig. 8.4 Decoding of visual ideas from temporo-occipital brain regions (Cichy et al. 2011b). **a** Subjects were shown pictures of different categories. **b** During an fMRI measurement, they had to imagine them pictorially. **c** Although specialized regions had the most information about their optimal category (e.g., faces in the fusiform face area FFA), it was also possible to some extent to read out which of various non-preferred objects had been imagined (e.g., chair versus clock in the face area). Other abbreviations: *EBA* extrastriatal body area, *FBA* fusiform body area, *OFA* occipital face area, *PPA* parahippocampal place area, *TOS* transverse occipital sulcus

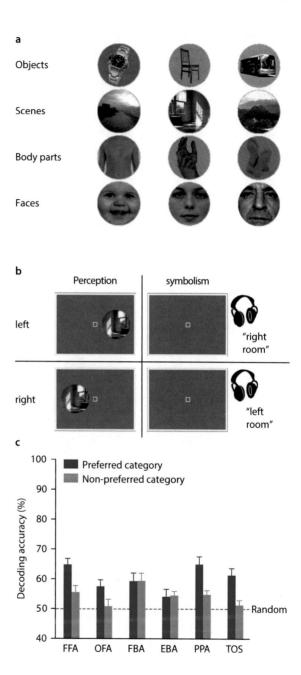

were found to be encoded in spatial brain activity patterns in specialized brain regions (Cichy et al. 2011b). However, it was also shown that other visual regions, apparently specialized in other ways, also represent the contents in question to a certain extent (☐ Fig. 8.4).

An important property of higher-level conscious perception could also be investigated, namely invariance to presentation conditions (Cichy et al. 2011b). Subjects had to imagine objects in either the left or the right visual field. This made it possible to train pattern recognition on images in one hemifield and test it on another hemifield. The logic here is that object identity, as a higher-level feature, should be invariant to changes in the details of the presentation. Despite the change in position, it was possible to decode object category from patterns in the ventral path, arguing for a spatially invariant representation. In the ventral path, invariances to other stimulus properties, such as colors or textures, have also been demonstrated (Sáry et al. 1993).

8.4   Structure of Consciousness

Another important question is how similarities between consciousness contents are neuronally encoded. Let us take as an example two brightness gradations H_1 and H_2 of a grey area and their neuronal correlates N_1 and N_2. Put simply, the aforementioned strict correlation between consciousness content and brain activity means that whenever a person sees a certain brightness, the same neuronal activity is observed in the brain. In other words, H_1 would always measure N_1, H_2 would always measure N_2, and so on. A broader question is whether the *relations* and similarity ratios between consciousness contents also translate into the same relations and similarity ratios of activity patterns in the brain.

In the simplest case, this could mean, for example, the following: If area 2 is brighter than area 1 (i.e., $H_2 > H_1$), then the neuronal activity N_2 is also greater than N_1. The advantage of such an encoding would be that not only would the individual contents be explained by specific brain activities, but the perceived **similarity relations would** thus be automatically explained as well. Evidence

for such an encoding exists in certain ases, however, according to imaging experiments, especially the brightness encoding in the visual system seems to be organized rather bipolar, namely in the form of deviations from an average gray value. Thus, if a surface is particularly dark or particularly bright, specialized neurons react more strongly to this than if the surface is gray (Haynes et al. 2004).

However, in other cases, particularly in object perception, there is strong evidence that the experienced similarities between perceptual states translate into similarities of neural representations. For example, in an early study Edelman et al. (1998) showed their subjects a series of images of different categories. They then asked the subjects to rate how similar they found the images. Based on the similarity judgments, the images were then arranged on a two-dimensional surface using multidimensional scaling in such a way that the relative positions of the objects in this space reflected their *subjectively perceived similarity to* each other (◻ Fig. 8.5a). In a second step, fMRI was used to measure brain activity patterns in the object perception cortex LOC for each object. Then, these activity patterns of different objects were compared with each other, and similarity was determined. Now, the objects were arranged with a multidimensional scaling such that the spatial proximity indicated the *similarity of their neural representation* (◻ Fig. 8.5b). It was found that the space of subjectively perceived similarities and the space of neural similarities reflect each other well. This means that the subjectively perceived similarity between objects can be explained by neural representations in this region, the LOC.

a

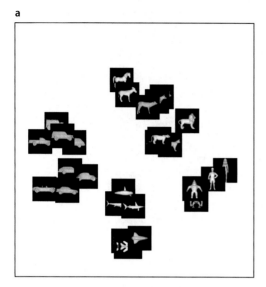

b

◻ Fig. 8.5 Psychophysical and neuronal perceptual spaces. A comparison of the perceived similarities of objects **a** with their neuronal similarity in the lateral occipital complex (**b** see text). (Adapted from Edelman et al. 1998)

8.5　Models of Consciousness

A central question in research on the neural correlates of consciousness is what specific difference in the neural processing of a stimulus determines whether it enters consciousness or not. There are a number of candidates, and some evidence can be found for all of them. Some representative theories are briefly outlined below:

- **Activity Level**

The core idea of this theory is that the awareness of a stimulus is tied to how strong the activity is in the brain that the stimulus triggers. Accordingly, a weak activation in a brain region may remain unconscious, but above a certain level of arousal, the neuronal representations it contains become conscious.

There is ample evidence to support this theory. The perceptual threshold for masked stimuli is directly correlated with the level of activity in object processing cortex, in monkeys (Kovács et al. 1995) and in humans (Grill-Spector et al. 2000). Already in primary visual cortex, there is evidence that activity level determines perception, even for small fluctuations in perception around threshold (Ress and Heeger 2003). However, work by Leopold and Logothetis (1996) shows that the link between activity and awareness may be more complex, at least in binocular rivalry. They found cells whose activity increased when the corresponding representation became conscious, but there were also cells whose activity dropped.

Activity theory can explain well an important property of conscious perception, namely its availability for access. We can respond to conscious stimuli intentionally, we can remember them, we can describe them verbally. This means that conscious stimuli undergo further processing in the brain. If the associated neural representations are "stronger" according to activity theory, it could also explain that they influence later regions more strongly.

- **Communication and Synchronization**

Another and very popular model states that neuronal stimulus representations reach consciousness when the distributed subrepresentations synchronize their firing rate (Engel and Singer 2001). The cortical activity associated with this is thought to be primarily in the high-frequency range (gamma band, 30–80 Hz). Accordingly, consciousness would not be reducible to individual brain regions, but would be a network phenomenon. This model can explain particularly well why, despite the distributed, quasi-modular processing of individual feature dimensions (color, shape, motion), an object is nevertheless perceived as integrated. The explanation would be that the uniformity of perception can be attributed to the synchronization of neuronal partial representations. This synchronization theory could also well explain the "access property" of consciousness, because neurons that become synchronously active have a higher probability of suprathreshold excitation of downstream regions (König et al. 1996). However, the neurophysiological processes by which long-range, high-frequency synchronization is thought to occur remain unclear. Although neurons that fire in the gamma band are found in narrowly defined populations of neurons, such as in the visual cortex (V1, V4), these are inhibitory interneurons that act on pyramidal cells as the respective output neurons of the population. However, apart from specialized pyramidal cells, the latter fire at low frequency in the 1–10 Hz range and thus cannot transmit gamma acivity over a long range (cf. Buszáki and Schomburg 2015; Ray and Maunsell 2015). However, they can be "forced" by high-frequency input from interneurons to fire with low-frequency pulses during specific time windows. Coupling of closely neighboring populations via interneurons could lead to synchronous firing of pyramidal cells in neighboring populations, thereby enhancing certain activities, such as attentional, within circumscribed cortex areas (see Taylor et al. 2005; Fries 2015; Ni et al.

2016). Longer-range synchronization phenomena in the gamma band region, however, cannot yet be explained by this.

■ **Recurrent Processing and Feedback**

Another theory also relates to the **dynamics of long-range neural processes** (Lamme et al. 2000; Pascual-Leone and Walsh 2001) and had previously been used to explain masking phenomena. Super et al. observed that conscious recognition of texture-defined shapes is expressed by increased activity at late stages of processing in primary visual cortex (Supèr et al. 2001). They interpreted this late activity as a "recurrent" phase of processing on which forward, backward, and lateral processing overlap. Further evidence for the importance of feedback signals after V1 was provided by a paper by Pascual-Leone and Walsh (2001). They showed that disrupting backward projection after V1 to a late stage of stimulus processing can prevent conscious awareness of motion stimuli. A related perceptual model is the reverse hierarchy theory (Hochstein and Ahissar 2002). If recurrence should also be a necessary condition for conscious perception, feedback after V1 might not be required for it, since in certain cases conscious experiences of motion can occur even without V1 (Zeki and Ffytche 1998).

■ **Specific Regions and Micro-awareness(es)**

Semir Zeki (2001) formulated a theory according to which different modules in the visual cortex can produce independent "*micro-consciousness*". This is attributed, among other things, to the fact that lesions can be so selective that only a certain category of consciousness content is precipitated (e.g., color or face recognition). Further evidence is that different sensory feature dimensions (e.g., motion and color) reach consciousness at different times (Moutoussis and Zeki 1997). Most notable are the rare cases of patients with Riddoch syndrome (Zeki and Ffytche 1998). These patients have a lesion in the primary visual cortex and are blind because of it. However, their MT+/V5 motor area is intact, and they report being able to see dimly fast motion and can discriminate motion stimuli correctly. They describe movement as being like a "black shadow against a dark background" (Zeki and Ffytche 1998). This presumably means that even when visual perception is largely absent, an isolated dimension can remain intact and be accompanied by activity in the corresponding cortex area. Moreover, a content of consciousness without V1 means that feedback to V1 cannot be a necessary condition for visual consciousness, which contradicts the aforementioned V1 feedback theories of consciousness (Pascual-Leone and Walsh 2001; Lamme et al. 2000).

■ **Dynamic Core**

Tononi and Edelman (1998) formulated a theory that consciousness is equivalent to a "*dynamic* core" of brain activity. This was based on two observations: First, we experience our consciousness as integrated or "bound." Thus, we experience a visual scene as a single entity rather than as an unconnected collection of colors, brightnesses, and edges, even though these features are processed in different places in the brain. Similarly, we find it difficult to perform more than a single task at a time unless the processing is highly automated. The second observation is that consciousness is differentiated despite integration. We can have a variety of experiences, hence our consciousness contains a lot of "information".

They bring these two aspects together in their theory of the dynamic core. According to this theory, there is an information-theoretically irreducible core in brain activity that exhibits a high degree of integration and recurrence between posterior and anterior thalamocortical loops, thus linking sensory categorization with action planning, evaluation, and memory. Evidence

for this comes from an experiment on binocular rivalry in which two images flickering at different frequencies are presented separately to the left and right eyes (Tononi et al. 1998). It showed a response strength and an increase in coherence between parietal and frontal MEG sensors at the frequency corresponding to the image when one image was conscious. A body of evidence, according to Tononi and Edelman (1998), suggests that the core process is dynamic, that is, it can alternately integrate different brain regions into a functional unit. This includes, in particular, the encapsulation of automated, unconscious processing, as expressed, for example, in the activation of extensive sensory cortex regions by unconscious stimuli.

The theory has been developed several times into a **theory of information integration** that gives explicit mathematical definitions of integration (Tononi 2004). In this, an attempt is made to also explain the diversity of qualitative dimensions of experience (qualia) within the same theoretical framework by stating that each dynamic unit spans its own coding space. The dynamic core theory explains the dynamism, integratedness and metastability of consciousness very well. One problem, however, is that the mathematical measures of integration are challenging to measure.

■ **Global Workspace**

An important property of conscious perception is that we can cognitively access conscious representations differently from unconscious representations. When we consciously recognize a visual stimulus, we know we have seen it and can remember it later, we can describe it verbally, or we can otherwise use it to control behavior, e.g., by pressing a response button. To explain this property, Baars formulated the theory of the global workspace (Baars 2002), which was then further developed by Dehaene and Naccache (2001). The idea is that visual stimulation that does not reach consciousness is informationally encapsulated in sensory cortex regions. Information about stimuli that do reach consciousness, on the other hand, is broadly "distributed" in the cortex, as if on a global worksurface where sensory information is available to multiple brain regions (such as for memory and action control) for readout. The *global workspace theory* (GWS; Baars 2002) is supported by numerous evidences. For example, there is direct evidence from Dehaene et al. (2001) that visually masked, unseen words activate only the sensory cortex. In contrast, consciously recognized words excite broad regions of the brain, including the prefrontal cortex (Dehaene et al. 2001). Baars (2002) lists a number of examples where it has been shown that conscious stimuli undergo more extensive processing. These include the extensive cortical distribution of information about binocular rivalry stimuli when they enter consciousness (Tononi et al. 1998), or the greater frontoparietal activation when small changes in dynamic visual displays are detected compared to the situation in which they are not detected (Beck et al. 2001).

▪ Specific Regions

A number of authors have discussed the importance of specific regions for visual consciousness. Crick and Koch (1995) argued for V1 not being directly involved in conscious processes. Among the main arguments for this is that V1 has no direct projections to the prefrontal cortex, which is incompatible with the availability of conscious information for complex behavioral control. Moreover, V1 is also activated by unconscious stimuli (He et al. 1996; Haynes and Rees 2005a, b). It would be difficult to explain why V1 should be involved in consciousness and yet these stimulus representations do not reach consciousness. Goodale and Milner (1992) argued against involvement of the parietal-dorsal visual processing pathway in visual consciousness. This was based on a double dissociation in the behavioral profile of two patients who had a lesion in the dorsal and ventral pathways, respectively. One patient with a dorsal visual lesion was able to describe a visual stimulus but was unable to use it to control behavior. In contrast, a patient with a lesion in the ventral visual pathway was able to incorporate the stimulus into behavioral programs but was unable to describe it. The authors interpreted this as evidence that only the ventral visual pathway is conscious.

Overview

An assessment of the various theories of consciousness is complicated by the fact that they are good at explaining different aspects of consciousness. Some theories focus on the availability of consciousness content for multiple complex behavioral performances. Activation theory and synchronization theory predict that subsequent regions are more amenable to suprathreshold arousal and explain this in terms of greater activity or synchronization of conscious neural representations. Both aspects are in principle compatible with global workspace theory, although other physiological realizations have been proposed by the authors. Moreover, in empirical data, activation and synchronization (and presumably recurrence) are strongly correlated in most cases (e.g., Siegel and König 2003), making it difficult to distinguish between the theories.

The theories mentioned in this section make very different predictions about what happens to content-specific neural representations during conscious versus unconscious processing (◘ Fig. 8.6). For example, according to the microconsciousness theory, it is sufficient for awareness to occur if a neural representation arises in a dedicated perceptual area. For example, content-specific activation in the fusiform face area would already automatically lead to face perception

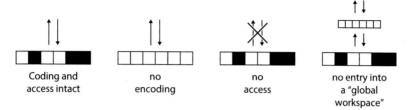

| Coding and access intact | no encoding | no access | no entry into a "global workspace" |

🔲 **Fig. 8.6** Contents of consciousness in different models of consciousness. (Adapted from Haynes 2009). From left: In the state of normal awareness, both the stimulus representation and the access to the stimulus representation are intact. If a stimulus is not consciously perceived, there could be several reasons for this: A lack of encoding of the content in content-specific brain regions, a lack of attional or executive access to the information despite an intact representation, no long-range distribution of the information in the sense of a "global workspace"

without requiring additional conditions such as attention. In contrast, for theories that focus on *access* to visual representations, a selection or distribution mechanism must be operative in addition to a representation. Accordingly, it could happen that a content is encoded in sensory regions but does not reach consciousness.

To distinguish between these two theories, it would be important to investigate whether encoding in sensory regions actually necessarily leads to awareness of the stimulus. Some experiments have shown that even unconscious stimuli can lead to content-specific activation in the visual system (Rees et al. 2002; Moutoussis and Zeki 2002; Haynes and Rees 2005a). Moreover, awareness leads to more widespread activations outside sensory brain regions as well. This suggests that additional processes are required to bring a representation into consciousness. However, whether these additional processes operate in the sense of a global workspace is currently unclear. This theory suggests that content must be

distributed widely in the brain in order to become conscious.

Accordingly, not only would one have to find widespread activity in the brain upon awareness of a stimulus, but these processes would have to be shown to be content-specific and, in fact, to encode *information* about the sensory process. More recently, advocates of the different positions on consciousness have started to engage in adversarial collaborations in order to define crucial experiments that will help decide between the different views (Melloni et al. 2023).

8.6 Awareness, Selection and Attention

According to a widely held view, selective attention regulates access to consciousness and is, in effect, the selective gatekeeper of consciousness. This would mean that we can only consciously recognize stimuli when we attend to them.

There is indeed some evidence for a close connection between consciousness and attention. The clearest evidence is provided by so-called cueing experiments (Sperling 1960; Posner et al. 1980; Hawkins et al. 1990). Experiments by Hawkins et al. (1990), for example, showed that the accuracy with which target stimuli presented at different positions could be recognized increased when subjects were given a cue indicating the most likely target position beforehand. Earlier, in a similar experiment, George Sperling (1960) had shown subjects random sequences of letters and asked them to recall as many letters as possible afterwards. It turned out that their hit rate was considerably better when the subjects were given a cue as to which rows they should report, even when this cue was given only *after the* row of letters had been removed. Accordingly, attention can even operate on representations in iconic memory and select its contents for conscious perception.

Another example of the close relationship between awareness and attention is provided by experiments on inattentional blindness (*Inattentional Blindness*; Mack and Rock 1998) and change blindness (*Change* Blindness; McConkie and Currie 1996; Rensink et al. 1997; Niedeggen et al. 2001; Beck et al. 2001). In inattentional blindness, subjects complete a few runs of a difficult fixation task, and in a final, crucial single run, a stimulus is presented in the periphery of the visual field during the fixation task. Few subjects report seeing the unexpected peripheral stimulus at all (Mack and Rock 1998). The lack of conscious recognition of the peripheral stimulus is attributed to lack of attention. In the very similar phenomenon of change blindness, even striking changes in visual images are not detected until they are processed with attention.

However, there is also evidence in these experiments that attention does not automatically lead to consciousness. Thus, a similar "blindness" can also occur with fully attended stimuli.

> ► **Example**
>
> In one experiment, subjects were engaged in fake conversations with a stranger (Simons and Levin 1998). The conversation was briefly interrupted by two construction workers carrying a door between them so that their view was briefly obscured. Unbeknownst to the subject, the interlocutor was exchanged, which was not noticed by numerous subjects. Since the subject had been looking at the interlocutor (with only one minor interruption) with full attention, this argues that the faulty recognition was not so much due to a lack of attention but to a lack of *expectation* that an interlocutor's identity might change. Other experiments show that attention can also affect the processing of unconscious stimuli (e.g., Martens et al. 2011). Together with the other experiments elaborated above, this would imply that attention is a necessary but not a sufficient condition for consciousness. ◄

However, in recent years some dissociations have been elaborated suggesting that attention is not necessary for consciousness either, because consciousness can occur without attention (Lamme 2003; Koch and Tsuchiya 2007; Srinivasan 2008). A number of experimental findings lead to this conclusion:

1. *Some stimulus properties can be recognized even without attention.* Even without attentional guidance (*Full Report*), the recognition rate in Sperling's experiments was well above chance (Sperling 1960), which could argue for residual processing, presumably by the subject's spontaneous attention. On the other hand, it could also mean that minor processing is possible even in the absence of attention. This is confirmed by studies with dual tasks (Braun and Julesz 1998). Here, subjects' attention is bound by a difficult primary discrimination task at

the fixation point. The extent to which additional peripheral stimuli can be detected without diverting attention from the central task is then investigated. It is first shown that simple stimulus features (such as color) can be detected without attentional cost, whereas complex stimulus features (such as shape) cannot. This suggests that conscious recognition of simple features is not dependent on attention. Later, it was shown that even the core content of extended and complex visual scenes can be reproduced without attention (Li et al. 2002).

2. *Attention can be a hindrance to awareness.* Awareness can in some cases be improved by paying less attention to the target stimulus (Yeshurun and Carrasco 1998; Olivers and Nieuwenhuis 2005). For example, Yeshurun and Carrasco (1998) have shown that foveal texture detection can be degraded by attention, presumably because processing of coarse-resolution shape information suffers from the increased spatial resolution induced by attention.

3. *Attention and consciousness can have opposite effects on visual processing.* Some experiments suggest a double dissociation between the effects that attention and conscious stimulus detection have on visual information processing. For example, a number of studies on afterimages showed that attention shortens their duration, whereas consciousness can increase their duration (see Koch and Tsuchiya 2007 for a review), although there are also opposite findings (Kaunitz et al. 2011).

4. *Attention can also be controlled by unconscious stimuli.* For example, McCormick (1997) directly compared attentional control by suprathreshold and subliminal stimuli in a study. It was shown that even invisible stimuli can trigger exogenous orienting of attention (see Mulckhuyse and Theeuwes 2010 for a detailed review). Attention can even be

guided by complex properties of invisible stimuli, which suggests that subliminal stimuli and can be processed deep enough to control attention (Jiang et al. 2006). Interestingly, the effect may depend on the unconscious stimulus that attracts attention being relevant to the subject's task (Ansorge et al. 2010). Taken together, then, the above points show a clear dissociability between attention and consciousness, even though the two processes are usually closely intertwined.

Summary

In this chapter, we first considered the psychological conditions of the occurrence of conscious awareness and found that it is not an all-or-nothing phenomenon, but that there is a gradual transition between completely unconscious and clearly conscious stimuli: Thus, some features of a stimulus may be recognized but not others, or seemingly completely unconscious "masked" stimuli may have a marked influence on choice decisions between alternative consciously presented stimuli (priming). There is evidence that certain stimuli are not consciously perceived at all, but nevertheless influence our behavior, for example in unconscious learning and conditioning tasks.

The crucial cognitive neuroscience question is whether there are stable differences in the brain activities evoked by unconscious or conscious stimuli. It has been established that unconscious stimuli can activate not only low-level but also high-level processing areas of the brain, e.g. the temporal face recognition area (fusiform gyrus), but that activation by conscious stimuli is always more extensive and involves a larger number of cortical areas. This mainly concerns the so-called fronto-parietal network, which extends between the superior frontal cortex (PFC) and the posterior parietal cortex (PPC). However, even in primary sensory areas, e.g. the primary visual cortex,

8

responses to stimuli that become conscious are stronger than those that remain unconscious.

Neuroscience has also addressed the question whether there are discrete single-unit representations, called "cardinal cells", not only for very simple stimuli, but also for certain complex contents such as faces, persons and situations, or whether such contents are encoded in a distributed code involving activity patterns of wide cortical populations of neurons. Recent research has confirmed the latter view for many domains as the rule using the method of multivariate decoding. It shows that even in the case of distributed coding, the same perceptual contents give rise to the same neuronal representations.

Another central question of neuroscientific consciousness research is to what extent differences and similarities between contents of consciousness are also expressed in differences and similarities of neuronal coding, i.e. whether the structure of our experience is reflected in the relational structure of our brain activity patterns. Subjectively perceived similarities are indeed expressed in similar neural responses. Thus, the link between subjective experience and brain activity can be established also at the level of detailed contents.

Consciousness and attention are usually closely linked: Attention acts to amplify the neural responses to sensory stimuli. Attention and awareness are not identical, however, because we can miss information in stimuli even when they are fully focused. This can happen, for example, when the information was not expected.

Neurotheorists and neurophilosophers have developed numerous models for the emergence of consciousness and its relationship to brain processes. Some authors assume that everything that affects the brain strongly enough and for long enough also penetrates into consciousness. Other authors assume that this requires "recurrent" activity between primary sensory areas and associative areas in the cortex. According to one older model such areas "bind together" in a specific way using high-frequency synchronization in the so-called gamma range (30–80 Hz) to produce meaningful conscious content. This can also be integrated with the idea that there is a "core mechanism" underlying all consciousness content that is active between the cortex and thalamus and constantly aligns with new current content. The various models can explain many aspects of consciousness well, but by no means all of them. According to a different view, consciousness can be understood as a "global workspace" involving special mechanisms of information processing, enabling current detailed perception, its memorability and linguistic reportability, and greater access to behavioral responses, including complex action planning.

Current neuroscience provides strong evidence against so-called dualistic positions, according to which there is an independence of brain and mind. Rather, the contents of consciousness are precisely related to neuronal processes in the brain.

References

Adams JK (1957) Laboratory studies of behavior without awareness. Psychol Bull 54:383–405

Ansorge U, Horstmann G, Worschech F (2010) Attentional capture by masked colour singletons. Vis Res 50(19):2015–2027. (Epub 2010 Jul 24)

Baars BJ (2002) The conscious access hypothesis: origin and recent evidence. Trends Cogn Sci 8:47–52

Beck DM, Rees G, Frith CD, Lavie N (2001) Neural correlates of change detection and change blindness. Nat Neurosci 4(6):645–650

Berti A, Rizzolatti G (1992) Visual processing without awareness: evidence from unilateral neglect. J Cogn Neurosci 4:345–351

Binder JR, Rao SM, Hammeke TA, Yetkin FZ, Jesmanowicz A, Bandettini PA, Wong EC, Estkowski LD, Goldstein MD, Haughton VM et al (1994) Functional magnetic resonance imaging of human auditory cortex. Ann Neurol 35(6):662–672

Blankertz B, Dornhege G, Schäfer C, Krepki R, Kohlmorgen J, Müller KR, Kunzmann V, Losch F, Curio G (2003) Boosting bit rates and error detection for the classification of fast-paced motor commands based on single-trial EEG analysis. IEEE Trans Neural Syst Rehabil Eng 11(2):127–131

Boring EG (1930) A new ambiguous figure. Am J Psychol 42:444

Braun J, Julesz B (1998) Withdrawing attention at little or no cost: detection and discrimination tasks. Percept Psychophys 60(1):1–23

Bullier J, Kennedy H (1983) Projection of the lateral geniculate nucleus onto cortical area V2 in the macaque monkey. Exp Brain Res 53(1):168–172

Buszáki G, Schomburg EW (2015) What does gamma coherence tell us about interregional neural communication? Nat Neurosci 18:484–489

Cannon MW (1984) A study of stimulus range effects in free modulus magnitude estimation of contrast. Vis Res 24(9):1049–1055

Cichy RM, Chen Y, Haynes JD (2011a) Encoding the identity and location of objects in human LOC. NeuroImage 54(3):2297–2307

Cichy RM, Heinzle J, Haynes JD (2011b) Imagery and perception share cortical representations of content and location. Cereb Cortex 22:372–380

Crick F, Koch C (1995) Are we aware of neural activity in primary visual cortex? Nature 375(6527):121–123

Crick F, Koch C (1998) Consciousness and neuroscience. Cereb Cortex 8(2):97–107

Debner JA, Jacoby LL (1994) Unconscious perception: attention, awareness, and control. J Exp Psychol Learn Mem Cogn 20(2):304–317

Dehaene S, Naccache L (2001) Towards a cognitive neuroscience of consciousness: basic evidence and a workspace framework. Cognition 79(1–2):1–37

Dehaene S, Naccache L, Cohen L, Bihan DL, Mangin JF, Poline JB, Rivière D (2001) Cerebral mechanisms of word masking and unconscious repetition priming. Nat Neurosci 4(7):752–758

Edelman S, Grill-Spector K, Kushnir T, Malach R (1998) Towards direct visualization of the internal shape space by fMRI. Psychobiology 26:309–321

Engel AK, Singer W (2001) Temporal binding and the neural correlates of sensory awareness. Trends Cogn Sci 5(1):16–25

Eriksen CW (1954) The case for perceptual defense. Psychol Rev 61:175–182

Eriksen CW (1960) Discrimination and learning without awareness: a methodological survey and evaluation. Psych Rev 67:279–300

Fang F, He S (2005) Cortical responses to invisible objects in the human dorsal and ventral pathways. Nat Neurosci 8(10):1380–1385. (Epub 2005 Sep 4)

Fechner GT (1860) Elemente der Psychophysik. Breitkopf & Härtel, Leipzig

Fries P (2015) Rhythmus for cognition: communication for coherence. Neuron 88:220–235

Gescheider GA (1988) Psychophysics: the fundamentals. Annu Rev Psychol 39:169–200

Gescheider G (1997) Psychophysics: the fundamentals. Erlbaum, New Jersey

Goebel R, Muckli L, Zanella FE, Singer W, Stoerig P (2001) Sustained extrastriate cortical activation without visual awareness revealed by fMRI studies of hemianopic patients. Vis Res 41(10–11):1459–1474

Goodale MA, Milner AD (1992) Separate visual pathways for perception and action. Trends Neurosci 15(1):20–25

Green DM, Swets JA (1966) Signal detection theory and psychophysics. Wiley, New York

Grill-Spector K, Kushnir T, Hendler T, Malach R (2000) The dynamics of object-selective activation correlate with recognition performance in humans. Nat Neurosci 3(8):837–843

Gur M, Snodderly DM (1997) A dissociation between brain activity and perception: chromatically opponent cortical neurons signal chromatic flicker that is not perceived. Vis Res 37(4):37782

Hadjikhani N, Liu AK, Dale AM, Cavanagh P, Tootell RB (1998) Retinotopy and color sensitivity in human visual cortical area V8. Nat Neurosci 1(3):235–241

Hawkins HL, Hillyard SA, Luck SJ, Mouloua M, Downing CJ, Woodward DP (1990) Visual attention modulates signal detectability. J Exp Psychol Hum Percept Perform 16:802–811

Haynes JD (2009) Decoding visual consciousness from human brain signals. Trends Cogn Sci 13(5):194–202

Haynes JD, Rees G (2005a) Predicting the orientation of invisible stimuli from activity in human primary visual cortex. Nat Neurosci 8(5):686–691

Haynes JD, Rees G (2005b) Predicting the stream of consciousness from activity in human visual cortex. Curr Biol 15(14):1301–1307

Haynes JD, Rees G (2006) Decoding mental states from brain activity in humans. Nat Rev Neurosci 7(7):523–534

Haynes JD, Lotto RB, Rees G (2004) Responses of human visual cortex to uniform surfaces. Proc Natl Acad Sci U S A 101(12):4286–4291

Haynes JD, Deichmann R, Rees G (2005a) Eye-specific effects of binocular rivalry in the human lateral geniculate nucleus. Nature 438(7067):496–499

Haynes JD, Driver J, Rees G (2005b) Visibility reflects dynamic changes of effective connectivity between V1 and fusiform cortex. Neuron 46(5):811–821

He S, Cavanagh P, Intriligator J (1996) Attentional resolution and the locus of visual awareness. Nature 383(6598):334–337

Hochstein S, Ahissar M (2002) View from the top: hierarchies and reverse hierarchies in the visual system. Neuron 36(5):791–804

Holender D (1986) Semantic activation without conscious identification in dichotic listening, parafoveal vision, and visual masking: a survey and appraisal. Behav Brain Sci 9:1–23

Jiang Y, Costello P, Fang F, Huang M, He S (2006) A gender- and sexual orientation-dependent spatial attentional effect of invisible images. Proc Natl Acad Sci U S A 103(45):17048–17052. (Epub 2006 Oct 30)

Kahnt T, Grueschow M, Speck O, Haynes JD (2011) Perceptual learning and decision-making in human medial frontal cortex. Neuron 70(3):549–559

Kanwisher N, McDermott J, Chun MM (1997) The fusiform face area: a module in human extrastriate cortex specialized for face perception. J Neurosci 17(11):4302–4311

Kaunitz L, Fracasso A, Melcher D (2011) Unseen complex motion is modulated by attention and generates a visible aftereffect. J Vis 11(13):10. https://doi.org/10.1167/11.13.10

Koch C, Tsuchiya N (2007) Attention and consciousness: two distinct brain processes. Trends Cogn Sci 11(1):16–22

König P, Engel AK, Singer W (1996) Integrator or coincidence detector? The role of the cortical neuron revisited. Trends Neurosci 19(4):130–137

Kouider S, Dupoux E (2004) Partial awareness creates the "illusion" of subliminal semantic priming. Psychol Sci 15:75–81

Kovács G, Vogels R, Orban GA (1995) Cortical correlate of pattern backward masking. Proc Natl Acad Sci U S A 92(12):5587–5591

Kunimoto C, Miller J, Pashler H (2001) Confidence and accuracy of near-threshold discrimination responses. Conscious Cogn 10:294–340

Lamme VA (2003) Why visual attention and awareness are different. Trends Cogn Sci 7(1):12–18

Lamme VA, Supèr H, Landman R, Roelfsema PR, Spekreijse H (2000) The role of primary visual cortex (V1) in visual awareness. Vis Res 40(10–12):1507–1521

Leopold DA, Logothetis NK (1996) Activity changes in early visual cortex reflect monkeys' percepts during binocular rivalry. Nature 379(6565):549–553

Leopold DA, Logothetis NK (1999) Multistable phenomena: changing views in perception. Trends Cogn Sci 3(7):254–264

Li FF, VanRullen R, Koch C, Perona P (2002) Rapid natural scene categorization in the near absence of attention. Proc Natl Acad Sci U S A 99(14):9596–9601

Lumer ED, Rees G (1999) Covariation of activity in visual and prefrontal cortex associated with subjective visual perception. Proc Natl Acad Sci U S A 96(4):1669–1673

Mach E (1886) Beiträge zur Analyse der Empfindungen. Fischer, Berlin

Mack A, Rock I (1998) Inattentional blindness. MIT Press, Cambridge

Marcel AJ (1983) Conscious and unconscious perception: experiments on visual masking and word recognition. Cogn Psychol 15(2):197–237

Martens U, Ansorge U, Kiefer M (2011) Controlling the unconscious: attentional task sets modulate subliminal semantic and visuomotor processes differentially. Psychol Sci 22(2):28291

Melloni L, Mudrik L, Pitts M, Bendtz K, Ferrante O, Gorska U, Hirschhorn R, Khalaf A, Kozma C, Lepauvre A, Liu L, Mazumder D, Richter D, Zhou H, Blumenfeld H, Boly M, Chalmers DJ, Devore S, Fallon F, de Lange FP, Jensen O, Kreiman G, Luo H, Panagiotaropoulos TI, Dehaene S, Koch C, Tononi G (2023) An adversarial collaboration protocol for testing contrasting predictions of global neuronal workspace and integrated information theory. PLoS One. 18(2):e0268577. https://doi.org/10.1371/journal.pone.0268577. PMID: 36763595; PMCID: PMC9916582

McConkie GW, Currie CB (1996) Visual stability across saccades while viewing complex pictures. J Exp Psychol Hum Percept Perform 22(3):563–581

McCormick PA (1997) Orienting attention without awareness. J Exp Psychol Hum Percept Perform 23(1):168–180

McGinnies E (1949) Emotionality and perceptual defense. Psychol Rev 56:244–251

Merikle PM, Joordens S, Stolz JA (1995) Measuring the relative magnitude of unconscious influences. Conscious Cogn 4(4):422–439

Moutoussis K, Zeki S (1997) A direct demonstration of perceptual asynchrony in vision. Proc Biol Sci 264(1380):393–399

Moutoussis K, Zeki S (2002) The relationship between cortical activation and perception investigated with invisible stimuli. Proc Natl Acad Sci U S A 99(14):9527–9532

Mulckhuyse M, Theeuwes J (2010) Unconscious attentional orienting to exogenous cues: a review of the literature. Acta Psychol 134(3):299–309

Necker LA (1832) Observations on some remarkable optical phaenomena seen in Switzerland; and on an optical phaenomenon which occurs on viewing a figure of a crystal or geometrical solid. Lond Edinb Philos Mag J Sci 1(5):329–337

Ni J, Wunderle T, Lewis CM, Desimone R, Diester I, Fries R (2016) Gamma-rhythmic gain modulation. Neuron 92:240–251

Niedeggen M, Wichmann P, Stoerig P (2001) Change blindness and time to consciousness. Eur J Neurosci 14(10):1719–1726

8

Olivers CN, Nieuwenhuis S (2005) The beneficial effect of concurrent task-irrelevant mental activity on temporal attention. Psychol Sci 16(4):265–269

Parker AJ, Newsome WT (1998) Sense and the single neuron: probing the physiology of perception. Ann Rev Neurosci 21:227–277

Pascual-Leone A, Walsh V (2001) Fast backprojections from the motion to the primary visual area necessary for visual awareness. Science 292(5516):510–512

Peirce CS, Jastrow J (1884) On small differences in sensation. Mem Natl Acad Sci 3:73–83

Pessiglione M, Petrovic P, Daunizeau J, Palminteri S, Dolan RJ, Frith CD (2008) Subliminal instrumental conditioning demonstrated in the human brain. Neuron 59(4):561–567

Posner MI, Snyder CR, Davidson BJ (1980) Attention and the detection of signals. J Exp Psychol Gen 109:160–174

Quiroga RQ, Reddy L, Kreiman G, Koch C, Fried I (2005) Invariant visual representation by single neurons in the human brain. Nature 435(7045):1102–1107

Ray RD, Maunsell JH (2015) Do gamma oscillations play a role in cerebral cortex? Trends Cogn Sci 19:78–85

Rees G, Wojciulik E, Clarke K, Husain M, Frith C, Driver J (2002) Neural correlates of conscious and unconscious vision in parietal extinction. Neurocase 8(5):387–393

Reingold EM, Merikle PM (1988) Using direct and indirect measures to study perception without awareness. Percept Psychophys 44:563–575

Rensink RA, O'Regan JK, Clark JJ (1997) To see or not to see: the need for attention to perceive changes in scenes. Psychol Sci 8:368–373

Ress D, Heeger DJ (2003) Neuronal correlates of perception in early visual cortex. Nat Neurosci 6(4):414–420

Sáry G, Vogels R, Orban GA (1993) Cue-invariant shape selectivity of macaque inferior temporal neurons. Science 260(5110):995–997

Schmidt T, Vorberg D (2006) Criteria for unconscious cognition: three types of dissociation. Percept Psychophys 68(3):489–504

Sidis B (1898) The psychology of suggestion. Appleton, New York

Siegel M, König P (2003) A functional gamma-band defined by stimulus-dependent synchronization in area 18 of awake behaving cats. J Neurosci 23(10):4251–4260

Simons DJ, Levin DT (1998) Failure to detect changes to people in a real-world interaction. Psychon Bull Rev 5:644–649

Sperling G (1960) The information available in brief visual presentations. Psychol Monogr 74:1–29

Srinivasan N (2008) Interdependence of attention and consciousness. Prog Brain Res 168:65–75

Supèr H, Spekreijse H, Lamme VA (2001) Two distinct modes of sensory processing observed in monkey primary visual cortex (V1). Nat Neurosci 4(3):304–310

Taylor K, Mandon S, Freiwald WA, Kreiter AK (2005) Coherent oscillatory activity in monkey area v4 predicts successful allocation of attention. Cereb Cortex 15:1424–1437

Tong F, Engel SA (2001) Interocular rivalry revealed in the human cortical blind-spot representation. Nature 411(6834):195–199

Tong F, Nakayama K, Vaughan JT, Kanwisher N (1998) Binocular rivalry and visual awareness in human extrastriate cortex. Neuron 21(4):753–759

Tononi G (2004) An information integration theory of consciousness. BMC Neurosci 2(5):42

Tononi G, Edelman GM (1998) Consciousness and complexity. Science 282(5395):1846–1851

Tononi G, Srinivasan R, Russell DP, Edelman GM (1998) Investigating neural correlates of conscious perception by frequency-tagged neuromagnetic responses. Proc Natl Acad Sci U S A 95(6):3198–3203

Tootell RB, Dale AM, Sereno MI, Malach R (1996) New images from human visual cortex. Trends Neurosci 19(11):481–489

Vuilleumier P, Sagiv N, Hazeltine E, Poldrack RA, Swick D, Rafal RD, Gabrieli JD (2001) Neural fate of seen and unseen faces in visuospatial neglect: a combined event-related functional MRI and eventrelated potential study. Proc Natl Acad Sci U S A 98(6):3495–3500

Vuilleumier P, Armony JL, Clarke K, Husain M, Driver J, Dolan RJ (2002) Neural response to emotional faces with and without awareness: event-related fMRI in a parietal patient with visual extinction and spatial neglect. Neuropsychologia 40(12):2156–2166

Watanabe T, Náñez JE, Sasaki Y (2001) Perceptual learning without perception. Nature 413(6858):844–848

Watanabe M, Cheng K, Murayama Y, Ueno K, Asamizuya T, Tanaka K, Logothetis N (2011) Attention but not awareness modulates the BOLD signal in the human V1 during binocular suppression. Science 334(6057):829–831

Watson AB (1986) Temporal sensitivity. In: Boff K, Kaufman L, Thomas J (eds) Handbook of perception and human performance. Wiley, New York

Weiskrantz L (2004) Roots of blindsight. Prog Brain Res 144:229–241

Weiskrantz L, Warrington EK, Sanders MD, Marshall J (1974) Visual capacity in the hemianopic field following a restricted occipital ablation. Brain 97(4):709–728

Wunderlich K, Schneider KA, Kastner S (2005) Neural correlates of binocular rivalry in the human lateral geniculate nucleus. Nat Neurosci 8(11):1595–1602

Yeshurun Y, Carrasco M (1998) Attention improves or impairs visual performance by enhancing spatial resolution. Nature 396(6706):72–75

Zeki S (2001) Localization and globalization in conscious vision. Annu Rev Neurosci 24:57–86

Zeki S, Ffytche DH (1998) The Riddoch syndrome: insights into the neurobiology of conscious vision. Brain 121(Pt 1):25–45

Zeki S, Watson JD, Lueck CJ, Friston KJ, Kennard C, Frackowiak RS (1991) A direct demonstration of functional specialization in human visual cortex. J Neurosci 11(3):641–649

8

Nature, Diagnosis and Classification of Mental Disorders

Henrik Walter

Contents

What is a mental disorder? This question must be addressed by any diagnostic and classification system in psychiatry. In this chapter, a brief systematic overview of theories of mental disorders is given and an integrative working definition is suggested. Furthermore, an overview of the prevalence of mental disorders in Europe is given and the current situation for psychiatric care in Germany is presented. Subsequently, some problems of the current classification systems (ICD and DSM), which are based on clinical criteria alone, are discussed. Finally, three more recent approaches to the understanding of mental disorders are presented, which follow an integrative approach, i.e. take clinical, neuroscientific and psychosocial aspects into account.

Learning Objectives

After reading this chapter, the reader should be able to explain how mental disorders can be defined, how they are diagnosed and classified, how common they are, and which recent integrative approaches to mental disorders are currently being discussed.

9.1 What Is a Mental Disorder?

9.1.1 Background and Historical Context

Psychiatry deals with mental disorders such as schizophrenia, depression, anxiety disorders, obsessive-compulsive disorders or post-traumatic stress disorders. In contrast, neurology deals with diseases such as stroke, brain tumors, multiple sclerosis, peripheral nerve damage and muscle diseases. However, there is also overlap between these two medical specialties, such as dementia, which is addressed by both. While it is relatively easy to give examples of mental disorders, it is much more difficult to say what a mental disorder actually *is*. Problems of demarcation arise on two sides. First, where does the

psychological normal end and the pathological begin? Where, for example, is the boundary between shyness and social phobia, between grief and depression, or between extraordinary experiences like hearing voices and schizophrenia? Secondly, where is the boundary between neurology and psychiatry, when both are apparently based on dysfunctions of the brain? In addition to these problems of demarcation, there is the specific issue that psychiatry has repeatedly been accused throughout its history—and in certain cases quite rightly—that the term "disease", "illness" or "disorder" serves only to pathologize deviation from social standards and thus serves to discipline society. It was not until 1990, for example, that the World Health Organization (WHO) removed homosexuality from the list of mental disorders. In Germany, homosexual acts were still considered a criminal offence until 1969, and the corresponding paragraph §175 was deleted from the Criminal Code only in 1994. Political dissidents in the Soviet Union were sometimes labeled as mentally disturbed and internalised in hospitals for "treatment" if they did obey to the insights into the truths of Marxist-Leninist doctrines. And in the Third Reich, based on racial hygiene doctrins, the persecution and murder of mentally ill patients was systematically organized within the T3 action.

Considering this background, it is not surprising that strong opposition to clinical psychiatry arose in the 1960s, which must also be regarded against the background that at that time many mentally ill patients were housed in large, isolated state hospitals under conditions that are unacceptable from today's standards and were treated inadequately, if at all. The anti-psychiatric movement emerged, which postulated that mental disorders did not exist, but were only socially constructed based on deviant behavior and thus ultimately an expression of social problems. In Western Germany, the Psychiatry Enquete Commision in 1975 drafted a report on the state of psychiatry. As a result, many of the state hospitals were dissolved, Departments of Psychiatry were created in normal hospitals, i.e. psychiatry was integrated into medicine, and many things improved.

These psychiatry-specific developments coincided with a boom of the neurosciences, a better understanding of the nervous system, the introduction of effective psychotropic drugs in the 1960s and 1970s, and a professionalization of psychotherapy with the development of new, effective treatment methods. Knowledge about mental disorders has increased considerably since then, although not to the same extent as, for example, in neurologgy.

In addition to its therapeutic mandate to treat disorders and reduce suffering, psychiatry today still has a public-legal function. When patients, due to a mental disorder, have a lack or strong impairment in their capacity for insight or in controlling their actions and at the same time therefore are a danger for themselves or others, they can be compulsory admitted and/or treated against their will in a psychiatric hospital, a procedure strictly regulated by specific laws. This may be the case for example while being in a delirium due to alcohol withdrawal, or being in a psychotic delusional state of mind, e.g. in schizophrenia, or being suicidal in a severe depressive episode. After the first effective psychotropic drugs were introduced to the market, the pharmaceutical industry gained a great deal of influence in the field of psychiatry. As we know today, drug effects were exaggerated, results presented far too positively, side effects were played down, and frequently all of this was done with criminal intent (cf. for example Hasler 2013). Against this background, it is not surprising that widespread mistrust in medication had developed as well as public reservations about the shere possibility of compulsory admission and treatment. Despite all of these negative aspects, psychiatric care in Germany has improved dramatically in recent decades. There is often very good inpatient care, a dense network for outpatients, excellent funding compared to many other countries, complementary facilities and care structures, a more modest use of drugs has developed with using lower doses, and a wealth of evidence-based psychotherapies for almost all mental disorders is now available. Nevertheless, one must be aware of this historical context when looking at classification and diagnosis in order to be able to properly understand some underlying controversies in the field of psychiatry.

9.1.2 Construction of a General Concept of Disease and Disorder from a Philosophy of Science Point of View

The question of what disease and health *are* arises not only in psychiatry, but in medicine in general. The medical ethicist Peter Hucklenbroich from Münster, Germany, provides a sound construction of the general concept of disease in biomedicine from a philosophy of science point of view (Hucklenbroich 2012). First, he distinguishes four levels of the concept of disease, disorder or pathology (in German: Krankheit). The first is personal and related to the life-world of a person (person X is sick). On the second level, a distinction can be made between healthy and pathological life processes (process X is pathological). On the third level, reference is made to a normal model of the human organism (X is pathologically altered), on which the pathodisciplines are based (pathophysiology, pathobiochemistry and psychopathology). Only at the fourth level disease entities and categories are postulated (X is a disease). These entities denote either individual diseases (influenza, myocardial infarction, femoral neck fracture, alcohol withdrawal delirium) or categories (e.g., cystic kidney disease, tachyarrhythmias). This distinction of levels is helpful in better understanding some of the discussions in the field. For example, it is popular to claim that there are no diseases at all, only sick people. This sounds good, but mixes up the first and fourth level and does not change the need to look for correlations, mechanisms and causes at level two and three.

Life processes to which the four criteria apply can be said to be pathological (in German: krank): They are conditions, processes or procedures

1. of or within individuals,
2. that can be attributed to the organism, not the environment,
3. which exist and develop independently of the will and knowledge of the individual organism, and
4. to which there exists at least one non-pathological alternative course.

But what exactly is pathological and what is not? Here it is important to distinguish between positive and negative disease criteria (◘ Table 9.1).

Clearly, this list of criteria is very general and includes some critical formulations that have been discussed again and again in the history of psychiatry, especially the fourth and fifth positive disease criteria. The first negative criterion also has been discussed again and again in view of the possibilities of modern medicine. But at least this set of criteria provides a blueprint for assessing the relevance of the biomedical model of disease in psychiatry.

Another helpful distinction is to distinguish secondary and tertiary pathological features from primary pathological ones. Secondary pathological features occur as a result of primary pathological processes and do not otherwise occur in the organism, e.g. fever, redness or swelling, scarring. Accordingly, in psychiatry, some symptoms may be only secondary pathological. Tertiary pathological features are not pathological in the first place, but can be considered pathological because they are a causal result of other pathological processes. An example would be pathological short stature compared to people who are simply short within the normal variance. In psychiatry, this may apply, for example, to certain forms of social behavior that may be very similar to normal variation but have different causes, for example "normal variants" (shyness) and social withdrawal (in the case of pronounced social phobia or in the context of schizophrenia).

Finally, the concept of disease entities postulates that there are individual diseases that can be distinguished from each other. The system of disease entitities is referred to as "nosology", whereas the doctrine of disease causes is called "etiology". In medicine, five dimensions of diseases are typically distinguished: the nature of the initial cause, the nature of the subjects potentially affected, the nature of the effect of the cause on those who are affected by the disease, a specific pathogenesis and pathophysiology, and a time characteristic of the course and signs of the disease. This can be well spelled

◘ **Table 9.1** Positive and negative disease criteria. (According to Hucklenbroich 2012 with changes)

Five positive disease criteria	Two negative disease criteria (= not sick)
1. Lethality 2. Pain, discomfort, suffering 3. Disposition for 1 or 2 4. Inability to reproduce 5. Inability to live together	1. Universal occurrence and inevitability, e.g. sex, intrauterine and ontogenetic phases, pregnancy, menopause, age, natural death 2. Knowingly and willingly self-induced behaviour (provided self-determination is not impaired), such as value judgements, risky behaviour, abstinence, deliberate lying, reflected suicide ("German: Freitod")

out by the example of a respiratory infection or the radius fracture *loco classico* (cf. in detail Hucklenbroich 2012). As a final aspect, it should be pointed out that, once established, treatment methods can of course also be used to treat non-pathological conditions in the above sense, such as age-related complaints, unavoidable pain conditions (childbirth, teething pain), prophylaxis or protection against social disadvantages due to physical stigmas, on request (cosmetic surgery) or for life and family planning (contraception, induction of childbirth, sterilization). Applied to psychiatry, this means that one can of course also "treat" life problems, which are not diseases or disorders, with psychotherapeutic techniques.

9.1.3 Current Definitions of Mental Disorders

- **From the Biomedical to the Biopsychosocial Model**

In the early days of psychiatry, the leading model was the biomedical one. Mental diseases (disorders), as the then 28-year-old Wilhelm Griesinger put it in his textbook "Die Pathologie und Therapie der psychischen Krankheiten, für Aerzte und Studirende" (The Pathology and Therapy of Mental Diseases, for Physicians and Students) as early as 1845, are brain diseases. This claim did not stand in contrast to his own, quite progressive social psychiatric approach. Subsequently, psychiatrist tried to find causes for mental disorders according to the respective state of knowledge and the available methods of their time by looking for infectious causes (prime example: Treponema pallidum as causative agent of progressive paralysis), histological changes of the brain (heyday of neuroanatomy at the beginning of the twentieth century), genetic factors (hereditary theory), neurotransmitter dysfunction in the brain (discovery of

psychotropic drugs) or in anatomical and functional connectivity changes of the brain (with the emergence of functional neuroimaging some decades ago). Interestingly, one of the main proponents of anti-psychiatry, Thomas Szasz, also follows the biomedical model of disease. In accordance with this model, he argues that mental illnesses would be "real" illnesses if one could identify a clear neuropathology, as in neurology. But since this is not the case, or so he argues, mental illness is a "myth" (Szasz 1961) that falsely categorizes common life problems as illnesses. In his 1961 book, however, he does not take on schizophrenia or depression, but discusses hysteria, as popularized by Charcot around 1900 (among other things, Freud attended Charcots clinical demonstrations), as a prime example. Today, the clinical picture of hysteria is categorized as a "dissociative disorder" in modern classification systems, and conceptualized as a mainly psychogenic disorder, and plays only a very marginal role in psychiatry. However, there are recent studies on dissociative disorders that attempt to identify the neurobiological mechanisms involved in the etiopathogenesis of "hysteria" (Boeckle et al. 2016; Schönfeldt-Lecuona et al. 2004). Nowadays, it seems self-evident to us that psychogenic diseases must also be manifest in the brain in some or the other way (cf. the preface in Walter 2005).

Furthermore, today we take it for granted that, in addition to genetic predispositions and neurobiological factors, life experiences, psychological processing and social factors also play a role in the development of mental illnesses and disorders. This was not always so clear, because the narrow biomedical model had no place for social, psychological and behavioural mechanisms. Their relevance was effectively postulated only by the historian of psychiatry George L. Engel in his biopsychosocial model, now cited in virtually all psychiatry textbooks (Engel

1977). It is closely related to the vulnerability-stress or stress-diathesis model. These models state that we are all endowed with a greater or lesser degree of vulnerability which, under the influence of 'stressors' on our experience and behaviour, can result in us becoming ill. Or to put it even more simply. Mental illness is always a combination of predisposition and environmental influences. However, this statement is so general that it is almost trivially true. Moreover, it says nothing about what a mental illness is, but rather something about its etiology, that is, how it comes about.

■ Definitions of Mental Disorders

How do the two major classification systems of psychiatry, the ICD (International Classification of Diseases) of the WHO (Dilling et al. 2015) and the DSM (Diagnostic and Statistical Manual of Mental Disorders) of the American Psychiatric Association (APA 2013) define mental disorders? The ICD classification of diseases evolved from the need to categorize diseases for death statistics. It is now of central administrative, and therefore financial and statistical, relevance in billing with health insurers in both Germany (ICD-10) and the USA (ICD 9). The fifth chapter of ICD-10 covers "Mental and behavioural disorders". The title itself suggests that it is apparently not always clear what exactly is mental, and that it is sometimes easier to simply classify behaviour. The DSM, also called the "bible" of psychiatry, is a manual that is much more comprehensive than the ICD-10 and contains detailed scientific explanations of the individual clinical conditions. It only attained a far-reaching significance in its third version of 1980 (DSM-I: 1952, DSM-II: 1968) On page 20 the following definition of a mental disorder can be found:

> A mental disorder is a syndrome characterized by a clinically relevant disturbance in an individual's cognition, emotion regulation, or behavior that indicates dysfunction in the psychological, biological, or developmental processes underlying mental functioning. Mental disorders are usually associated with significant distress or disability in social, occupational, or other important activities. An expected or culturally recognized reaction to a common stressor or loss, such as the death of a loved one, is not a mental disorder. Socially deviant behavior (e.g., political, religious, or sexual) or conflict that exists primarily between the individual and society is not a mental disorder unless the deviance or conflict results from dysfunction in the individual as described above. DSM-III, 1980.

This definition is not as bad as its reputation. Similar to Hucklenbroich, it contains both positive and negative criteria. However, it is very broad, so that it is not surprising that the demarcation of non-pathological psychological problems from disordes is not always easy.

A rather narrow definition of the term *clinically relevant* mental disorder can be found in Heinz (2015). He distinguishes three aspects associated with different concepts of "disorder":

– the medical aspect, i.e. in the broadest sense pathophysiological, objectifiable functional disorders (*disease*),
– the subjective feeling of being *ill* (*illness*), i.e. the aspect of suffering
– finally, impairments in the way of life or social participation (*sickness*).

Only when all three aspects are present, it is suggested, can there be a relevant mental illness. This definition is compatible with Wakefield's (1992) classical theory of illness, which focuses on the concept of "harmful dysfunction," although Wakefield takes a more evolutionary approach. For Heinz, only dysfunctions that affect mental functions relevant to surviving should be classified as diseases, such as dysfunctions of alertness, orientation, reasoning ability, memory, delusion, or the loss of affective vibratory capacity. The advantage of such a narrow definition is that it covers all severe mental disorders such as dementia, delirium, paranoid-hallucinatory schizophrenia or severe depression; however, many other disorders, in particular personality disorders, can no longer be regarded as diseases without further ado.

The meaning of dysfunction as well as the consideration of the clinical relevance or severity of a disease makes it clear that normative aspects play a role in the classification of a dysfunction as a disease. At this point, it should be noted that the concept of "normality" can occur in at least three meanings. Often, it is used in a prescriptive sense, i.e. as a setting or social norm. However, there are also norms and normality in a statistical sense (cf. the biostatistical theory of disease by Boorse 2011) and thirdly in a biological sense (function for the organism, so-called Cummins-functions) i.e. evolutionary normality, i.e. with reference to the history of the coming into existence of a function, also called *proper function*.

A very useful definition of mental illness has been provided by the philosopher Georg Graham. According to him, a mental disorder is "a disability, dysfunction, or impairment in one or more basic mental or psychological *faculties* (in the original: "*mental faculties or psychological capacities*") of a person that has harmful or potentially harmful consequences for the person concerned." (Graham 2010, p. 28). What is important in his theory is that the affected person does not necessarily recognise or acknowledge the harmful consequences themselves, she cannot simply control the condition and the condition cannot be made to go away simply by using additional psychological resources, e.g. by simply 'pulling oneself together'. Graham also makes a distinction from typical neurological diseases such as Huntington's disease (genetic defect), Alzheimer's disease (neurodegenerative disease), or encephalitis (inflammation of the brain). Whereas in these cases the disease, which may well also include mental symptoms, is caused by a direct affection of the brain, i.e. through the "brute forces of the neurological", according to Graham in mental disorders the mental is always involved in the genesis of the disease via intentional or conscious processes.

Of course, the question arises here as to what is the nature of intentional or conscious processes—but we can put that question aside here as long as we assume that what is meant by this is not some ominous substance that cannot be grasped scientifically, but a particular kind of natural process that constitutes the mental and for which the brain plays a central and indispensable role.

In the box: Working Definition of Mental Disorder, a working definition for mental disorders is proposed that attempts to preserve the insights of the above theories and closely follows Hucklenbroich's general theory of illness.

Working Definition of Mental Disorder

A mental disorder is a (P1) mental dysfunction, i.e. a disability, disorder or impairment of one or more of a person's basic mental faculties that (P2) results in clinically relevant subjective distress or discomfort and thus (P3) impairs everyday skills of living in a clinically relevant way. It is (N1) not controllable by simple volition or reasonable effort, (N2) not an unavoidable universally occurring process (such as exhaustion, separation stress, or fear of pain), (N3) not an expectable and culturally recognized response to stressors or loss (such as grief), and (N4) not simply a deviation from social values, preferences, or behaviors (e.g., political, religious, or sexual) unless it is secondary caused by one or more independent mental dysfunctions.

This working definition contains three positive (P) and four negative criteria (N). The latter are mainly used to demarcate much discussed "simple" problem cases such as "reasonable fears" grief, homosexuality or political and religious beliefs. The definition does not contain a clear demarcation between disease or simple dysfunction. The simplest demarcation can probably be made by severity. It also does not contain a clear demarcation between a mild mental disorder and severe life problems. The reason for this is simple: there simply is no clear boundary, even though there are clear examples at the ends of the spectrum, i.e. of severe illness on the one hand and clear mental health on the other. Normative and societal factors play an elusive role in drawing the boundary, as will become clear from the discussion of pathological grief below. The exact demarcation is determined by too many theoretical and social factors that make it impossible to give clear boundaries in a definition.

9.2   How To Diagnose a Mental Disorder?

The clinical diagnosis of a mental disorder is made in a similar way as with other diseases and disorders, i.e. by taking a medical history, objective additional tests (laboratory values, brain imaging—in psychiatry, however, usually only to exclude "organic" causes of the complaints, such as neurological or medical diseases), the systematic assessment of the psychopathological status, the consideration of the family history and the observation of the clinical course. With the exception of Alzheimer's dementia, there is no single mental disorder for which objective additional diagnostic findings from the laboratory or imaging exist to objectively confirm or prove a clinical diagnosis. Many patients believe that neurotransmitter deficiencies can be measured, but this is not the case. The major diagnostic tool of the psychiatrist therefore is the psychopathological status, cross-sectionally as well as longitudinally.

Today, standardized procedures for assessing psychopathology exist (i.e., alertness, orientation, memory, formal and content-related thought disorder, affective symptoms, etc.), which will not be presented in detail here; see the respective chapters in this book. Instead, two central concepts for a theory of disease will be briefly discussed: Validity and Reliablity. The validity of a diagnosis means that what is diagnosed actually exists. The problem here, of course, is where the *ground truth* is, that is, how we know that a disease is present. In general medicine, the validity often only could be confirmed by an autopsy, i.e. the (histo-) pathological findings. Nowadays, outside of psychiatry, modern medicine has a variety of objective parameters or biomarkers, laboratory tests, biopsies, or imaging techniques. As already mentioned, these objective measures usually do not exist in psychiatry in such a way that they could be used in clinical routine (or not yet, see below).

Therefore, since the publication of the DSM-III, great emphasis has instead been placed on reliability, i.e., the reliability of a diagnosis (independent of its validity)—in other words, whether two independent examiners arrive at the same diagnosis for the same patient. Whereas prior to the DSM-III, psychiatry was dominated by the triadic system (exogenous ("organic") disorders with known physical causes, endogenous ("internal") disorders with as yet unknown physical causes, and psychogenic disorders) based on a theoretical nosology, the DSM-III marked a clear shift to a descriptive approach. Experts sat together and formulated diagnostic criteria (see the individual disease chapters) so that a diagnosis could be made reliably by determining how many symptoms of a syndrome had to be present over what amount of time for a diagnosis to be made. This approach is descriptive in that it is neutral to the question of what gives rise to such a defined syndrome or what its causes are (exception: neurological and internal causes of an "organic" mental disorder). Thus, in the past, a distinction was made between endogenous depression (comes from within, without an external cause, must be treated with drugs), neurotic depression (has its cause in early conflicts, requires psychotherapy) and reactive depression (is caused by an external event such as death, divorce, job loss, or partner problems). Nowadays, depression is diagnosed only on the basis of number of symptoms and course, and the severity rather than the (assumed) cause is relevant for the type of treatment. We will discuss related problems below.

It was this approach that allowed for reliable diagnosis, epidemiological studies and comparisons between regions and countries, as we will discuss in the next section. It should be noted that a close co-evolution of DSM and ICD has occurred at least since the publication of DSM-III.

9.3 What Mental Disorders Are There and How Common Are They?

Due to the diagnostic criteria of the ICD-10, which have been trimmed for reliability, there are now very reliable studies on the frequency of mental disorders. Probably the most comprehensive and highest-quality study currently available for Europe was conducted in 2011 (Wittchen et al. 2011). In this study, systematic reviews, reanalysis of existing datasets, national surveys and consultations with experts were used to determine the frequency of diagnoses based on the ICD-10 within one year in 27 EU Member States (EU-27) as well as Switzerland, Norway and Iceland, together accounting for a good 500 million people. Accordingly, in 2010, more than one third of the European population met the diagnostic criteria for at least one of 27 mental disorders, more precisely 38.2% or roughly 165 million people. The absolute and percentage frequencies for the most important individual disorders according to ICD-10 are shown in ◻ Table 9.2.

Of great interest here is the question of whether mental disorders have increased in frequency, as it repeatedly has been claimed. In 2005, Wittchen and Jacobi (2005) published a study that used the same methodology. At that time, only 13 diseases were examined. Five years later, there was no increase in the frequency of diagnosis for these disorders (2005: 27.4%. 2010 27.1%). The increase in mental disorders often reported in the media usually is due to other factors, such as the number of sick leaves counted by health insurance companies due to mental disorders. These have indeed increased (Wittchen and Jacobi 2005). However, in practice the diagnostic criteria are generally not checked as strictly as in the surveys on which the Wittchen study is based. At best, therefore, it can be stated that the number of sick leaves has increased,

◘ **Table 9.2** Frequency of mental illnesses in Europe 2010 (EU-27 plus Switzerland, Iceland, Notwergen) from Wittchen et al. 2011. The three disorders printed in bold are core disorders of clinical psychiatry (schizophrenia, bipolar disorder, major depression). Disorders that were only included in the 2010 survey but not in 2005 (Wittchen and Jacobi 2005) are printed in italics

ICD-10	Category	Diagnosis	Frequency	Absolute number
F00-09	Organic, including symptomatic mental disorders	*Dementia*	1.2%	6.3 million
F10-19	Mental and behavioural disorders due to use of psychotropic substances	Alcohol Opioids Cannabis	**3.4%** 0.1–0.4% 0.3–1.8%	**14.6 million** 1.0 million 1.4 million
F20-29	**Schizophrenia**, schizotypal and delusional disorders	**Psychotic disorders**	**1.2%**	**5.0 million**
F30-39	Affective disorders	**Depression** **Bipolar disorder**	**6.9%** **0.9%**	**30.3 million** **3.0 million**
F40-48	Neurotic, stress related and somatoform disorders	Panic disorder Agoraphobia Social phobias Generalised anxiety disorder, Obsseve-compulsive disorder, somatoform disorder, Post-traumatic stress disorder	1.8% 2.0% 2.3% 1.7–3.4% 0.7% 4.9% 1.1–2.9%	7.9 million 8.8 million 10.1 million 8.9 million 2.9 million 20.4 million 7.7 million
F50-59	Behavioural syndromes associated with physiological disturbances and physical factors	Anorexia Bulimia *Insomnia* *Hypersomnia* *Narcolepsy* *Sleep apnea*	0.2–0.5% 0.1–0.9% 3.5% (7%) 0.8% 0.02% 3.0%	0.8 million 0.7 million 14.6 (29.1) million 3.1 million 0.1 million 12.5 million
F60-69	Personality and behavioural disorders	*Borderline PD* *Dissocial PD*	0.7% 0.6%	2.3 million 2.0 million
F70-79	Mental retardation	*Mental retardation*	1.0%	4.2 million
F80-89	Developmental disorders	*Autism*	0.6%	0.6 million
F90-F98	Behavioural and emotional disorders with onset in childhood and adolescence	*ADHD* *Behavioral disorders*	5.0% 3.0%	3.3 million 2.1 million
F99-F99	Unspecified mental disorders	Remainder category		

9

but not the number of scientifically established diagnoses. This probably has to do both with the increased willingness to recognise and diagnose mental illness in the first place, and possibly also with the overly hasty attribution of a diagnosis.

However, the number of diagnoses alone says nothing about their clinical relevance, since all degrees of severity are combined in one category here. Gallinat et al. (2017) have presented the clinical reality in Germany much more realistically in terms of severity. According to these authors, 90% of all mental disorders are mild to moderate and only 10% of all mental disorders are severe, of which, surprisingly and depressingly, half are among adolescents between 13 and 17 years of age. The reality of care for these groups is as follows. The mild and moderate disorders are dominated by older people with depression, anxiety, stress and somatoform disorders. About 95% of them are in outpatient psychotherapy. Treatment is poorly managed, there are waiting times for psychotherapy of 3–9 months and thus a backlog, including those in need of an inpatient treatment. Patients in this group account for 90% of all days of incapacity to work, with direct costs (2014) of €8.3 billion and €13.1 billion in gross value added.

In contrast, the severe disorders are dominated by the schizophrenia spectrum, bipolar disorders, borderline personality disorder and psychotic depression. The risk factor migration plays a major role there. Only 3–5% of these patients are in outpatient psychotherapy, and they cause only 5–10% of all days of incapacity to work, since most of those affected are not employed or not able to work. Many of the patients are revolving door inpatients, meaning that they come back again and again or they live in therapeutic and long-term facilities or are housed in forensic institutions. This group accounts for 60% of all emergencies and 80% of all compulsory admissions. It shows high morbidity and mortality with an average life expectancy of only 55 years.

The direct medical costs caused per case are high, averaging around €45,000 per year. This reality of care makes it clear, among other things, why there are such different views, even within professionals, on the reality of psychiatry and the incidence of mental illness, where everyone feels confirmed by their everyday experience in their own field of daily work.

9.4 Problems with the Classification of Mental Disorders

In the following, the problems with the classification of mental illnesses will be shortly explained. I will primarily refer to the DSM, since most of the literature refers to it, but the problems described apply equally to the ICD-10.

9.4.1 Heterogeneity

As explained, the current classifications are descriptive, i.e. a diagnosis is based on characteristic syndromes and is made when a number of specific symptoms are present for a certain period of time. For example, a diagnosis of depression according to DSM-5 is made when at least five of nine symptoms are present for most of the day for a period of at least 2 weeks, and one of the symptoms must be (1) or (2). At first, this sounds very plausible to any clinician. However, the question arises as to whether the diagnosis of depression is really a single clinical picture. Thus, it is possible for two individuals to be given the same diagnosis (depression) without having a single symptom in common. (◻ Table 9.3).

Theoretically, there are 227 unique symptom combinations that all lead to the same diagnosis; if one also takes into account that there can be too much or too little of sleep, appetite or psychomotor function, there are

◼ **Table 9.3** Possibilities for making a diagnosis of depression according to DSM-5 without a single overlapping symptom. (After Pawelzik, unpublished, with kind permission)

Mr. Miller	Mrs. Schmidt
(1) Depressed mood	(2) Loss of interest or pleasure
(3) Loss of appetite or weight	(3) Increase in appetite or weight gain
(4) Insomnia	(4) Hypersomnia
(5) Psychomotor agitation	(5) Psychomotor retardation
(7) Feelings of worthlessness or inappropriate guilt	(6) Loss of energy
(9) Thoughs of death or suicide	(8) Decreased concentration

even 945, and if one takes into account the sub-symptoms, there are even 16,400. Now, one could assume that this is just a theoretical consideeration, but that most depressions are very similar. This was examined empirically by Fried and Nesse (2015) using one of the largest treatment studies of depression (*n* = 3703), the so-called Star*D study. Using a symptom list (QIDS 16), 1030 unique symptom profiles emerged empirically with an average of only 3.6 individuals per profile. 501 symptom profiles (48.6%) existed in only one patient and 864 profiles (83.9%) included only 2–5 individuals. Thus, it is empirically apparent that there is a great deal of heterogeneity in depressive disorder. Time and again, it has been investigated whether there might be distinct subtypes that can be characterized on the basis of common symptom profiles, but all these attempts have so far not been supported by convincing evidence.

9.4.2 Demarcation Problems

For mental disorders, there are at least two demarcation problems. First, "normal" depression must be distinguished from so-called "organic" depression, e.g. depression after stroke, in Parkinson's disease or in medical diseases, e.g. thyroid disorders, or as side effects of medication (e.g. cortisone) or drugs (e.g. after esctasy consumption). This so-called exclusion diagnosis of other primary diseases that secondarily lead to a psychiatric syndrome is an obligatory part of every diagnostic process. It also includes the distinction between neurological and psychiatric disorders, which is obsolete for some disorders (dementia) but useful for others (▶ Sect. 9.1). Second, a much more difficult problem is the demarcation from normal psychological processes or life problems. A much discussed example is grief following the death of a significant other (Wakefield 2015). Thus, after the death of a life partner or even a child, it is not surprising to feel despair, to cry, to doubt the meaning of life, to feel no more pleasure, to have reduced drive, in short, to grieve. Looking at the symptoms alone, a diagnosis of depression can be made easily. But of course it is normal and natural to grieve after a significant other dies; indeed, not to grieve would be rather unnatural or even pathological. The authors of the DSM were well aware of this life problem. That is why in the DSM-IV (1994) the so-called *bereavement exclusion existed*. After a bereavement, it was only allowed to diagnose depression only 2 months after the event at the earliest. With the introduction of the DSM-5, this *bereavement exclusion* was dropped. Why? The argument for this decision was that this exception would make it impossible for people who were grieving and in the process developed depression to be diagnosed, and thus impossible to receive treatment, since therapy would only be paid for by health insurance companies if a diagnosis was made. Another argument was that if one sticks with the *bereavement exclusion*, it does not seem plausible to define only death as an exception. After all, isn't it normal to have depressive symptoms when your partner leaves you, you lose your job, or your home is destroyed by a fire? So either the exception should be extended to include such other cases or it should be consistently dropped. Opponents of the abolition argued that the possibility of diagnosing depression only 2 weeks after a death would pathologise normal psychological processes and lead to an unjustified inflation of diagnoses.

This discussion also has a scientific part. First there have been attempts for a long time to establish so-called prolonged or complicated grief as an independent clinical condition (Wagner 2014). Second, empirical studies exist, e.g., by Wakefield, the theorist of mental disorder as *harmful dysfunction*, that there are uncomplicated depressions, i.e., conditions that although meeting the diagnostic criteria of depression cross-sectionally do not show an increased likelihood for future depressive episodes longitudinally, and thus should be considered to be benign depressions (Wakefield and Schmitz 2014). These are characterized as single episodes that resolve within 6 months, do not cause severe impairments, and are not associated with psychotic symptoms, suicidal ideation, psychomotor slowing, or feelings of worthlessness. Now, is this a "benign" depression or a "normal" psychological process? This question is difficult to answer or to decide by definition.

Fortunately, however, a discussion of this issue can now be based on empirical data.

A third demarcation problem is to distinguish between between different types of mental disorders. According to DSM-5, mental disorders are defined categorically, i.e. there is a disorder or not. But this creates the problem of comorbidity. Often several disorders are present at the same time. For example, there is a close comorbidity of depression and anxiety disorders or of addictive disorders and depression. Is this really a case of the presence of two different disorders? Or is there not rather a connection between both disorders, or even a causal relationship? Someone who suffers from an addictive disorder could, for example, become secondarily ill with depression because he suffers from the consequences of his addictive disorder. An inverse relationship is also conceivable. For this reason, it was also considered in the DSM-5, especially for the personality disorders, to introduce a dimensional approach to mental disorders instead of a categorical approach. This means that a mentally ill person may present with symptoms in different dimensions that are more or less pronounced, instead of being diagnosed with different disorders. Such an approach has been already elaborated in the field of personality disorders, but has not yet gained wide acceptance.

9.4.3 The Problem of Biomarkers

The heterogeneity of purely symptomatically defined mental disorders has always been an argument for including neurobiological findings in the definition or diagnosis, as has now been achieved with cerebrospinal fluid diagnostics for the diagnosis of Alzheimer's disease, which shows high sensitivity and specificity. This is the promise of biological psychiatry. And indeed, it was a promise for the transition from DSM-IV to DSM-5. Yet neurobiological criteria found virtually no entry into DSM-5 diagnoses. Why? The answer is simply that despite the wealth of neurobiological research and knowledge, there are virtually no clinically useful biomarkers—with a few exceptions, such as Alzheimer's disease. Take schizophrenia, for example: with such a severe and relatively uniform clinical picture worldwide, one would think that the chance of finding one or more reliable biomarkers should be quite good. Prata et al. (2014) investigated this empirically by performing a detailed analysis of all papers on biomarkers in psychosis (n = 3200). About half of the studies were related to diagnostic biomarkers, a quarter were reviews, and fewer than 200 papers were longitudinal studies. For the latter, the authors examined whether genetic, metabolic, or imaging markers were predictive of treatment outcome. They assessed the quality of the biomarkers based on quality (positive outcome, controlled trial, a priori definition of biomarker, sufficient statistical power, independent replication) and effect size with a maximum score of 8. The result?

» The only biomarker with a final score above 6 from the total of 362 predictive & monitoring biomarkers in the 114 studies was a pharmacogenetic biomarker that scored 7: the C allele of the 6672 G > C single nucleotide polymorphism (SNP) in the HLA-DQB1 region (Athanasiou et al. 2011) predicted risk for clozapine-induced agranulocytosis with an O.R. 16.8, was defined a priori, and its effect replicated in an independent sample. (Prata et al. 2014, p. 138).

In other words, the results of decades of biomarker research are very disappointing. Critics of biological psychiatry see this as an argument to stop doing this kind of research. Biologically oriented psychiatrists, on the other hand, argue that this merely shows that it is unlikely to find consistent biomarkers for purely clinically defined, heterogeneous syndromes as found in current classification systems (cf. on this ► Sect. 9.5.1).

9.4.4   Non-medical Interests

Another problem in the diagnosis of mental disorders is non-medical interests. On the one hand, this refers to financial conflicts of interest (cf. Hasler 2013) and the resulting distortions of nosology. Many co-authors of the DSM had consulting contracts with the pharmaceutical industry. The industry has an interest in defining, some say inventing, novel disorders in order to create new markets for drugs. Finally, there are individual, not financial, interests. For example, in the committees that drafted the DSM, some researchers fought for their own favorite disorder that was object of their own research activities for years or decades to be officially included in the DSM. Which, of course, is easier than in other medical disciplines when there are no objective biomarkers anyway. However, there are also stakeholders who have non-medical interests, e.g. because they fear losing financial advantages (e.g. through the omission of Asperger's syndrome from the DSM-5), or because they insist that a disorder has biological causes and not psychological ones (e.g. chronic fatigue syndrome), or simply because, from the perspective of those affected, the respective clinical syndrome naturally has a very high priority and, with ever more limited resources, lobbying for it is of high importance for them.

9.4.5   The Lock-in Syndrome

As already mentioned, the ICD forms the basis for the medical care system. Only with an official diagnosis do health insurance companies pay for therapies, you are eligible for an official notification of illness for the workplace, or have a chance of having an occupational illness recognised or receiving a pension. That is why, according to philosopher Rachel Cooper, it is almost impossible to fundamentally reform the DSM or the ICD (Cooper 2015). This is because any change would have existential consequences for those in the existing care system and would destabilise a complex, constantly used and deeply embedded system. This, she argues, is similar to the QWERTY keyboard on a computer. Originally, the arrangement of the keys was due to the fact that the levers of a typewriter where supposed not to get stuck while typing. To establish a different arrangement of the keys, which would be more suited from a purely technical point of view on the computer, which has no more mechanical levers, is today however practically impossible, since a change of the system would be such a large expenditure that this will never happen: The system is in a "lock-in state". The only option would be a radical system change. In the computer realm, for example, this might be achieved by the development of voice input, i.e. speaking replacing manual input, i.e. typing. A similar radical change is proposed by the RDOC system (▶ Sect. 5.1).

9.4.6 The Mentalism Problem

Another fundamental problem that can only be outlined here is the mentalism problem (cf. Walter and Pawelzik 2018). It can be seen as a successor problem to the mind-body problem. As we saw in ▶ Chap 1, dualism has proved increasingly dispensable in the course of (Western) history. At the same time, the rejection of dualism by no means clarifies what the mental actually is, for instance in the biopsychosocial model. Hardly any of the theories of mental disorders address this question, but simply assume without further explanation that the mental or phenomenal experience is another level or aspect of reality. But what exactly is it? Is it simply identical with the neural, that is, with brain processes, as it would be natural to assume? A whole range of non-reductionist theories argue against this, e.g. by pointing out that brain states should not be ascribed properties on the personal level,

that an "animate" organism also has a body, that mental states always develop in social interaction, are involved in a social and cultural context, are closely tied to language, at least in humans, and therefore mental processes must be regarded as complex, emergent phenomena. In the context of psychiatric controversies, however, three aspects of the mentalism problem are often mixed and not distinguished. The first aspect is the question of what exactly mental processes are. This is an *ontological* question about the nature of mental states. The second is the question of how we acquire knowledge about the nature of our experience and the content of normal (and pathological) mental processes. This is an *epistemiological* question. Thus, there is no contradiction in assuming that mental states are nothing more than brain states plus x (e.g., bodily states plus other external factors; on the externalism of the mental, see Walter 1997, 2018), but that we have access to the content of these mental states mainly and inevitably through subjective experience and socially grounded language. Such an assumption would explain why both, those who hold reductionist intuitions and those who point to the irreduciblity of personal experience feel justified. The third and ultimately most important aspect for psychiatry as a science is how mental dysfunction comes about. This question of the genesis of mental disorder and the causal factors relevant to it is neither identical with the ontological nor with the epistemological question alone. Although all three problems are related, they are conceptually independent of each other. For example, it is now generally assumed that a mix of causes, including genetic predisposition, neurobiological influences during the development of the organism, subjective experiences, stressors, and their processing, is relevant to most common mental disorders. Thus, when discussing mental disorders, we should always be careful to distinguish between ontological, epistemological, and causal issues, i.e., whether we are talking about the nature ("essence") of mental phenomena, how we can recognize them, and how they arise.

9.5   Recent Approaches

The difficulties with current classification systems and the lack of success of biological approaches to psychiatry have led to new proposals for changing and/or theorizing about the nature, study, and classification of mental disorders in the long run. Three such approaches will be presented here.

9.5.1   Research Domain Criteria (RDoc)

The RDoC initiative was launched in 2009, at the world's largest psychiatric research institute, the NIMH (National Institute of Mental Health) in Bethesda (Insel 2013; Kozak and Cuthbert 2016). Briefly, the issue was this (adapted from Walter 2017): Thomas Insel, then director of the NIMH, himself a researcher in the field of social neurobiology (functions of oxytocin and vasopressin), had always promoted neuroscientific research on mental disorders, from molecular biology to neuroimaging. With the introduction of the DSM-5 (2013), it was planned to also incorporate neurobiological findings into classification and diagnosis and to move from a categorical to a dimensional system. However, neither intention could be realized. This was for a variety of reasons, not least that the neurobiological findings were not robust enough to be incorporated into a clinically useful classification. This was unsatisfactory to many scientists, since much more was known about neural circuitry, including in humans, since the publication of the DSM-IV (1994), not least through non-invasive neuroimaging. For a long time, science-minded doctors had been dissatisfied with diagnosing mental illness only at the symptom level. But it is no wonder, says Thomas Insel, if biomarker research does not lead to success:

» But it is critical to realize that we cannot succed if we use the DSM categories as the "gold standard." The diagnostic system has to be based on the emerging research data, not on the current symptom-based categories. Imagine deciding that ECGs were not useful because many patients with chest pain did not have ECG changes. That is what we have been doing for decades when we reject a biomarkes because it does not detect a DSM category. We need to begin collecting genetic, imaging, physiologic, and cognitive data to see how all the data—not just the symptoms—form clusters and how these clusters relate to treatment response. (Insel 2013)

The RDoC system, which was made public in April 2013, therefore proposes that research into mental disorders should not be oriented towards (superficial) symptoms and syndromes, but should start from domains of neurocognitive functions based on the function of specific circuits, and map these at different levels (from gene to behaviour), in order to then use this database to classify the heterogeneous disorders into more specific disorders, independent of DSM criteria. Only constructs for which there was independent evidence of validity and for which knowledge about association with neural circuits were accepted as candidates for RDoC. The resulting constructs (currently 25) were sorted into six higher-level domains (◻ Fig. 9.1) and also have subconstructs. Domains and constructs are not fixed and unmutable, but have been tentatively selected based on current knowledge. They can and should be refined, modified and extended by new empirical findings. For example, the sixth, sensorimotor domain has only recently been added,

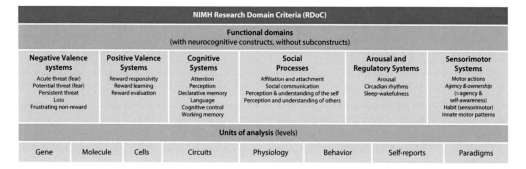

NIMH Research Domain Criteria (RDoC)					
Functional domains (with neurocognitive constructs, without subconstructs)					
Negative Valence systems	**Positive Valence Systems**	**Cognitive Systems**	**Social Processes**	**Arousal and Regulatory Systems**	**Sensorimotor Systems**
Acute threat (fear) Potential threat (fear) Persistent threat Loss Frustrating non-reward	Reward responsivity Reward learning Reward evaluation	Attention Perception Declarative memory Language Cognitive control Working memory	Affiliation and attachment Social communication Perception & understanding of the self Perception and understanding of others	Arousal Circadian rhythms Sleep-wakefulness	Motor actions Agency & ownership (≈agency & self-awareness) Habit (sensorimotor) innate motor patterns
Units of analysis (levels)					
Gene	Molecule	Cells	Circuits	Physiology	Behavior	Self-reports	Paradigms

◘ Fig. 9.1 The functional domains and neurocognitive constructs of the RDoC (as of June 30, 2019, retrieved from ► https://www.nimh.nih.gov/research/research-funded-by-nimh/rdoc/constructs/rdoc-matrix.shtml). Currently, there are six domains with a total of 25 constructs (see table) and 28 subconstructs (not listed here)

and a further domain "impulsivity" is being discussed.

All constructs can and should be systematically investigated at different levels (the RDoC authors prefer the term: units of analysis): from genes to molecules and cells to circuits, the physiological level (e.g. heart rate, cortisol level), observable behaviour and self-reports. To do this, different paradigms are used. Therefore, the RDoC matrix can be thought of as a two-dimensional table in which knowledge is collected. On the constantly updated RDoC homepage, one can look up what we currently know about each field in this table. Two other dimensions that are included in RDoC, though not in the matrix, and are relevant to all domains, are the developmental and the environmental dimension. This is because the functions mentioned above all emerge only in the course of an individual's development, and are shaped and modified by environmental influences (paradigmatic: epigenetic and learning effects).

The RDoC matrix provides a grid for the systematic study of mental disorders that is not fixed to previously defined disease categories. In the long term, the RDoC system is intended to contribute to a differentiated, and in some cases also novel, classification of mental disorders. Above all, however, it is intended to create the basis for better, tailor-made, ideally individualised therapy in the sense of *precision psychiatry*. Of course, the RDoC initiative has not been without criticism (cf. Walter 2017). What remains to be said is that it has had a significant impact on psychiatric research in recent years, as there is now a substantial body of work exploring and redesigning categories of mental disorders using this model. The RDOC approach can be succinctly summarized as "psychiatric research as applied cognitive neuroscience" (Walter 2017). A major extension of cognitive neuroscience is computational neuroscience, which has expanded into computational psychiatry within psychiatry (Friston et al. 2014; Heinz 2017, see for example ► Chap. 11 on schizophrenia in this book).

9.5.2 Network Theories of Mental Disorders

Another theory that has been discussed in recent years, not coincidentally, is the network theory of mental disorders (Borsboom 2017; Borsboom et al. 2019). It opposes the essentialist notion that mental health symptoms are "surface features" of an underlying pathological process, as in other diseases. Measles, for example, is clinically manifested by certain symptoms (rash, Koplik's spots, fever), all of which are caused by a common cause, infection with the measles virus, and can thus be explained. The assumption that it is the same with depression, i.e. that the clinical symptoms of depression have a common cause (such as a lack of serotonin or a loss experience in childhood) is a misleading idea. Rather, the disease of depression consists (an ontological statement) of being a network of symptoms that are causally connected. For example, the symptoms of sleep disturbance, difficulty concentrating, ruminating, and self-worth problems are causally related. Those who sleep little are not rested the next morning, cannot concentrate, and have more time to ruminate, which can lead to thinking about their worthlessness. Conversely, a lot of ruminating can lead to insomnia. The pathological nature of depression is shown by the fact that the causal links between symptoms are so strong that a full-blown clinical picture is rapidly formed by an external event and causal interactions between symptoms, remains stable, and is difficult to become deactivated. A resilient network is characterised by the fact that although symptoms may develop, they quickly fade—due to the only weakly developed causal interconnections between the symptoms—and do not lead to a full-blown clinical picture of depression. A vulnerable network, on the other hand, only needs a trigger at some point, and then the symptoms spread by themselves, so to speak, and persist even if the external trigger is removed (◘ Fig. 9.2). In network terminology, then, someone is mentally healthy with high resilience when the symptom network is weakly connected and there are few external stressors. With a weakly connected network and strong external stressors, there are increased symptoms, but the person is still healthy. A strongly connected network with few stressors indicates high vulnerability and with a strongly connected network with strong stressors, mental illness occurs.

Such a network approach differs from conventional theories of disorders in several aspects: For example, no common cause of mental illness is assumed; symptoms are given a crucial role. Symptoms are not merely counted independent of each other, but a causal nexus between them is postulated; a mental disorder is defined and described in terms of networks. Further, at

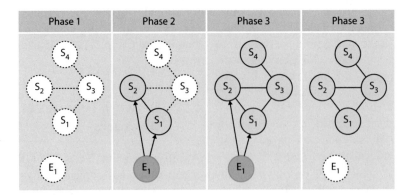

◘ **Fig. 9.2** Development of a mental disorder according to the network theory: In the first phase there are no symptoms (S), the network is "asleep" (phase 1). Under the influence of one (E1) or even several external events (stressors), individual symptoms are then activated (phase 2), which causes further symptoms to develop (phase 3). If the symptom network has strong connections, then recovery does not occur due to the omission of the external event. The external event has acted as a trigger and the network of symptoms keeps itself active and stuck in its active state. Similarly, it is conceivable that the activity of the network as a whole decreases as a result of therapy on individual symptoms. (After Borsboom 2017)

least according to the proponents of the network theory (Borsboom et al. 2019), neither disorders nor individual symptoms or their causal nexus can be attributed to causal processes in brain circuits, because the connections are too complex, because mental processes can be realized multiply, and because the focus on such details loses sight of what defines a mental disorder, namely the network properties of the symptoms, in the process. Thus, neurobiological reductionism is blocked. However, it remains a mystery how the different levels, such as those of neurobiology and mental symptoms and behavior, are connected. It is probably helpful to know that the founders of network theory are neither neurobiologically trained researchers nor clinical practitioners. Rather, they have their roots in psychiatric epidemiology and statistics, that is, in a field that deals with large, primarily clinical, data sets (symptom questionnaires) in large studies. It was mentioned at the outset that it is no coincidence that network theories are popular right now. Network theories are currently enjoying a lot of interest in various fields (social networks, neural networks, networks in physics). To date, however, there has been little contact or collaboration between epidemiological statisticians and neural network theories (Braun et al. 2018; Waller et al. 2018). Thereby, the attempt to bring together neurobiological and clinical levels using a common terminology and mathematical tools could be of high interest.

9.5.3 The New Mechanism

We have introduced two approaches that are in principle complementary to each other: While RDoC starts from basic neurocognitive processes and their brain circuits and neglects the level of symptoms, network theories focus on the symptom level and declare the underlying processes negligible. What both approaches have in common is that they want to move away from a purely descriptive approach and—in different ways—bring causal conditions into play. After all, identifying causal factors seems to be the fundamental hallmark of science, and both approaches emphasize this. In recent years, a new approach to understanding the brain has emerged in the field of philosophy of science in biology and neuroscience that is also of interest to psychiatry. This is the "new mechanism". For a long time, the philosophy of science was primarily concerned with theories in physics, in the form of quantifiable laws. One goal was to find as few as possible but fundamental general laws of nature to which other laws could be reduced. Another area of application for scientific theorists was the theory of evolution, since here, too, a general theory is available that claims to be able to explain the origin of life with a few general laws (mutation, selection, population dynamics). In the neurosciences, on the other hand, there is (as yet) no generally accepted "theory of the brain". Therefore, the new mechanism is devoted to the question of how concrete explanations in the neurosciences work. The result is a diverse mosaic of individual explanations rather than a general theory. The concept of mechanism has been identified as central to this (Craver 2007; Glennan 2017; Machamer et al. 2000). Phenomena are explained by identifying in detail the mechanisms that produce and sustain the phenomena, involving multiple levels and integrating different fields of science.

This approach is an approach of philosophy of science because it defines the concept of mechanism in an abstract and general way and then applies it in detail to specific phenomena in the field of neuroscience. However, mechanisms can also be found for mental (Bechtel 2008) and social phenomena (Hedström and Ylikoski 2000). A mechanism is defined as "a set of entities and activities organized to produce the phenomenon that is to be explained" (Craver 2007, p. 5). Entities are parts or components

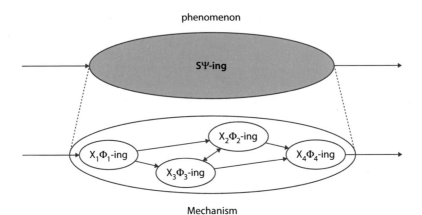

phenomenon

S𝛹-ing

Mechanism

◻ Fig. 9.3 The figure shows a phenomenon 𝛹 (pronounced: Psi) explained by a mechanism (S) (upper part of the figure). The lower part shows (abstract) details of the mechanism, namely entities (circles) and activities (arrows) that have a specific spatio-temporal organization. (After Craver 2007, p. 7); t, S = mechanism as a whole, 𝛹 (Psi) = phenomenon, X = entity or component, 𝛷 (Phi) = activity

of the mechanism that exhibit activities that produce causal effects. Crucially, the organisation of these active components in space and time and in a hierarchy is what ensures that the mechanism produces the phenomenon. Craver (2007) goes through this extensively using the example of explaining neurotransmitter release and long-term potentiation. His general scheme for a mechanism has now become a canonical account (◻ Fig. 9.3).

Already the abbreviation "Psi" in this scheme, which in philosophy often stands for psychological, indicates that the mechanism scheme is also meant to be applied to mental processes. However, there is much more about the brain than about the mind in Craver's writings, and nowhere is it stated exactly what the mental actually is (cf. ▶ Sect. 9.4.6 on the mentalism problem). For a detailed account of the new mechanism as an explanatory approach in cognitive neuroscience, see Kästner (2017). There, various problems with the approach are discussed, such as the question of the difference between the constitution of a mental process and its causal effects, how causal effects must be conceptualized across different levels, and what notion of causality is

experimentally and conceptually relevant to neuroscience. Crucially for us at this point is, that the new mechanism is designed from the outset to move away from the physics centered approach in philosophy of science, and rather is concerned with real-world explanations of relevant phenomena in neuroscience, and has been conceived from the outset as a multilevel and integrative approach. Moreover, it refrains from the widespread correlationism that all too easily leads back to dualism, but claims to describe causal processes in the production of phenomena.

In a 2011 paper, psychiatrist and geneticist Kenneth Kendler, along with philosopher of science Carl Craver and philosopher of psychiatry Peter Zachar, made an attempt to use the new mechanism for a contemporary theory of mental disorders (Kendler et al. 2011; for another recent attempt, see Kästner 2019). In doing so, the authors take their cue from evolutionary biology approaches to explaining what a species is and apply them to types of disease. They somewhat awkwardly call their approach *mechanistic property cluster theory*, or MPC. According to MPC, disorders are like biological species, i.e. fuzzy "populations" with paradigmatic central and more marginal specimens. One could also say that disorders are accumulations of (causally relevant) properties. In this context, different specimens of a species (disorders) do not share all properties with each other, but rather exhibit a family resemblance. In the high-dimensional property space of all (causally relevant) properties that contribute to their emergence, they are found close to each other, they "cluster". However, these properties are not surface properties (such as only symptoms), but the "co-occurrence of these properties between individuals is explained by causal mechanisms that regularly ensure that these properties are realized together." (Kendler et al. 2011, p. 1147). In this context, different levels can interact with each other to causally produce specimens of a species in the first place. Examples include epigenetic effects from stress, effects of substance abuse on the brain, or the effect of insomnia on depressed mood. Here, then, the authors combine elements of the new mechanism (mechanisms, that is, spatially and temporally organized active components causally producing a phenomenon, a disorder) and elements of network theories (symptoms may causally influence each other and this interplay may be part of the mechanism). Moreover, MPC is a multilevel theory and is able to integrate different theoretical approaches (biological, sociological, phenomenological) without giving up the claim to causal explanations.

In the introduction to their article, Kendler et al. (2011) categorize the types of diseases that have been theoretically explained as essentialist (a disease has an underlying essence as gold does, i.e., a single biological cause), socially constructed (invented for extra-medical reasons), and pragmatic (useful for the practice of medicine, such as the DMS-5). The MPC approach allows all kinds of mechanisms (biological, psychological, social factors, societal) as parts of a mechanistic explanation. Some disorders might be more essen-

tialist (schizophrenia), others more socially constructed (hysteria) and some more constituted by social circumstances (anorexia nervosa). Only time will tell whether the MPC theory, which has not yet been worked out in great detail, will become accepted; it certainly will not under its complicated name. However, the new mechanistic approach will probably play an important role in any scientifically based psychiatry in the future.

Summary and Outlook

In the present chapter we have seen how difficult it is to define, diagnose and distinguish mental disorders from neurological diseases, life problems and from each other. Any theory of disorder in psychiatry will inevitably have to draw boundaries and live with the fact that those are fuzzy. Given the historical background of psychiatry, it is understandable why there are so many controversies about mental disorders. Any definition must contain both positive and negative criteria and must include the personal level refering to the person's ability to function and the subjective suffering component of the postulated dysfunction. At present, a system trimmed to reliability (DSM-5 and ICD-10 or -11) dominates the practice of psychiatry with all its advantages (no commitment to possibly incorrect etiological theories) as well as disadvantages (inflation of diagnoses, demarcation problems aggravated, a non-adaptive system with a "lock-in syndrome"). On the scientific and philosophical level, there are new approaches, all of which are more or less pluralistic and include different levels of explanation, but—in contrast to the DSM and ICD-10—place great emphasis on including causal mechanisms in the definition and explanation of mental disorders. None of these has yet become generally accepted, and it is possible that for different disorders different theories may be appropriate. In practice, however, the established diagnostic systems will persist for a long time. In the future, the inclusion of biological findings will become increasingly standard, although it is hard to imagine that the central role of subjective experience in the diagnosis of most mental disorders will change.

References

APA—American Psychiatric Association (2013) Diagnostic and statistic manual of mental disorders, 5. Aufl. Amercian Psychiatric Association Publishing, Arlington, VA

Athanasiou MC, Dettling M, Cascorbi I, Mosyagin I, Salisbury BA, Pierz KA, Zou W, Whalen H, Malhotra AK, Lencz T, Gerson SL, Kane JM, Reed CR (2011) Candidate gene analysis identifies a polymorphism in HLA-DQB1 associated with clozapine-induced agranulocytosis. J Clin Psychiatry 72(4):458–463

Bechtel W (2008) Mental mechanisms. Philosophical perspectives on cognitive neuroscience. Routledge, London

Boeckle M, Liegl G, Jank R, Pieh C (2016) Neural correlates of conversion disorder: overview and metaanalysis of neuroimaging studies on motor conversion. BMC Psychiatry 16:195

Boorse C (2011) Concepts of health and disease. In: Gifford F (ed) Philosophy of medicine. North Holland, Oxford, pp 13–64

Borsboom D (2017) A network theory of mental disorders. World Psychiatry 16:5–13

Borsboom D, Cramer AOJ, Kalis A (2019) Brain disorders? Not really. Why network structures block reductionism in psychopathology research. Behav Brain Sci 42(e2):1–63

Braun U, Schaefer A, Betzel RF, Tost H, Meyer-Lindenberg A, Bassett DS (2018) From maps to multi-dimensional network mechanisms of mental disorders. Neuron 97:14–31

Cooper R (2015) Why is the diagnostic and statistical manual of mental disorders so hard to revise? Path-dependence and "lock-in" in classification. Stud Hist Philos Sci Part C 51:1–10

Craver CF (2007) Explaining the brain. Mechanisms and the mosaic unity of neuroscience. Oxford, Oxford University Press

Dilling H, Mombour W, Schmidt MH (2015) Internationale Klassifikation psychischer Störungen: ICD-10 Kapitel V (F)—Klinisch-diagnostische Leitlinien. Hogrefe, Göttingen

Engel GL (1977) The need for a new medical model: a challenge for biomedicine. Science 196:129–136

Fried EI, Nesse RM (2015) Depression is not a consistent syndrome: an investigation of unique symptom patterns in the STAR*D study. J Aff Disord 172:96–102

Friston KJ, Stephan KE, Montague R, Dolan RJ (2014) Computational psychiatry: the brain as a phantastic organ. Lancet Psychiatry 1(2):148–158

Gallinat J, Karow A, Lambert M (2017) Psychiatrie der Zukunft. Vortrag in Greifswald. http://www2.medizin.uni-greifswald.de/psych/fileadmin/user_upload/veranstaltungen/2017/15.-17.02.2017__Die_Subjektive_Seite_der_Schizophrenie_/Vortraege/Vortrage_17.02.2017/Gallinat_Psychiatrie_der_Zukunft.pdf. Accessed 19. Jul 2019

Glennan S (2017) The new mechanical philosophy. Oxford University Press, Oxford

Graham G (2010) The disordered mind. An introduction to philosophy of mind and mental illness, 2 Aufl. Routledge, London

Hasler F (2013) Neuromythologie: Eine Streitschrift gegen die Deutungsmacht der Hirnforschung. Transcript, Bielefeld

Hedström P, Ylikoski P (2000) Causal mechanisms in the social sciences. Ann Rev Sociol 36:49–67

Heinz A (2015) Krankheit vs. Störung. Medizinische und lebensweltliche Aspekte psychischen Leidens. Nervenarzt 86:36–41

Heinz A (2017) A new understanding of mental disorders: computational models for dimensional psychiatry. MIT Press, Cambridge

Hucklenbroich P (2012) Der Krankheitsbegriff der Medizin in der Perspektive einer rekonstruktiven Wissenschaftstheorie. In: Rothhaar M, Frewer A (eds) Das Gesunde, das Kranke und die Medizinethik. Moralische Implikationen des Krankheitsbegriffs. Steiner, Stuttgart, pp 33–63

Insel T (2013) Transforming diagnosis. Directors Blog an April 29, 2013. http://www.nimh.nih.gov/about/director/2013/transforming-diagnosis.shtml

Kästner L (2017) Philosophy of cognitive neuroscience: causal explanations. Mechanisms and experimental manipulations. De Gruyter, Berlin

Kästner L (2019) Identifying causes in psychiatry. Paper submitted to Philosophy of Science Association

Kendler KS, Zachar P, Craver C (2011) What kind of things are psychiatric disorders? Psychol Med 41:1143–1150

Kozak MJ, Cuthbert BN (2016) The NIMH research domain criteria initiative: background, issues, and pragmatics. Psychophysiology 53(3):286–297

Machamer D, Darden L, Craver CF (2000) Thinking about mechanisms. Philos Sci 67:1–25

Prata D, Mechelli A, Kapur S (2014) Clinically meaningfulbiomarkers for psychosis: a systematic and quantitativereview. Neurosci Biobehav Rev 45:134–141

Schönfeldt-Lecuona C, Connemann BJ, Höse A, Spitzer M, Walter H (2004) Konversionsstörungen.

Von der Neurobiologie zur Behandlung. Nervenarzt 75:619–627

Szasz T (1961) The myth of mental illness. Hoeber-Harper, New York

Wagner B (2014) Komplizierte Trauer: Grundlagen, Diagnostik und Therapie. Springer, Berlin

Wakefield JC (1992) Disorder as harmful dysfunction: a conceptual critique of DSM-III-R's definition of mental disorder. Psychol Rev 99(2):232–247

Wakefield JC (2015) The loss of grief: sciene and pseudoscience in the debate over DSM-5's elimination of the bereavement exclusion. In: Singy P, Demazeux S (eds) The DSM-5 in Perspective. Philosophical reflections on the psychiatric bible. Springer, Heidelberg, pp 157–178

Wakefield JC, Schmitz MF (2014) Predictive validation of single-episode uncomplicated depression as a benign subtype of unipolar major depression. Acta Scand Acta Psychiatr Scand 129:445–457

Waller L, Brovkin A, Dortschmidt L, Bzdok D, Walter H, Kruschwitz J (2018) GraphVar 2.0: a user-friendly toolbox for machine learning on functional connectivity measures. J Neurosci Meth 308:21–33

Walter H (1997) Neurophilosophie der Willensfreiheit. Von libertarischen Illusionen zum Konzept natürlicher Autonomie. Mentis, Paderborn

Walter H (2017) Research Domain Criteria (RDoC). Psychiatrische Forschung als angewandte kognitive Neurowissenschaft. Nervenarzt 88(5):538–548

Walter H (2018) Über das Gehirn hinaus. Aktiver Externalimus und die Natur des Mentalen. Nervenheilkunde 37(07/08):479–488

Walter H (ed) (2005) Funktionelle Bildgebung in Psychiatrie und Psychotherapie: Methodische Grundlagen und klinische Anwendungen. Schattauer, Stuttgart

Walter H, Pawelzik M (2018) Die Mentalismus-Frage in der Nervenheilkunde. Nervenheilkunde 37(07/08):466

Wittchen H-U, Jacobi F (2005) Size and burden of mental disorders in Europe—a critical review and appraisal of 27 studies. Eur Neuropsychopharmacol 15(4):357–376

Wittchen H-U, Jacobi F, Rehm J et al (2011) The size and burden of mental disorders and other disorders of the brain in Europe 2010. Eur Neuropsychopharmacol 21:655–679

Psyche and Mental Illness: Addiction

Stefan Gutwinski and Andreas Heinz

Contents

© The Author(s), under exclusive license to Springer-Verlag GmbH, DE, part of Springer Nature 2023
G. Roth et al. (eds.), *Psychoneuroscience*, https://doi.org/10.1007/978-3-662-65774-4_10

In this chapter we deal with the topic of addiction: first with the question of the occurrence of addiction, the diagnostic classification and the frequency of addictive diseases. We present which clinical symptoms are present in people with addictive disorders and how these are related to neuronal changes. We will discuss functional as well as structural changes of the brain and the processes on the level of neuronal messenger substances. We want to relate these points to the therapeutic understanding derived from them and possible future therapies.

Learning Objectives

After reading this chapter, the reader should be familiar with the clinical classification and neurobiological models of addictive disorders.

▶ Example

Mr. J. (44 years) reported that he had drunk alcohol for the first time when he was 8 years old. His father had been "alcoholic" and had actually always drunk when he was not working. His father was a teacher and Mr. J. recalls that his father was particularly cheerful and cordial to him and his siblings when he was drunk. Mr. J. reports that he often fetched beer and wine for his father and that he once tasted it when he was 8 years old.

When he was 12, he would occasionally drink from his father's open wine in the afternoon with his brother, who was 3 years older. He liked the "cosy" feeling. At the age of 14, he had also drunk larger amounts on weekends, when he went to football matches and later they drank together in the club house. He continued to do this throughout his high school years and at the beginning of his law studies. Looking back, he could hardly remember a weekend during his time as a young adult when he did not drink.

In the later course of his studies, he liked to drink two or three beers alone during the week and to get "really" drunk at the weekends. He had then continued this rhythm for many years, whereby he had also "ended up" with a bottle of wine in the evening at some point during the week. However, these had been good years, in which he had changing relationships and was also recognized professionally, as a lawyer in an authority.

Actually, since the birth of his two children, there were permanent quarrels with his then partner and the births were actually just attempts to save the relationship. He had then somehow crossed a threshold and recalled that during this time he was also regularly drunk during the week for the first time. With the divorce, he said, he lost control and also drank alcohol during the day during the week. He felt he had somehow gone down a slippery slope which he was slowly "sliding down" and consumption was taking place more and more frequently without any real pleasure in it. Then in the last 2 years he often had a slight restlessness in the morning, which he could calm down with a small bottle of brandy.

The detoxification treatment took place due to increasing difficulties at work. Mr. J. reported that he had survived the withdrawal well, but since then he had felt a strong urge to drink alcohol several times a day. He did not experience this as a desire for alcohol at all, but rather like the feeling one has when wearing a "scratchy sweater that one wants to take off". Several times he had already stood at the checkout in the supermarket and wanted to pay for several bottles of wine, but so far he had always been able to bring them back. He no longer goes to the kiosk near his home because he can no longer trust himself. The discussions in the self-help group would help him a lot—nevertheless, he was often dissatisfied and did not really have the feeling that he fit in with these people in the self-help group. But there was no one else who could understand him so well. ◀

10.1 The Most Important Facts in Brief

The consumption of substances that alter the physical and mental states of humans is part of all human cultures. Consumption is shaped by historical, regional and religious backgrounds, but also by a variety of other factors such as gender, age and socioeconomic status (von Heyden et al. 2018).

When substance use is considered from the perspective of a disorder, the term "addiction" usually emerges, which is usually understood as strong desire for substances (drugs, alcohol, etc.) or certain behaviors (such as excessive gambling, etc.) which is often pursued despite negative consequences for the individual, such as physical, psychological and social consequences. Scientifically and clinically, the term has long been used differently and usually limited to the effects of psychoactive substances (including alcohol, opiates, nicotine, etc.). In contrast, "behavioral addictions" have long been understood by professionals as impulse control disorders, and thus have also been classified elsewhere in the classification system, for example in the International Classification of Disease Version 10 (Dilling and Freyberger 2013). However, since certain behavioural addictions, such as pathological gambling—also referred to as gambling disorder—are now assumed to have a partly similar origin and neurobiological background, some of the models described, for example those of learning theory, can also be partially applied to non-substance-related addictions (Penka et al. 2018).

10.2 Diagnosis and Classification of Substance Use Disorders

Heavy use of a substance does not automatically equate to the presence of a dependence disorder. In order to establish the diagnosis of a dependence disease, further criteria must be fulfilled. The diagnostic guidelines for dependence disorders are laid down in both the ICD-10 (Dilling and Freyberger 2013) and the DSM-5 (DSM-5 2013) classification systems. In the ICD-10 classification, the distinction between "dependence" and "harmful use" plays a crucial role. Harmful use here refers to demonstrable physical or mental harm caused by sustained use over at least 1 month or repeatedly within 12 months. Dependence refers to disorders of prolonged use of at least 12 months, which comprise six criteria, of which at least three must be fulfilled for the diagnosis to be made (◘ Table 10.1).

> Classification system ICD-10: The diagnosis of addiction diseases is carried out according to clearly defined criteria and is carried out internationally according to the classification system ICD-10.

While the American classification system, the DSM-IV, also distinguished between "harmful use" and "dependence", the version revised since 2013, the DSM-5, does without the distinction and rather distinguishes the severity of *"substance use disorder"* into mild, moderate and severe (DSM-5 2013; ◘ Table 10.1).

◻ Table 10.1 Classification of addictive disorders, comparison of DSM-5 and ICD-10 criteria. (Penka et al. 2018)

DSM-5	ICD-10

DSM-5 criteria Substance Use Disorder
At least two of the following criteria must be present within a 12-month period:
1. Substance is frequently consumed in larger quantities or for longer than intended
2. Persistent desire or unsuccessful attempts to reduce or control substance use
3. High expenditure of time to obtain, consume or recover from the effects of the substance
4. Craving or strong desire to use the substance
5. Repeated substance use that results in failure to meet important responsibilities at work, school, or home
6. Continued substance use despite ongoing or repeated social or interpersonal problems caused or exacerbated by the effects of the substance
7. Important social, professional or recreational activities are abandoned or restricted due to substance use
8. Repeated substance use in situations where the use leads to physical danger
9. Continued substance use despite knowledge of a persistent or recurrent physical or mental problem that is likely to have been caused or exacerbated by the substance
10. Development of tolerance, defined by one of the following criteria:
  • Dose increase to induce intoxication state or a desired effect
  • Significantly reduced effect with continued use of the same amount of substance
11. Withdrawal symptoms manifested by one of the characteristic withdrawal symptoms (defined for each substance)
Current severity:
Mild: 2–3 symptom criteria
Means: 4–5 symptom criteria
Severe: 6 or more symptom criteria
	Dependence
	At least three of the following criteria in the last 12 months:
	1. Consumption despite evidence of harmful effects
	2. Development of tolerance
	3. Withdrawal syndrome
	4. Reduced ability to control onset, cessation, or quantity
	5. Strong desire or compulsion (craving) to use the substance
	6. Neglect of other interests in favour of substance use, increased time spent on acquisition or consumption
	Harmful use: substance-induced mental or physical problems

10.3 Symptoms

In the case of harmful use, the symptoms of addiction primarily include physical disorders, such as increased liver values in the case of alcohol consumption, and psychological disorders, such as depressive states and anxiety disorders in the case of v consumption. In addition, further symptoms occur in the development of dependence, such as the development of tolerance and the associated increase in dosage, withdrawal symptoms, loss of control or craving.

Important clinical symptoms and terms are listed in Box: Important clinical terms in substance use disorders and in ◻ Table 10.1.

10

Important Clinical Terms in Substance Use Disorders

Tolerance: Another word for **tolerance** is **habituation** and means the decrease of the substance effect after repeated intake of the same dose. A classic example is the decrease in the effect of morphine after repeated administration. Tolerance occurs both centrally in the brain through desensitization of various target regions and peripherally through accelerated degradation of substances, for example via enzyme induction in the liver.

Withdrawal symptoms: Withdrawal symptoms result from the reduction or sudden cessation of the regularly used substance. Withdrawal symptoms are often the opposite symptoms of the original substance effect, so that withdrawal from euphoric substances, such as cocaine, tend to cause depressive symptom complexes and withdrawal from sedative substances, such as opioids, tend to be accompanied by physical restlessness and increased vigilance. These symptoms often develop in the case of a pre-existing development of tolerance and are thus an expression of the disturbance of the homeostasis (state of equilibrium) formed in the brain during substance consumption. Withdrawal symptoms can develop after only a few weeks in the case of opioids, and in the case of other substances usually only after several months to several years of high consumption. In adolescents, they can occur much more quickly.

Automatisms and reduction of control: The term "automatisms" stands for little or no conscious, motorically automated processes, which can, for example, lead to a relapse (Tiffany and Carter 1998). An example of this is the habitual, automated consumption of alcohol supplies found and consumed in the home despite a desire to abstain. If such a habitual routine is interrupted, for example when all remaining stocks have been used up, consumption cannot continue in the usual way and is then frequently interrupted. Now the person concerned usually first notices his craving for the substance. Conscious craving—but also non-conscious motor automatisms—are considered by many authors to be a central variable in the development of addiction disorders and relapse (Everitt and Robbins 2016). Reduced control in this context describes the lack of control over these habitual, motor processes and thus the amount consumed. An example is the "after-work beer", which can not be stopped and ends in massive alcohol consumption.

Craving: **Craving** is a term used to describe a compulsive, conscious desire for a substance that may exist even when the use of the substance itself is no longer enjoyed (Everitt and Robbins 2016; Tiffany and Carter 1998; Wetterling et al. 1996). However, the place of **craving** in people with substance dependence is largely unresolved. It is true that casuistics report sometimes excruciating cravings in people with dependence disorders, sometimes associated with pronounced physical reactions, which can occur in people with substance dependence even after years of abstinence and can lead to relapse in use. However, the occurrence of cravings can also prevent relapse in situations in which those affected are aware of these cravings and can thus interrupt automated processes.

Substance reactivity (cue reactivity): Situations that were associated with the use of substances in the past, such as places of consumption, persons in whose presence consumption took place, certain times of day, memories or moods, can make the occurrence of a relapse more likely in substance-dependent individuals. This is due to classical (linking of stimulus and response) and operant (linking of action and effect) conditioning processes, which can lead to the fact that confrontation with these situations activates a neuronal cascade in the behaviour-reinforcing system of the brain. This can result in an activation of automatic behaviour patterns, which leads to renewed consumption.

10.4 Incidence of Addictive Disorders

The recording of the occurrence and frequency of addictive disorders in the population is not easy, because in addition to difficulties in diagnosing and defining problematic consumption quantities, the data situation is also often only of limited significance due to the tabooing of addictive disorders. In representative surveys, there is often a selection problem, since among those who do not participate in the surveys, there are more persons with dependence diseases or problematic substance use than among those who do participate in the surveys. Despite this systematic problem, population surveys are regularly conducted, such as the Epidemiological Survey on Addiction (ESA 2012) or the Drug Affinity Study of the Federal Centre for Health Education (BZgA), which uses a representative sample to determine the development of young people's use (Orth 2016). National health surveys such as the DEGS (Study on Adult Health in Germany) and the annual drug report for the EU also provide information on the development of substance use in the population (Burger and Mensink 2003; EMCDDA 2016; Jacobi et al. 2014). In addition, for licit substances, consumption in the general population is estimated indirectly on the basis of sales figures, and for illicit drugs, such as methamphetamine, heroin and cocaine, and consumption trends are estimated on the basis of police information such as seizures.

10.4.1 Prevalence of Alcohol Consumption

According to the WHO, the average per capita consumption of pure alcohol in Germany was 11.8 L in 2014 (WHO 2016). This put Germany in 16th place in a European comparison after countries such as Poland, Lithuania, Croatia, the Czech Republic and Ireland. The limits for alcohol consumption that poses a health risk are currently set by the WHO at <40 g of pure alcohol per day for men and <20 g per day for women. Bühringer et al. (2000) also distinguish alcohol consumption levels into low-risk (men <30 g/women <20 g), high-risk (>30 g/>20 g), hazardous (>60 g/>40 g) and high consumption (>120 g/>80 g). According to the Epidemiological Survey on Addiction (ESA 2012) published in 2012, 28.5% of the persons surveyed in Germany lived abstinent from alcohol in the last month. For the last year, 9.8% of the respondents indicated abstinence from alcohol. About 12.0% of the respondents had a risky consumption, 1.8% a hazardous consumption and 0.4% a high consumption. The criteria for alcohol dependence were met by 3.4% of the respondents between 18 and 64 years of age, with men being affected more frequently than women (Kraus et al. 2013).

The diagnosis of alcohol dependence affects about 3.4% of the German population between the ages of 18 and 64.

Alcohol consumption in different countries varies greatly internationally. In Turkey, for example, the annual prevalence for alcohol consumption is only 25.8%, which is much lower than in all other European countries (WHO 2016). The proportion of people with an alcohol dependence is even estimated at only 0.8% (WHO 2016), which can be explained, among other things, by the religious ostracism of alcohol consumption. Alcohol consumption in the population in other Muslim countries, however, varies widely from 6.5% in Egypt to 20.1% in Lebanon and prevalence rates of alcohol dependence from 0.1% in Egypt to 0.3% in Lebanon (WHO 2016). In Germany, according to WHO, these rates are 80.3% and 2.9% which is approximately the same as the ESA figures above (WHO 2016). In Russia, the annual prevalence for alcohol consumption is only 67.8%, which is significantly lower than in other European countries, but the proportion of people with alcohol dependence is 9.3%, which is significantly higher than in most European countries (WHO 2016).

10.4.2 Prevalence of Use of Non-legal Substances

Among the non-legalised drugs in Germany, the use of cannabis is the most widespread. While the lifetime prevalence of cannabis use among adults is currently around 25%, only around 5–6% of adults have ever used another illicit substance in their lives (BMG 2016; EMCDDA 2016). Accordingly, annual prevalence also differs at 4.5% for cannabis and 1.5% for other drugs and monthly prevalence at 2.3% for cannabis and 0.8% for other drugs. For individual other drugs, much lower monthly prevalences were determined, for example for heroin 0.1%, other opiates 0.2% and for cocaine use 0.3% (ESA 2012).

The use of illicit drugs varies greatly worldwide and, in addition to regional differences, is also subject to strong fluctuations over time. For example, the number of estimated heroin users internationally has been relatively constant for around 15 years at 17 million, with phases of increased use in certain regions, such as the USA, which already had and currently again has a high number of heroin users (UNODC 2017). However, the number of users of cannabis internationally has been increasing for years, with now an estimated 183 million users worldwide and an estimated 30 million people with harmful use, although again the increase relates to North and South America rather than Europe, which have had relatively stable rates of use for a number of years (GDS 2018). Of concern, however, are increasing production rates for most illicit drugs, which for opium, for example, increased by 30% in 2016 and may precede increasing international consumption (UNODC 2017).

The use of drugs that can lead to addiction, such as benzodiazepines and opiates, can now hardly be separated from the use of non-legal substances, as the drugs are often not prescribed by a doctor but traded on the black market. In this context, there has been an increase in non-medical use of opioid painkillers in particular (UNODC 2017). Reliable figures on dependence rates among people with regular non-medical use of medicines are not yet available in Germany, but are currently estimated at around 2.3 million people (BMG 2016) and would thus be comparable to alcohol dependence in terms of the incidence of dependence.

However, nicotine dependence remains the most socially relevant substance use disorder, with both high morbidity and mortality. In Germany, 31% of men and 24% of women reported being current smokers in 2016 (BMG 2016), with the proportion of adult smokers falling significantly between 1995 (14.7%) and 2006 (9.6%) (Baumeister et al. 2008). This is also the case for adolescents, where the proportion of smokers in the population has fallen from 27.5% in 2001 to 7.8% in 2015 (BMG 2016).

Rates of dependent smoking vary widely internationally, with some marked gender differences, such as in Turkey, where male smokers account for 50% of smokers, and 10% of women, or Russia, where 63% are male and 16% female (Bobak et al. 2006).

10.5 Risk Factors and Development Models

Risk factors for the development of dependence disorders are assessed differently in different disciplines (medicine, psychology, neurosciences, social sciences). What the models have in common, however, is the view of dependence disorders as a disease that cannot be explained by individual misconduct or weakness of will. So far, no isolated cause could be proven, such as particularly addiction-producing social situations or the proof of an "alcoholism gene". In view of the complexity, a multifactorial model currently seems most suitable for describing the development and maintenance of dependence disorders. The neurobiological basis of dependence diseases must be seen in the context of psychological and social influencing variables, which leads to concordance of the development models.

10.5.1 Biographical and Individual Factors

In clinical observation, traumatic events, such as experiences of abuse or stress in the parental home, are more frequent in persons with dependence disorders than in persons without dependence experiences. Furthermore, addiction disorders are often accompanied by comorbid mental illnesses such as depression or personality disorders (e.g. emotionally unstable and narcissistic personality disorders; von Heyden et al. 2018).

Genetic factors are of particular importance in the development of dependence disorders: It is estimated that about 50% of the risk of developing substance dependence can be explained by genetic factors. An important predictor for the later development of alcohol dependence is a presumably genetically determined reduced acute response to the consumption of alcohol (Heinz et al. 2013).

Learning theory approaches use classical learning psychology models to explain the development of addictive disorders. Of particular importance here are models of imitation learning and classical and instrumental conditioning. In animal models and human studies with neurobiological, partly imaging procedures, attempts are being made on the basis of these models to better understand

the neuronal correlates of the learning of dependent behaviour (▶ Sects. 10.7 and 10.8, below).

Neurobiological approaches focus primarily on the biochemical correlates of the leading symptoms of addiction diseases, such as the development of tolerance, withdrawal symptoms and addiction craving, through animal experiments and human models. The corresponding adaptation of neuronal transmission and the significance of important neuronal centres, for example the ventral striatum including the nucleus accumbens, are discussed in Sect. 10.8.1.

Sociological factors can be important additional factors influencing the development of addiction diseases. In this context, sociological theories (e.g. Schmidt et al. 1999) are to be understood as complementary rather than contradictory to neurobiological explanatory models. For example, it has been shown in animal models that social exclusion processes can be associated with stress reactions, which in turn are associated with altered serotonergic neurotransmission and can result secondarily, for example, in increased aggressive actions in the case of acute and chronically increased alcohol consumption (Heinz et al. 2013).

In addition, there are a number of other models, such as *socio-social* theories, **which** discuss processes of social exclusion and individualization and dissolving traditional family forms, structures and relationships (Hurrelmann and Bründel 1997), or *psychodynamic models*, which address individual triggering events such as situations of grievance and family constellations (Rost 2009).

10.6 Treatment of Substance Use Disorders

For most substances, such as alcohol dependence, the treatment of substance dependence is primarily aimed at abstinence from the respective substances. For some substances, such as heroin, abstinence is a secondary goal and the focus is on ensuring survival, for example by administering long-acting opioids to replace the procurement of heroin (Gutwinski et al. 2014).

10.6.1 Treatment of Substance Use Disorders Distinguishes Between Different Phases

■ **Contact and Motivation Phase**
It usually takes place in general practitioners' surgeries and addiction counselling centres with the aim of building up a relationship with the person concerned, consolidating the decision to change (for example through Motivational Interviewing/ Motivierende Gesprächsführung) and initiating therapy.

■ **Detoxification Treatment**
This is the acute treatment, in which the affected person leaves out the substance and there are usually vegetative symptoms, which are typically opposite to the substance effect. For example, with benzodiazepines, which have a sedative effect, a restless, agitated syndrome occurs during withdrawal.

In this phase, two strategies for action are usually pursued:
— Treatment with agonists, substances that replace the effect of the substance. In the simplest case, this is the same substance but in a different form of application, for example nicotine patches for nicotine withdrawal. Furthermore, substances of the same type with different half-lives are used, such as the long-acting opioid methadone for withdrawal from heroin, which has a shorter effect. Agonists on the GABA receptor, for example benzodiazepines, are also used for alcohol dependence. The aim of treatment with agonists is first to reduce withdrawal

symptoms and then to gradually reduce agonists with the aim of reducing the risk of dangerous side effects, such as epileptic seizures, and to reduce discontinuation rates due to severe side effects (Muller et al. 2014).
— Symptomatic therapy: use of substances that alleviate the side effects of withdrawal, for example, sleep-inducing substances, blood pressure-lowering drugs, drugs against nausea and vomiting, etc.

■ **Weaning Treatment**
Long-term therapeutic treatment in specialised psychotherapeutic clinics to acquire abstinence maintenance skills and identify individual risk factors (e.g. problems at work).

■ **Follow-Up Phase**
Maintaining abstinence through therapeutic support in addiction counselling centres, by general practitioners and self-help groups. The focus is on building a life worth living, in which substance use is replaced by other, positive activities. Outpatient psychotherapies are useful in this phase, as many of those affected use substance use as "self-therapy" for mental disorders, for example in the presence of trauma or an emotionally unstable personality disorder. Unfortunately, many psychotherapists still have reservations about treating formerly "addicted" patients.

■ **Longer-Term Pharmacological Treatment After the End of Detoxification**
In the case of heroin addiction, this is primarily achieved by substitution treatment with long-acting opioids such as methadone or buprenorphine. In the case of alcohol dependence, substances can be used which reduce the risk of relapse, for example naltrexone (opioid antagonist) and acamprosate (effect on glutamate receptors; Muller et al. 2014).

10.7 Progression of Addictive Disorders

Dependence diseases are often accompanied by consumption relapses. In the case of alcohol dependence, this amounts to 50–80% in the course of 5 years, and to 80–90% in the case of heroin dependence. In today's therapies, relapses are regarded as part of the therapeutic process, from which those affected can draw important conclusions about their own susceptibility factors for future abstinence phases.

Substance-specific physical damage is also frequently found after years of consumption. A frequent finding after years of significantly increased alcohol consumption is, for example, brain atrophy, which in turn can lead to an impairment of cognitive performance (e.g. Czapla et al. 2016; Loeber et al. 2009). However, studies also indicate that in addition to chronic consumption of alcohol, its withdrawal can also lead to brain damage (Duka and Stephens 2014). Presumably, the increased glutamate level during withdrawal plays a role in this, which can have neurotoxic effects (Tsai and Coyle 1998). Thus, patients with a history of frequent detoxification treatments show significantly worse cognitive performance than patients who have had no or only one detoxification treatment (Loeber et al. 2009).

Long-term excessive alcohol consumption can also lead to a variety of other consequential damages affecting almost all organ systems, such as liver cirrhosis, polyneuropathy and pancreatitis (see Charlet and Heinz 2016), and is therefore associated with a significantly increased mortality.

The consequential harm caused by other substances is again substance-specific and can be caused by the substance itself, for example through intoxication, or by the lifestyle associated with substance use, for example through malnutrition in the case of alcohol consumption or poor hygiene in the case of intravenous use of heroin (Häbel and Gutwinski 2018).

10.7.1 Neurobiological Basis of Substance Dependence

To date, there is no conclusive neurobiological theoretical conception of substance dependence. However, a large body of research suggests that pre-existing and/or substance-induced differences in the neurobiological structure of the behavioural reinforcement system in the brain (the so-called reward system) may be important for the development of substance dependence. In this regard, important scientific results are emerging from animal experiments and, not least, findings from neurobiological imaging in humans, from which the main models presented below are derived. However, the individual model assumptions are in part very complex, so that a simplification is necessary in the following presentation.

10.7.1.1 Learning Mechanisms

Learning mechanisms play an essential role in the development of addictive disorders. These mechanisms can be interpreted in terms of classical conditioning processes (association of cue stimulus and behavioral response). These were first described in animal experiments (Heinz et al. 2013; Wikler 1948), in which withdrawal symptoms from opioids were associated with box placement in rats, and withdrawal symptoms also occurred when they were placed in the box again, without being related to opioid administration. In this case, one would assume a conditioned response in which the location is already linked in learning theory with occurrence of withdrawal symptoms. This reaction presumably also has a protective function, since withdrawal symptoms also represent an opposite reaction to substance use and increase tolerance to the addictive substance. This opposing response would be described as conditioned withdrawal symptomatology and may pave the way for substance abusers to use again or relapse, as it may increase craving for the substance (Heinz et al. 2013; Verheul et al. 1999). Another learning mechanism in addiction disorders is the so-called positive reinforcement. Here, the dopamine release in the ventral striatum caused by substance intake seems to be of primary importance for the development of dependence (Wise 1998). The positive conditioning results from this dopamine release and the pleasant effect after the stimulus in the form of drug intake (Wise 1998). However, some authors discuss whether it is not so much the experience of feelings of pleasure or happiness (liking) that plays the important role, but rather the craving (wanting) for the substance that arises. This could also explain the fact that substance use in the case of addiction is continued even when the substance use is no longer associated with a euphoric effect (Everitt and Robbins 2005, 2016).

10.7.2 Development of Tolerance

Dependence diseases are characterized, among other things, by adaptation processes with a development of tolerance: Increasing amounts of the substances are needed to achieve the desired effect (Dilling and Freyberger 2013; Gutwinski et al. 2016). The decreasing analgesic efficacy of morphine after repeated administration is a clinical example of an adaptation process and can be described pharmacologically as a rightward shift of the dose-response curve (Battegay 1966; Reidenberg 2011). In this context, the development of tolerance is an expression of counter-regulatory changes in the body in response to ongoing substance use through pharmacodynamic, pharmaco-kinetic, and behavioral mechanisms (Battegay 1966; Bespalov et al. 2016; Reidenberg 2011). A pharmacodynamic example is the altered subunit composition of the GABA-A receptor after chronic alcohol consumption (Clapp et al. 2008) (GABA or γ-aminobutyric acid is the major inhibitory neurotransmitter in the central nervous system). After chronic alcohol consumption, there may be a decrease in the production of the α1-subunit of the GABA-A receptor (Clapp et al. 2008), which likely contributes to alcohol tolerance (Blednov et al. 2003; Clapp et al. 2008). Interestingly, it has been shown that in people with alcohol dependence, unlike in healthy individuals, a decrease in the number of GABA-A receptors has been demonstrated in terms of tolerance formation, which had not normalized even several weeks after stopping alcohol (Abi-Dargham et al. 1998). This finding is also of essential importance for the psychotherapeutic and medical further treatment of patients in the first weeks after detoxification, because it can be assumed that precisely these changes in the brain initially persist and only regress over time. Longer-lasting, not immediately visible symptoms such as temporary cognitive performance losses or an emotional dysbalance with behavioural disorders are presumably an expression of associated restrictions in the brain's performance, which only regresses after some time of abstinence and also only gradually. This process of regeneration must always be taken into account in psychotherapeutic treatment in order to avoid excessive demands, and has so far received little recognition, for example by health insurers, who primarily regard the obvious withdrawal symptoms such as muscle tremor, ataxia, irritability or epileptic withdrawal attacks and other vegetative symptoms as acute markers of substance withdrawal.

Another effect of many dependence-producing substances is the blockade of a subtype of the excitatory glutamate receptors (NMDA receptors), which leads to an inhibition of the glutamatergic system. This mechanism has been particularly extensively studied in chronic alcohol consumption, where it appears to be associated with an increase in these receptors as part of tolerance formation. This counter-regulatory process seems to allow the brain to regain some of the capacity it has in the sober state despite the persistent effects of alcohol, and it partly explains the sometimes prima facie good functioning of some people despite marked levels of blood alcohol.

However, this balance becomes problematic during acute withdrawal: If the alcohol supply is suddenly missing, the NMDA receptor for glutamate is released and the excitatory transmitter glutamate meets the still increased number of receptors. Since at the same time too few GABA-A receptors are present, which otherwise ensure counter-regulatory inhibition of the activity of the nerve cells, disinhibition or over-stimulation of the so-called locus coeruleus in the brain stem can now occur, which is responsible for the development of vegetative withdrawal symptoms (increased pulse, blood pressure, sweating, etc.). Furthermore, epileptic withdrawal seizures can be caused by these changes in neuronal homeostasis (Gutwinski et al. 2016).

Problematic for users is the experience that withdrawal symptoms are often perceived as very stressful and can be quickly overcome by using substances again. This process of negative reinforcement then in turn leads to a continuation of substance use.

10.8 NMDA Receptors and Learning Mechanisms

NMDA receptors are thought to play an important role in the mediation and storage of learning content. A current hypothesis assumes that memories are "stored" in the brain as specific spatiotemporal excitation patterns within neuronal networks. These networks consist of groups of neurons that are synaptically connected. In order for such enduring synaptic circuitry to emerge, the new incoming information must be capable of influencing the pre-existing synaptic connections. Hebb (1949) and Konorski (1948) studied essential mechanisms for this and described the process of coincidence: synaptic connections between two neurons are strengthened when they are activated simultaneously ("*Neurons that fire together, wire together.*"). Such processes are found in various brain regions and have also been described in the hippocampus, which plays a central role in memory formation. It is now interesting to note that high-frequency activation of neurons with an excitatory function in these brain regions can lead to a long-term sustained consolidation of synaptic connections and thus increase the probability of activation (Bliss and Collingridge 1993). This mechanism is referred to as Long Term Potentiation (LTP). Both glutamatergic NMDA receptors and other presynaptic processes are thought to be crucially involved in the generation of LTP. It is assumed that the nerve cells can form an engram of a simultaneous activation pattern via synaptic connections. Since alcohol can impair the function of NMDA receptors, for example by influencing glycine binding, which in turn is thought to facilitate magnesium blockade, there is then presumably also a reduced formation of LTP and associated disturbances in memory function (Tsai et al. 1995). Since the glutamatergic system still seems to be impaired in the period after discontinuation of substances, this may lead to a reduced ability to learn in the first weeks after detoxification and, for example, to impaired learning processes in psychotherapy.

The mechanism described above and the associated excessive calcium influx are probably associated with neurotoxic cell damage in the case of excessive substance use, which further impairs learning ability. Furthermore, the upregulated NMDA receptors may cause activation of glutamatergic neurotransmission, leading to sensitization of the motivationally important dopaminergic system. This process is thought to be elementally linked to the development of addiction memory, which in turn may be associated with behavioral-level changes such as decreased control, disruption of motivation, excessive craving, or automatic behaviors that lead to repeated use of a dependence-producing substance. The addiction memory term describes the assumption that this behaviour seems to be reactivatable quickly even after long periods of abstinence and that it must be actively suppressed to some extent by the affected person even during abstinence. However, the biological basis for this has not yet been clarified and cannot be attributed solely to the process described above. Studies indicate, among other things, that a disturbance in the dopaminergic system, especially in the Ncl. accumbens, (Heinz et al. 2013), as well as influences of the dopaminergic system on the hippocampus seem to play an essential role (Lisman and Grace 2005).

10.9 The Nucleus Accumbens and the Importance of the Dopaminergic Reinforcement System

A central structure of the behavior-reinforcing system is the core area of the nucleus accumbens and the dopaminergic neurotransmission there. The intake of virtually all common dependence-producing substances leads to an increased availability of dopamine in this region and thus to a change in neuronal processes. This is presumably one of the central mechanisms by which dependence-producing substances produce a more pronounced behaviour-reinforcing effect than normal rewarding

stimuli ("natural" reinforcers such as food, play or sex), which in turn leads the brain to prefer consumption of the substance over everyday rewards (for a review, see e.g. Charlet et al. 2013).

In studies of the dopaminergic system with radioactive markers (positron emission tomography; e.g., Kienast and Heinz 2005), it was further observed that people with alcohol dependence have a lower number of dopamine D2 receptors in the striatum compared to healthy individuals (see also Heinz et al. 1998; Koob and Volkow 2016). There is evidence that in humans, low dopamine D2 receptor availability in the corpus striatum, the region where the ncl. accumbens is also located, is associated with impairments in emotional responsiveness and experience, and therefore this type of receptor plays a role in mediating emotional responses (Heinz et al. 2013). Also, the number of available dopamine D2 receptors appears to affect the effects of dependence-producing substances and contribute in the sense of a disposition to dependence disorders (Volkow et al. 1999). However, evidence of a genetic disposition has so far only been found in animal experiments (Koob and Volkow 2016).

Since the availability of dopamine D2 receptors in the corpus striatum seems to be reduced in some individuals during pronounced environmental stress, this could be one of the molecular mechanisms underlying the association of environmental stress on the development of addictive disease (Heinz et al. 2013).

In addition to the dopaminergic system, other neurotransmitters play an essential role in the development of dependence disorders, including GABAergic and serotonergic neurotransmitters (Heinz et al. 2013). For example, there is evidence that disturbances in serotonergic neurotransmission influence the responsiveness of the centrally depressing GABAergic system and could thus influence the tolerance of substances, for example alcohol (Heinz et al. 1998).

10.10 **Sensitization**

Another important mechanism in the development of addictive disorders is sensitization, which is associated with far-reaching consequences at the behavioral-biological level. The neurobiological model of sensitization as the basis for dependent behavior replaces the concepts of "psychological dependence" and the assumption of a "weakness of will of the patient", which are unfortunately widespread. The model of sensitization can be used to explain to a large extent why people, after repeated use of a dependence-producing substance, continue to use it repeatedly and in a dependent form against their own wishes.

The model of sensitization is derived from animal experiments in which it was shown that animals exhibited enhanced psychomotor reactions after repeated administration of stimulants. The cause is presumably attributable to altered neuronal transmission in the ncl. accumbens, but can also be understood as a principal mechanism. Sensitization is highly substance-specific, so that, for example, sensitivity to nicotine does not automatically lead to sensitivity to alcohol or cocaine.

In very simplified terms, the process of sensitization can be described as the process opposing the formation of tolerance. This is apparently associated with a pronounced activity in the ncl. accumbens, so that certain neurons are activated even by the smallest amounts of substances or by individual stimuli that an affected person associates with the substance (cues). This causes an increased urge (wanting) to consume the substance again. This mechanism is called incentive sensitization and is a possible explanation for the frequent action of people with substance dependence to continue using, even without experiencing the effect positively and even using against their own declared intention (Robinson and Berridge

2008). In extreme cases, these processes are completely automatic and need not involve conscious action.

10.11 **Automatisms**

Another mechanism, which has already been mentioned several times and seems to play a role in addiction, are so-called automatisms. There are indications that these processes take place via the corpus striatum, in which this serves as a memory for routinized courses of action (*habits*), and templates for courses of action are activated by conditioned stimuli (cues). Transferred to the learning of dance steps, in which no conscious thought about the step sequence is necessary after multiple repetitions, similar automatisms would take place in substance dependence: For example, the presence of a familiar brand of cigarette (cue 1) in the backyard (cue 2) with colleagues (cue 3) during breakfast break (cue 4), etc., would lead to the use of nicotine without the person consciously planning the sequence. One can speak here of a transition from goal-directed to automated behaviour (Tiffany and Carter 1998). At the moment when there is no more nicotine and the sequence of automated action is interrupted, a conscious craving (craving) for the substance arises. It is only at this moment of action interruption that most people with dependence disorders realize and evaluate their behavior (orbito-frontal cortex), and then consciously decide on the further course of the behavior (frontal cortex and anterior cingulate, among others; Tiffany and Carter 1998) At this moment it is easier for the affected people to interrupt the automatisms, and is therefore the focus of therapeutic measures (development of alternative action strategies). However, the fact that individuals recognise these automatisms and wish to resist them is no guarantee of successful behavioural change, because some patients report, for example in the case of heavy heroin use, that although they perceive these unwanted courses of action in themselves, they are practically unable to influence them in the process. A possible explanation for this would be that the activation of the reinforcement system of the nucleus accumbens, once triggered, is so pronounced that it cannot be sufficiently modulated by other brain regions (e.g. frontal cortex and anterior cingulate) (Kienast and Heinz 2006).

Importance of Neurobiology for Psychotherapy

Neurobiological knowledge is of relevance for therapies in the field of addictive disorders when psychotherapeutic procedures aim to exert specific influence on certain brain regions of the persons concerned, for example on the reinforcement system with the aim of reducing pathological activity or on prefrontal control functions with the aim of increasing activity. In recent years, individualized techniques have been developed, such as computer-assisted training elements, which aim to reduce automatisms of attentional function to alcoholic stimuli and resulting automated behaviors. The effectiveness of such alcohol avoidance training has been shown in several studies (e.g., Eberl et al. 2013; Wiers et al. 2015). Such avoidance training could therefore be an effective adjunct to existing psychotherapeutic procedures with people with dependence disorders. Current studies also show that these methods also lead to the desired reduction in neuronal reactions to alcoholic stimuli at the neurobiological level (Vollstädt-Klein et al. 2011).

The neurobiological model of disorders in the dopaminergic system and the assumption of a reduced response to natural reward stimuli is also relevant for the clinical attitude of psychotherapists in addiction treatment in the first weeks after detoxification, as the establishment of natural reinforcers is often more difficult for those affected. Similarly, the toward response to the new stimuli is diminished (Schultz 2006; Schultz et al. 1997), often despite the sufferer's desire to maintain abstinence. In this tension between the intention to maintain abstinence and remaining impairments in brain performance, unexpected incidents of use may occur, which can often be associated with feelings of discouragement and shame reactions for the individual. The therapist can prepare this knowledge with the affected person during the therapy process in order to reduce the risk of renewed consumption. In this neurobiologically sensitive phase, accompanying pharmacological treatment (e.g. with substances that reduce the risk of relapse, such as acamprosate, naltrexone, etc.) in addition to psychotherapeutic treatment can also be useful. In the case of additional stress factors, such as limited availability of social resources (e.g. debts, stressful social environment, unemployment) or persistent one-sided occupational stress (e.g. police officers, managers), there may be an indication for withdrawal treatment.

For more information, see ▶ Chap. 5, Psychotherapy Procedures and Their Effects.

Concluding Remarks in the Sense of an Attempt at a Neurobiological-integrative Approach: Arguments Against the Stigmatisation of Addicted People

The neurobiological and behavioural findings of recent years make it clear that addictive disorders can occur in a wide variety of people and are not merely an expression of "weakness of will" or an "addictive character". The evidence of neurobiological correlates underlines the now common assumption that these are diseases which are also characterised by a frequently high level of suffering. This is also true for the individuals who are sometimes referred to as "type 2 addicts", who are not a clearly delineable or clinically unambiguous group of people with addictive disorders, but rather individuals who have often experienced violence at an early age and may themselves respond to experiences of threat with aggression and rash (impulsive) actions (Cloninger 1981; Gutwinski et al. 2018). Especially in these individuals, neurobiological dysfunction, for example in serotonergic neurotransmission, seems to be modulated to a particular extent by environmental factors (Gutwinski et al. 2018; Heinz et al. 2013). Aggression is often the result of feelings of isolation or threat, which can be caused by experiences of social isolation.

Neurobiological research, if not interpreted in a biologically pessimistic way, could therefore help to link other perspectives, for example on characteristics such as stress or trauma, which are associated with the development of addiction, and could thus counteract the myths of guilt and weakness of character, in the sense of a model of "addicted people" who can be reached by specific therapies.

References

Abi-Dargham A, Krystal JH, Anjilvel S, Scanley BE, Zoghbi S, Baldwin RM et al (1998) Alterations of benzodiazepine receptors in type II alcoholic subjects measured with SPECT and [123I]iomazenil [Research Support, U.S. Gov't, Non-P.H.S. Research Support, U.S. Gov't, P.H.S.]. Am J Psychiatry 155(11):1550–1555. https://doi.org/10.1176/ajp.155.11.1550

Battegay R (1966) Drug dependence as a criterion for differentiation of psychotropic drugs [Clinical Trial Comparative Study Controlled Clinical Trial]. Compr Psychiatry 7(6):501–509

Baumeister S, Kraus L, Stonner T, Metz K (2008) Tabakkonsum, Nikotinabhängigkeit und Trends. Ergebnisse des Epidemiologischen Suchtsurveys 2006. Sucht 54(Sonderheft 1):26–S35

Bespalov A, Muller R, Relo AL, Hudzik T (2016) Drug tolerance: a known unknown in translational neuroscience [Review]. Trends Pharmacol Sci 37(5):364–378. https://doi.org/10.1016/j.tips.2016.01.008

Blednov YA, Jung S, Alva H, Wallace D, Rosahl T, Whiting PJ, Harris RA (2003) Deletion of the alpha1 or beta2 subunit of GABAA receptors reduces actions of alcohol and other drugs [Research Support, Non-U.S. Gov't Research Support, U.S. Gov't, P.H.S.]. J Pharmacol Exp Ther 304(1):30–36. https://doi.org/10.1124/jpet.102.042960

Bliss T, Collingridge G (1993) A synaptic model of memory: long-term potentiation in the hippocampus. Nature 361:31–39

BMG (2016) Bundesministerium für Gesundheit. Sucht und Drogen. http://www.bmg.bund.de/themen/praevention/gesundheitsgefahren/sucht-und-drogen.html

Bobak M, Gilmore A, McKee M, Rose R, Marmot M (2006) Changes in smoking prevalence in Russia, 1996–2004 [Research Support, Non-U.S. Gov't]. Tob Control 15(2):131–135. https://doi.org/10.1136/tc.2005.014274

Bühringer G, Augustin R, Bergmann E, Bloomfield K, Funk W, Junge B, Töppich J (2000) Alkoholkonsum und alkoholbezogene Störungen in Deutschland. Nomos, Baden-Baden

Burger M, Mensink G (2003) Bundes-Gesundheitssurvey: Alkohol. Konsumverhalten in Deutschland Robert-Koch-Institut, Beiträge zur Gesundheitsberichterstattung des Bundes. Berlin

Charlet K, Heinz A (2016) Harm reduction-a systematic review on effects of alcohol reduction on physical and mental symptoms. Addict Biol. https://doi.org/10.1111/adb.12414

Charlet K, Beck A, Heinz A (2013) The dopamine system in mediating alcohol effects in humans. Curr Top Behav Neurosci 13:461–488

Clapp P, Bhave SV, Hoffman PL (2008) How adaptation of the brain to alcohol leads to dependence: a pharmacological perspective [Research Support, N.I.H., Extramural Research Support, Non-U.S. Gov't Review]. Alcohol Res Health 31(4):310–339

Cloninger C (1981) A systematic method for clinical description and classification of personality variants: a proposal. Arch Gen Psych 38:861–868

Czapla M, Simon JJ, Richter B, Kluge M, Friederich HC, Herpertz S et al (2016) The impact of cognitive impairment and impulsivity on relapse of alcohol-dependent patients: implications for psychotherapeutic treatment [Research Support, Non-U.S. Gov't]. Addict Biol 21(4):873–884. https://doi.org/10.1111/adb.12229

Dilling H, Freyberger H (2013) Taschenführer zur ICD-10-Klassifikation psychischer Störungen. Hogrefe, vorm. Verlag Hans Huber, Göttingen

DSM-5 (2013) American Psychiatric Association Publishing; 5 Revised edition

Duka T, Stephens D (2014) Repeated detoxification of alcohol-dependent patients impairs brain mechanisms of behavioural control important in resisting relapse. Curr Addict Rep 1:1–9

Eberl C, Wiers R, Pawelczack S, Rinck M, Becker ES, Lindenmeyer J (2013) Approach bias modification in alcohol dependence: do clinical effects replicate and for whom does it work best? Dev Cogn Neurosci 4:38–51

EMCDDA (2016) European monitoring centre on drugs and drug addiction. European drug report 2016. http://www.emcdda.europa.eu/edr2016

ESA (2012) Epidemiologischer Suchtsurvey. http://esa-survey.de/fileadmin/user_upload/Literatur/Berichte/ESA_2012_Drogen-Kurzbericht.pdf

Everitt BJ, Robbins TW (2005) Neural systems of reinforcement for drug addiction: from actions to habits to compulsion [Research Support, Non-U.S. Gov't Review]. Nat Neurosci 8(11):1481–1489. https://doi.org/10.1038/nn1579

Everitt BJ, Robbins TW (2016) Drug addiction: updating actions to habits to compulsions ten years on [Research Support, Non-U.S. Gov't Review]. Annu Rev Psychol 67:23–50. https://doi.org/10.1146/annurev-psych-122414-033457

GDS (2018) Global Drug Survey 2018. https://www.globaldrugsurvey.com/

Gutwinski S, Bald LK, Gallinat J, Heinz A, Bermpohl F (2014) Why do patients stay in opioid maintenance treatment? Subst Use Misuse 49(6):694–699. https://doi.org/10.3109/10826084.2013.863344

Gutwinski S, Kienast T, Lindenmeyer J, Löb M, Löber S, Heinz A (2016) Alkoholabhängigkeit—Ein Leitfaden zur Gruppentherapie. Kohlhammer, Stuttgart

Gutwinski S, Heinz A, Heinz A (2018) Alcohol-related agression and violence. In: Beech A, Carter A, Mann R, Rotshtein P (eds) The Wiley Blackwell Handbook of Forensic Neuroscience. Wiley, Blackwell, NJ

Häbel T, Gutwinski S (2018) Opioide. In: Von Heyden M, Jungaberle H, Majic T (eds) Handbuch Psychoaktive Substanzen. Springer, Heidelberg

Hebb D (1949) The organisation of behaviour. Wiley, New York

Heinz A, Higley JD, Gorey JG, Saunders RC, Jones DW, Hommer D et al (1998) In vivo association between alcohol intoxication, aggression, and serotonin transporter availability in nonhuman primates. Am J Psychiatry 155:1023–1028

Heinz A, Batra A, Scherbaum N, Gouzoulis-Mayfrank E (2013) Neurobiologie der Abhängigkeit. Grundlagen und Konsequenzen für Diagnose und Therapie von Suchterkrankungen. Kohlhammer, Stuttgart

Hurrelmann K, Bründel H (1997) Drogengebrauch–Drogenmißbrauch. Eine Gratwanderung zwischen Genuß und Abhängigkeit. Primus, Darmstadt

Jacobi F, Hofler M, Strehle J, Mack S, Gerschler A, Scholl L, Wittchen HU (2014) Mental disorders in the general population: study on the health of adults in Germany and the additional module mental health (DEGS1-MH). Nervenarzt 85(1):77–87. https://doi.org/10.1007/s00115-013-3961-y

Kienast T, Heinz A (2005) Suchterkrankungen. In: Walter H (ed) Funktionelle Bildgebung in Psychiatrie und Psychotherapie. Schattauer, Stuttgart

Kienast T, Heinz A (2006) Dopamine and the diseased Brain. Curr Drug Targets-CNS and Neurol Disord 5:109–131

Konorski J (1948) Conditioned reflexes and neuron organisation. Cambridge University Press, Cambridge

Koob G, Volkow N (2016) Neurobiology of addiction: a neurocircuitry analysis. Lancet Psych 3:760–773

Kraus L, Pabst A, Piontek D, Gomes de Matos E (2013) Substanzkonsum und substanzbezogene Störungen in Deutschland im Jahr 2012. SUCHT 59(6):333–345

Lisman J, Grace A (2005) The hippocampal-VTA loop: controlling the entry of information into long-term memory. Neuron 45(5):703–713

Loeber S, Duka T, Welzel H, Nakovics H, Heinz A, Flor H, Mann K (2009) Impairment of cognitive abilities and decision making after chronic use of alcohol: the impact of multiple detoxifications [Research Support, Non-U.S. Gov't]. Alcohol Alcohol 44(4):372–381. https://doi.org/10.1093/alcalc/agp030

Muller CA, Geisel O, Banas R, Heinz A (2014) Current pharmacological treatment approaches for alcohol dependence [Review]. Expert Opin Pharmacother 15(4):471–481. https://doi.org/10.1517/14656566.2014.876008

Orth B (2016) Die Drogenaffinität Jugendlicher in der Bundesrepublik Deutschland 2015. Rauchen, Alkoholkonsum und Konsum illegaler Drogen: aktuelle Verbreitung und Trends. BZgA-Forschungsbericht. Köln: Bundeszentrale für gesundheitliche Aufklärung

Penka S, Gutwinski S, Heinz A (2018) Abhängigkeit und Sucht. In: Machleidt W, Kluge S, Sieberer M, Heinz A (eds) Praxis der Interkulturellen

10

Psychiatrie und Psychotherapie. Urban & Fischer Verlag & Elsevier, München

Reidenberg MM (2011) Drug discontinuation effects are part of the pharmacology of a drug [Lectures Research Support, N.I.H., Extramural Research Support, U.S. Gov't, P.H.S.]. J Pharmacol Exp Ther 339(2):324–328. https://doi.org/10.1124/jpet.111.183285

Robinson T, Berridge K (2008) The incentive sensitization theory of addiction: some current issues. Philos Trans R Soc Lond Ser B Biol Sci 363:3137–3146

Rost D (2009) Psychoanalyse des Alkoholismus: Theorie Diagnostik, Behandlung. Psychosozial-Verlag, Gießen

Schmidt B, Alte-Teigeler A, Hurrelmann K (1999) Soziale Bedingungsfaktoren von Drogenkonsum und Drogenmissbrauch. In: Gastpar M, Mann K, Rommelspacher H (eds) Lehrbuch der Suchterkrankungen. Thieme, Stuttgart, pp 50–69

Schultz W (2006) Behavioral theories and the neurophysiology of reward. Annu Rev Psychol 57:87–115

Schultz W, Dayan P, Montague P (1997) A neural substrate of prediction and reward. Science 275:1593–1599

Tiffany ST, Carter BL (1998) Is craving the source of compulsive drug use? J Psychopharmacol 12(1):23–30. https://doi.org/10.1177/026988119801200104

Tsai G, Coyle J (1998) The role of glutamatergic neurotransmission in the pathophysiology of alcoholism. Annu Rev Med 49:173–184

Tsai G, Gastfriend D, Coyle J (1995) The glutamatergic basis of human alcoholism. Am J Psychiatry 152:332–340

UNODC (2017) World Drug Report 2017. https://www.unodc.org/wdr2017/index.html

Verheul R, van den Brink W, Geerlings P (1999) A three-pathway psychobiological model of craving for alcohol. Alcohol Alcohol 34:197–222

Volkow N, Fowler J, Wang G (1999) Imaging studies on the role of dopamine in cocaine reinforcement and addiction in humans. J Psychopharmacol 13:337–345

Vollstädt-Klein S, Loeber S, Kirsch M, Bach P, Richter A, Bühler M, von der Goltz C et al (2011) Effects of cue-exposure treatment on neural cue reactivity in alcohol dependence: a randomized trial. Biol Psychiatry 69:1060–1066

von Heyden M, Jungaberle H, Majic T (2018) Handbuch Psychoaktive Substanzen. Springer, Berlin

Wetterling T, Veltrup C, Junghanns K (1996) [Craving–an adequately defined concept?] [Review]. Fortschr Neurol Psychiatr 64(4):142–152. https://doi.org/10.1055/s-2007-996380

WHO (2016) Global status report on alcohol and health. file:///C:/Users/Gutwinski/Downloads/msbgsruprofiles.pdf and http://www.who.int/substance_abuse/publications/global_alcohol_report/profiles/en/

Wiers C, Ludwig V, Gladwin T, Park S, Heinz A, Wiers R, Bermpohl F (2015) Effects of cognitive bias modification training on neural signatures of alcohol approach tendencies in male alcohol-dependent patients. Addict Biol 20:990–999

Wikler A (1948) Recent progress in research of the neurophysiological basis of morphin addiction. Am J Psychiatry 105:329–338

Wise R (1998) The neurobiology of craving: implication for the understanding of addiction. J Abnorm Psychol 97:118–132

Psychotic Disorders ("Schizophrenia")

Florian Schlagenhauf and Philipp Sterzer

Contents

In this chapter, the psychotic experience is presented using paranoid schizophrenia as an example.

Learning Objectives

After reading the chapter, the reader should be able to name the symptoms of schizophrenia as well as its treatment options and to classify different neurobiological explanations.

▶ **Example**

A 22-year-old mechanical engineering student, accompanied by his flatmates, presents at the emergency department. They had persuaded him to do so because they were increasingly worried about his changed behaviour. About one and a half years ago, after he had passed the first phase of exams, he had let the university slide. In the last two semesters he had not gone at all and had withdrawn more and more. The reasons he gives for this are concentration problems and lack of interest. His flatmates report that there was a clear change 6 weeks ago. He had hardly come out of his room, had neglected eating and personal hygiene. In addition, he had repeatedly said strange and sometimes incomprehensible things. For example, he reported that he could sometimes clearly hear neighbors in his room making nasty comments about him, speaking directly to him. He had noticed suspicious black cars outside the house, indicating the machinations of the neighbours, and was convinced that they had accessed his smartphone to use it to manipulate his thoughts. Finally, he turned off the TV in a rage today, because the anchorwoman of the daytime news was constantly making allusions to him.

Lately he had hardly drunk alcohol and had not used drugs. Only at the end of his school years had he occasionally smoked a joint. Once he had had "strange" experiences (everything around him had felt strangely changed and threatening), whereupon he had decided to "keep his hands off the stuff".

His mother had recurrent depressions and a paternal uncle suffered from schizophrenia.

The patient is offered inpatient admission for psychiatric diagnosis and treatment. He is initially ambivalent, but allows himself to be persuaded with the offer of a thorough medical clarification. Laboratory tests of blood and cerebrospinal fluid (CSF) as well as magnetic resonance imaging of the brain remain without pathological findings.

After initial mistrust, the patient agrees to drug treatment with an antipsychotic (aripiprazole) and receives group psychotherapy with a psychoeducational focus as well as metacognitive training and occupational therapy. The fears described on admission increasingly recede into the background and the perceptual disturbances cease within a week. After 3 weeks, the patient can be discharged home. At this time, he still states that he is suspicious of the neighbours, but is now more relaxed in this respect. He wants to spend a few weeks with his parents in Bavaria first and then resume his studies in the next semester. ◀

11.1 Psychotic Disorders: Overview and Incidence

Nowadays, the diagnosis of "schizophrenia" follows operationalized criteria that describe the occurrence of certain symptoms of thinking, self-reference and emotional experience and behavior over a defined period of time while organically tangible causes have been excluded. The focus is on a disorder of thinking with false perceptions (hallucinations), fixed beliefs (delusions) and often disorganised thought processes. Such psychotic experiences, i.e. those associated with a loss of the previously self-evident "reality", were already described in various traditions before the introduction of our current psychiatric classification systems.

The **concept of schizophrenia** on which today's classification systems (ICD and

DSM) are based represents a mixture of various historical concepts and draws on different theories. Important influences can be traced back to Hecker (1871) and his concept of "juvenile insanity" (hebephrenia) as well as to Kahlbaum (1874) with his description of motor phenomena in the form of "Spannungsirresein (tension insanity)" (catatonia). Emil Kraeplin (1856–1926) established the dichotomy that has persisted to this day, although he incorrectly formulated it as a distinction between "manic-depressive insanity" (today's bipolar disorders), which has a cyclic and more benign course, and "dementia praecox", which develops progressively and less favourably, as a premature loss of cognitive abilities in early adulthood (as mentioned in ▶ Sect. 11.3, and in contrast to Kraepelin's postulation, the courses are very heterogeneous). Accordingly, Eugen Bleuler (1857–1939) already distanced himself from Kraepelin's classification with regard to the unfavourable prognosis in his paper "Dementia praecox or the group of schizophrenias". Bleuler postulated instead a disturbance in the association of thoughts as the central characteristic of schizophrenia and contrasted the so-called basic symptoms such as association looseness, affect disorders, autism and ambivalence ("the four As") with what he regarded as more marginal (accessory) symptoms such as perceptual disturbances, delusions and catatonic symptoms. Important for today's concept of schizophrenia were Kurt Schneider (1887–1967) and his emphasis on symptoms based on the phenomenal experience of the patient, such as commenting voices or thought insertion. The variety of phenomena subsumed under the term "schizophrenia" is thus on the one hand historically determined and on the other hand due to the symptom variablility in the course of the illness—from the prodromal stage over the florid psychotic episode to the possible residual state. In the following, we will concentrate on paranoid schizophrenia.

Info Box Schizophrenia	
Annual incidence[a]	**15/100.000**
Lifetime prevalence[a]	1%
Gender ratio	w = m
Age of onset	Mean age of first episode in men 21 years, in women 26 years 90% before the age of 30 First manifestation after 40 years of age rare
Major psychiatric comorbidities	Dependence on nicotine, alcohol or illicit drugs (lifetime prevalence 50%) Depression Obsessive compulsive disorder
Hereditary factor	Concordance in monozygotic twins 40–60%.

[a] Average information

11.2 Epidemiology, Symptoms and Diagnosis

11.2.1 Epidemiology

Schizophrenia is a mental disorder which occurs worldwide. The lifetime prevalence, i.e. the proportion of people suffering from schizophrenia in the lifetime up to the time of the survey, is given on average as 1% and ranges between 0.7% and 1.4% in the 15–60 age group worldwide—depending on the breadth of the diagnostic criteria. Approximately 19 new cases are diagnosed per 100,000 inhabitants per year in Germany, so that with a population of 82.3 million in Germany, approximately 15,600 newly diagnosed schizophrenia cases can be expected each year (Gaebel 2010). Recent epidemiological studies suggest that there are significant variations in incidence (McGrath et al. 2008). According to this study, the global median annual incidence is 15.2 per 100,000 persons with a range from 7.7 to 43.0 per 100,000 persons. Increased incidence rates have been described for people with a migration background and low socioeconomic status, as well as in urban settings (Heinz et al. 2013). Interestingly, in epidemiological studies, individual psychotic symptoms are found significantly more frequently than the disorders from the schizophrenia spectrum itself. The prevalence of single psychotic symptoms such as delusions or hallucinations is estimated at 7.2% in the general population and the annual incidence rate at 2.5% (Linscott and van Os 2013). In the vast majority of individuals, such psychotic symptoms are transient and have no medical significance; in approximately 20% of individuals, symptoms may persist and be associated with psychiatric disorders.

11.2.2 Symptoms

The symptoms of schizophrenic disorders affect areas of thinking, self-reference and emotional experience and behaviour. A common classification is that of positive symptoms, which are added to normal experience, and negative symptoms, which describe a pathological loss of mental functions. Positive symptoms, which describe florid psychotic experience, include delusions, formal thought disorder, perceptual disturbances, and self-disturbances.

Delusion is defined as a false fixed belief and refers to thought content. A delusional belief represents a rigid misjudgment of reality that is held to with subjective certainty, even when the beliefs contradict observations that refute the belief. Delusional interpreted events are experienced as "centered" on the affected person, who sees herself at the center of the imagined events. Typical delusional contents are ideas of reference, for example that television news refer to one's own person, or persecutory delusions, where the patient experiences himself as the target of hostility or surveillance.

Various symptoms associated with delusional experience can be distinguished. Before delusions are fully developed a so-called delusional mood is often observed, which is characterized by an unspecific feeling of being alarmed and the experience that something unusual and threatening is going on. Delusional beliefs may occur suddenly (delusional idea) or perceptions of environmental events may be misinterpreted (delusional perception). Delusional thoughts can greatly determine the patient's experience and develop into a system where different areas of experience are linked. The emotional involvement associated with the delusional experience is referred to as delusional dynamics and can vary profoundly during the course of the disorder as well as through antipsychotic therapy.

Self-disturbance refers to the loss of ego boundaries and thus to the self-reference of the affected person. Patients experience their own thoughts as coming from outside and being manipulated (thought insertion) or describe that other persons would have access to their thoughts and could, for example, read them (thought broadcasting) or take them away (thought withdrawal). Self-disturbance is conceived in the Anglo-American tradition as specific delusional content (*delusions of control*) rather than as a discrete symptom complex, ignoring the fundamental difference between self-disturbance and delusional symptoms: Self-disturbance refers to the experience of one's own thoughts, delusions to the external world. Delusional systems can then connect all these phenomena by complex explanations and secondarily "rationalize" them, for example by explaining that they are technically complicated interventions and manipulations of a secret service.

Formal thought disorders, on the other hand, describe the process rather than the content of thought and manifest themselves, among other things, as disorganized speech. Disorganized thinking is a condition in which the logical coherence and coherence of thought or verbal utterances no longer exist and other people can no longer follow the patient's train of thought. Other formal thought disorders include the interruption or blocking of the train of thought or the use of terms in other meanings up to the formation of new words (neologisms).

Hallucinations are perceptions without a corresponding stimulus source and are therefore also called "objectively false" perceptions. Hallucinations can affect all sensory qualities. In schizophrenia spectrum disorders, auditory hallucinations are most common, followed by tactile hallucinations. Visual hallucinations are rare and if present should give reason to rule out an acute brain disorder such as delirium.

For schizophrenia the hearing of voices (phonemes) are characteristic, which often comment on the patient's actions or talk about the patient in dialogue form.

Negative symptoms refer to a reduction in functioning such as a general reduction in motivated and goal-directed behavior, social contact, or emotional experience and expression. Avolition refers to a reduction in goal-directed activities and in efforts to carry out a resolution, and anhedonia refers to a reduction in positive emotional experience and/or decreased interest in activities that are perceived as pleasurable. In addition, there are symptoms that also affect emotional expression, such as affective flattening or decreased speech production. Negative symptomatology contributes strongly to impairments in social and occupational functioning and quality of life.

Cognitive symptoms include disturbances in cognitive performance and are an important feature of schizophrenic psychosis, described by Bleuler and Kraeplin (Green and Harvey 2014). Neuropsychological testing procedures allow for measurement of various cognitive domains, with many of the tests used capturing more than a single isolated domain. However, a substantial proportion of patients exhibit only minor impairments at best. The affected cognitive domains include: Working memory, processing speed, attention, verbal as well as visual learning and memory, problem solving, and social cognition. Both chronic schizophrenia patients and patients with a first psychotic episode may exhibit cognitive deficits. In fact, cognitive deficits are often detectable in the prodromal stage (before the onset of full-blown disorder), and their magnitude may be predictive of the transition to psychosis. Although cognitive deficits are not part of the diagnostic criteria, they are important for prognosis and level of functioning (Kahn and Keefe 2013). Since chronic neuroleptic administration may contribute to a low-grade reduction in brain volume, drug effects should be ruled out (Aderhold et al. 2015).

Furthermore, the symptoms of schizophrenia include disorganized behavior and psychomotor symptoms, which are referred to as catatonia. Catatonic symptoms are very diverse. Both hyperkinetic and hypokinetic states can occur. In a characteristic catatonic state, patients fall completely silent (mutism), while speaking abilities are preserved, and may take on bizarre-looking postures, sometimes for hours.

11.2.3 Diagnostics

The **diagnosis of schizophrenia** is made using operationalized criteria according to ICD-10 or DSM-5. As shown in ◼ Table 11.1, a minimum number of certain symptoms is required for a specified time. The two systems are largely in agreement. Differences between the classification systems include the required symptom duration, which is only 4 weeks in ICD-10 but 6 months in DSM-5, and the emphasis on specific symp-

◘ Table 11.1 Diagnosis of schizophrenia according to ICD-10 and DSM 5

ICD-10 (F20)	DSM 5 (295.90)
1. Thought echo, thought insertion, thought withdrawal, thought broadcasting 2. Delusions of control, influence or passivity, delusional perception 3. Commenting or dialoguing voices 4. Persistent delusion 5. Persistent other hallucinations 6. Formal thought disorders 7. Catatonic symptoms 8. Negative symptoms Symptoms: 1 from 1 to 4 2 from 5 to 8 Time criterion: >1 month	1. Delusions 2. Hallucinations 3. Disorganized speech 4. Severely disorganized or catatonic behavior 5. Negative symptoms (e.g. reduced emotional expression, avolition) Symptoms: 2 out of 5 (including 1, 2, or 3) Time criterion: >1 resp. 6 months

toms. In ICD-10, eight symptom groups are distinguished, with the first four being considered particularly characteristic. Thus, only one of these symptoms is sufficient for the diagnosis, such as ego disturbances (which, according to Kurt Schneider, belong to the so-called first-rank symptoms). In contrast, the DSM-5 distinguishes only five psychopathological domains: Delusions, hallucinations, disorganized thinking and speech, abnormal psychomotor behavior, and negative symptoms. At least two symptom domains are required for diagnosis (one of which must involve domains 1–3).

Psychotic symptoms such as delusions and hallucinations may also be present in other psychiatric and neurological disorders, for example, autoimmune disorders and affective disorders with psychotic symptoms, as well as schizoaffective disorder (▶ Chap. 12). Schizophrenia must be distinguished from brief psychotic disorders that do not meet the required time criteria, from delusional disorder in which other psychotic symptoms are absent, and from schizotypal personality disorder in which symptomatology is less severe and personality traits are enduring. In severe forms of obsessive-compulsive disorder, differentiation can be difficult. Psychotic experiences may also occur in the context of post-traumatic stress disorder. Autism spectrum disorders are characterized by early onset and are not characterized by marked delusions or hallucinations. Psychotic experience may also be due to acute drug effects. Drug-induced psychosis resolves after abstinence, but a clear distinction may be difficult in some cases because of the high comorbidity between schizophrenia and addiction.

The currently available diagnostic systems follow a categorical approach. However, as can be seen from the brief description of symptoms above, the clinical presentation of schizophrenia is very heterogeneous. This is true between individual patients cross-sectionally as well as intraindividually in terms of the clinical course.

A better understanding of the underlying neurobiological mechanisms might be promoted by a dimensional approach. Thus, it has been proposed to describe symptomatology along eight dimensions to map the individual symptom constellation (Heckers et al. 2013). In addition to the five psychopathological domains of the DSM-5, these dimensions include impaired cognition as well as depressive and manic symptoms.

11.2.4 Exclusion Diagnostics

The diagnosis of a schizophrenic disorder requires to exclude brain-organic diseases such as inflammatory processes (e.g. encephalitis, syphilis, multiple sclerosis, Huntington's disease, etc.), epilepsy, delirium or drug intoxications. In addition to a physical examination, necessary additional diagnostic procedures include structural brain imaging by means of magnetic resonance tomography, electroencephalography, and cerebrospinal fluid analysis.

Patients with schizophrenia often have other mental illnesses (comorbidity) such as addiction, depression or obsessive-compulsive disorders. About half of those with schizophrenia meet diagnostic criteria for dependency on alcohol or illicit drugs such as cannabis at some point in their lives (lifetime prevalence), and about 80% of schizophrenia patients are smokers.

11.3 Therapy and Prognosis of "Schizophrenia"

The treatment of schizophrenic disorders is based on the interaction of different therapeutic approaches and is multimodal and multiprofessional. In principle, treatment should be relationship-oriented and adapted to the needs of the patient. The involvement of relatives and participative decision-making play an important role. Stability and consistency of therapeutic relationships with good networking between outpatient and inpatient settings with a preference for outpatient treatment approaches should be of high importance. Therapy is essentially based on three pillars: pharmacotherapy, psychotherapy and sociotherapy.

A basic component of every therapy and an important link between the medical-psychotherapeutic treatment and the social environment is sociotherapy. Sociotherapeutic approaches include measures of structuring the social environment, occupational and work therapy, as well as occupational and social rehabilitation. An important role play integrated treatment concepts aimed at establishing long-term treatment continuity and avoiding or shortening hospitalization.

The importance of **psychotherapy** for schizophrenic disorders was neglected for a long time, but has increased significantly in recent years. In particular, the effectiveness of cognitive behavioural therapy and psychoeducational oriented family interventions is very well documented and has now found its way into most guideline recommendations (National Collaborating Centre for Mental Health 2014).

Pharmacotherapy is mainly based on antipsychotics, the efficacy of which has been well documented in randomized controlled trials, particularly for positive symptoms in all phases of the disease. In addition, depending on individual symptoms and comorbidities, other psychotropic drugs may be indicated, such as anxiolytic agents or antidepressants. The antipsychotic drugs represent a chemically heterogeneous group of substances with different side effect profiles. They influence a variety of neurotransmitter systems, so that it is assumed that clinical efficacy results from an interplay of different mechanisms of action. The antipsychotic mechanism of action is essentially related to the blockade of postsynaptic dopamine receptors of the D2 type. All currently approved neuroleptics block these D2 receptors to a greater or lesser extent. Receptor occupancy of 60–80% is thought to be therapeutically effective. This degree of receptor occupation is already achieved with relatively low dosages (e.g. 3 mg haloperidol per day; Farde et al. 1992; Heinz et al. 1996). Consistent with these findings, higher dosing is not more effective for treating positive symptomatology, but only increases the risk of side effects (Donnelly et al. 2013). Daily doses of 10 mg haloperidol or more, which were commonplace practice in psychiatric emergency treatment for many years, are no longer justifiable in light of these findings. In particular, D2 receptor blockade frequently leads to undesirable extrapyramidal motor side effects such as parkinsonian symptoms, dyskinesia, and agitation (akathisia). These side effects are particularly pronounced with first generation antipsychotics, as their action is mainly due to D2 blockade. Many second generation antipsychotics (also called "atypical antipsychotics") act more strongly by blocking the receptors of other neurotransmitter systems, such as serotonergic, noradrenergic, histaminergic, and muscarinic acetylcholine receptors. Second generation antipsychotics have fewer overall extrapyramidal motor side effects than first generation antipsychotics. They have also been attributed superior antipsychotic efficacy as well as better efficacy on negative symptomatology, although this superiority is controversial and has not been convincingly demonstrated (Lieberman et al. 2005). In particular, the metabolic side effects of some second generation antipsychotics, such as obesity and an increased risk of diabetes mellitus, should not be underestimated and should be taken into account when choosing an antipsychotic.

A major challenge in the treatment of people with schizophrenia is that psychotic symptoms are often not interpreted as symptoms of illness by those affected, and medical concepts are not perceived as helpful without a detailed explanation and the establishment of a relationship of trust. Even if affected persons are assessed as incapable of giving consent, i.e. the nature, significance and scope of a medical measure (or the omission thereof) cannot be properly grasped due to the mental illness, even medically justified treatment against the will is not lawful in most cases. Such compulsory treatment would constitute a serious violation of the fundamental right to physical integrity. Under German law, it is only possible in exceptional situations, namely when there is an acute or chronic danger to life or health.

The course of schizophrenic disorders is very heterogeneous. About 20% of patients experience only one episode and 30% several episodes with complete remission (reduction of symptoms) in the interval. In about half of the patients, the course is unfavorable with incomplete remission between episodes and increasing social and occupational limitations (Watts 1985). People suffering from schizophrenia die on average about 15 years earlier than healthy comparison persons, which is a particularly high mortality compared to other mental disorders. Approximately 5–10% of patients take their own lives. Other reasons for the increased mortality are somatic comorbidities such as metabolic and cardiovascular diseases, which are partly related to the adverse side effect profile of antipsychotic drugs.

A number of prognostically relevant factors have been identified, although individual parameters hardly allow a reliable prognosis of the course in individual cases (Moller 2004). Male gender, positive family history, early first manifestation and comorbid substance use or dependence are associated with a rather unfavourable course. Psychopathological predictors of an unfavorable course include marked negative symptomatology, cognitive deficits, residual delusions after treatment, and the presence of auditory hallucinations or obsessive-compulsive symptoms. Regarding the course, a long prodromal phase, long untreated episodes, an insidious onset, and a slow response indicate an unfavorable prognosis. Prognostically unfavourable social factors include, above all, a low level of education and functioning and the absence of a partnership.

11.4 Emergence of "Schizophrenia": A Spectrum of Theories

Today, it is assumed that schizophrenia is a brain development disorder with multifactorial etiopathogenesis, in which there is an increased vulnerability to environmental influences according to a **vulnerability-stress model** based on genetic factors. The genetic basis for the development of schizophrenia is considered certain due to the familial clustering of schizophrenia with concordance rates of 40–60% in identical twins, which has been proven in numerous studies (Häfner 1995). Genome-wide association studies (GWAS), which have now identified over a hundred genetic risk variants for schizophrenia, suggest a polygenetic etiology (Ripke et al. 2013). The so-called "three-hit" hypothesis specifies the vulnerability-stress model in that genetic predisposition results in increased vulnerability during particularly critical periods of brain development (Keshavan 1999). Harmful influences during early brain development, such as viral infections during pregnancy and perinatal hypoxia, lead to changes in brain development that contribute to an increased risk of disease (*first hit*). However, if detectable at all, such neuropathological findings are very heterogeneous and do not provide a diagnostically useful picture. Environmental factors that exist during childhood also increase the risk of developing schizophrenia (*second hit*). These include early separation from parents, childhood abuse or neglect, and probably also a family communication style with *High-Expressed Emotions*, characterized by strong criticism, hostility, and overprotectiveness (Cechnicki et al. 2013). Under the influence of further factors such as drug use (especially cannabis) or psychosocial stress (e.g. due to migration) in adolescence or early adulthood (*third hit*), the disorder then manifests. While the influence of each

of the above-mentioned psychosocial risk factors on their own is relatively small, it is now assumed that these factors together play a significant role (Kirkbride et al. 2010).

Some influential earlier theories on the etiopathogenesis of schizophrenia are no longer considered relevant today. Not least due to the influence of Emil Kraepelin, who coined the term "dementia praecox", a progressive brain disease was long assumed to be the basis of schizophrenic disorders, which is characterized by progressive neurodegenerative processes comparable to dementia. However, neither longitudinal imaging or neurocognitive studies nor clinical outcome studies have provided clear evidence for the hypothesis of a progressive brain disease (Zipursky et al. 2013).

A central question in contemporary **schizophrenia research** is how the now well-documented genetic and psychosocial factors are reflected in neurobiological changes that underlie schizophrenic disorders. An important clue is provided by findings that point to a neuro-developmental dysregulation in the balance between excitatory and inhibitory neurotransmitter systems (excitation-inhibition balance or "E/I balance") as early as adolescence (Rapoport et al. 2012). Genetic and epigenetic factors are thought to cause changes in glutamatergic neurotransmission early in brain development, which in turn, together with other mechanisms, such as immunological ones, influence further brain development. Such changes could lead to homeostatic adjustments in the finely tuned interplay of neuronal systems. For example, it has been proposed that hypoactivity of (excitatory) glutamatergic neurotransmission counter-regulates a reduction in (inhibitory) GABAergic activity (Krystal and Anticevic 2015). The balance between excitation and inhibition that is thus restored could be the reason why schizophrenic disorders manifest with a delay, despite the presumably early onset of the brain developmental disorder. However, reduced activity of GABAergic interneurons will lead to long-term disinhibition of both cortical glutamatergic neurons and subcortical dopaminergic neurons in the mesolimbic system. Under the influence of stress, such disinhibition could contribute to the increased dopaminergic neurotransmisison in the striatum, which is directly related to the development of psychotic symptoms. This brief summary illustrates that the different neurotransmitter hypotheses of schizophrenia—glutamate, GABA and dopamine—are not incompatible with each other, but can be embedded together in an overarching theory about the disruption of mechanisms at the level of neuronal regulatory circuits (■ Fig. 11.1; Heinz and Schlagenhauf 2010).

Theories about the role of neuronal control circuits raise the question of how their disturbances are related to altered subjective experience and observable symptoms of schizophrenic psychoses. The methods of **computational neuroscience** offer a promising possibility to relate changes at the level of behavior and experience not only qualitatively-descriptively, but also quantitatively with specific neuronal processes. This still young scientific discipline deals with the information-processing properties of the nervous system and uses mathematical modelling to describe and simulate neuronal processes. Such modelling can describe processes on different levels. Thus, individual subprocesses on the level of local neuronal control circuits (e.g., E/I balance in the prefrontal cortex) or on the level of cognitive functions (e.g., reward learning) can be captured in mathematical models. The individual estimation of model parameters offers the possibility of assigning disease-

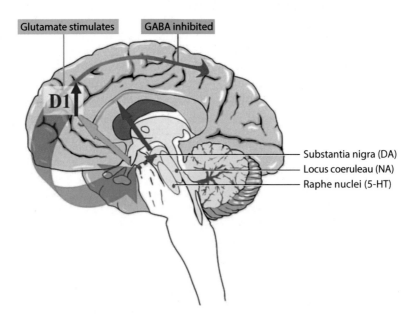

11

Glutamate stimulates   GABA inhibited

D1

Substantia nigra (DA)
Locus coeruleau (NA)
Raphe nuclei (5-HT)

■ **Fig. 11.1** Possible relationship between changes in glutamatergic, GABAergic and dopaminergic neurotransmission according to Heinz and Schlagenhauf (2010). Hypofunction of glutamatergic prefrontal-subcortical projections (green arrow) leads to decreased activation of GABAergic interneurons and thus to disinhibition of dopaminergic neurons in the midbrain. In contrast, there is a decreased dopamine release in the prefrontal cortex with upregulation of dopamine D1 receptors, which in turn results in a disruption of the function of prefrontal glutamatergic neurons (*DA* dopamine; *NA* norepinephrine; *5-HT* 5-hydroxytryptamine). (After Heinz and Schlagenhauf 2010, with kind permission)

relevant behavioral or symptom dimensions to neuronal dysfunctions in a quantitative manner, thus bridging the gap between neurobiological and symptom levels (Friston et al. 2014).

11.5 Neurobiological Foundations

The neurobiological basis of schizophrenia is poorly understood. No circumscribed anatomical or functional abnormality has been identified as specific for this disorder. Serious brain changes, as described in neurological diseases (e.g. dementias, inflammations), do not seem to be associated with the disease (Kahn et al. 2015).

As described in ▶ Sect. 11.4, most of today's explanatory models of schizophrenia assume a complex interplay of genetic factors and environmental conditions that affect brain development and influence the course of neurobiological adaptation processes to further (stressful) life events, which is referred to as the **neurodevelopmental model**. This is matched by the fact that the onset of the disease occurs in early adulthood, but that important risk factors relate to prenatal (e.g. infections) or perinatal events (e.g. hypoxia at birth). Thus, early environmental exposures may increase the risk of developing psychotic symptoms in response to later social stressors. However, the exact molecular mechanisms are not clear.

Subtle changes in specific cell populations, such as GABAergic interneuron populations, are postulated. Most importantly, current theories assume pathological alteration of functional networks. Several of the domains proposed in the RDoC approach (▶ Sect. 9.5.1) and the neural circuits underlying them appear to be involved in schizophrenic disorders. Alterations in perception and cognition in schizophrenia are associated with dysfunctions of fronto-parietal networks as well as fronto-striatal-hippocampal circuits under the influence of neuromodulatory systems.

11.6 Genetics

In genome-wide association studies, schizophrenia has been linked to genes predominantly expressed in the brain and immune system (Ripke et al. 2013). The majority of identified single nucleotide polymorphisms (SNPs) are non-protein coding and appear to be involved in gene regulation. Individually, such SNPs have a negligible effect on schizophrenia risk, and collectively they explain only a moderate proportion. However, these findings are crucial for a pathophysiological understanding of the mechanisms involved in the disease. Thus, associations were found with genes encoding the dopamine receptor (DRD2), or with genes associated with glutamatergic neurotransmission as well as synaptic plasticity and interneuron function.

In addition to the involvement of gene variations frequently found in the population with low impact, rarely occurring gene variants with higher risk are of importance. These include *copy number variances* (CNVs), which are structural variants of DNA where the number of copies of larger DNA segments differs greatly between individuals. A gene can occur more than once (duplication) or be completely absent (deletion). An example is the microdeletion syndrome 22q11, in which changes exist on the long arm of chromosome 22 at position 11, which is associated with a greatly increased risk of psychosis. In addition, some de novo mutations have been identified, but these findings are still inconsistent.

In summary, there are not just a few genes that can be associated with schizophrenia. Rather, the disorder is polygenetic, i.e. it is caused by the interaction of numerous genes. Many of the identified genes are also involved in the occurrence of other psychiatric disorders and are therefore not specific to schizophrenia (pleiotropic). Many genetic markers for schizophrenic psychosis overlap with risk factors for bipolar disorder. In addition, gene-gene interactions and gene-environment interactions are likely to be crucial for a more detailed understanding of etiopathogenesis.

For further reading, please refer to Psychiatric Genomic Consortium: ▶ https://www.med.unc.edu/pgc/ and Avramopoulos (2018).

11.7 Neurotransmitter

Several neurotransmitter systems are involved in schizophrenia. However, the neuromodulatory dopaminergic system is of particular importance in theories of schizophrenia. The so-called dopamine hypothesis of schizophrenia is based primarily on pharmacological evidence. Initially, a global overactivity of the dopaminergic system was assumed. A modified version of the dopamine hypothesis postulates hyperactivity subcortically and hypoactivity in the mesocortical system. The most important evidence for an involvement of dopaminergic neurotransmission comes from the fact that all currently available drugs have an antagonistic effect on dopamine receptors, especially of the D2 type. D2-antagonistic drugs such as the first, accidentally discovered antipsychotic chlorpromazine act on positive symptoms but not on negative symptoms or cognitive deficits, and antipsychotic medication at very high doses may even increase avolition and apathy. A further argument for involvement of the dopaminergic system is the observation that dopamine agonistic substances such as amphetamine can trigger psychotic experience, i.e. have a psychotomimetic effect. Direct studies of the dopaminergic system in individuals with psychosis are crucial: in humans, in vivo measurements of the dopaminergic system are possible by nuclear medicine techniques. When a weakly radiolabeled precursor of the transmitter such as ^{18}F-fluoro-3,4-dihydroxyphenyl-l-alanine (F-DOPA) is applied, its metabolites accumulate in the presynaptic vesicles of dopaminergic neurons (◻ Fig. 11.2). The emitted alpha radiation can be measured by positron emission tomography to provide a measure of presynaptic dopamine synthesis. A well-

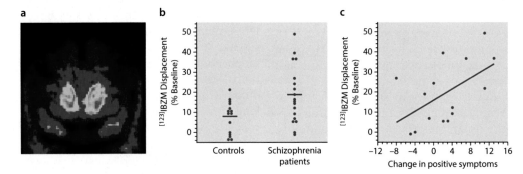

◻ **Fig. 11.2** Nuclear medicine techniques for in vivo measurement of dopaminergic neurotransmission in patients with schizophrenia. **a** Imaging of dopamine synthesis in the striatum using positron emission tomography (PET) with ^{18}F-DOPA as the radiolabeled precursor of dopamine. **b** Increased striatal dopamine release after amphetamine administration in schizophrenia patients compared to healthy controls according to Laruelle et al. (1996; © National Academy of Sciences); **c** Correlation between the extent of amphetamine-induced dopamine release and the increase in positive symptoms under amphetamine administration according to Laruelle et al. (1996; © National Academy of Sciences)

replicated finding is that schizophrenia patients show increased subcortical dopamine synthesis capacity primarily in the associative part of the striatum compared to healthy individuals (Howes et al. 2012). An increased striatal dopamine release has also been shown after amphetamine administration (◘ Fig. 11.2a), whereby the extent of the increased release was related to the increase in positive symptoms (◘ Fig. 11.2b; Laruelle et al. 1996).

Pharmaceuticals with antagonistic action at the glutamatergic N-methyl-d-aspartate (NMDA) receptor, such as the anesthetic ketamine or the designer drug phencyclidine, can induce psychotic experience, negative symptoms, and cognitive deficits. Therefore, ketamine is used as a pharmacological model for psychosis, and a deficit in NMDA receptor function is postulated. NMDA receptors are important for synaptic plasticity and learning through processes such as long-term potentiation (LTP). Together with the inhibitory neurotransmitter GABA, glutamate is crucial for the E/I balance of cortical networks that show alterations in schizophrenia spectrum disorders.

There is **no independent serotonin hypothesis** for schizophrenia. However, serotonin agonists such as the hallucinogens lysergic acid diethylamide (LSD) or mescaline can trigger psychotic symptoms, especially perceptual disturbances. However, blocking serotonin receptors alone has no antipsychotic effect. However, many second generation antipsychotics also have an antagonistic effect at the 5-HT2 receptor, which is associated with their better tolerability with regard to extrapyramidal side effects.

11.7.1 Excitation-Inhibition Balance

Neuronal information processing is based on a functional balance of excitatory and inhibitory networks, the E/I balance. At the level of a single neuron, such a balance consists of an appropriate ratio of excitatory glutamatergic and inhibitory GABAergic synaptic inputs. Glutamatergic and GABAergic neurons form most cortical synapses and are targets of cortical and subcortical modulatory connections. When an excitatory synapse is activated by glutamate, this leads to depolarization of the neuron, increasing the likelihood of triggering an action potential, whereas inhibitory GABAergic synapses have an opposite effect (Gao and Penzes 2015). GABAergic interneurons account for approximately 10% of cortical neurons and can be divided into several subtypes. One type of GABAergic neurons expresses the calcium-binding protein parvalbumin (PV+). This class includes basket cells, which inhibit pyramidal cells perisomatically, and chandelier cells, which form their inhibitory synapses at the axon initial segment of pyramidal cells. Therefore, these PV+ interneurons have a central influence on the formation of pyramidal cell action potentials and are crucial for the synchronized activity of neuronal assemblies. Dysfunction of these PV+ interneurons has been postulated in the cortex as well as in the hippocampus of patients with schizophrenia. E/I imbalance could result from hypofunction of NMDA-type glutamate receptors on the dendrites of the inhibitory PV+ interneurons. This would result in less activation of the inhibitory PV+ interneurons, leading to decreased inhibition of the excitatory pyramidal cells (Gonzalez-Burgos et al. 2015). Due to the reduced GABAergic influence, the inhibitory influences on the pyramidal cells are reduced, resulting in their disinhibition and thus increased excitability.

The branched connections of the basket cells lead to a simultaneous, coordinated inhibition of numerous pyramidal cells and thereby enable a synchronized activity of the pyramidal cells. Through synchronized activity, cortical neurons dynamically connect to form functional networks. This rhythmic activity occurs in different frequency ranges and can be measured electrophysiologically, for example, as temporal coherence between anatomically distributed areas. In schizophrenia patients, impairments of gamma oscillations (frequency range between 30 and 80 Hz) have been repeatedly found and associated with cognitive deficits such as working memory impairment. PV+-GABAergic interneurons are critical for the generation of gamma oscillations and associated synchronized cortical network states. NMDA-R dysfunction could lead to reduced inhibition by PV+ interneurons and overstimulation of excitatory neurons with a reduction in gamma oscillations and dysconnectivity observed in schizophrenia patients (Uhlhaas and Singer 2015, see also there for further information).

11.7.2 Aberrant Salience

The **theory of aberrant salience** relates psychotic symptoms to a stress-related or chaotic hyperfunction of the subcortical dopaminergic system in order to explain the subjective experience of patients (Heinz 2002; Kapur 2003). From a clinical perspective, environmental stimuli that are actually neutral and insignificant often acquire an extraordinary importance and significance for people with psychotic experiences. For example, in the patient description at the beginning of the chapter, the patient notices black cars in front of his house that seem particularly significant and suspicious to him. According to the theory, the patient tries to explain this aberrant salience experience by delusional explanations, for example, the particular importantance of the cars is explained by the thought that the patient is being persecuted, thus forming the starting point for the development of a persecutory delusion.

Salience refers to the property of a stimulus or event to attract attention and *arousal*, which favors its neural processing and a behavioral response to such a salient stimulus. Several properties of salience can be distinguished: the physical properties, the novelty or unexpectedness of a stimulus, and motivational salience. The latter is referred to as *incentive salience* and describes the motivational component in the response to a stimulus, which has been linked to the activity of the dopaminergic system (Berridge 2012; Robinson and Berridge 1993). In the healthy state, context- or stimulus-related dopamine release mediates that motivational salience, and thus meaningfulness, is attributed to that particular context or stimulus. In the context of psychotic experience, dysregulated dopamine release, which occurs independently of stimulus and context due to biological dysregulation, leads to erroneous salience attribution to what should be neutral and insignificant environmental stimuli or internal representations (Heinz 2002; Kapur 2003). Accordingly, a dysregulated, chaotic dopamine release could lead to a subtle change in experience. Its persistence can then lead to the formation of delusional beliefs, through which the constant experience of aberrant salience can be explained away. The administration of antipsychotics then leads to a reduction in the salience experience associated with psychosis, so that a gradual distancing from, for example, the experience of persecution may occur (◘ Fig. 11.3). If antipsychotics are overdosed, however, general motivational aspects of environmental stimuli may also be blocked, and avolition and apathy may result.

In addition to encoding salience, dopamine is also involved in other functions. Animal studies show that dopaminergic neurons are also activated when an unexpected reward arrives. However, if the reward is predicted as part of a learning process, the dopaminergic signal remains absent and may even decrease if the reward that arrives is less than expected. Accordingly, the dopaminergic signal follows a prediction error that encodes the difference between

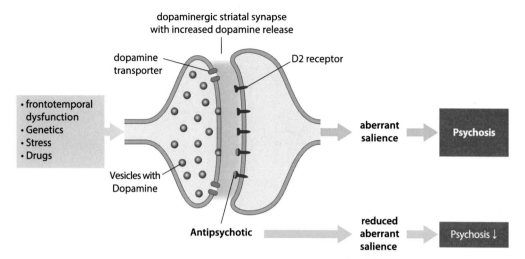

■ **Fig. 11.3** Under the influence of multiple factors (e.g. genetics, stress, drugs), there is an increased striatal dopamine release, which leads to "aberrant salience" as the basis for the development of psychotic symptoms. Most currently available antipsychotics interfere with this process by blocking postsynaptic dopamine receptors. (Adapted from Howes and Kapur 2009)

the expectation and the actual event and can be used to adjust future expectation (Schultz 2017). Accordingly, only an unexpected stimulus that indicates reward leads to a short-term (phasic) dopamine release and thus to the attribution of salience, so again the—in this case temporal—surprise effect and thus a prediction error is crucial. Accordingly, it has been postulated that a reduction in adaptive (dopaminergic) prediction error signals due to disturbed encoding of relevant stimuli and events contributes to the emergence of the motivational negative symptoms, whereas an increase in aberrant, chaotic error signals is involved in the emergence of the psychotic positive symptoms (Maia and Frank 2017).

11.8   The Bayesian Brain and Predictive Processing

An influential general theory about how the human brain works is the "**Bayesian Brain Hypothesis**". This term stands for the idea of inference (Helmholtz 1867), first formulated by Helmholtz, which states that the brain uses learned prior assumptions to infer their causes from sensory input data. This process can be formulated as a Bayesian inference process in which probabilistic predictions (*priors*) are combined with the probability of the presence of sensory data, the sensory evidence (*likelihood*), to calculate an a *posterio* probability (*posterior*) of the cause of a sensory event (Friston 2005). The posterior thus corresponds to the perception that is most likely based on the combination of prior and likelihood (Hohwy 2012). This Bayesian inference might be implemented in the brain in the form of a hierarchical prediction model (*Hierarchical Predictive Coding*), in which increasingly abstract predictions are encoded at higher levels of the cortical hierarchy (Friston 2005; Lee and Mumford 2003). If the predictions do not match the sensory data, prediction error signals are generated to correct the predictive model. The precision with which predictions and sensory data are encoded plays an important role. In the Bayesian formulation, precision corresponds to the inverse variance of the probability distributions representing predictions and sensory data. If the precision of the sensory data is high, this also results in a stronger prediction error. How much this prediction error in turn corrects the predictions also depends on the precision of the predictions. Imprecise predictions are more strongly corrected by the prediction error than precise predictions.

A leading hypothesis for schizophrenia within this theory is that changes in this interplay of predictions and sensory data lead to erroneous conclusions (inferences) that form the basis for the development of psychotic symptoms (◻ Fig. 11.4). For example, it has been suggested that reduced precision of prior assumptions may lead to sensory data being weighted more heavily, resulting in stronger prediction errors (Adams et al. 2013; Sterzer et al. 2018). This reasoning is based on numerous empirical findings, such as the lower susceptibility of people with schizophrenia to visual illusions (Notredame et al. 2014). An example that well illustrates this putative change in Bayesian inference is the so-called hollow-face illusion. When healthy subjects are shown the mask of a face from the inside, it is not perceived as concave, but erroneously as a convex face, i.e., curved outward. According to Bayesian theory, extensive experience with faces in healthy subjects leads to a very accurate prior about the configuration of faces, namely that they are convex. When the prior is integrated with the sensory evidence (likelihood), the prior is weighted so heavily because of its high precision that the perceptual outcome corresponds to a convex face despite stimulus properties that indicate a concave configuration of the face. In contrast, people with schizophrenia are more likely to actually perceive the concave face as concave (Schneider et al. 2002), indicating a lower precision of the prior and a stronger weighting of the likelihood. These theories may also explain why psychosis is clustered among people with immigrant backgrounds and experiences of social exclusion: Priors may be less precise compared to accurate

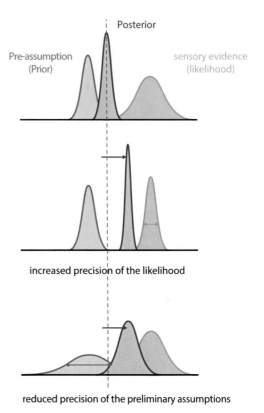

◻ **Fig. 11.4** Schematic representation of the changes in Bayesian inference. Prior, likelihood and posterior are shown as probability distributions. The larger the variance of a distribution (the wider it is), the lower its precision. The calculation of the posterior depends on the respective precision of the prior and the likelihood. Both an increased precision of the likelihood (middle) and a decreased precision of the prior (bottom) can lead to a shift of the posterior towards the likelihood. This results in a stronger weighting of sensory evidence over prior assumptions. (Adapted from Adams et al. 2013)

observation of the threat-experienced environment, resulting in clustered dopaminergic encoded prediction errors (Heinz et al. 2018).

The neuronal basis of altered Bayesian inference is currently the subject of research. A central mechanism may lie in the NMDA-R dysfunction discussed above. The NMDA receptor is thought to play an important role in the transmission of excitatory feedback signals (Bastos et al. 2012), so NMDA-R dysfunction may be a mechanism of the reduced precision of feedback prediction signals (Corlett et al. 2009). In addition, increased dopaminergic neurotransmission could contribute to increased precision (not in the sense of correctness, but in the sense of relevance) of prediction error signals (Galea et al. 2012). As a result of these changes, individuals with schizophrenia experience a greater weighting of sensory information in the context of Bayesian inference, which, similar to the mechanism discussed above, leads to aberrant salience and, subsequently, the development of delusional mood and delusional thoughts. Aberrant salience theory, which in its original form was limited to motivational salience, can thus be embedded in the more general theory of Bayesian inference (Heinz et al. 2018; Maia and Frank 2017; Sterzer et al. 2018).

11.9 Outlook: A Neurobiological Integrative Approach

Neurobiological research on the pathophysiology of schizophrenia has revealed a multitude of findings at different levels of observation. Integrating these findings into a coherent picture that has clinical relevance beyond theoretical insight is a major challenge for current schizophrenia research. As with other mental disorders (▶ Chap. 12), the vulnerability-stress model described above remains the standard model for the development of psychotic disorders such as schizophrenia. This model has the advantage of establishing a link between the genetic predisposition to psychosis, which is now regarded as undoubted, and the recognition that environmental factors also have a significant influence on the development of the disorder. Thus, on the one hand, the vulnerability-stress model provides a helpful framework, but on the other hand, it must be filled with specific mechanisms in order to fulfill the claim of a neurobiologically

integrative approach. At least two major challenges arise: on the one hand, the integration of different psychosocial and neurobiological findings into a comprehensive neurodevelopmental disorder model, and on the other hand, the still existing explanatory gap between neurobiological mechanisms and the subjective experience of those affected.

Currently, we have only a rudimentary understanding of the neurobiological processes and mechanisms underlying disease predisposition and its interaction with environmental factors, and how this might integrate different neurobiological findings into a comprehensive neurodevelopmental disorder model. For example, little is known about how genetic factors are related in detail to neurotransmitter alterations (such as glutamatergic NMDA-R hypofunction or dopaminergic hyperfunction) and what role these relationships play in brain development and vulnerability to stressors occurring later in life. Further neurobiological research using a variety of methodological approaches, from animal models to brain imaging in humans, will be required to address these questions. However, an important step towards an improved understanding will also lie in the mathematical ("computational") modeling of neuronal processes or data measured as proxies (fMRI, EEG, etc.) with regard to their function or dysfunction. Basically, two categories of such computational models can be distinguished (Valton et al. 2017): top-down models provide algorithms for describing behavioral phenomena and then relating model parameters to neural signals. Bottom-up models, on the other hand, describe how computational processes are implemented on the neuronal level and how this results in behavior. Although these two categories involve different approaches, they are not mutually exclusive and can be combined. For example, the Bayesian brain theory described above is a top-down approach, but it can also be combined with models about implementation at the level of neural control circuits (e.g., Bastos et al. 2012). Such "predictive processing" models therefore have the potential to link different levels of observation, from the dysfunction of specific transmitter systems to psychopathology, and thereby provide the basis for a comprehensive neurodevelopmental disorder model.

With regard to the explanatory gap between neurobiology and subjective experience, previous theories already provide some promising starting points. For example, the theory of aberrant salience (Heinz 2002; Kapur 2003) aimed to establish a link between subcortical dopamine hyperfunction and psychopathological phenomena such as the experience of meaning and delusion via the functional relevance of dopaminergic signals and their stress-related alteration (which may also differ with regard to neurodevelopmental risk factors). This approach offers people with psychotic experiences the possibility of a plausible explanation of their subjective experience, which may well contribute to the depathologization and thereby destigmatization of psychosis, especially when the influence of stressful experiences on dopaminergic transmission is taken into account (Heinz et al. 2018). However, aberrant salience theory, like other previous models explaining psychotic phenomena (see also, for example, the comparator model; Frith and Done 1989), remains limited to a single mechanism for explaining a particular symptom. It is therefore an important task for the future to develop more comprehensive disorder models that allow to derive plausible neurobiological models for the different and variable symptom domains of schizophrenia spectrum disorders and thus make psychotic experience more understandable for the affected individuals and their environment.

The goal of integrating neurobiology into a clinical approach is therefore by no means merely to develop better neurobiological methods (e.g. psychosocial interventions or medications) that act more specifically on certain symptoms and their neurobiological correlates. Rather, a transparently and critically communicated, neurobiologically integrative approach can also lead to a better understanding of the disease and its psychosocial risk factors. On the one hand, this can enable the development of new therapeutic approaches, but on the other hand it can also lead to an open "trialogical" discussion (between affected persons, their relatives and professionals) and thus contribute to reducing the stigmatization to which mentally ill people are still exposed.

References

Adams RA, Stephan KE, Brown HR, Frith CD, Friston KJ (2013) The computational anatomy of psychosis. Front Psych 4:47. https://doi.org/10.3389/fpsyt.2013.00047

Aderhold V, Weinmann S, Hagele C, Heinz A (2015) Frontal brain volume reduction due to antipsychotic drugs? Nervenarzt 86(3):302–323. https://doi.org/10.1007/s00115-014-4027-5

Avramopoulos D (2018) Recent advances in the genetics of schizophrenia. Mol Neuropsychiatry 4(1):35–51. https://doi.org/10.1159/000488679

Bastos AM, Usrey WM, Adams RA, Mangun GR, Fries P, Friston KJ (2012) Canonical microcircuits for predictive coding. Neuron 76(4):695–711. https://doi.org/10.1016/j.neuron.2012.10.038

Berridge KC (2012) From prediction error to incentive salience: mesolimbic computation of reward motivation. Eur J Neurosci 35(7):1124–1143. https://doi.org/10.1111/j.1460-9568.2012.07990.x

Cechnicki A, Bielanska A, Hanuszkiewicz I, Daren A (2013) The predictive validity of expressed emotions (EE) in schizophrenia. A 20-year prospective study. J Psychiatr Res 47(2):208–214. https://doi.org/10.1016/j.jpsychires.2012.10.004

Corlett PR, Frith CD, Fletcher PC (2009) From drugs to deprivation: a Bayesian framework for understanding models of psychosis. Psychopharmacology 206(4):515–530. https://doi.org/10.1007/s00213-009-1561-0

Donnelly L, Rathbone J, Adams CE (2013) Haloperidol dose for the acute phase of schizophrenia. Cochrane Database Syst Rev 8:Cd001951. https://doi.org/10.1002/14651858.cd001951.pub2

Farde L, Nordstrom AL, Wiesel FA, Pauli S, Halldin C, Sedvall G (1992) Positron emission tomographic analysis of central D1 and D2 dopamine receptor occupancy in patients treated with classical neuroleptics and clozapine. Relation to extrapyramidal side effects. Arch Gen Psychiatry 49(7):538–544

Friston K (2005) A theory of cortical responses. Philos Trans R Soc Lond Ser B Biol Sci 360(1456):815–836. https://doi.org/10.1098/rstb.2005.1622

Friston KJ, Stephan KE, Montague R, Dolan RJ (2014) Computational psychiatry: the brain as a phantastic organ. Lancet Psychiatry 1(2):148–158. https://doi.org/10.1016/S2215-0366(14)70275-5

Frith CD, Done DJ (1989) Experiences of alien control in schizophrenia reflect a disorder in the central monitoring of action. Psychol Med 19(2):359–363

Gaebel WW (2010) Schizophrenie, vol 50. Robert Koch-Institut in Zusammenarbeit mit dem Statistischen Bundesamt, Berlin

Galea JM, Bestmann S, Beigi M, Jahanshahi M, Rothwell JC (2012) Action reprogramming in parkinson's disease: response to prediction error is modulated by levels of dopamine. J Neurosci 32(2):542–550. https://doi.org/10.1523/jneurosci.3621-11.2012

Gao R, Penzes P (2015) Common mechanisms of excitatory and inhibitory imbalance in schizophrenia and autism spectrum disorders. Curr Mol Med 15(2):146–167

Gonzalez-Burgos G, Cho RY, Lewis DA (2015) Alterations in cortical network oscillations and parvalbumin neurons in schizophrenia. Biol Psychiatry 77(12):1031–1040. https://doi.org/10.1016/j.biopsych.2015.03.010

Green MF, Harvey PD (2014) Cognition in schizophrenia: past, present, and future. Schizophr Res Cogn 1(1):e1–e9. https://doi.org/10.1016/j.scog.2014.02.001

Häfner H (1995) Was ist Schizophrenie? In: Häfner H (ed) Was ist Schizophrenie? Fischer, Stuttgart, pp 1–56

Hecker (1871) Die Hebephrenie. Ein Beitrag zur klinischen Psychiatrie. Arch Pathol Anat Physiol Klin Med 52:394–429

Heckers S, Barch DM, Bustillo J, Gaebel W, Gur R, Malaspina D, Carpenter W (2013) Structure of the psychotic disorders classification in DSM-5. Schizophr Res 150(1):11–14. https://doi.org/10.1016/j.schres.2013.04.039

Heinz A (2002) Dopaminergic dysfunction in alcoholism and schizophrenia–psychopathological and behavioral correlates. Eur Psychiatry 17(1):9–16

Heinz A, Schlagenhauf F (2010) Dopaminergic dysfunction in schizophrenia: salience attribution revisited. Schizophr Bull 36(3):472–485. https://doi.org/10.1093/schbul/sbq031

Heinz A, Knable MB, Weinberger DR (1996) Dopamine D2 receptor imaging and neuroleptic drug response. J Clin Psychiatry 57(11):84–88; discussion 89–93

Heinz A, Deserno L, Reininghaus U (2013) Urbanicity, social adversity and psychosis. World Psychiatry 12(3):187–197. https://doi.org/10.1002/wps.20056

Heinz A, Murray GK, Schlagenhauf F, Sterzer P, Grace AA, Waltz JA (2018) Towards a unifying cognitive, neurophysiological, and computational neuroscience account of schizophrenia. Schizophr Bull. https://doi.org/10.1093/schbul/sby154

Helmholtz H (1867) Handbuch der physiologischen Optik. Voss, Leipzig

Hohwy J (2012) Attention and conscious perception in the hypothesis testing brain. Front Psychol 3:96. https://doi.org/10.3389/fpsyg.2012.00096

Howes OD, Kapur S (2009) The dopamine hypothesis of schizophrenia: version III–the final common pathway. Schizophr Bull 35(3):549–562. https://doi.org/10.1093/schbul/sbp006

Howes OD, Kambeitz J, Kim E, Stahl D, Slifstein M, Abi-Dargham A, Kapur S (2012) The nature of dopamine dysfunction in schizophrenia and what this means for treatment. Arch Gen Psychiatry 69(8):776–786. https://doi.org/10.1001/archgenpsychiatry.2012.169

Kahlbaum K (1874) Die Katatonie oder das Spannungsirresein. Eine klinische Form psychischer Krankheit. A. Hirschwald, Berlin

Kahn RS, Keefe RS (2013) Schizophrenia is a cognitive illness: time for a change in focus. JAMA Psychiat 70(10):1107–1112. https://doi.org/10.1001/jamapsychiatry.2013.155

Kahn RS, Sommer IE, Murray RM, Meyer-Lindenberg A, Weinberger DR, Cannon TD, Insel TR (2015) Schizophrenia. Nat Rev Dis Primers 1:15067. https://doi.org/10.1038/nrdp.2015.67

Kapur S (2003) Psychosis as a state of aberrant salience: a framework linking biology, phenomenology, and pharmacology in schizophrenia. Am J Psychiatry 160(1):13–23

Keshavan MS (1999) Development, disease and degeneration in schizophrenia: a unitary pathophysiological model. J Psychiatr Res 33(6):513–521

Kirkbride J, Coid JW, Morgan C, Fearon P, Dazzan P, Yang M, Jones PB (2010) Translating the epidemiology of psychosis into public mental health: evidence, challenges and future prospects. J Public Ment Health 9(2):4–14. https://doi.org/10.5042/jpmh.2010.0324

Krystal JH, Anticevic A (2015) Toward illness phase-specific pharmacotherapy for schizophrenia. Biol

Psychiatry 78(11):738–740. https://doi.org/10.1016/j.biopsych.2015.08.017

Laruelle M, Abi-Dargham A, van Dyck CH, Gil R, D'Souza CD, Erdos J, Innis RB (1996) Single photon emission computerized tomography imaging of amphetamine-induced dopamine release in drug-free schizophrenic subjects. Proc Natl Acad Sci 93(17):9235–9240. https://doi.org/10.1073/pnas.93.17.9235

Lee TS, Mumford D (2003) Hierarchical Bayesian inference in the visual cortex. J Opt Soc Am A Opt Image Sci Vis 20(7):1434–1448

Lieberman JA, Stroup TS, McEvoy JP, Swartz MS, Rosenheck RA, Perkins DO, Hsiao JK (2005) Effectiveness of antipsychotic drugs in patients with chronic schizophrenia. N Engl J Med 353(12):1209–1223. https://doi.org/10.1056/NEJMoa051688

Linscott RJ, van Os J (2013) An updated and conservative systematic review and meta-analysis of epidemiological evidence on psychotic experiences in children and adults: on the pathway from proneness to persistence to dimensional expression across mental disorders. Psychol Med 43(6):1133–1149. https://doi.org/10.1017/s0033291712001626

Maia TV, Frank MJ (2017) An integrative perspective on the role of dopamine in schizophrenia. Biol Psychiatry 81(1):52–66. https://doi.org/10.1016/j.biopsych.2016.05.021

McGrath J, Saha S, Chant D, Welham J (2008) Schizophrenia: a concise overview of incidence, prevalence, and mortality. Epidemiol Rev 30:67–76. https://doi.org/10.1093/epirev/mxn001

Moller HJ (2004) Course and long-term treatment of schizophrenic psychoses. Pharmacopsychiatry 37(Suppl 2):126–135. https://doi.org/10.1055/s-2004-832666

National Collaborating Centre for Mental Health (2014) National Institute for Health and Clinical Excellence: guidance. In: Psychosis and schizophrenia in adults: treatment and management: updated edition 2014. National Institute for Health and Care Excellence (UK), London

Notredame CE, Pins D, Deneve S, Jardri R (2014) What visual illusions teach us about schizophrenia. Front Integr Neurosci 8:63. https://doi.org/10.3389/fnint.2014.00063

Rapoport JL, Giedd JN, Gogtay N (2012) Neurodevelopmental model of schizophrenia: update 2012. Mol Psychiatry 17(12):1228–1238. https://doi.org/10.1038/mp.2012.23

Ripke S, O'Dushlaine C, Chambert K, Moran JL, Kahler AK, Akterin S, Sullivan PF (2013) Genome-wide association analysis identifies 13 new risk loci for schizophrenia. Nat Genet 45(10):1150–1159. https://doi.org/10.1038/ng.2742

Robinson TE, Berridge KC (1993) The neural basis of drug craving: an incentive-sensitization theory of addiction. Brain Res Brain Res Rev 18(3):247–291

Schneider U, Borsutzky M, Seifert J, Leweke FM, Huber TJ, Rollnik JD, Emrich HM (2002) Reduced binocular depth inversion in schizophrenic patients. Schizophr Res 53(1–2):101–108

Schultz W (2017) Reward prediction error. Curr Biol 27(10):R369–R371. https://doi.org/10.1016/j.cub.2017.02.064

Sterzer P, Adams RA, Fletcher P, Frith C, Lawrie SM, Muckli L, Corlett PR (2018) The predictive coding account of psychosis. Biol Psychiatry. https://doi.org/10.1016/j.biopsych.2018.05.015

Uhlhaas PJ, Singer W (2015) Oscillations and neuronal dynamics in schizophrenia: the search for basic symptoms and translational opportunities. Biol Psychiatry 77(12):1001–1009. https://doi.org/10.1016/j.biopsych.2014.11.019

Valton V, Romaniuk L, Steele D, Lawrie S, Series P (2017) Comprehensive review: computational modelling of schizophrenia. Neurosci Biobehav Rev. https://doi.org/10.1016/j.neubiorev.2017.08.022

Watts CA (1985) A long-term follow-up of schizophrenic patients: 1946–1983. J Clin Psychiatry 46(6):210–216

Zipursky RB, Reilly TJ, Murray RM (2013) The myth of schizophrenia as a progressive brain disease. Schizophr Bull 39(6):1363–1372. https://doi.org/10.1093/schbul/sbs135

Affective Disorders Using the Example of Unipolar Depression

Stephan Köhler and Henrik Walter

Contents

© The Author(s), under exclusive license to Springer-Verlag GmbH, DE, part of Springer Nature 2023
G. Roth et al. (eds.), *Psychoneuroscience*, https://doi.org/10.1007/978-3-662-65774-4_12

The following chapter focuses on the topic of affective disorders using the example of unipolar depression, one of the most common mental disorders and one of the largest contributors to disability due to mental illness worldwide.

Learning Objectives

After reading this chapter, the reader should be able to name the symptoms of depression as well as different treatment options and to classify the different neurobiological explanations for the development of depression.

Case Report

The 49-year-old Mr. S. is admitted as an inpatient on a psychiatric ward for diagnosis and treatment of a severe anxious-depressive syndrome. On admission, the patient reports a decrease in performance for about 2 months. His mood became increasingly deteriorating and was characterized by hopelessness and despair. He was constantly worried about the future of his family. In addition, he was increasingly brooding and had concrete suicidal thoughts. Mr. S. describes a pronounced loss of pleasure and interest, there was currently nothing that could give him pleasure. He bemoans a loss of daily structure, although he had always been "well organised" and order and performance always had been very important to him. He had lost 8 kg in weight and no longer felt any appetite. He found it very difficult to concentrate on certain things, and therefore had strong feelings of guilt and self-doubt, as he was no longer able to organise many things. The symptoms were particularly pronounced in the morning. He often woke up at 4 a.m. and then had the feeling of being completely exhausted. Mr. S. reports a first depressive episode at the age of 33, which had completely subsided after a few weeks without treatment.

First of all, a detailed psychiatric and medical diagnosis is made, which confirms the presumptive diagnosis of a severe depressive episode without any other underlying disease being identified as the cause of the current symptoms. A staged treatment approach is discussed with the patient based on the severity of the symptoms and his treatment experience. Initially, antidepressant medication is started with the antidepressant venlafaxine, dosed up to 300 mg/day. After 2 weeks, there is a slow improvement in drive and mood. Because of pronounced sleeping problems, sleep-inducing medication is also used temporarily. At the same time, an individual psychotherapeutic model of the disorder is developed with the patient already at this this early stage using cognitive behavioural therapy. In addition, psychotherapeutic treatment starts with working on typical depressive cognitions (e.g. "nothing ever works out", "I am a failure") and the development of positive activities. After 8 weeks of inpatient treatment, with significant stabilization, the patient is discharged to further outpatient treatment.

12.1 Affective Disorders: Overview and Incidence

Major depressive disorder (MDD) is one of the most common mental disorders with a lifetime prevalence of up to 20% (H. U. Wittchen et al. 2011). Typically, these disor-

ders progress in episodes. The designation "unipolar" distinguishes the disorder from the much rarer "bipolar" depression in bipolar disorder (lifetime prevalence 2–3%), in which manic phases with increased drive and pathological elevated mood also occur. Depressive disorders lead to a high level of suffering for patients and their relatives. Correspondingly, according to the World Health Organization (WHO), they are among the diseases that lead to the greatest health impairments and most Disability Adjusted Life Years (DALYs) worldwide (Murray et al. 2013). When characterizing depressive disorders, it is important to distinguish them from common, "normal" human feelings such as sadness, disappointment, and hopelessness that everyone experiences in their lives. The term (unipolar) depression is used, when these basic experiences persist in a certain form over a certain period of time (◘ Table 12.2), although the distinction between mild depression and "normal" low mood naturally remains blurred. The term "burnout" should also be strictly separated from clinical depression, as is not regarded as an illness in its own right now but rather is understood as a risk condition for the development of depression. However, it is more readily accepted in society. This is important not only for diagnostic reasons, but also to counteract the stigmatization of the diagnosis of depression. Depression can manifest itself in a variety of ways that can vary widely. Throughout the history of depressive disorders, there have been and continue to be efforts to distinguish subtypes of depressive disorders, e.g., reactive versus endogenous, anxious, masked, melancholic, atypical depression, and several more. However, only very few subtypes of depression are described in current diagnostic manuals (e.g., persistent depressive disorder in the DSM-5), as the various subtypes have not been sufficiently delineated neither diagnostically nor therapeutically (see also the *specifiers* in ▶ Sect. 2.2, Diagnostics). A major factor that may contribute to this is the multifactorial aetiopathogenesis (causal development) of depressive disorders. In this context, the problem of the threshold for diagnosis and of over- as well as underdiagnosis of depression should be mentioned: On the one hand, depressive disorders are still not sufficiently recognized, diagnosed and treated (Trautmann and Beesdo-Baum 2017). This is especially true for certain patient groups, such as patients with geriatric depression. On the other hand, nowadays the diagnosis of depression is often made lightly and treated directly with medication, without distinguishing it from moods in the context of adjustment disorders, for example. Even if the diagnosis is correct, the treatment depends on the degree of severity. This is particularly true with regard to the recommendation of antidepressant medication, which must be carefully weighed. For years, the S3 guideline Unipolar Depression has recommended a wait-and-see approach for mild depressive episodes and explicitly no pharmacotherapy (DGPPN 2015). This contradicts the data of prescribing antidepressants in everyday clinical practice. The need for a critical and indication-based perspective is also evident from various studies on the efficacy of antidepressants, some of which question the effects for entire antidepressant classes (Jakobsen et al. 2017), or were able to demonstrate an effect only for a major depressive episode in contrast to placebo (Fournier et al. 2010). Furthermore, as in any field of medicine, drug treatment also involves possible side effects, so that every treatment decision must be the result of a weighing of risk and benefits.

In the ICD-10, the episodic character of depression is emphasized and recurrent episodes are referred to as recurrent depressive disorder. Since 30% of all depressive episodes take a chronic course (duration of the episode longer than 2 years; (Murphy and Byrne 2012), the diagnosis of persistent depressive disorder as described in the DSM-5 should be mentioned here again. In addition, dysthymia must also be distinguished from unipolar depression. It usually begins early and is long-lasting (even longer than 2 years) and resembles depression, but the diagnostic criteria for a depressive episode (◘ Table 12.1) are not met. A clinically relevant and difficult-to-treat subtype of

■ **Table 12.1** Information: unipolar depression. (According to Hasin et al. 2005, 2018; Wittchen et al. 2011)

12-month prevalence	5–10%
Lifetime prevalence	13–20%
Gender ratio	W > m, approx. 2:1
Age of onset	Any age 50% before the age of 30 First manifestations also possible after 60 years of age
Major psychiatric comorbidities	Anxiety disorders (approx. 30%) Addictive disorders (30–60%) Personality disorders Eating disorders, obsessive-compulsive disorders, somatoform disorders
Hereditary factor	Two to threefold increased risk in relatives of patients

chronic depression has been identified as double depression, which is defined as a combination of a persistent dysthymia with additional depressive episodes (Köhler et al. 2015).

12.2 Epidemiology, Symptoms and Diagnosis

The incidence of unipolar depression peaks around the age of 30, with half of patients becoming ill before this age. The likelihood of becoming affected decreases with age: Only 10% of patients become ill after the age of 60. The disease is typically episodic, with severe depressive episodes in particular recurring frequently (50–85%). There is a tendency towards a younger age of onset and an increasingly chronic course (approx. 30% of all illnesses; Angst et al. 2009). In the USA, unipolar depression is diagnosed more frequently in the last decade with a doubling of the 12-month prevalence from 5.28% in 2001/2002 (Hasin et al. 2005) to 10.4% in 2012/2013 (Hasin et al. 2018); in Europe, the prevalence is unchanged from 2005 to 2010 at 6.9%, although there is a significant increase in depression-related sick leave in Germany. Before the medication era, the average episode duration was about 6 months; today, 50–75% of all patients are no longer depressed after 8 and 16 weeks, respectively, with the first episode usually lasting longer. The total annual economic costs and consequential costs are estimated atEUR 16 billion (Krauth et al. 2014) and, next to cardiovascular diseases, cause the highest health care costs (Lopez et al. 2006). Risk factors are female gender, first-degree family members with depression, *stressful life events* such as the loss of a partner or job, but also marriage or the birth of a child, chronic stress and excessive demands, a lack of a social network, or living alone.

12.2.1 Symptoms

The three core symptoms of depression are a marked depressed mood, a reduction in drive, and anhedonia (loss of interest and pleasure), which lasts most of the day and is prevalent for at least 2 weeks. Common symptoms include sleep disturbances, reduced self-esteem and self-confidence, guilt, nonspecific pain, suicidal thoughts or behavior, reduced concentration and attention, loss of appetite or increased appetite with weight changes. In addition, there is often a strong anxious symptomatology, which can be undirected or expressed in often inappropriate fears of, for example, job loss, impoverishment, or illness. Furthermore, there are formal thought disorders such as rumination, inhibition and slowing down of thinking, as well as content-related thought disorders that are mood-congruent, such as thoughts of illness, impoverishment, sinfulness or delusions of sinfulness. Often there are diurnal fluctuations with increased symptoms in the morning and improvement in the evening. In addition, psychomotor changes in the sense of a slowing down or signs of a strong restlessness are often noticed. Although depression occurs in various degrees of severity, it can be a fatal illness. Patients with depression have a 1.8-fold increased risk of dying and, on average, a 10.6 (men) and 7.2 (women) year shorter life expectancy (Otte et al. 2016). In addition to the increased risk of other diseases, suicidality is a significant contributing factor. Worldwide, about 800,000 people die by suicide each year, about half of which are completed in the context of a depressive episode. In Germany, about 10,000 people die by suicide each year, but only 3500 die from traffic accidents and 400 from AIDS (Bundesamt 2015). Recognition of suicidality and active targeting it in treatment are therefore essential aspects in the diagnosis of depression. In the treatment of suicidality in the context of depressive disorders, lithium salts in particular appear to have a positive effect in the long term (Smith and Cipriani 2017), while no significant effects on reducing the suicide rate could be demonstrated for antidepressants (Braun et al. 2016).

12.2.2 Diagnostics

When making a diagnosis according to ICD-10, a distinction is made between three main and various secondary symptoms and thus different degrees of severity. According to DSM-5, a distinction is made between nine symptom groups and various specifiers (◘ Table 12.2). These describe the severity (mild, moderate, severe) and the course (single or recurrent episode, in remission, with or without psychotic features). The other specifiers can be thought of as successors of subtypes, for whose diagnosis one must in

turn meet certain criteria. For example, the specifier anxious distress means that at least two out of five symptoms (feeling keyed up or tense, feeling unusually restless, difficulty concentrating because of worry, fear that something awful might happen, feeling of losing control of oneself) are present in the depressive episode (for further specifiers see DSM-5).

Ideally, the diagnosis should be supplemented by questionnaires that allow the severity of depression to be quantified. These include short questionnaires such as the PHQ-9 (often used in epidemiological studies), self-report questionnaires such as the Beck Depression Inventory (BDI) or third-person questionnaires such as the Hamilton Depression Scale (HAMD).

12.2.3 Exclusion Diagnostics

As with any psychological syndrome, a depressive syndrome can also be the result of another underlying disease. Therefore, common somatic diseases must be excluded as a cause. Treatment of an somatic disorder causing depressive symptoms is mandatory and often leads to the disappearance of depressive symptoms. With regard to important somatic differential diagnoses, hypothyroidism should be mentioned in particular, but kidney disease, anaemia or cancer are also frequent causes. At the same time, however, patients with depression often present to their general practitioner primarily because of physical complaints (Simon et al. 1999). Therefore, in addition to a medical and neurological physical examination, a blood test is always obligatory in the basic diagnosis of depressive disorders. Furthermore, especially in the case of a first episode, brain scans are necessary to exclude direct affections of the brain (tumor, hemorrhage, stroke, inflammation) (overview: DGPPN 2015). But of course, depression can also occur together and independently

◻ **Table 12.2**   Diagnosis of an episode of unipolar depression according to ICD-10 and DSM 5

DSM-5	ICD-10
A. Five (or more) of the following symptoms have been present during the same 2-week period and represent a change from previous functioning; at least one of the symptoms is either (1) depressed mood or (2) loss of interest or pleasure Note: Do not include symptoms that are clearly attributable to a another medical disease condition. 1. Depressed mood most of the day, nearly every day, as indicated by either subjective report (e.g., feels sad, empty, or hopeless) or observation made by others (e.g., appears close tearful). (Note: Iin children and adolescents, can be irritable mood). 2. Markedly diminished interest or pleasure in all or almost all activities, most of the day, nearly every day (as indicated by by either subjective account or observation) 3. Significant weight loss when not dieting or weight gain (e.g.,a change of more than 5% of body weight in a month) or decreased or increase in appetite on almost all days. (Note: In children, consider failure to make expected weight gain) 4 .Insomnia or hypersomnia nearly every day 5. Psychomotor agitation or retardation nearly every day (observable by others, not merely subjective feelings of restlessness or being slowed down) 6. Fatigue or loss of energy nearly every day. 7. Feelings of worthlessness or excessive or inappropriate guilt (which may be delusional) nearly every day (not just self-reproach or guilt about being sick) 8. Diminished ability to think or concentrate or indecisiveness, nearly everyl days (either by subjective account or asbobserved by others) 9. Recurrent thoughts of death (not just fear of dying), recurrent suicidal ideation without a specific plan, or a suicide attempt or a specifi plan for commiting suicide B. The symptoms cause clinically significant distress or impairment in social, occupational, or other important areas of functioning C. The episodes is not attributable not to the physiological effects of a substance or to a another medical condition Note: Criteria A–C represent a major depressive episode **Specifiers:** • with anxious distress • with mixed features • with melancholic features • with atypical features • with mood-congruent psychotic features • with mood-incongruent psychotic features • with catatonia • with peripartum onset • with seasonal pattern	Simultaneous presence of at least two of the following three core symptoms for at least 2 weeks 1. Lowering of mood to a degree that is clearly unusual for the affected person for most of the day 2. Loss of interest or pleasure in normally enjoyable activities 3. Decreased energy or increased fatigability Additionally at least two symptoms up to a total of 4 (mild episode) to 8 (severe episode) from the following group 1. Reduced concentration and attention 2. Reduced self-esteem and self-confidence 3. Ideas of guilt and worthlessness 4. Bleak and pessimistic views of the future 5. Ideas or acts of self-harm or suicide 6. Disturbed sleep 7. Diminshed appetite Possible somatic syndrome that is characterized by the following typical features: Loss of pleasure or interest in acivites that are normally enjoyable, lack of emotional reactivity to positive events, waking in the morning at least 2 h before the usual time, morning low, psychomotor retardation or agitation, loss of appetite, weight loss, loss of libido F32.0 Mild depressive episode **Info:** Two core and two additional symptoms are present. The affected patient is generally impaired by them, but often able to continue most daily activities F32.1 Moderate depressive episode **Info:** Two core symptoms and three to four additional symptoms are present, and the affected patient usually has great difficulty in continuing everyday activities F32.2 Major depressive episode without psychotic symptoms **Info:** A depressive episode with three core symptoms and at least four additional symptoms that are experienced as overall distressing. Typically, there is a loss of self-esteem and feelings of worthlessness and guilt. Suicidal thoughts and acts are common, and some somatic symptoms are usually present

12

of somatic diseases. In fact, additional illnesses (comorbidity) are common in patients with depression. Severe physical diseases such as myocardial infarction, stroke or traumatic brain injury are associated with increased occurrences of depressive disorders (e.g. Gan et al. 2014; Pan et al. 2011), as are other neuropsychiatric disorders such as Parkinson's disease or Alzheimer's disease. In terms of differential diagnosis, episodic unipolar depression must be distinguished from normal phases of grief and mood swings, on the one hand, and from more chronic forms (persistent depressive disorder, dysthymia (▶ Sect. 12.2)) and from depression in bipolar disorder, schizophrenia and schizoaffective disorder, on the other. The distinction from anxiety disorders (▶ Chap. 13) is also not always easy. In addition, patients with depression often have other mental illnesses ("psychiatric comorbidity") that must be considered, especially for the specification of treatment. Most common and important are comorbid anxiety disorders (30–50%), addictive disorders (30–60%), as well as somatoform and personality disorders (DGPPN 2015).

12.3 Therapy and Prognosis of Unipolar Depression

A variety of strategies are available for the treatment of unipolar depression. The therapeutic "portfolio" includes pharmacological, psychotherapeutic, physical and psychosocial treatment options. In addition, further treatment strategies such as sleep deprivation or light therapy can be used if indicated. The effectiveness of all therapy methods, with the exception of ketamine treatment and sleep deprivation therapy, is usually a few weeks. The treatment is divided into acute treatment, maintenance treatment and relapse prevention. With regard to pharmacotherapy, it is important to note that antidepressants, unlike painkillers or tranquilizers, do not show effects immediately after ingestion; rather, an antidepressant effect does not occur before 1–3 weeks, so that medication generally must be administered at a standard dosage for 2–4 weeks until efficacy can be assessed. The effects of pharmacotherapy are particularly proven for severe depressive episodes (Fournier et al. 2010). Therefore, the different therapy recommendations are also depending on the severity of the depressive episode (DGPPN 2015). Furthermore, as mentioned above, psychotherapy represents an essential pillar in the treatment of depressive disorders with the best evidence for cognitive behavioural therapy and interpersonal psychotherapy (DGPPN 2015). In addition, there are other therapeutic methods such as electroconvulsive treatment (▶ Sect. 12.5.6, Stimulation methods), transcranial magnetic stimulation (rTMS), which has fewer side effects, deep brain stimulation (still experimental), treatment with ketamine (a glutamate antagonist), and methods such as light therapy, sleep deprivation therapy, and sports therapy, all of which have different indications and effectiveness. What can be stated is that depressive disorders generally have a good prognosis due to the available treatment methods, but that initial treatment often leads to an insufficient therapy response. The proportion of treatment failure (lack of remission) for a first treatment attempt is 30–50% in both pharmacotherapy (Rush 2007) and psychotherapy. A total of 20–30% of all depressive disorders take a treatment-resistant and/or chronic course.

12.4 Theories on the Development of Depression

As with the development of mental disorders in general, depression can also be assumed to be a multifactorial process. Biological (e.g. genetic predisposition), psychological (e.g. negative thoughts) and social (e.g. unemployment) factors always interact. One integrative model is the so-called vulnerability-stress model (Wittchen 2011), also known as the diathesis-stress model (diathesis roughly means "readiness or disposition to develop a disease"). It conceptualizes the genesis (development) of depressive disorders through the interaction of current or chronic stresses with neurobiological and psychological changes as well as other modifying variables (previous mental disorders, etc.) against the background of a predisposition (vulnerability; ◘ Fig. 12.1).

The theories on the origin of depression are so numerous and varied that an overall presentation is not possible here. Apart from historical theories (obsession, the black bile of melancholy), psychological constructs dominated for a long time. These include, on the one hand, psychoanalytic theories which, greatly abbreviated, consider a real or only in the inner imagination experienced objects loss on the basis of a certain personality structure as the causal factor. Behavioral explanations focus either on the reduced experience of positive reinforce-

ment (Peter M. Lewinsohn) or on the presence of negative cognitions with a bias for negative information, rumination, and thought patterns associated with self-deprecation and negative future expectations (Aaron Beck). The learned helplessness theory, based on animal experiments, states that people who have been exposed to negative experiences in the past, the cause of which they could not influence, do not change their situation even if they could (Martin Seligman). Sociologically inspired theories emphasize external factors, interpreting a depressive syndrome as a "healthy" response to a "sick" society with too much pressure to achieve success and stress (Alain Ehrenberg). Evolutionarily inspired theories are similarly broad. Social competition theory, for example (John Price), sees depressive syndromes as an evolutionarily meaningful pattern of refection that protects individuals from wasting resources or suffering loss and injury in a in a competition they cannot win. Instead, it is evolutionarily more sensible to retreat into a

12

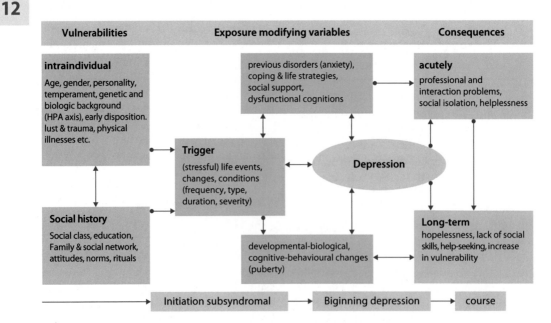

◘ **Fig. 12.1**   Vulnerability-stress model of depression. (According to Wittchen 2011)

protected space in order to develop new behavioural strategies.

However, neurobiological factors and mechanisms will be emphasized in the following presentation. These do not directly compete with learning theory, evolutionary, or sociological explanations. Neurobiological mechanisms do not necessarily have to be understood as independent causes. They simply can be understood as the causal mechanisms by which psychological functions are realised or on which other, e.g. social, contributing factors exert their effects. Due to space limitations, we cannot discuss all theories, such as the chronobiological factors of depression (Zaki et al. 2018). Rather, we will refer to particularly general ones and on those that are thoroughly researched and current neurobiological mechanisms such as genetics, neurotransmitters, the HPA stress axis, inflammatory processes, neurocognitive network models, models of neuroplasticity, and epigenetics. At the end will we return to psychological functions.

12.5 Neurobiological Mechanisms

12.5.1 Genetics of Depression

As with all mental illnesses, genetic factors are important contributing factors in depression. The higher familial incidence of depression is well documented. For example, children of depressive patients have a twofold increased risk of developing an affective disorder (Mattejat and Remschmidt 2008). Heritability is estimated at 37% based on twin and adoption studies. This is, compared to schizophrenia, bipolar disorder or autism (81, 75 and 80% respectively) only about half (Sullivan et al. 2012). As with all mental disorders, a polygenic model is assumed, involving many individual small genetic variants with small effects. Until recently, however, there were no specific

gene variants associated with depression in genome-wide analyses with sufficient statistical probability, i.e., corrected for multiple comparisons. This has changed. In the current largest study from 2018 with 135,438 patients (more than half of them from 23andme) and nearly 344,910 controls, 44 genome-wide significant SNPs (*single nucleotide polymorphisms*) were found (Wray et al. 2018a). The so-called SNP heritability (heritability calculated based on all known SNPs) of 8.7% (as usual) was lower than that estimated via twin and adoption studies. Of further interest is that there are gene variants associated with personality traits such as neuroticism that play a role in the genesis of depression (Nagel et al. 2018).

In addition to genetics, it has long been known that the environment, especially negative experiences, are important aspects in the development of depression. Research used to focus on events that occurred in the year prior to the development of depression (separation, unemployment, financial problems, bereavement, health problems, experience of violence, etc.) as patients usually attribute causal relevance to them for the occurrence of their depression. More recent research has tended to focus on traumatic events in childhood (lack of caregivers, abuse, neglect). Not least because of the **monoamine deficiency hypothesis** (▶ Sect. 12.5.2), a finding from the so-called Dunedin Longitudinal Study in New Zealand has long been cited as a prime example of the interaction of genetics and environment. In 2003, Caspi et al. described that the probability of developing depression after experiencing abuse in childhood was increased, especially in carriers of the so-called short variant of the serotonin transporter gene (Caspi et al. 2003). Subsequent meta-analyses, however, have cast doubt on the existence or at least the strength of the effect. What is certain today is that the strength was overestimated in the Caspi paper and that the effect, if present, is rather small (for the current state of the debate cf. (Bleys et al. 2018; Culverhouse et al. 2018). Nowadays, research focuses more on the influences of early negative experiences on the regulation of gene expression, as described in ▶ Sect. 12.5.8 on epigenetics.

12.5.2 Neurotransmitters and the World of Antidepressants

In addition to the monoaminergic neurotransmitters (serotonin, norepinephrine and dopamine), glutamate, as the most important neurotransmitter in the CNS, is probably also involved in the development of depression (cf. ▶ Sect. 12.5.6 on ketamine). This leads to the monoamine (deficiency) hypothesis and the glutamate (toxicity) hypothesis.

The **monoamine deficiency hypothesis** of depression has been derived primarily from the clinical finding that the first substances that led to an improvement in mood have an influence on the metabolism of the monoaminergic neurotransmitters serotonin, norepinephrine and dopamine that can be clearly demonstrated in animal experiments. Their immediate pharmacological effect is an increase in neurotransmitter concentration in the synaptic cleft. Conversely, it has been concluded that in depressive disorders the activity of monoaminergic neurotransmission should be decreased and constitutes the core neurobiological mechanism of depression (Nutt 2008; Prins et al. 2011; Schildkraut and Kety 1967). Accordingly, in human experiments, it was found that an acute reduction in the availability of tryptophan (a precursor of serotonin) by a tryptophan-reduced diet led to a mood deterioration in patients with a previous or currently major depressive episode. Reductions in serotonin, norepinephrine, and dopamine worsen mood even in healthy subjects with a concurrent positive family history of MDD

(Ruhe et al. 2007). Consistent with preclinical studies as well as post-mortem studies, positron emission tomography (PET) studies have demonstrated that serotonin receptors 1A, 2A, 1B, as well as serotonin transporters, play an important role in MDD in the context of serotonergic dysfunction (Savitz and Drevets 2013). Another indication of the central role of serotonin are the effects of ecstasy (MDMA): Immediately after intake, MDMA leads to a general release of serotonin and a peaceful feeling of happiness (so-called entactogenic effects). A few days after ingestion, however, a depressive mood often develops, as serotonin capacities have been depleted and are only slowly replaced.

Today, however, the monoamine hypothesis in the sense of a simple deficiency hypothesis ("patients with depression lack serotonin like patients with diabetes lack insulin, which must therefore be substituted") is considered outdated. Even if the effects mentioned are experimentally demonstrable, they are only one aspect of the underlying neurobiological processes. A major argument against the simple deficiency hypothesis is that increase of neurotransmitter concentration in the synaptic cleft are measurable immediately after AD administration, but the clinical effects of AD occur with a latency of 1–3 weeks. One explanation is that downstream molecular adaptation processes (upregulation of receptors, influence on intracellular signaling cascades) are responsible for the antidepressant effects. However, it is also possible that completely different pathomechanisms are involved in the development of depressive disorders. For example, recent research has identified the importance of ceramides (fatty acids from the sphingolipid group) in the development of depression. Stress and inflammation can lead to an increased release of sphingolipids, which is directly related to depressive symptoms (Kornhuber et al. 2014).

12.5.3 Pharmacological Classification and Significance of Antidepressants

A variety of substances from the group of **antidepressants** as well as other substances—e.g. lithium, St. John's wort, quetiapine—are available for the treatment of depression. The primary pharmacological effect of antidepressants is to enhance serotonergic, noradrenergic and/or dopaminergic neurotransmission by various mechanisms such as inhibition of reuptake, inhibition of degradative enzymes or indirect effects on various pre- and postsynaptic receptors. Antidepressants are grouped into five major classes of action according to chemical structure and function: Tricyclic antidepressants (TCAs), monoamine oxidase inhibitors (MAO inhibitors), selective serotonin reuptake inhibitors (SSRIs), selective serotonin/norepinephrine reuptake inhibitors (SRNIs), and autoreceptor blockers. MAO inhibitors inhibit the mitochondrial degradation of monoamine oxidase after reuptake into the presynaptic nerve terminal (Köhler et al. 2014). This increases the amount of neurotransmitter in the synaptic cleft (□ Fig. 12.2). This in turn causes changes at the post- and presynaptic receptor level (upregulation) as well as the intracellular second messenger signaling cascades (Kraus et al. 2017). These include changes in gene expression as well as other adaptive processes, such as an increase in brain-derived neurotrophic factor (BDNF; Polyakova et al. 2015). Recent research attempts to optimize the diagnosis of unipolar depression by measuring various metabolites of neurotransmitters and, if necessary, to develop parameters for the course of treatment and the response to antidepressants (Pan et al. 2018).

The individual substances within the different classes of antidepressants are clinically described as rather homogeneous and similar in their mode of action. In general, there is a lack of efficacy studies demonstrating clinically relevant differences between antidepressants (Gartlehner et al. 2011). However, in everyday clinical practice it often becomes clear that antidepressants and their individual characerics are not equal. Thus, certain effects and especially side effects (increased appetite and sedation) of individual substances can have a positive impact on comorbid mental illnesses. In the most recent network meta-analysis by Cipriani and colleagues, four antidepressants (amitriptyline, mirtazapine, duloxetine, and venlafaxine) were found to be slightly superior to the other substances (Cipriani et al. 2018). Moreover, there are individual differences in the degradation and transport of AD. Thus, depending on the expression of the CYP2D6 gene located on chromosome 22 (it codes for an enzyme from the cytochrome P-450 group), a distinction is made between (ultra)slow and (ultra)fast metabolizers. Another influencing factor is genetically determined difference in blood-brain barrier function. There are special transporter mechanisms that actively transport molecules from the nervous system back into the blood. Glycoprotein P, for example, transports certain AD from the CNS back into the blood. In patients with a rather unfavourable genotype of the ABCD gene (it codes for glycoprotein P), certain AD are rapidly transported back (e.g. citalopram, escitalopram, venlafaxine, paroxetine and amitriptyline), so that only low concentrations are reached in the CNS, whereas this does not affect other AD (e.g. mirtazapine and fluoxetine). Knowledge of these gene variants is in principle helpful for optimising therapy, but is (still) rarely applied in practice and is subject to critical discussion (Bschor et al. 2017).

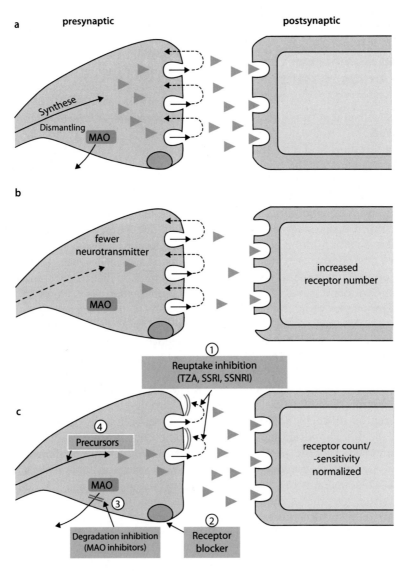

◻ Fig. 12.2 Model of the mode of action of antidepressants: **a** normal state; **b** depression; **c** normalization by various antidepressant mechanisms. (Adapted from Laux and Dietmaier 2018)

12.5.4 HPA Axis Disorder: Stress, Stress, Stress

It seems immediately obvious to everyday experience and common sense that extreme and chronic strains in everyday life ("stress") can lead to depressive mood. Scientifically, there is no doubt that psychosocial stress is a significant and validated factor in the

onset, worsening, relapse and chronification of depressive disorders (Gilman et al. 2013; Harkness et al. 2014).

Patients with depression exhibit 2.5 times more *stressful life events* prior to the manifestation of their disorder than corresponding control groups (Hammen 2005). What is the biological basis for this relationship? A central regulator is located in a dysfunction of the hypothalamic-pituitary-adrenocortical axis (HPA) (Stetler and Miller 2011).

In summary, the "stress hormone" cortisol (a glucocorticoid) released from the adrenal cortex under acute stress dampens inflammatory processes, which makes neurobiological sense. The release is controlled by a hormonal signaling cascade from the hypothalamus to the pituitary to the adrenal cortex. This HPA axis (from *hypothalamus—pituitary—adrenal cortex*) is regulated by negative feedback inhibition. However, chronic inflammatory processes can then cause glucocorticoid resistance in immune cells by inducing various signaling cascades (e.g. MAP kinases, c-jun N-terminal kinase JNK, p38). Cytokine signal transduction (e.g., *nuclear factor-kB*, NF-B) disrupts glucocorticoid receptor and expression function, which in turn perpetuates inflammatory processes. Chronic inflammation can subsequently itself often lead to core depressive symptoms such as reduced drive, anhedonia and depressed mood (Kiecolt-Glaser et al. 2015).

Cytokine-dependent glucocorticoid receptor resistance in turn leads to disinhibition of the HPA axis, i.e. the system "tries" to produce more cortisol. There is an increased release of corticotropin-releasing hormone (CRH) from the hypothalamus and an increase in cytokines, which leads to a further amplification of the body's stress response (Dantzer et al. 2008).

The increased cortisol release due to chronic stress presumably has a neurotoxic effect on the very sensitive neurons of the hippocampus: These possess glucocorticoid receptors, and their permanent activation by cortisol leads to a decrease in synaptogenesis (formation of cell connections) and dendrite formation (branching of neurons) and simultaneously to increased apoptosis (cell death). The consequence is a disinhibition of the HPA axis, which leads in the sense of a *circulus vitiosus* (vicious circle) to the

amplification of the effects mentioned. Additionally changes such as a volume change in the hippocampus and cortex, which is presumably caused by cell death (Anacker et al. 2013) can be found. With the aforementioned alterations, also a reduced release of BDNF has been detected, which is considered a marker for synaptogenesis. Furthermore, chronic stress causes changes in the neurotransmitter system, such as an increase in the release of glutamate, the main excitatory neurotransmitter in the prefrontal cortex and hippocampus, which is associated with neurotoxic effects (Sanacora et al. 2012). Furthermore, chronic stress leads to reduced release of serotonin and dopamine in mesocortical monoaminergic circuits (Dillon et al. 2014; Mahar et al. 2014).

In patients with depression, meta-analyses have shown that there is an average—albeit moderate—increase in cortisol levels (Stetler and Miller 2011). Furthermore, it is known that treatment with cortisone increases the risk of developing depression by a factor of 2 and the risk of suicide by a factor of 7 (Otte et al. 2016). However, only about 50% of successfully treated patients show a change in cortisol levels after treatment. And the dexamethasone inhibition test, which was promoted for some time as a diagnostic test for depression (in depressed individuals dexamathone should not lead to a decrease in cortisol levels as it does in healthy individuals), has not proven to be effective in practice. The hope that CRH antagonists (CRH is released from the hypothalamus and stimulates ACTH release, which in turn stimulates cortisol production in the adrenal cortex) could be used as drugs to treat depression has unfortunately not been fulfilled either. Thus, although there is clear evidence for disruption of the HPA axis from decades of research, this is likely only one factor in the multifactorial genesis of depression, and, moreover, may play an important role only in a subgroup of patients.

12.5.5 Inflammation: Lifestyle and the Influence on Depressive Symptoms

The immune system is closely linked to the physiological stress system and the HPA axis, as it can be clearly concluded from animal experiments. The connection of depressive illnesses with inflammatory processes has been a topic of intensive research for many years. It is no coincidence that the feeling of sickness in infections such as influenza resembles a depressive syndrome. In systemic diseases such as rheumatoid arthritis, there is an increased incidence of depression. Inflammatory molecules in particular are thought to be a common mechanism. In several large meta-analyses, patients with depression had elevated levels of proinflammatory cytokines (including IL-6, TNF-a, CRP) compared to healthy control subjects (Goldsmith et al. 2016; Haapakoski et al. 2015). Overall, there appears to be a reciprocal causal relationship: Depression and inflammation mutually reinforce each other (Bauer and Teixeira 2018). Yet, only about one-third of depressed patients actually show elevated levels of inflammation (Raison and Miller 2011). Therefore, inflammation is neither necessary nor sufficient to trigger or maintain depression, but seems to play an important role in certain groups of depressed patients (Kiecolt-Glaser et al. 2015).

What is the detailed pathophysiological relationship between inflammation and depression? Peripherally released cytokines can lead to changes in the production, metabolism and transport of neurotransmitters via signalling cascades in the CNS (Capuron and Miller 2011). For example, cytokines can inhibit tryptophan metabolism via enzyme induction in terms of reduced serotonin production (Dantzer 2017). In addition, cytokines lead to oxidative stress and glial damage, a reduction in the production of neurotrophins (e.g., BDNF), and thus impaired neuroplasticity and neurogenesis (Eyre and Baune 2012; Zhang et al. 2016). Cytokines also lead to alterations of the HPA axis function already presented above and hereby provide an important link to the theory of impaired HPA axis function (Stetler and Miller 2011). Increased inflammation can lead to symptoms of depression, which is also illustrated by the high incidence of depression as a side effect in the context of cytokine treatment, such as INF-alpha therapy (Udina et al. 2012).

In the context of the **inflammation hypothesis of depression**—for a recent introduction cf. (Bullmore 2018)—the gut microbiome has recently become the focus of scientific interest. Via neuroendocrine and endocrine signaling cascades, tight feedbacks exist on the so-called "gut-brain axis". Physical and psychological processes can alter the gut microbiome, leading to altered gut endocrine responses, which in turn can have a direct impact on brain and neurophysiological processes (Mayer et al. 2014; Slyepchenko et al. 2017). In the context of depressive disorders and chronic stress, for example, it has been found that there can be an increase in gut permeability (*leaky gut*) as well as a change in the initial microbiome composition, which was associated with reduced BDNF release and expression of 5HT1a receptors in animal models (Bercik et al. 2011). Several studies have found differential gut bacterial composition in patients with depression and healthy control subjects (Jiang et al. 2015; Naseribafrouei et al. 2014). However, it currently remains questionable to what extent these findings can be replicated and to what extent therapeutic targets can be derived, such as specific dietary approaches, administration of certain probiotics or even the so-called *microbiota transfer therapy* (MTT; "transplantation" of microbiomes). The overall evidence however, is very heterogeneous (Cepeda et al. 2017; Romijn et al. 2017).

Further therapeutic principles can be derived from the inflammation hypothesis: For example, certain diets are associated with prevention of depressive symptoms. On the one hand, there is a clear link between obesity, inflammation (increased levels of IL-6, TNF-a and CRP) and increased risk of developing depression (Luppino et al. 2010). At the same time, a Mediterranean diet can lead to a reduction in inflammatory processes (IL-6 reduction) and a decrease in depressive symptoms in depressed patients (Milaneschi et al. 2011). Here, too, the pro-inflammatory processes appear to play the decisive role, which may also explain the antidepressant effects of anti-inflammatory substances such as omega-3 fatty acids (Schefft et al. 2017).

The anti-inflammatory and antidepressant effects of sports interventions (Gleeson et al. 2011) in the prevention and intervention of depressive disorders might also relate to this context (Azevedo Da Silva et al. 2012). Their effects were demonstrated primarily in the acute treatment of depression (Kvam et al. 2016). Nevertheless, a cautious interpretation of the findings is necessary, as the findings are highly dependent on the quality of the studies (Krogh et al. 2017). Overall, patients who already had increased evidence of inflammation before a specific diet seem to benefit most (Rethorst et al. 2013), which could play an important role in the diagnosis of inflammatory processes in the future.

12.5.6 Neuroplasticity and Neurogenesis: From Electroconvulsive Treatment to Ketamine

Neurostimulation procedures in the treatment of depressive disorders have a long history. Electroconvulsive therapy (ECT), for example, has been used for many decades to treat psychiatric disorders, but there has also been a critical discussion of this procedure for just as long ECT is known to the majority of the population only through the lurid and shocking film "One Flew Over the Cuckoo's Nest". (A more realistic portrayal, on the other hand, can be found in the ninth episode of the seventh season of the popular Netflix series "Homeland.") Yet ECT is now an established, safe and highly effective procedure, particularly for treatment-resistant depression, with response rates of up to 70% (UK ECT Review Group 2003). The aim of ECT is to induce a (time-limited) generalized seizure. The exact mechanism of action of ECT remains unclear, yet all postulated mechanisms of action are good examples of neuroplasticity and neurogenesis and their impairment in the context of depressive disorders: One theory is the so-called neurogenic theory. For a long time it was considered dogma that no new neurons are formed during life. However, it is now known that neurogenesis can occur in the hippocampus, which is promoted by antidepressants, among other things (Olesen et al. 2017). Various studies have shown that depressive symptoms in patients were correlated with impaired hippocampal neurogenesis as well as reduced hippocampal volume. At the same time, animal studies have shown that an increase in brain-derived neurotrophic factor (BDNF) as well as increased synaptogenesis and an overall increase in hippocampal volume can be detected in the hippocampus after ECT treatment. Increased BDNF levels have also been found in humans following ECT series (Rocha et al. 2016). A second mechanistic theory of ECT is the neuroendocrine theory, which relates to the theory of impaired HPA axis activity (▶ Sect. 12.5.4). HPA axis overactivity, particularly with the lack of negative feedback inhibition, appears to be changed by ECT. Thus, it has been demonstrated that ECT leads to stimulation of the diencephalon and excessive release of various hormones and neuropeptides, such as ACTH, prolactin and vasopressin (Bolwig 2011).

Further evidence for the importance of neurogenesis in the context of depressive disorders is provided by the almost spectacular effect of ketamine in depression: ketamine has long been known as an anesthetic. Pharmacologically, it is a glutamate antagonist at the NMDA receptor, thus blocking the action of the excitatory neurotransmitter glutamate. Therefore, by simple inversion, one conjecture has been that too much glutamate is related to depression. Infused at low doses, ketamine has produced dramatic improvements in patients with depressive disorders in controlled clinical trials in recent years (Kishimoto et al. 2016). In particular, the rapid onset of action within 24 h represents a novelty in the treatment of depression. As a reminder, conventional drugs for depression take 1–3 weeks to take effect. Meanwhile, it is assumed that many of the positive effects of ketamine are probably associated with the release of BDNF and an onset of synaptogenesis and neurogenesis. For example, it has been shown microscopically that there is an increase in the density of dendrites in layer V of the pyramidal cells in the prefrontal cortex (PFC) as early as 24 h after ketamine administration (Sattar et al. 2018).

The molecular mechanisms underlying these neurobiological remodeling processes are essentially based on the activation of protein biosynthesis and the activation of the enzyme mTOR (*mammalian target of rapamycin*). Ketamine, via blockade of the NMDA receptor, leads to reduced activity of eEF2 (eukaryotic elongation factor-2) kinase and thus of eEF2 proper (◘ Fig. 12.3). Ketamine increases BDNF release, furthermore, via tyrosine kinase B (TrkB), ERK and protein kinase B (PKB/Akt) are subsequently activated and glycogen synthase kinase-3 (GSK-3) is inhibited, processes that lead to synaptogenesis. At the same time, in addition to NMDA blockade,

ketamine increases signal transduction via AMPA, which in turn itself conditions an enhancement of BDNF translation and subsequent processes, and may even represent a more central role in ketamine's antidepressant effects (Sattar et al. 2018; Zanos et al. 2016).

The mechanism of action of ketamine and its antidepressant properties provide the opportunity to gain new insights into pathophysiology and develop specific new compounds for treatment (Köhler and Betzler 2015; Wray et al. 2018b). Meanwhile, ketamine itself is also being further developed as a nasal application for the treatment of depressive disorders and has also shown very positive results in studies (Canuso et al. 2018). Despite all the euphoria regarding efficacy, the side effects of ketamine treatment (short-term dissociative or psychotic symptoms) should nevertheless be mentioned (albeit with very low frequency and severity), as well as the question of possible dependence due to iatrogenic ketamine administration, which has not yet been adequately answered. In a recent paper, the antidepressant effects of ketamine were abolished with an opiate antagonist, which at least supports the concern of long-term addictive effects of ketamine (Williams et al. 2018).

12.5.7 Disturbance of Neuronal Networks

An important extension of neuroscientific methods for the study of mental processes and their disorders is structural and functional imaging with MRI, a subfield of cognitive neuroscience. Current approaches to the study of mental disorders, such as the Research Domain Criteria (RDoC) initiative, assume that all mental processes rely on the activity of specific neural *circuits* that realize certain classes of basic cognitive, emotional, and social functions. Early stud-

□ Fig. 12.3 The mode of action of ketamine. (According to Köhler and Betzler 2015)

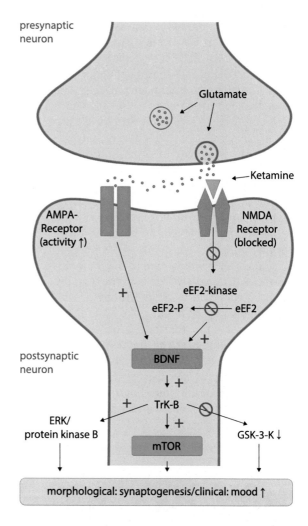

ies in psychiatry focused on volume changes or altered activation of individual brain regions during cognitive tasks or emotional stimulation in relatively small groups of usually less than 20 patients (see e.g. Vasic et al. 2007). Later, connectivity of brain regions was increasingly studied (e.g., Erk et al. 2010) and group sizes were expanded. Nowadays, it is assumed that all brain regions are parts of larger networks. Mathematical analysis of networks (*connectomics*) plays a role in many fields of science, e.g., the study of social networks, and can also be applied to neural networks (Bassett et al. 2018; Zalesky et al. 2012). Also, the large-scale and relatively simple approach of studying the brain in MRI while subjects and patients simply do nothing at all (so-called resting-state studies) has led to the availability of relatively large datasets now, with group sizes of hundreds (Walter et al. 2011), thousands (Quinlan et al. 2017), or even up to 10,000 subjects (Thiebaut de Schotten et al. 2018). This is highly beneficial, because due to the relatively poor signal-to-noise ratio, small group sizes, and methodological shortcomings, many, especially early findings, in the field of neuroimaging are not replicable and probably not valid. Using connectomics methods, it is now possible to identify a number of definable networks in the resting state that are known to be involved in specific cognitive functions. Therefore, these networks are often functionally characterized, e.g. as "cognitive control network", "salience network" or "reward network". Strictly speaking, these terms are not entirely correct, since the networks are involved in different functions.

A number of integrative review papers are now available that attempt to summarize the most important findings on altered neural networks in depression and bring them together with the symptoms of depression (Li et al. 2018; Williams et al. 2016). Perhaps the most detailed overview of depression has been compiled by Stanford-based neuroscientist Lianne Williams, who also makes a strong case for a transdiagnostic perspective that accounts for the frequent comorbidity of anxiety and depression. She postulates a set of "biotypes" in the anxiety-depression spectrum and links the main findings of "over-" and "under-activation"

in network regions, as well as findings of connectivity within networks, to the main symptoms of the anxiety-depression spectrum (◘ Fig. 12.4).

Of course, this assignment is not exhaustive. This can be seen, for example, in the fact that the default mode network (◘ Fig. 12.4, top left in blue) is only associated with rumination. In fact, however, the default mode network is also associated with the ability to mentalize (to put oneself in the thoughts and feelings of others). What is still missing in Williams' approach are more recent findings on the interaction of networks.

In the following, two current research areas will be briefly outlined that can be counted as network approaches in a broader sense. *Deep brain stimulation* (DBS) for severe, treatment-resistant depression is based on early neuroimaging findings. In the field of neurology, DBS is an established therapy for Parkinson's disease and certain forms of tremor. An electrode is neurosurgically implemented, usually in the subtha-

12

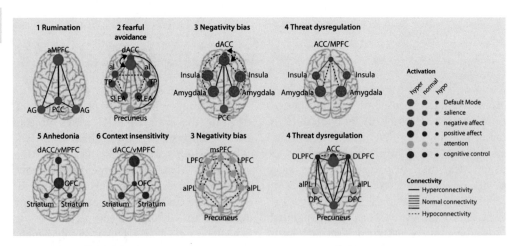

◘ **Fig. 12.4** A hypothesis for eight "biotypes" of the anxiety-depression spectrum. Shown here, based on a review of existing evidence from imaging studies, is a speculative hypothesis of how six known brain circuits relate to eight postulated clinical subtypes within the anxiety-depression spectrum, not found in the DSM-5 in this way, in terms of their functional connectivity. *ACC* anterior cingulate cortex, *DLPFC* dor-

solateral prefrontal cortex, *DPC* dorsal parietal cortex, *MPFC* medial prefrontal cortex, *OFC* orbitofrontal cortex, *PCC* posterior cingulate cortex, *PFC* prefrontal cortex, *SLEA* sublenticular extended amygdala, *TP* temporal pole, *aIPL* anterior inferior parietal lobule. (From Walter 2017, adapted from Williams et al. 2016)

lamic nucleus, and continuous stimulation at 100 Hz (which is not directly felt, because the brain is paradoxically insensitive) leads to an often dramatic, clinically relevant improvement in the movement disorder. It is thought that this effect is not a purely local one, but that stimulation in the right spot normalises disturbed networks or improves their function. Encouraged by this, Helen Maybergs group performed deep brain stimulation on patients with severe depression whose condition did not improve with medication, psychotherapy or electro-convulsive therapy (Crowell et al. 2015). But where to stimulate? It was known from early positron emission tomography (PET) studies that patients with depression often show increased glucose metabolism in the sub-genual ACC (Brodman area 25), which nor-malizes after successful therapy. Therefore, she chose this place as the stimulation site. The initial findings from open studies in a few patients were spectacular. More than 50% of completely refractory patients improved significantly and this, as follow-up observations show, over a longer period of time. There are now three double-blind therapeutic trials of DBS of the subgenual ACC, which were unfortunately disappoint-ing, showing no significant differences between verum and placebo conditions (Holtzheimer et al. 2017). Mayberg herself attributes this to the fact that it very much depends on exactly where you stimulate. She is currently developing protocols that use a different technique (DTI, *diffusion tensor imaging*) to map the course of fiber tracts before implantation in order to be able to stimulate more precisely. A group led by German psychiatrist Thomas Schläpfer and neurosurgeon Volker Coenen has also recently presented impressive results with the stimulation of a different structure (Bewernick et al. 2017), which are currently being tested in blinded studies. Based on imaging studies and theoretical consider-ations, they stimulate a tract, called the medial forebrain bundle, that connects the brainstem with the "reward center", the nucleus accumbens. It remains to be seen whether the very encouraging initial results will be confirmed in randomized, controlled and double-blind studies. What is clear, however, is that these forms of therapy could not have been developed at all with-out modern neuroimaging (current over-view in Kisely et al. 2018).

Another research direction is **computa-tional neuroscience** (Heinz 2017; Huys et al. 2016; Stephan et al. 2016). Instead of rely-ing solely on activation strength or connec-tivity in simple tasks or even at rest, it investigates networks using computational models of cognitive processes. This is usu-ally based on theoretical models that describe in mathematical terms how certain information processing steps are computed, for example in reward learning. If one now examines a behavior in the scanner, one can model the behavior, make model compari-sons, and analyze the extent to which certain parameters of the model (e.g., learning rate, reward sensitivity) correlate with brain activity or connectivity. One could also say that instead of studying many patients with simple tools and little theory (like large scale studies with Resting State), computational neuroscience studies complex models in rather few patients very closely. It remains to be seen how far this approach will bear fruit in the field of depression (cf. e.g. Huys et al. 2015).

12.5.8 Epigenetics and Early Childhood: Scars of Life

Severe stressful *life events* are of particular importance in individuals with a family his-tory of depressive disorders. The heritability for MDD is just under 40%. Viewed the other way round, this means that almost two thirds of the variance cannot be explained by genetic factors. If stressful events and cir-cumstances occur in persons with a genetic predisposition to the disease, this can lead to the occurrence of the disorder (gene-environment interaction).

Early childhood trauma is a significant risk factor for the development of depression in adulthood, especially if additional stressful life events occur later (Nanni et al. 2012; Nemeroff 2016). Trauma in childhood can also lead to chronic inflammatory processes (Fagundes et al. 2013; Tursich et al. 2014), which thus maintain the above-mentioned processes, especially the increased cytokine levels and hypercortisolism. Overall, early childhood neglect and abuse lead to a significantly increased risk for various psychiatric disorders (additionally PTSD, substance abuse, bipolar disorders) and various persistent neurobiological changes (Nemeroff 2016).

Both early childhood trauma and other stressful life events (job loss, separation, etc.) can lead to changes at the epigenetic level. The term "epigenetics" refers to all structural and molecular factors that alter gene expression without changing the structure of the DNA (Egger et al. 2004). These include methylation of DNA (usually leading to silencing of the corresponding gene), modification of histones (acetylation and methylation of the proteins around which DNA wraps), chromatin remodelling (movement of histones along DNA), influences of non-coding microRNA on gene expression and finally, a very new area of research, alteration of chromatin structure (changing the 3D arrangement, e.g. in the form of loops; Pena and Nestler 2018). Epigenetic processes are necessary for genes to be expressed in an organ-specific manner and for non-organ-specific genes to be silenced. They are important primarily in cancer research, but increasingly also in psychiatric research. Many of the findings have been obtained in animal models, especially in the social defeat stress model in mice. In living humans, the problem is that there is no direct access to brain tissue, but epigenetic changes are quite organ-specific and can usually only be determined peripherally in blood and saliva.

A good overview of the now abundant research on **epigenetics and depression is** provided by the reviews of the New York brain researcher Eric Nestler (Nestler 2014; Pena and Nestler 2018). In the following, some results on DNA methylation will be outlined by way of example (Pena and Nestler 2018): for example, an increase in de novo methyltransferase 3 (DMNT3a) was found in the nucleus accumbens of depressed people post mortem. Through animal experiments, it has been shown to be associated with depression-like behavior produced by social-defeat stress. In patients, however, post-mortem studies show not global but rather regional hypermethylation in genes related to nervous system development, mitochondria and immune regulation. Furthermore, there are a number of studies of candidate genes. For example, more than 20 studies have now examined methylation of the neurotrophin BNDF in blood or saliva. Most found increased methylation in its promoter region, which in some studies was reduced after antidepressant therapy. More than ten studies examined the gene for the serotonin transporter (SCL6A4) and generally found hypermethylation that normalized after treatment in some studies. Finally, there are several studies that found hypermethylation of genes relevant to the HPA axis, such as the CRF promoter in mice (reversible by treatment with antidepressants) or the glucocorticoid receptor in the hippocampus (NR3C1) in both mice and humans, especially with early childhood stress.

There are a number of studies on microRNAs, also in humans, that provide evidence for their involvement in the aetiology of depression, but could also act as potential biomarkers for diagnosis and response to drug treatment (especially MiR 1202, but also miR-146b-5p, miR-425-3p, miR-24-3p, miR-185, and miR-491-3p (Pena and Nestler 2018). It remains to be seen whether some of these findings can also be confirmed as robust in large epigenome-wide and expression studies. However, epigenetics is a another fresh example of why research in humans, which is often only correlative and subject to limitations, must be viewed together with animal experiments, as only here are manipulations possible that allow causal statements to be made.

12.6 Outlook: Integration or Pluralism?

As described in ▶ Sect. 12.4, the standard model that still is valid today is the vulnerability-stress model. This can now be underpinned in many places by neurobiological mechanisms. The more fundamental question, however, is: Can there be an integrative theory at all, if "the" depression in a DSM-5 sense may not exist? Recall that it has been empirically shown that among 3703 patients in the largest depression study ever conducted (STAR*D study), there were 1030 different profiles of depression, shared on average by only 3.6 patients (Fried and Nesse 2015). There is much overlap with other disorders, particularly anxiety disorders. Uncomplicated depressive episodes that were shorter than 2 months and did not include suicidal ideation, psychotic ideations, psychomotor retardation, or feeling worthless were not different in prognosis from normal grief episodes (Wakefield and Schmitz 2014). An alternative to the standard assumption that depressive symptoms have a common underlying cause, i.e. depression, is the network theory. It postulates instead that depressive disorders exist in the symptoms themselves, which are linked to each other by causal relationships that can be quite individual (Borsboom et al. 2016). Another pluralistic approach, the so-called mechanistic cluster property theory

(Kendler et al. 2011), conceives depression as syndromes with a family resemblance conditioned by causal factors at different levels, similar to how species are defined in biology: The different expressions of depression are related to each other (they "cluster"), i.e., they are more or less similar, but one always finds "relatives" that are quite different (Kendler et al. 2011). Family resemblance is explained by causal mechanisms that cause specimens with (family) similar characteristics to be produced. What these critics of the model of "the" depression have in common is that they consider diagnosis based solely on counting symptoms without theories of their origins to be the wrong approach. But what are the alternatives? One of them has already been presented, the cross-diagnostic characterization based on basic neurocognitive domains at different levels (RDoC).

A new interesting proposal comes from evolutionary psychology, namely to take the evolutionarily justifiable genesis ("trigger") of depressive dysfunction as a starting point for subtyping (Rantala et al. 2018). These authors assume that a depressive response pattern is an evolutionarily adaptive response to specific, fundamental problems of social organisms. Some of these subtypes are already known clinically (see below), and others are already formulated in older evolutionary theoretical approaches (the theory of social competition, ▶ Sect. 12.4). However, Rantala et al. (2018) choose a comprehensive systematics and identify the following 12 triggers: infections, chronic stress ("melancholia"), loneliness, traumatic events, hierarchical conflict, grief, romantic rejection, postpartum depression, seasonal depression, chemical-induced depression, depression due to somatic illness, and starvation-induced depression. From an evolutionary theoretical perspective, no principled distinction is made between physical and "psychological" causes. Depressive symptoms are initially seen as adaptations to the precipitating triggers, much as a fever may be a response to a pathogen, or a hypomanic state an adaptive response to infatuation. They should only be treated if they take on a magnitude or duration that is (too much) mal-adaptive. From an evolutionary theoretical perspective, depressive symptoms are ultimate functions and neurobiological processes are their proximate mechanisms.

Can a conclusion be drawn from all these approaches? Probably yes, and this could be as follows: The DSM-5 logic of diagnosing depression allows for a fairly reliable classification that is independent of causes. However, subtypes will emerge that can be distinguished by their triggering mechanisms, the mechanisms involved and the functions fulfilled by them, also in terms of course and prognosis. RDoC-like characterization will potentially lead to mechanistically based subtypes that can be related to functional subtypes. It can be speculated that three major subcategories may emerge in the process: First, functional depressions that serve adaptive functions (e.g., grief reactions following losses, depressive syndromes associated with infections and illnesses, capitulatory depressions associated with hierarchical conflicts), for which treatment, if needed at all, must address the original adaptive functions. Second, subtypes that are more likely to be seen as illnesses in which certain mechanisms dominate (e.g., inflammation, melancholia). Third, a trajectory that represents a kind of terminal course, i.e., chronic, frequently recurrent, and severe depression that is more similar and in which physiologically and psychologically ingrained behavioral routines or physiological patterns exist that are decoupled from their triggers. Of course, the boundaries of these subtypes should not be considered rigid. Nevertheless, it remains to be concluded: An integrative theory of depression will only succeed if it is pluralistic enough to bring together descriptive symptoms, their trigger-specific functions, and proximate mechanisms of neurobiology in a coherent approach, and to abandon the idea of a singular disorder that is generalized about.

References

Anacker C, Cattaneo A, Luoni A, Musaelyan K, Zunszain PA, Milanesi E, Pariante CM (2013) Glucocorticoid-related molecular signaling pathways regulating hippocampal neurogenesis. Neuropsychopharmacology 38(5):872–883. https://doi.org/10.1038/npp.2012.253

Angst J, Gamma A, Rossler W, Ajdacic V, Klein DN (2009) Long-term depression versus episodic major depression: results from the prospective Zurich study of a community sample. J Affect Disord 115(1–2):112–121. https://doi.org/10.1016/j.jad.2008.09.023

Azevedo Da Silva M, Singh-Manoux A, Brunner EJ, Kaffashian S, Shipley MJ, Kivimaki M, Nabi H (2012) Bidirectional association between physical activity and symptoms of anxiety and depression: the Whitehall II study. Eur J Epidemiol 27(7):537–546. https://doi.org/10.1007/s10654-012-9692-8

Bassett DS, Zurn P, Gold JI (2018) On the nature and use of models in network neuroscience. Nat Rev Neurosci. https://doi.org/10.1038/s41583-018-0038-8

Bauer ME, Teixeira AL (2018) Inflammation in psychiatric disorders: what comes first? Ann N Y Acad Sci. https://doi.org/10.1111/nyas.13712

Bercik P, Denou E, Collins J, Jackson W, Lu J, Jury J et al (2011) The intestinal microbiota affect central levels of brain-derived neurotropic factor and behavior in mice. Gastroenterology 141(2):599–609, e591–593. https://doi.org/10.1053/j.gastro.2011.04.052

Bewernick BH, Kayser S, Gippert SM, Switala C, Coenen VA, Schlaepfer TE (2017) Deep brain stimulation to the medial forebrain bundle for depression- long-term outcomes and a novel data analysis strategy. Brain Stimul 10(3):664–671. https://doi.org/10.1016/j.brs.2017.01.581

Bleys D, Luyten P, Soenens B, Claes S (2018) Gene-environment interactions between stress and 5-HTTLPR in depression: a meta-analytic update. J Affect Disord 226:339–345. https://doi.org/10.1016/j.jad.2017.09.050

Bolwig TG (2011) How does electroconvulsive therapy work? Theories on its mechanism. Can J Psychiatr 56(1):13–18. https://doi.org/10.1177/070674371105600104

Borsboom D, Rhemtulla M, Cramer AO, van der Maas HL, Scheffer M, Dolan CV (2016) Kinds versus continua: a review of psychometric approaches to uncover the structure of psychiatric constructs. Psychol Med 46(8):1567–1579. https://doi.org/10.1017/S0033291715001944

Braun C, Bschor T, Franklin J, Baethge C (2016) Suicides and suicide attempts during long-term treatment with antidepressants: a meta-analysis of 29 placebo-controlled studies including 6,934 patients with major depressive disorder. Psychother Psychosom 85(3):171–179. https://doi.org/10.1159/000442293

Bschor T, Baethge C, Hiemke C, Muller-Oerlinghausen B (2017) Genetic tests for controlling treatment with antidepressants. Nervenarzt 88(5):495–499. https://doi.org/10.1007/s00115-017-0310-6

Bullmore E (2018) The inflamed mind: a radical new approach to depression. Short Books, London

Bundesamt S (2015) Todesursachenstatistik. https://www.gbe-bund.de/oowa921-install/servlet/oowa/aw92/dboowasys921.xwdevkit/xwd_init?gbe.isgbetol/xs_start_neu/&p_aid=3&p_aid=10324223&nummer=670&p_sprache=D&p_indsp=-&p_aid=66408671

Canuso CM, Singh JB, Fedgchin M, Alphs L, Lane R, Lim P, Drevets WC (2018) Efficacy and safety of intranasal esketamine for the rapid reduction of symptoms of depression and suicidality in patients at imminent risk for suicide: results of a double-blind, randomized. Placebo-controlled study. Am J Psychiatry 175(7):620–630. https://doi.org/10.1176/appi.ajp.2018.17060720

Capuron L, Miller AH (2011) Immune system to brain signaling: neuropsychopharmacological implications. Pharmacol Ther 130(2):226–238. https://doi.org/10.1016/j.pharmthera.2011.01.014

Caspi A, Sugden K, Moffitt TE, Taylor A, Craig IW, Harrington H, Poulton R (2003) Influence of life stress on depression: moderation by a polymorphism in the 5-HTT gene. Science 301(5631):386–389. https://doi.org/10.1126/science.1083968

Cepeda MS, Katz EG, Blacketer C (2017) Microbiome-gut-brain axis: probiotics and their association with depression. J Neuropsychiatry Clin Neurosci 29(1):39–44. https://doi.org/10.1176/appi.neuropsych.15120410

Cipriani A, Furukawa TA, Salanti G, Chaimani A, Atkinson LZ, Ogawa Y, Geddes JR (2018) Comparative efficacy and acceptability of 21 antidepressant drugs for the acute treatment of adults with major depressive disorder: a systematic review and network meta-analysis. Lancet 391(10128):1357–1366. https://doi.org/10.1016/S0140-6736(17)32802-7

Crowell AL, Garlow SJ, Riva-Posse P, Mayberg HS (2015) Characterizing the therapeutic response to deep brain stimulation for treatment-resistant depression: a single center long-term perspective. Front Integr Neurosci 9:41. https://doi.org/10.3389/fnint.2015.00041

Culverhouse RC, Saccone NL, Bierut LJ (2018) The state of knowledge about the relationship between 5-HTTLPR, stress, and depression. J Affect Disord 228:205–206. https://doi.org/10.1016/j.jad.2017.12.002

Dantzer R (2017) Role of the kynurenine metabolism pathway in inflammation-induced depression: preclinical approaches. Curr Top Behav Neurosci 31:117–138. https://doi.org/10.1007/7854_2016_6

Dantzer R, O'Connor JC, Freund GG, Johnson RW, Kelley KW (2008) From inflammation to sickness and depression: when the immune system subjugates the brain. Nat Rev Neurosci 9(1):46–56. https://doi.org/10.1038/nrn2297

DGPPN, B., KBV, AWMF, AkdÄ, BPtK, BApK, DAGSHG, DEGAM, DGPM, DGPs, DGRW für die Leitliniengruppe Unipolare Depression (2015) S. 3-Leitlinie/Nationale Versorgungs Leitlinie Unipolare Depression—Langfassung, 1. Aufl. Version 5. 2009, zuletzt verändert: Juni 2015. https://www.leitlinien.de/themen/depression

Dillon DG, Rosso IM, Pechtel P, Killgore WD, Rauch SL, Pizzagalli DA (2014) Peril and pleasure: an rdoc-inspired examination of threat responses and reward processing in anxiety and depression. Depress Anxiety 31(3):233–249. https://doi.org/10.1002/da.22202

Egger G, Liang G, Aparicio A, Jones PA (2004) Epigenetics in human disease and prospects for epigenetic therapy. Nature 429(6990):457–463. https://doi.org/10.1038/nature02625

Erk S, Mikschl A, Stier S, Ciaramidaro A, Gapp V, Weber B, Walter H (2010) Acute and sustained effects of cognitive emotion regulation in major depression. J Neurosci 30(47):15726–15734. https://doi.org/10.1523/JNEUROSCI.1856-10.2010

Eyre H, Baune BT (2012) Neuroplastic changes in depression: a role for the immune system. Psychoneuroendocrinology 37(9):1397–1416. https://doi.org/10.1016/j.psyneuen.2012.03.019

Fagundes CP, Glaser R, Kiecolt-Glaser JK (2013) Stressful early life experiences and immune dysregulation across the lifespan. Brain Behav Immun 27(1):8–12. https://doi.org/10.1016/j.bbi.2012.06.014

Fournier JC, DeRubeis RJ, Hollon SD, Dimidjian S, Amsterdam JD, Shelton RC, Fawcett J (2010) Antidepressant drug effects and depression severity: a patient-level meta-analysis. JAMA 303(1):47–53. https://doi.org/10.1001/jama.2009.1943

Fried EI, Nesse RM (2015) Depression is not a consistent syndrome: an investigation of unique symptom patterns in the STAR*D study. J Affect Disord 172:96–102. https://doi.org/10.1016/j.jad.2014.10.010

Gan Y, Gong Y, Tong X, Sun H, Cong Y, Dong X, Lu Z (2014) Depression and the risk of coronary heart disease: a meta-analysis of prospective cohort studies. BMC Psychiatry 14:371. https://doi.org/10.1186/s12888-014-0371-z

Gartlehner G, Hansen RA, Morgan LC, Thaler K, Lux L, Van Noord M, Lohr KN (2011) Comparative benefits and harms of second-generation antidepressants for treating major depressive disorder: an updated meta-analysis. Ann Intern Med 155(11):772–785. https://doi.org/10.7326/0003-4819-155-11-201112060-00009

Gilman SE, Trinh NH, Smoller JW, Fava M, Murphy JM, Breslau J (2013) Psychosocial stressors and the prognosis of major depression: a test of Axis IV. Psychol Med 43(2):303–316. https://doi.org/10.1017/S0033291712001080

Gleeson M, Bishop NC, Stensel DJ, Lindley MR, Mastana SS, Nimmo MA (2011) The anti-inflammatory effects of exercise: mechanisms and implications for the prevention and treatment of disease. Nat Rev Immunol 11(9):607–615. https://doi.org/10.1038/nri3041

Goldsmith DR, Rapaport MH, Miller BJ (2016) A meta-analysis of blood cytokine network alterations in psychiatric patients: comparisons between schizophrenia, bipolar disorder and depression. Mol Psychiatry 21(12):1696–1709. https://doi.org/10.1038/mp.2016.3

Haapakoski R, Mathieu J, Ebmeier KP, Alenius H, Kivimaki M (2015) Cumulative meta-analysis of interleukins 6 and 1beta, tumour necrosis factor alpha and C-reactive protein in patients with major depressive disorder. Brain Behav Immun 49:206–215. https://doi.org/10.1016/j.bbi.2015.06.001

Hammen C (2005) Stress and depression. Annu Rev Clin Psychol 1:293–319. https://doi.org/10.1146/annurev.clinpsy.1.102803.143938

Harkness KL, Theriault JE, Stewart JG, Bagby RM (2014) Acute and chronic stress exposure predicts 1-year recurrence in adult outpatients with residual depression symptoms following response to treatment. Depress Anxiety 31(1):1–8. https://doi.org/10.1002/da.22177

Hasin DS, Goodwin RD, Stinson FS, Grant BF (2005) Epidemiology of major depressive disorder: results from the national epidemiologic survey on alcoholism and related conditions. Arch Gen Psychiatry 62(10):1097–1106. https://doi.org/10.1001/archpsyc.62.10.1097

Hasin DS, Sarvet AL, Meyers JL, Saha TD, Ruan WJ, Stohl M, Grant BF (2018) Epidemiology of adult DSM-5 major depressive disorder and its specifiers in the United States. JAMA Psychiat 75(4):336–346. https://doi.org/10.1001/jamapsychiatry.2017.4602

Heinz A (2017) A new understanding of mental disorders. Computational models for dimensional psychiatry. The MIT Press, Cambridge

Holtzheimer PE, Husain MM, Lisanby SH, Taylor SF, Whitworth LA, McClintock S, Mayberg HS (2017) Subcallosal cingulate deep brain stimulation for treatment-resistant depression: a multisite, randomised, sham-controlled trial. Lancet Psychiatry 4(11):839–849. https://doi.org/10.1016/S2215-0366(17)30371-1

Huys QJ, Daw ND, Dayan P (2015) Depression: a decision-theoretic analysis. Annu Rev Neurosci 38:1–23. https://doi.org/10.1146/annurev-neuro-071714-033928

Huys QJ, Maia TV, Frank MJ (2016) Computational psychiatry as a bridge from neuroscience to clinical applications. Nat Neurosci 19(3):404–413. https://doi.org/10.1038/nn.4238

Jakobsen JC, Katakam KK, Schou A, Hellmuth SG, Stallknecht SE, Leth-Moller K et al (2017) Selective serotonin reuptake inhibitors versus placebo in patients with major depressive disorder. A systematic review with meta-analysis and trial sequential analysis. BMC Psychiatry 17(1):58. https://doi.org/10.1186/s12888-016-1173-2

Jiang H, Ling Z, Zhang Y, Mao H, Ma Z, Yin Y, Ruan B (2015) Altered fecal microbiota composition in patients with major depressive disorder. Brain Behav Immun 48:186–194. https://doi.org/10.1016/j.bbi.2015.03.016

Kendler KS, Zachar P, Craver C (2011) What kinds of things are psychiatric disorders? Psychol Med 41(6):1143–1150. https://doi.org/10.1017/S0033291710001844

Kiecolt-Glaser JK, Derry HM, Fagundes CP (2015) Inflammation: depression fans the flames and feasts on the heat. Am J Psychiatry 172(11):1075–1091. https://doi.org/10.1176/appi.ajp.2015.15020152

Kisely S, Li A, Warren N, Siskind D (2018) A systematic review and meta-analysis of deep brain stimulation for depression. Depress Anxiety 35(5):468–480. https://doi.org/10.1002/da.22746

Kishimoto T, Chawla JM, Hagi K, Zarate CA, Kane JM, Bauer M, Correll CU (2016) Single-dose infusion ketamine and non-ketamine N-methyl-d-aspartate receptor antagonists for unipolar and bipolar depression: a meta-analysis of efficacy, safety and time trajectories. Psychol Med 46(7):1459–1472. https://doi.org/10.1017/S0033291716000064

Köhler S, Betzler F (2015) Ketamine–a new treatment option for therapy-resistant depression. Fortschr

12

Neurol Psychiatr 83(2):91–97. https://doi.org/10.1055/s-0034-1398967

Köhler S, Stover LA, Bschor T (2014) [MAO-inhibitors–a treatment option for treatment resistant depression: application, efficacy and characteristics]. Fortschr Neurol Psychiatr 82(4), 228–236; quiz 237–228. https://doi.org/10.1055/s-0034-1365945

Köhler S, Fischer T, Brakemeier EL, Sterzer P (2015) Successful treatment of severe persistent depressive disorder with a sequential approach: electroconvulsive therapy followed by cognitive behavioural analysis system of psychotherapy. Psychother Psychosom 84(2):127–128. https://doi.org/10.1159/000369847

Kornhuber J, Muller CP, Becker KA, Reichel M, Gulbins E (2014) The ceramide system as a novel antidepressant target. Trends Pharmacol Sci 35(6):293–304. https://doi.org/10.1016/j.tips.2014.04.003

Kraus C, Castren E, Kasper S, Lanzenberger R (2017) Serotonin and neuroplasticity—links between molecular, functional and structural pathophysiology in depression. Neurosci Biobehav Rev 77:317–326. https://doi.org/10.1016/j.neubiorev.2017.03.007

Krauth C, Stahmeyer JT, Petersen JJ, Freytag A, Gerlach FM, Gensichen J (2014) Resource utilisation and costs of depressive patients in Germany: results from the primary care monitoring for depressive patients trial. Depress Res Treat 2014:730891. https://doi.org/10.1155/2014/730891

Krogh J, Hjorthoj C, Speyer H, Gluud C, Nordentoft M (2017) Exercise for patients with major depression: a systematic review with meta-analysis and trial sequential analysis. BMJ Open 7(9):e014820. https://doi.org/10.1136/bmjopen-2016-014820

Kvam S, Kleppe CL, Nordhus IH, Hovland A (2016) Exercise as a treatment for depression: a meta-analysis. J Affect Disord 202:67–86. https://doi.org/10.1016/j.jad.2016.03.063

Laux G, Dietmaier O (2018) Antidepressiva. In: Laux G, Dietmaier O (eds) Psychopharmaka. Springer, Berlin, p 89

Li BJ, Friston K, Mody M, Wang HN, Lu HB, Hu DW (2018) A brain network model for depression: from symptom understanding to disease intervention. CNS Neurosci Ther. https://doi.org/10.1111/cns.12998

Lopez AD, Mathers CD, Ezzati M, Jamison DT, Murray CJ (2006) Global and regional burden of disease and risk factors, 2001: systematic analysis of population health data. Lancet 367(9524):1747–1757. https://doi.org/10.1016/s0140-6736(06)68770-9

Luppino FS, de Wit LM, Bouvy PF, Stijnen T, Cuijpers P, Penninx BW, Zitman FG (2010) Overweight, obesity, and depression: a systematic review and meta-analysis of longitudinal studies. Arch Gen Psychiatry 67(3):220–229. https://doi.org/10.1001/archgenpsychiatry.2010.2

Mahar I, Bambico FR, Mechawar N, Nobrega JN (2014) Stress, serotonin, and hippocampal neurogenesis in relation to depression and antidepressant effects. Neurosci Biobehav Rev 38:173–192. https://doi.org/10.1016/j.neubiorev.2013.11.009

Mattejat F, Remschmidt H (2008) Kinder psychisch kranker Eltern. Dtsch Arztebl 105(23):413–418

Mayer EA, Knight R, Mazmanian SK, Cryan JF, Tillisch K (2014) Gut microbes and the brain: paradigm shift in neuroscience. J Neurosci 34(46):15490–15496. https://doi.org/10.1523/JNEUROSCI.3299-14.2014

Milaneschi Y, Bandinelli S, Penninx BW, Vogelzangs N, Corsi AM, Lauretani F, Ferrucci L (2011) Depressive symptoms and inflammation increase in a prospective study of older adults: a protective effect of a healthy (Mediterranean-style) diet. Mol Psychiatry 16(6):589–590. https://doi.org/10.1038/mp.2010.113

Murphy JA, Byrne GJ (2012) Prevalence and correlates of the proposed DSM-5 diagnosis of Chronic Depressive Disorder. J Affect Disord 139(2):172–180. https://doi.org/10.1016/j.jad.2012.01.033

Murray CJ, Richards MA, Newton JN, Fenton KA, Anderson HR, Atkinson C, Davis A (2013) UK health performance: findings of the global burden of disease study 2010. Lancet 381(9871):997–1020. https://doi.org/10.1016/S0140-6736(13)60355-4

Nagel M, Jansen PR, Stringer S, Watanabe K, de Leeuw CA, Bryois J, Posthuma D (2018) Meta-analysis of genome-wide association studies for neuroticism in 449,484 individuals identifies novel genetic loci and pathways. Nat Genet 50(7):920–927. https://doi.org/10.1038/s41588-018-0151-7

Nanni V, Uher R, Danese A (2012) Childhood maltreatment predicts unfavorable course of illness and treatment outcome in depression: a meta-analysis. Am J Psychiatry 169(2):141–151. https://doi.org/10.1176/appi.ajp.2011.11020335

Naseribafrouei A, Hestad K, Avershina E, Sekelja M, Linlokken A, Wilson R, Rudi K (2014) Correlation between the human fecal microbiota and depression. Neurogastroenterol Motil 26(8):1155–1162. https://doi.org/10.1111/nmo.12378

Nemeroff CB (2016) Paradise LOST: the neurobiological and clinical consequences of child abuse and neglect. Neuron 89(5):892–909. https://doi.org/10.1016/j.neuron.2016.01.019

Nestler EJ (2014) Epigenetic mechanisms of depression. JAMA Psychiat 71(4):454–456. https://doi.org/10.1001/jamapsychiatry.2013.4291

Nutt DJ (2008) Relationship of neurotransmitters to the symptoms of major depressive disorder. J Clin Psychiatry 69(E1):4–7

Olesen MV, Wortwein G, Folke J, Pakkenberg B (2017) Electroconvulsive stimulation results in long-term survival of newly generated hippocampal neurons in rats. Hippocampus 27(1):52–60. https://doi.org/10.1002/hipo.22670

Otte C, Gold SM, Penninx BW, Pariante CM, Etkin A, Fava M, Schatzberg AF (2016) Major depressive disorder. Nat Rev Dis Primers 2:16065. https://doi.org/10.1038/nrdp.2016.65

Pan A, Sun Q, Okereke OI, Rexrode KM, Hu FB (2011) Depression and risk of stroke morbidity and mortality: a meta-analysis and systematic review. JAMA 306(11):1241–1249. https://doi.org/10.1001/jama.2011.1282

Pan JX, Xia JJ, Deng FL, Liang WW, Wu J, Yin BM, Xie P (2018) Diagnosis of major depressive disorder based on changes in multiple plasma neurotransmitters: a targeted metabolomics study. Transl Psychiatry 8(1):130. https://doi.org/10.1038/s41398-018-0183-x

Pena CJ, Nestler EJ (2018) Progress in epigenetics of depression. Prog Mol Biol Transl Sci 157:41–66. https://doi.org/10.1016/bs.pmbts.2017.12.011

Polyakova M, Stuke K, Schuemberg K, Mueller K, Schoenknecht P, Schroeter ML (2015) BDNF as a biomarker for successful treatment of mood disorders: a systematic & quantitative meta-analysis. J Affect Disord 174:432–440. https://doi.org/10.1016/j.jad.2014.11.044

Prins J, Olivier B, Korte SM (2011) Triple reuptake inhibitors for treating subtypes of major depressive disorder: the monoamine hypothesis revisited. Expert Opin Investig Drugs 20(8):1107–1130. https://doi.org/10.1517/13543784.2011.594039

Quinlan EB, Cattrell A, Jia T, Artiges E, Banaschewski T, Barker G, Consortium I (2017) Psychosocial stress and brain function in adolescent psychopathology. Am J Psychiatry 174(8):785–794. https://doi.org/10.1176/appi.ajp.2017.16040464

Raison CL, Miller AH (2011) Is depression an inflammatory disorder? Curr Psychiatry Rep 13(6):467–475. https://doi.org/10.1007/s11920-011-0232-0

Rantala MJ, Luoto S, Krams I, Karlsson H (2018) Depression subtyping based on evolutionary psychiatry: proximate mechanisms and ultimate functions. Brain Behav Immun 69:603–617. https://doi.org/10.1016/j.bbi.2017.10.012

Rethorst CD, Toups MS, Greer TL, Nakonezny PA, Carmody TJ, Grannemann BD, Trivedi MH (2013) Pro-inflammatory cytokines as predictors of antidepressant effects of exercise in major depressive disorder. Mol Psychiatry 18(10):1119–1124. https://doi.org/10.1038/mp.2012.125

Rocha RB, Dondossola ER, Grande AJ, Colonetti T, Ceretta LB, Passos IC, da Rosa MI (2016) Increased BDNF levels after electroconvulsive therapy in patients with major depressive disorder: a meta-analysis study. J Psychiatr Res 83:47–53. https://doi.org/10.1016/j.jpsychires.2016.08.004

Romijn AR, Rucklidge JJ, Kuijer RG, Frampton C (2017) A double-blind, randomized, placebo-controlled trial of Lactobacillus helveticus and Bifidobacterium longum for the symptoms of depression. Aust NZ J Psychiatry 51(8):810–821. https://doi.org/10.1177/0004867416686694

Ruhe HG, Mason NS, Schene AH (2007) Mood is indirectly related to serotonin, norepinephrine and dopamine levels in humans: a meta-analysis of monoamine depletion studies. Mol Psychiatry 12(4):331–359. https://doi.org/10.1038/sj.mp.4001949

Rush AJ (2007) STAR*D: what have we learned? Am J Psychiatry 164(2):201–204. https://doi.org/10.1176/ajp.2007.164.2.201

Sanacora G, Treccani G, Popoli M (2012) Towards a glutamate hypothesis of depression: an emerging frontier of neuropsychopharmacology for mood disorders. Neuropharmacology 62(1):63–77. https://doi.org/10.1016/j.neuropharm.2011.07.036

Sattar Y, Wilson J, Khan AM, Adnan M, Azzopardi Larios D, Shrestha S, Rumesa F (2018) A review of the mechanism of antagonism of N-methyl-D-aspartate receptor by ketamine in treatment-resistant depression. Cureus 10(5):e2652. https://doi.org/10.7759/cureus.2652

Savitz JB, Drevets WC (2013) Neuroreceptor imaging in depression. Neurobiol Dis 52:49–65. https://doi.org/10.1016/j.nbd.2012.06.001

Schefft C, Kilarski LL, Bschor T, Kohler S (2017) Efficacy of adding nutritional supplements in unipolar depression: a systematic review and meta-analysis. Eur Neuropsychopharmacol 27(11):1090–1109. https://doi.org/10.1016/j.euroneuro.2017.07.004

Schildkraut JJ, Kety SS (1967) Biogenic amines and emotion. Science 156(3771):21–37

Simon GE, VonKorff M, Piccinelli M, Fullerton C, Ormel J (1999) An international study of the relation between somatic symptoms and depression. N Engl J Med 341(18):1329–1335. https://doi.org/10.1056/NEJM199910283411801

Slyepchenko A, Maes M, Jacka FN, Kohler CA, Barichello T, McIntyre RS, Carvalho AF (2017) Gut microbiota, bacterial translocation, and interactions with diet: pathophysiological links

12

between major depressive disorder and non-communicable medical comorbidities. Psychother Psychosom 86(1):31–46. https://doi.org/10.1159/000448957

Smith KA, Cipriani A (2017) Lithium and suicide in mood disorders: updated meta-review of the scientific literature. Bipolar Disord 19(7):575–586. https://doi.org/10.1111/bdi.12543

Stephan KE, Bach DR, Fletcher PC, Flint J, Frank MJ, Friston KJ, Breakspear M (2016) Charting the landscape of priority problems in psychiatry, part 1: classification and diagnosis. Lancet Psychiatry 3(1):77–83. https://doi.org/10.1016/S2215-0366(15)00361-2

Stetler C, Miller GE (2011) Depression and hypothalamic-pituitary-adrenal activation: a quantitative summary of four decades of research. Psychosom Med 73(2):114–126. https://doi.org/10.1097/PSY.0b013e31820ad12b

Sullivan PF, Daly MJ, O'Donovan M (2012) Genetic architectures of psychiatric disorders: the emerging picture and its implications. Nat Rev Genet 13(8):537–551. https://doi.org/10.1038/nrg3240

Thiebaut de Schotten M, Walter H, Sabuncu MJ, Holmes AJ, Gramfort A, Varoquaux GP et al. (2018) Subspecialization within default mode nodes in 10,000 UK Biobank participants. PNAS, in revision

Trautmann S, Beesdo-Baum K (2017) The treatment of depression in primary care. Dtsch Arztebl Int 114(43):721–728. https://doi.org/10.3238/arztebl.2017.0721

Tursich M, Neufeld RW, Frewen PA, Harricharan S, Kibler JL, Rhind SG, Lanius RA (2014) Association of trauma exposure with proinflammatory activity: a transdiagnostic meta-analysis. Transl Psychiatry 4:e413. https://doi.org/10.1038/tp.2014.56

Udina M, Castellvi P, Moreno-Espana J, Navines R, Valdes M, Forns X, Martin-Santos R (2012) Interferon-induced depression in chronic hepatitis C: a systematic review and meta-analysis. J Clin Psychiatry 73(8):1128–1138. https://doi.org/10.4088/JCP.12r07694

Uk, ECT Review Group (2003) Efficacy and safety of electroconvulsive therapy in depressive disorders: a systematic review and meta-analysis. Lancet 361(9360):799–808. https://doi.org/10.1016/S0140-6736(03)12705-5

Vasic N, Wolf RC, Walter H (2007) [Executive functions in patients with depression. The role of prefrontal activation]. Nervenarzt 78(6):628, 630–622, 634–626 passim. https://doi.org/10.1007/s00115-006-2240-6

Wakefield JC, Schmitz MF (2014) Predictive validation of single-episode uncomplicated depression as a benign subtype of unipolar major depression.

Acta Psychiatr Scand 129(6):445–457. https://doi.org/10.1111/acps.12184

Walter H (2017) Research domain criteria (RDoC): psychiatric research as applied cognitive neuroscience. Nervenarzt 88(5):538–548. https://doi.org/10.1007/s00115-017-0284-4

Walter H, Schnell K, Erk S, Arnold C, Kirsch P, Esslinger C, Meyer-Lindenberg A (2011) Effects of a genome-wide supported psychosis risk variant on neural activation during a theory-of-mind task. Mol Psychiatry 16(4):462–470. https://doi.org/10.1038/mp.2010.18

Williams LM, Goldstein-Piekarski AN, Chowdhry N, Grisanzio KA, Haug NA, Samara Z, Yesavage J (2016) Developing a clinical translational neuroscience taxonomy for anxiety and mood disorder: protocol for the baseline-follow up research domain criteria anxiety and depression ("RAD") project. BMC Psychiatry 16:68. https://doi.org/10.1186/s12888-016-0771-3

Williams NR, Heifets BD, Blasey C, Sudheimer K, Pannu J, Pankow H et al (2018) Attenuation of Antidepressant Effects of Ketamine by Opioid Receptor Antagonism. Am J Psychiatry. https://doi.org/10.1176/appi.ajp.2018.18020138

Wittchen H-U (2011) Klinische Psychologie & Psychotherapie. Springer, Berlin, ISBN 978-3-642-13017-5, Kap. 2, S. 21–23, 833f

Wittchen HU, Jacobi F, Rehm J, Gustavsson A, Svensson M, Jonsson B, Steinhausen HC (2011) The size and burden of mental disorders and other disorders of the brain in Europe 2010. Eur Neuropsychopharmacol 21(9):655–679. https://doi.org/10.1016/j.euroneuro.2011.07.018

Wray NR, Ripke S, Mattheisen M, Trzaskowski M, Byrne EM, Abdellaoui A, Major Depressive Disorder Working Group of the Psychiatric Genomics, C (2018a) Genome-wide association analyses identify 44 risk variants and refine the genetic architecture of major depression. Nat Genet 50(5):668–681. https://doi.org/10.1038/s41588-018-0090-3

Wray NH, Schappi JM, Singh H, Senese NB, Rasenick MM (2018b) NMDAR-independent, cAMP-dependent antidepressant actions of ketamine. Mol Psychiatry. https://doi.org/10.1038/s41380-018-0083-8

Zaki NFW, Spence DW, BaHammam AS, Pandi-Perumal SR, Cardinali DP, Brown GM (2018) Chronobiological theories of mood disorder. Eur Arch Psychiatry Clin Neurosci 268(2):107–118. https://doi.org/10.1007/s00406-017-0835-5

Zalesky A, Fornito A, Bullmore E (2012) On the use of correlation as a measure of network connectivity. NeuroImage 60(4):2096–2106. https://doi.org/10.1016/j.neuroimage.2012.02.001

Zanos P, Moaddel R, Morris PJ, Georgiou P, Fischell J, Elmer GI, Gould TD (2016) NMDAR inhibition-independent antidepressant actions of ketamine metabolites. Nature 533(7604):481–486. https://doi.org/10.1038/nature17998

Zhang JC, Yao W, Hashimoto K (2016) Brain-derived Neurotrophic Factor (BDNF)-TrkB signaling in inflammation-related depression and potential therapeutic targets. Curr Neuropharmacol 14(7):721–731

12

Anxiety Disorders

Jens Plag and Andreas Ströhle

Contents

The chapter highlights "anxiety" as a continuum from a normal psychological and functional phenomenon to a pathological state, which in the form of categorical anxiety disorders leads to a pronounced psychosocial impairment of those affected. In addition to the disorder-specific symptoms, important epidemiological data and the current treatment options for anxiety disorders, the psychological and biological basis and correlates of this group of disorders are presented, as well as the role of stress as an important symptom-triggering and symptom-maintaining factor. The aim is to provide an overview of the multifactorial conditions in the development of pathological anxiety and, in particular, to clarify the reciprocal relationships between the individual aspects.

Learning Objectives

The purpose of this chapter is to familiarize the reader with the respective symptomatology, clinical characteristics, and disease-promoting as well as symptom-triggering and symptom-maintaining factors of anxiety disorders, and then to enable the reader to comprehend the rationale of current therapeutic procedures.

13

▶ **Example**

34-year-old Luise T., accompanied by her partner, presents at the outpatient consultation of a psychiatrist. She reports to the doctor that about half a year ago, for the first time, in a pedestrian zone, "as if out of nowhere", a symptomatology consisting of palpitations, sweating, trembling, dizziness, a strong urge to urinate, nausea, an experience of unreality ("I perceived the environment as if through a pane of frosted glass"), a feeling of loss of control over the situation as well as fear of death had occurred. At the time, she had gone shopping with her husband and daughter in preparation for the family holiday. Due to a tight schedule and the child's related "whining", the patient had been

"quite stressed" situationally. At this time, she had also already been under a lot of pressure in the medium term, mainly due to an additional workload. Although the symptoms had subsided after about a quarter of an hour, her husband had driven her immediately to the emergency room of a nearby hospital to have her "checked out" ("we didn't know what was going on, and I wanted to rule out something serious like a heart attack or stroke"). There, however, an ECG and a blood test did not show any abnormal results. However, due to the description of the symptoms, the doctor already suspected a panic attack and recommended a consultation with a psychiatrist or psychotherapist in case the symptoms should occur again. After the patient had a few days of rest, the symptoms had finally reappeared; this time she had just been standing in the kitchen preparing dinner. It had been very hot and she had already had the feeling of not being able to breathe properly. With a comparable constellation of symptoms, however, the heart palpitations and the dizziness in particular had been much more pronounced this time and Ms. T. had feared going mad. After the symptoms had subsided again after about 20 min, she had taken a sleeping pill and gone to bed. For the next few days, she had first called in sick and took it easy on herself physically so as not to "provoke" the symptoms. In the following days, the anxiety attacks did not occur; however, a persistent fear of their recurrence had set in. After just under a week, Ms. T. had finally set off for work again with a "somewhat queasy feeling". As she did every morning, she took the underground. On this morning, however, the subway was unusually crowded due to a surprising cold day. After she had entered the carriage, however, this time she had immediately become very nervous and within a few minutes the full picture of an anxiety attack had developed again. She left the train at the next stop and returned home. Since then, the patient has no longer been able to use public transport, as she is afraid of not being able to get off in time in the event of a panic or of embarrassing herself in front of the other passengers. In the course of time, it had also become increasingly difficult or impossible for her to cope with other situations from which she could not immediately remove herself for spatial or social reasons or in which she could not expect immediate help. For example, she is currently no longer able to drive actively in city traffic or visit cinemas, and long train journeys, the use of lifts or professional meetings are only possible if she distracts herself (e.g. by listening to music or "mind games") or takes medication beforehand which reduces her heart rate. For commuting to work or shopping, she currently mainly uses her bicycle or is driven by her husband in bad weather. Together with the restrictions in the other areas of (social) life, the symptoms result not only in a massive burden for the patient, but also for the entire family. Accordingly, she is now seeking professional help, as she no longer knows what to do.

After (repeated) organic diagnostics (which in particular served to exclude thyroid dysfunction and higher-grade cardiac arrhythmia), Ms. T. was diagnosed with agoraphobia with panic disorder on the basis of her medical history and psychopathological findings. After an explanation of the nature, development, prognosis and treatment options, a pharmacotherapy with the selective serotonin reuptake inhibitor escitalopram was started in accordance with the patient's wish to improve her symptoms as quickly as possible, and this was increased to a daily dose of 15 mg in the medium term. This resulted in a gradual decrease of symptoms and thus in a relief of Mrs. T. already after about 10 weeks. Parallel to the drug treatment, the patient began an outpatient cognitive behavioural therapy, in which exposure treatment was carried out in addition to psychoeducation, cognitive restructuring and relaxation procedures. In the medium term, this initially led to a further progressive improvement and finally to a complete reduction of the symptoms, which could be maintained in a stable manner even after the gradual discontinuation of the medication. ◄

13.1 Fear: From a Protective Mechanism to a Pathological Phenomenon

Just like joy or anger, fear is an evolutionarily determined basic emotion that initially has an important function in the (survival) of humans as a normal psychological phenomenon. Fear helps us to better assess dangers or risks in everyday life and to react adequately in corresponding situations. If, for example, we are confronted with an overpowering opponent or feel "trapped" or defenceless because threatening or protective environmental conditions have suddenly appeared or disappeared, changes occur on various levels: *Perception* "narrows" with focus on the (potential) source of danger and *thinking* and *behavior* become dominated by the desire to escape or defend. In addition, symptoms occur on a *physical level*, which are caused in particular by an activation of the vegetative nervous system and enable the organism to initiate flight or to take up the fight. In order to be able to guarantee the energy for the increased muscular work necessary for this, the blood flow "centralises" (i.e. it is redistributed in favour of the muscles and at the expense of the central nervous system), and the heartbeat and blood pressure increase. At the same time, there is an increase in respiratory rate and metabolic rate to ensure adequate oxygenation of the blood and adequate energy supply, respectively, as part of the physical activation. Alternatively to this activating reaction, however, a fear reaction can lead to an acute drop in blood pressure and consecutive loss of consciousness due to a sudden dilation of the vessels. Within the framework of phylogenetic "logic", the organism can also escape from the threat through this—it "plays dead" and the attacker lets go of it. After a successful escape or a successful fight, however, the described psychophysical changes regress in the short to medium term, and the biological system returns to its original functionality. What has been described makes it clear that a "normal" fear reaction is situation-specific and self-limiting. Furthermore, its probability of occurrence and its extent regularly correlate with the real or assumed dangerousness of the trigger.

In this context, the latter aspect in particular marks the distinction between functional and pathological anxiety. Pathological anxiety is characterised by an anxiety reaction that occurs for no apparent reason or in relation to an objectively harmless trigger, or whose severity clearly exceeds an adequate reaction to the (potential) threat. An example of this is the *panic attack*, which is characterised by the occurrence of a psychovegetative symptom complex comparable to a normal anxiety reaction, but at the same time by the absence of an objectifiable trigger. The presence of such a trigger, on the other hand, is the hallmark of *phobic anxiety*. In this case, however, the nature of the fear-inducing situation or object (e.g. the elevator or the spider) or the expression of the fear reaction in relation to the stimulus is not comprehensible from an objective point of view. Pathological anxiety, however, is not only characterized by a sudden occurrence in the absence of or inadequate stimuli; it can also manifest itself in the form of a permanently increased "apprehension" in relation to various potential dangers of daily life (e.g., illness, loss of a caregiver or job, accidents, or violent crimes). Pathological characteristics of this *generalized anxiety* are again the lack of an objectifiable correlate of the fears, the extent as well as the more or less permanent presence of the fears. The physical aspect of generalized anxiety is also usually persistent (although qualitatively less pronounced than in a panic attack or a phobic reaction) and often manifests itself in the form of gastrointestinal complaints or muscular pain or tension (e.g. Plag and Hoyer 2019).

Regardless of its form, psychosocial *stress* plays a central role in the development and maintenance of pathological anxiety. The relationship between individual susceptibility (vulnerability) and stress can be illustrated particularly vividly for anxiety disorders using the transdiagnostically well-documented "diathesis-stress model" (◘ Fig. 13.1). People who are particularly

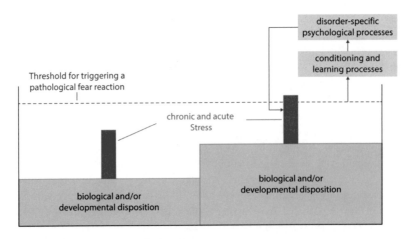

◘ **Fig. 13.1** The "diathesis-stress model" of anxiety

susceptible (predisposed) to the development of pathological anxiety due to biological and/or developmental characteristics (see also ► Sects. 13.2 and 13.5) develop symptoms more quickly under the influence of stress. In this context, both chronic and acute stress play an important role; both forms can develop a synergism. Once symptoms have developed, *conditioning* and further *learning processes* lead to the development of disease-specific psychological processes (► Sect. 13.4), which then contribute to the maintenance and exacerbation of symptoms, partly through a further increase in stress. The example of our patient Luise illustrates this well: With presumably increased sensitivity, the first panic attack occurred here under the influence of (sub)acute stress. Subsequently, the patient developed a strong anticipatory anxiety regarding a recurrence of the anxiety symptomatology. Although she took care of herself physically to prevent an "activation" of panic-associated bodily symptoms, a new panic attack occurred promptly. Due to the perceived loss of control in relation to symptom development, this resulted in a further increase in her stress level. This in turn favoured the phobic reaction to situations from which escape is difficult or impossible in the event of a panic attack, or in which no immediate help is available (e.g. public transport or lifts). Characteristically for phobic fears, in addition to the physical protective behaviour, a situational safety and avoidance behaviour is now also established, which reduces the patient's autonomy and radius of action more and more and thus leads into the "vicious circle" of ever increasing stress and impairment. In the development of generalised anxiety, comparable cognitive processes also intertwine; at the behavioural level, reassurance behaviour (e.g. regular visits to the doctor in the case of actually harmless bodily symptoms or illnesses among acquain-

tances) is regularly established here, by means of which the anticipated catastrophe (e.g. illness) is to be ruled out.

13.2 Symptoms, Diagnosis and Epidemiology of Anxiety Disorders

The ICD-10 distinguishes phobic disorders (specific phobia, agoraphobia and social phobia) from panic disorder and generalized anxiety disorder ("other anxiety disorders") on the basis of the presence or absence of an identifiable and immediate trigger. The DSM-5 does not make this separation and groups phobic and nonphobic anxiety disorders into a common category. Furthermore, in addition to the disorders listed in the ICD-10, the DSM-5 also lists "selective mutism" and "separation anxiety" as anxiety disorders (of adulthood). All anxiety disorders have in common both the presence of physical correlates of a stress or anxiety reaction that occurs suddenly (panic disorder), in response to the respective trigger (phobic disorders, selective mutism and separation anxiety) or in a persistent but then usually attenuated form (generalised anxiety disorder), and an avoidance or safety behaviour in relation to the triggers that trigger the respective symptom. The individual clinical pictures are described in more detail below.

According to the ICD-10 and the DSM-5, panic *disorder* is characterized by "unpredictable" and recurrent anxiety attacks (panic), which in turn are characterized by pronounced physical and psychological symptoms. Typically, many patients fear "going crazy" or dying during a panic attack. The ICD-10 and DSM-5 criteria for panic disorder are listed in ◘ Table 13.1. Both classification systems largely coincide in their syndromal description of the disor-

▫ Table 13.1   Criteria for panic disorder according to ICD-10 and DSM-5

ICD-10 (code: F41.0)	DSM-5 (code: 300.01)

1. Repeated panic attacks that are not related to a specific situation or object and often occur spontaneously (i.e. the attacks are not predictable). The panic attacks are not associated with particular exertion, dangerous or life-threatening situations
2. A panic attack has **all the** following characteristics:
  • It's a single episode of intense anxiety
  • It begins abruptly
  • It reaches a maximum within a few minutes and lasts at least a few minutes
  • **At least four symptoms from** the list below, *one of* which must be *from symptoms 1 to 4*, must be present:
    1. Palpitations or increased heart rate
    2. Sweating
    3. Fine/coarse tremor
    4. Dry mouth
    5. Respiratory symptoms
    6. Feeling of anxiety
    7. Chest pain or discomfort
    8. Nausea or abdominal discomfort
    9. Feeling of dizziness, unsteadiness, weakness or lightheadedness
    10. Feeling that objects are unreal or oneself is far away or "not really here" (depersonalisation), unreal (derealization) or fear of loss of control, going crazy
    11. Fear of losing control, going crazy, or "freaking out"
    12. Fear to die
    13. Hot flushes or cold shivers
    14. Numbness or tingling sensations
3. Exclusion: The panic attacks are not due to a physical disorder, an organic mental disorder or another mental disorder such as schizophrenia and related disorders, an affective disorder or a somatoform disorder
	A. Repeated unexpected panic attacks. A panic attack is a sudden surge of intense fear or discomfort that peaks within minutes
	  During this time, **at least four of** the following symptoms occur:
	  • Palpitations
	  • Perspiration
	  • Tremor/shaking
	  • Shortness of breath/breathlessness
	  • suffocating sensations
	  • Chest pain/cramping
	  • Nausea/gastrointestinal complaints
	  • Dizziness, unsteadiness, drowsiness
	  • Chills/Heat Feelings
	  • Paresthesias (numbness, tingling sensations)
	  • Derealization/depersonalization
	  • Fear of losing control or "going crazy".
	  • Fear of death
	B. At least one of the attacks was followed by a month (or longer) with at least one of the following symptoms:
	  • Persistent worry/concern about the occurrence of further panic attacks or their consequences (e.g., losing control, having a heart attack, or "going crazy")
	  • A markedly maladaptive change in behaviour as a result of the attacks (e.g. avoidance of physical activity or unfamiliar situations in order to avoid panic attacks)
	C. The disturbance is not the result of a physiological effect of a substance or a medical disease factor (e.g. hyperthyroidism, cardiovascular or pulmonary disease).
	D. The disorder cannot be better explained by another mental disorder

In the **ICD-10**, *two degrees of severity of panic disorder* can be distinguished:
Moderate: At least four panic attacks in 4 weeks (code: F41.00)
Severe: At least four panic attacks per week within 4 weeks (code: F41.01)

13

der. In contrast to the DSM-5, however, the ICD-10 allows differentiation of the severity of panic disorder; the DSM-5, on the other hand, emphasises the presence and importance of anticipatory anxiety and disorder-specific avoidance behaviour (subitem B in ▫ Table 13.1) as important elements of all anxiety disorders (American Psychiatric Association 2013; Dilling et al. 2015).

Patients with panic disorder often develop agoraphobia or the latter precedes the development of panic disorder (Bienvenu et al. 2006); the ICD-10 has taken this high comorbidity rate into account by allowing agoraphobia to be diagnosed with or without panic disorder as separate disorder entities. *Agoraphobia* itself is characterised by a (panic-like) fear of or in situations which the affected person cannot leave immediately, either in reality or for social reasons, or in which no immediate help is available. Typical situational triggers include using public transport or lifts, being in crowds, visiting cinemas and theatres, standing in a queue or attending a meeting. *Social phobia* (ICD-10) or *social anxiety disorder* (DSM-5), in turn, involves the fear of being negatively evaluated by third parties in a social context. Accordingly, those affected develop a fear reaction, for example, while giving a presen-

tation, eating in public or making new contacts. In contrast to the ICD-10, the DSM-5 also defines a variant of social anxiety disorder that occurs "only in performance situations" (*performance only type*). This subtype would correspond, for example, to "performance anxiety" in musicians or actors. A fear of circumscribed situations (not corresponding to social or agoraphobic triggers) or specific objects is the central element of *specific phobias*. Specific phobias represent the most common group of anxiety disorders (see also ◻ Table 13.2), which is explained by the fact that in principle any object or situation can become a trigger of a specific-phobic reaction. Both the ICD-10 and the DSM-5 distinguish between five subgroups, namely the animal, the environ-

◻ **Table 13.2** Epidemiology and diagnosis of anxiety disorders

Anxiety disorder	Initial manifestation	Prevalence[a]	Diagnosis possible from a symptom duration of (ICD-10/DSM-5):	Diagnostic tool (examples)
Panic disorder	3rd–4th decade of life	2%	Each not further defined	PAS[b]
Agoraphobia	3rd–4th decade of life	4%	Duration not defined/6 months	PAS[b]
Social phobia	1st–2nd decade of life	2.3%	Duration not defined/6 months	LSAS[c]
Specific phobia	1st decade of life	10.3%	Duration not defined/6 months	FSS-III[d]
Generalized anxiety disorder	1. and from 5. decade of life	2.2%	6 months/6 months	PSWQ[e]
Selective mutism	1st decade of life	<1%	Disorder is not defined/1 month	FEM[f]
Separation anxiety	1st decade of life	1.4%	Disorder is not defined/6 months	ASA-27[g]

[a] 12-month or point prevalence (literature: Jacobi et al. 2014; American Psychiatric Association 2013)
[b] Panic and Agoraphobia Scale (Bandelow 1995)
[c] Liebowitz Social Anxiety Scale (Heimberg et al. 1999)
[d] Fear Survey Schedule-III (Beck et al. 1998)
[e] Penn State Worry Questionnaire (Meyer et al. 1990)
[f] Selective Mutism Questionnaire (Steinhausen 2010)
[g] Adult Separation Anxiety-27 (Cyranowski et al. 2002)

mental, the blood-injection-injury and the situational type, as well as a residual category, into which common (e.g. arachnophobia or phobia of heights) and comparatively rarely occurring syndromes (e.g. fear of thunderstorms) can be classified. Finally, according to both classification systems, *generalized anxiety disorder (GAD)* is characterized by persistent and excessive worries and fears about one or more aspects of daily life. For example, one's own health, social or security-related situation or perspective or that of close relatives may be affected. Patients with GAD regularly suffer from somatic correlates of an increased level of mental stress (muscle tension, musculoskeletal pain, gastrointestinal complaints) as well as sleep disturbances, the latter being mainly due to worry-associated rumination. As part of the conceptualization of the DSM-5, two new anxiety disorders of adulthood have been defined, *selective mutism* and *separation anxiety*, which are already long established in the child and adolescent psychiatric field. Neither has yet been listed in the ICD; however, a corresponding addition is planned for ICD-11 (Kogan et al. 2016). Selective mutism is characterized by an inability to speak in specific situations where participation is socially expected. In this, it has a high comorbidity rate with social phobia, and it has been debated whether selective mutism represents an extreme variant of this very disorder (e.g., Yeganeh et al. 2003). An inappropriate fear of separation from close attachment figures (e.g., children, partners) characterizes the separation anxiety of adulthood. The psychophysical fear reaction manifests itself regularly in the case of spatial distance, but can already occur in the case of its expectation and is closely connected with worries regarding a vital endangerment of the reference persons. At present, however, there are only a few findings on the psychological and biological genesis of these "new anxiety disorders". Accordingly, they will only be mentioned in the following if a robust evidence base is available.

With regard to the time of first manifestation and frequency of occurrence of anxiety disorders, there are sometimes large differences between the individual disorders (important *epidemiological aspects* are shown in ◘ Table 13.2); however, women are affected by anxiety disorders up to twice as often as men. As *developmentally associated risk factors* for the development of anxiety disorders, studies in recent years have identified a number of sociobiographical aspects (e.g. divorce of parents, "overprotective" parental home or immigration), increased susceptibility to negative emotions ("neuroticism"), increased "anxiety sensitivity" (assumption that fear is harmful) and "behavioral inhibition" (excessive fear of the unknown) have been identified in particular (American Psychiatric Association 2013; Lahat et al. 2011). Anxiety disorders also frequently co-occur with other mental and physical illnesses. In this context, it has been shown that depression, dysthymia and alcohol dependence in particular, as well as pulmonary, gastrointestinal and cardiovascular disorders, are common *comorbidities* (Merikangas and Swanson 2010). With respect to depression in particular, the presence of shared biological and psychological risk factors is suggestive; however, often comorbidities may be a consequence of anxiety (e.g., alcohol use as an anxiety-relieving "self-medication") or the anxiety disorder may develop "reactively" in the context of increased stress from another disorder. Anxiety disorders should be *diagnosed* using standardised and validated diagnostic instruments. Established questionnaires for external and self-assessment are available for the individual anxiety disorders, and their use significantly increases diagnostic certainty and reliably measures the severity of symptoms (◘ Table 13.2).

However, before an anxiety disorder can be diagnosed, other causes of a pathological experience of anxiety must always be ruled out. These include, first of all, substance-induced anxiety, such as can be triggered by the use of caffeine, cocaine and amphetamines, but also by thyroid medication or asthma medication. In addition, other psychological causes, for example psychotically induced anxiety, as well as organic diseases (e.g. cardiac arrhythmia, thyroid or adrenal dysfunction) must also be taken into account, whereby the latter can lead "into anxiety" in particular by triggering anxiety-associated bodily symptoms. This so-called "differential diagnosis" is of great importance, not only in order to provide the patient with treatment appropriate to the cause, but also because corresponding factors can significantly complicate the treatment of anxiety disorders.

13.3 Therapy and Prognosis of Anxiety Disorders

Pharmacotherapy and psychotherapy are derived from the "diathesis-stress model" as the first-rate treatment methods for anxiety disorders. Through these, a reduction of the biological and developmentally associated sensitivity as well as a correction of the learning-associated psychological conditions should be achieved, resulting in a relearning, an improved stress tolerance and finally a reduction of symptoms.

In the context of pharmacotherapy, the German S3 treatment guideline for anxiety disorders (Bandelow et al. 2014) and numerous other international treatment guidelines (e.g. Katzman et al. 2014) recommend antidepressants from the groups of "selective serotonin (norepinephrine) reuptake inhibitors" (SS[N]RI) as drug treatment options of first choice. It has been shown for numerous agents of these substance classes that their use leads to an improvement in the respective disease-specific symptoms of panic disorder, agoraphobia, generalized anxiety disorder or social phobia (e.g. Plag and Ströhle 2012; Perna et al. 2016), and there are correspondingly extensive treatment approvals by the Federal Institute for Drugs and Medical Devices (BfArM). With regard to the drug treatment of specific phobia, there is no (convincing) evidence of efficacy to date; here, only psychotherapy has been recommended as a treatment method. The anxiolytic mechanism of SS(N)RI appears to be complex and is (still) less well researched than its antidepressant effect. The studies carried out to date can best be summarised as showing that the changes in serotonergic neurotransmission induced by SS(N)RI compensate for pathological hyper- or hypoactivity in various anxiety-associated brain areas, thereby (re)establishing homeostasis in the so-called "fear network" (see also ▶ Sect. 13.5). In addition to antidepressants, benzodiazepines are another relevant class of substances in the context of the treatment of anxiety disorders. In numerous clinical studies, they have been shown to be rapidly and reliably effective in patients with panic disorder, agoraphobia, generalized anxiety disorder and social phobia. Examples for this group of drugs are lorazepam, diazepam or clonazepam (Offidani et al. 2013). However, despite their effectiveness, benzodiazepines are obsolete in the medium- and long-term treatment of anxiety disorders due to their unfavourable spectrum of side effects (especially their high dependence potential) (see also Bandelow et al. 2014), and even their short-term use should be reserved exclusively for special situations (e.g., the acute "breakthrough" of a severe panic attack). Another potential pharmacotherapeutic approach is the modulation of glutamate and thus the reduction of the effect of the most excitatory neurotransmitter system in the central nervous system on the "fear network". Although a relatively large number of studies have been conducted in this regard in recent years, almost none of the agents developed in this context (e.g., glutamate receptor modulators) has yet reached clinical application (Plag et al. 2012). Only the substance pregabalin, which was originally developed for the treatment of epilepsy and which, as a modulator of voltage-dependent calcium channels, inhibits the release of glutamate and other excitatory neurotransmitters, has proven effective in the treatment of generalized anxiety disorder and has been approved and recommended accordingly (Plag and Ströhle 2012; Bandelow et al. 2014).

In the context of psychotherapy, cognitive behavioural therapy (CBT) is the method of first choice (Bandelow et al. 2014). In comparison with the other psychotherapy methods approved by the health insurance funds in Germany (psychodynamic psychotherapy and psychoanalytic therapy), there is the most evidence of effectiveness for CBT with regard to panic disorder, agoraphobia, generalized anxiety disorder, social phobia and specific phobia (Kaczkurkin and Foa 2015). In recent years, treatment manuals or guidelines have been developed for the individual anxiety disorders, each of which contains the "classic" elements of CBT, such as psychoeducation, cognitive restructuring and confrontation methods, taking into account the specific features of the disorder. This structure for the treatment of panic disorder with agoraphobia will be briefly described as an example: In the psychoeducational section, the patient is first informed about the function of (normal psychological) anxiety as well as about the distinction from and the development of pathological anxiety. Subsequently, the individual components of a panic attack or the agoraphobic-triggered fear reaction (especially the perception of agoraphobic environmental conditions and/or fear-associated body symptoms as well as their catastrophizing evaluation) as well as their mutual interaction in the sense of an "escalation" of the fear reaction ("vicious circle

of fear") are worked out. Within the framework of the creation of an individual disease model, this is then transferred to the patient's individual situation and fear-triggering or fear-maintaining perceptions, thoughts and behaviours (e.g. specific safety or avoidance behaviour) are first identified and finally changed. In this context, disorder-specific exposure therapy plays an essential role (see also Lang et al. 2012). Here, symptom provocation (seeking out fearful situations in agoraphobia or experimentally triggering panic-associated bodily symptoms such as palpitations or dizziness in panic disorder) initially triggers a fear response or panic attack. The (increasing) fear is then consciously endured with the deliberate elimination of any safety and avoidance behaviour until it finally spontaneously decreases. This "habituation" of fear gives the patient the "corrective experience" that the fear will subside on its own after a certain point and not continue to increase, as is usually feared. Particularly in patients with panic disorder with or without agoraphobia, as well as specific phobias, it has been repeatedly shown that the performance of exposure therapy alone was similarly effective to complete, disorder-specific CBT (e.g. Öst et al. 2004; Sanchez-Meca et al. 2010); these findings therefore clearly underline the importance of the real experience of fear reduction for therapeutic success and thus the prominent role of exposure treatment within the overall treatment plan.

Similar to psychopharmacotherapy, after the confirmation of the efficacy of CBT, there have been efforts to elucidate its underlying mechanisms. Especially in recent years, therefore, a relatively large number of studies have been conducted with the aim of objectifying the neurobiological effects of CBT or its specific components. In this context, it has been shown, particularly in the case of phobic disorders, that (exposure-based) CBT leads to a reduction in the activity of the amygdala, insula, thalamus and hippocampus, as well as to an increase in the activity of the orbitofrontal cortex (Galvao-de Almeida et al. 2013), and that it thus exerts similar effects on the "fear network" as serotonergic pharmacotherapy (▶ Sect. 13.5). Various psychotherapy methods of the so-called "third wave of behaviour therapy", which have been developed since the early 2000s on the basis of CBT, are also effective in the treatment of anxiety disorders. Compared to the "classical" disorder-oriented CBT programs, these procedures address so-called "neuromental functional areas" (e.g., metacognitions, cognitive [de]fusion, mindfulness) that are relevant in the pathogenesis of both anxiety disorders and other mental illnesses (▶ Chap. 5). Against this background, it has been shown so far within randomized-controlled or open studies that "acceptance and commitment therapy" (ACT), "metacognitive therapy" (MKT), and "mindfulness-based cognitive therapy" (MBCT) led to an improvement of the respective disorder-specific symptomatology in generalized anxiety disorder, social phobia, and panic disorder, respectively (e.g., Hayes-Skelton et al. 2013; Lakshmi et al. 2016; Wells et al. 2010). Accordingly, these procedures may become relevant as treatment options in patients for whom CT has not resulted in improvement of symptoms. Empirical evidence of efficacy also exists for depth-focused psychotherapy (TFP; usually referred to as "psychodynamic therapy" in international parlance) in the area of anxiety disorders; however, studies that addressed neurobiological bases of clinical effect are not yet available. Scientific papers that investigated TFP in patients with panic disorder, agoraphobia, generalized anxiety disorder, and social and specific phobia were able to demonstrate its efficacy against various active or non-active control conditions (Keefe et al. 2014). Compared to CBT, however, the studies conducted to date in this regard are (still) numerically clearly in the minority and are characterized by a high degree of heterogeneity in terms of content

and structure (especially due to a lack of manualization of TFP to date). In addition, direct comparisons between the two methods have repeatedly shown that TFP is not superior to CBT in the area of anxiety disorders with regard to its effectiveness, or that the respective disorder-specific anxiety symptoms could be reduced more by CBT (e.g. Beutel et al. 2013; Leichsenring et al. 2009, 2013). Against this background, depth psychology-based psychotherapy is recommended by the current national treatment guideline as a secondary psychotherapy method and should be used "if CBT has not proven effective, is not available or if there is a preference of the informed patient in this regard" (Bandelow et al. 2014). Unlike TFP, no studies exist to date for psychoanalytic therapy in relation to its clinical effect on anxiety disorders. Traditionally, the procedure-specific literature here is predominantly characterized by narratively written case reports of individual therapy courses and process-oriented questions, which is why psychoanalytic therapy has so far hardly been accessible to a scientific evaluation of its clinical effects. For this reason, its use is not currently recommended in the field of anxiety disorders.

With regard to the choice of treatment method (psychotherapy vs. pharmacotherapy), meta-analyses indicate that both methods develop a synergism and their combination is superior to a respective monotherapy (e.g. Cuijpers et al. 2014). An exception is specific phobia; here, only CBT is the method of choice. In clinical practice, however, psychotherapy can be carried out first, particularly in the case of mild to moderate symptom expression, and only from a moderate to severe burden or impairment of the patient should a combination treatment be considered initially. Compared with drug therapy, CBT has the advantage that it allows the patient to experience "self-efficacy" and teaches techniques that make it possible to maintain the therapeutic success independently even after the end of treatment. Clinically relevant prognostic factors predicting a particularly good or poor response to the respective therapy method have only been identified to date. However, individual study results indicate that, for example, the expression of "anxiety sensitivity", "neuroticism", heart rate variability and the activity of certain areas of the frontal brain could be predictors of the effectiveness of pharmaceutical or psychotherapy (e.g. Lueken et al. 2016).

Based on positive findings in the context of depression, the effect of physical activity in various anxiety disorders has also been studied relatively intensively in recent years. In this context, it has been shown in particular for panic disorder, GAD, and social phobia that aerobic training (e.g., running or circuit training) of moderate intensity, performed up to five times a week and lasting up to 30 min each, had moderately strong effects on the respective disorder-specific symptoms (Stubbs et al. 2017). Exercise-induced changes in various anxiety-associated psychological and biological parameters are discussed as moderators of anxiety-reducing efficacy, such as a reduction in anxiety sensitivity, an increase in self-efficacy, or an improvement in serotonergic neurotransmission (e.g., Asmundson et al. 2013). In any case, the study results to date underline the relevance of physical exercise as a complementary therapeutic option to established pharmacological and psychotherapeutic treatments and should stimulate further research in this field.

13.4   Disease-Specific Psychological Theories of Anxiety Disorders

Until the 1950s, conflict-centered psychoanalytic models of illness dominated the hypotheses for the development of anxiety disorders. Here, pathological anxiety was seen as the result of a "defense" symboli-

cally expressed within the framework of a conflict within the intrapsychic instances of the "ego", "id", and "superego". However, from the second half of the last century onwards, the importance of conditioning and learning processes was demonstrated, the relevance of which to individual anxiety disorders was well established empirically. These finally formed the basis for the development of disorder-specific disease models, which are presented below in an overview. In contrast to GAD, panic disorder and phobic disorders, no scientifically elaborated psychological models currently exist for selective mutism and separation anxiety in adulthood. Accordingly, these "new" anxiety disorders are not yet listed here.

13.4.1   Generalised Anxiety Disorder (GAD)

With regard to the psychological mechanisms of GAD, various concepts now exist, of which those with the highest evidence base to date will be presented here (a detailed overview is provided by Behar et al. 2009). The most established model of GAD is the Avoidance Model of Worry and GAD (AMW) by Thomas B. Borkovec, which was developed in the 1990s. The AMW postulates that "worrying" is a variant of an avoidance or safety behavior through which anxiety-associated imaginaries and bodily responses are suppressed. However, this leads to the psychological processing of anxiety-inducing stimuli being impeded or blocked, which in turn promotes the progression and chronification of the disorder. The cognitive-behavioural therapeutic approach of GAD is derived primarily from AMW, in which, in addition to cognitive elements and relaxation exercises, "worry exposure" plays a particularly important role. The aim of this is to achieve a habituation of the disorder-specific anxiety and associated bodily symptoms through confrontation with worry-related thoughts or contextual conditions (e.g. hospital, cemetery). Other well-studied pathogenetic concepts represent the Uncertainty Intolerance Model (IUM) and the Metacognitive Model (MCM), which were formulated primarily by Michel J. Dugas and Adrian Wells, respectively. The IUM emphasizes the role of *intolerance of uncertainty in* the disease process. According to this, new situations or those with unclear perspectives are highly stressful for GAD patients, which is why "worrying" should lead to an increase in control over future processes ("If I think about everything, nothing can happen to me"). Through this, worrying establishes itself as a dysfunctional problem-solving strategy and eventually leads to cognitive and situational avoidance, similar to AMW. Accordingly, addressing uncertainty intolerance has become an integral component of disorder-specific cognitive behavioral therapy, particularly in recent years (Van der Heiden et al. 2012). The MCM, in turn, highlights the importance of metacognitive processes through which patients enter a "vicious cycle" of worry. According to this, GAD patients develop worry (*type 2 worry*) in a multistep process in relation to the harmful effects of the actual topic-related worry (*type 1 worry*). This results not only in an indirect cognitive or situational avoidance behaviour towards worry-associated topics or situations (in order to directly prevent the occurrence of *type 1 worry*), but also in the belief that the worries combine to form an uncontrollable "chain reaction". Accordingly, metacognitive therapy (MCT) first attempts to reduce *type 2 worry* in order to enable patients to exit the escalating dynamic between the two forms of worry (Wells 2011).

13.4.2 **Panic Disorder**

In a psychophysiological model of panic disorder, a positive feedback loop is postu-

lated between physical symptoms, their association with danger and the development of panic attacks (Ehlers and Margraf 1989). Both internal and external stressors can lead to physical or cognitive changes, thus initiating the "vicious cycle" of the build-up process. Empirically, increased anxiety sensitivity is found in patients with panic disorder (Reiss and McNally 1985), so that panic attacks can also be triggered by interoceptive conditioning and the misinterpretation of physical symptoms. The basis of the developmental models of phobic disorders as well as agoraphobia is Mowrer's (1960) two-factor theory, which assumes classical conditioning in the development and operant conditioning in the maintenance of avoidance. Seligman (1971) assumed that certain stimulus-response associations are learned more easily because there is a biological or evolutionarily significant readiness (*preparedness*).

13.4.3 **Social Phobia**

According to Beck (1985), dysfunctional cognitive schemas are central to the development of mental illness. In the case of social phobia, these are in particular, for example, the evaluation of the self as incompetent or failing, the excessive weighting of the evaluation by others, the view that other people are always very critical in their evaluation, and perfectionist standards of evaluation of one's own behavior. According to Clark and Wells (1995), the maintenance of social anxiety occurs through a particular form of information processing in social situations with

1. increased self-awareness and distorted self-perception,
2. safety behaviour, and
3. problematic anticipatory and retrospective processing of social interactions.

Many patients also report socially "traumatising" experiences. Social competence deficits can play a role in the development, but can also be a consequence of the inhibition of socially adequate behavior by the fear.

13.4.4 Specific Phobias

Current research results show that in the development of specific phobias, immediate aversive learning experiences in relation to the phobic object or situation play a role in almost half of those affected. Beyond that, however, imitation learning and fear acquisition through communication with third parties ("I heard that…") are also important pathogenetic aspects, through which almost identical neuronal activations as in classical conditioning are triggered in patients (Hamm 2017). However, evidence from the literature suggests that learning theory and genetic factors play different roles in the development of different specific-phobic subtypes. Twin studies, for example, indicate that phobias of the situational type are more likely to be based on individual conditioning or learning processes, and animal-related fears and blood-injection-injury phobia show an increased familial—and thus probably also genetic—clustering (▶ Sect. 13.5, for a contribution to the discussion see also Fanselow 2018). Regardless of the type of phobia and its genesis, Michael S. Fanselow developed the *"threat-imminence model"* in the 1990s, which defines different psychophysical reaction patterns depending on the patient's spatial proximity to the phobic stimulus: First, in the context in which the phobic stimulus was previously experienced or in which it is suspected or expected on the basis of reports by others, a strongly increased vigilance (hypervigilance) sets in. This serves to quickly sensory the potential threat should it occur. If the feared object (e.g. the mouse or the syringe) is then perceived concretely or if there is an immediate approach to the situation perceived as dangerous (e.g. a viewing platform in the case of height phobia), a "movement rigidity" and

an attentional focus on the stimulus set in. On further approach to the phobic trigger, depending on the specific situation, there is finally an activation of the sympathetic nervous system or a vasovagal reaction, which can manifest itself in a panic-like reaction (phobia of the animal or situational type) or a syncope (often in the blood-injection-injury type or in dental fear).

13.5 Neurobiological Mechanisms of Fear

In recent decades, the number of studies dealing with the biological basis of pathological anxiety or anxiety disorders has increased steadily (Garcia 2017). Many findings, such as the interlocking of different conditioning processes during fear acquisition (see also ▶ Sect. 13.5.1), were initially collected in animal experiments and subsequently verified in humans, and currently often form the basis for psychotherapeutic and pharmacotherapeutic treatment approaches (Harro 2018). In the following description of the individual systems or alterations associated with anxiety disorders, it becomes clear at many points that there is a relatively high degree of overlap between the individual disorders in terms of findings and that the specific phenotypes "blur" into one another neurobiologically to a certain degree. In fact, this observation fits well with the efforts of the Research Domain Criteria (RDoC) of the National Institute of Mental Health (NIMH), presented in ▶ Sect. 9.5.1, to describe the biological basis of individual neurocognitive functional units and deficits independently of nosology and to assess their significance for individual psychiatric disorders. In the area of various anxiety disorders, numerous findings (e.g., at the level of functional brain anatomy or neurotransmitter systems) coincide well with the already defined units of analysis within the RDoC constructs of "fear" and "anxiety," which correspond to patterns of response to an active threat or potential danger, respectively. Thus, it seems possible that for the categorically defined anxiety disorders there is a relatively broad common biological basis on which the disorder-specific symptomatology of the individual clinical pictures develops on the basis of distinct "profiles" within the biological changes (▶ Sect. 13.5.3) as well as individual psychological and environmentally associated factors.

13.5.1 Learning

Basic emotions are innate behavioral patterns for communicating and regulating or coping with a situation. Fear is primarily associated with the behavioral inhibition system or the fight-flight system. The association of stimuli with fear is shaped by evolutionary preparatory learning (*preparedness* according to Seligman 1971). Going further, Öhman and Mineka (2001) distinguish a defense system against attackers in animal phobias, whereas a biological submission system in social phobias. Both systems can elicit a fear response. While specific phobias are more likely to result from conditioning to a concrete or circumscribed key stimulus (e.g., the spider), for anxiety disorders with longer-lasting fear, context seems to be more important as a learning-theoretical condition.

The following is a description of the forms of learning that are significant for our experience and behaviour and for the development, maintenance and treatment of anxiety disorders. The neurophysiological and molecular basis of learning has been described in particular in animal studies (e.g. Kandel and Hawkins 1992). Learning is the fundamental prerequisite for the adaptation of the organism to changing environmental conditions or demands on the individual. Associative learning processes are distinguished from non-associative learning processes. Non-associative learning processes include the orientation response (preparation for the reception of new information), habituation (familiarization with known information) and sensitization (increase in readiness to react). Pairing stimuli and responses results in associative learning processes such as classical and operant conditioning. In classical conditioning, a biologically relevant unconditioned stimulus/stimulus (UCS) elicits an unconditioned response (UCR). Repeated pairing of a neutral (conditioned) stimulus (CS) with the UCS causes the CS to elicit a conditioned response (CR) that corresponds to the UCR (e.g., Pavlov 1927). An often cited example of classical conditioning is Pavlov's dog, which is, so to speak, the "archetype" of this form of conditioning. This dog was given its food (UCS) over a long period of time exclusively in connection with the ringing (CS) of a bell. When the bell was finally rung without the simultaneous administration of a meal, the dog immediately developed an increased salivation (CR)—a physiological response previously associated exclusively with food intake (UCR). Classical conditioning can also occur through observational learning (model learning) (Askew and Field 2007). In operant conditioning (Skinner 1938; Thorndike 1911), the probability of occurrence of behaviour is altered by reinforcement. This can be positive reinforcement, e.g., a pleasant stimulus following a behavior, or negative reinforcement, i.e., the cessation or absence of an aversive stimulus. Negative reinforcement is particularly effective and resistant to extinction. Punishment is said to occur when the consequences of the behavior reduce the probability of occurrence. Repeated presentation of a stimulus (CS) without UCS prior to conditioning results in slower learning of classical conditioning or weaker elicited response (CS)—latent inhibition. Conditioned responses can also be triggered by stimuli that are similar to the conditioned stimuli, this is called stimulus generalization.

Patients with anxiety disorders may discriminate more poorly between conditioned fear stimuli and similar stimuli. This tendency to generalize may be a risk factor for the development of anxiety disorders (Lissek et al. 2005). Extinction occurs when the (conditioned) stimulus (CS) is presented repeatedly without UCS, and as a consequence the conditioned response is elicited less and less and then not at all. Conditioned responses are not extinguished by extinction; rather, learning other associations leads to the presentation of other responses. Recurrence of conditioned responses can occur through spontaneous recovery (presentation of CS), reinstatement (repeated presentation of UCS), or renewal (different context).

13.5.2 Structural and Functional Brain Anatomy

The biological system most frequently studied in the context of anxiety disorders is the so-called "*fear network*". The fear network is a functional network of different brain areas that are involved in the perception and processing of fear-associated stimuli as well as the triggering and control of the corresponding psychological and physiological reactions (Bandelow et al. 2016). A schematic representation of the fear network can be found in ◘ Fig. 13.2).

The "cascade" of processes within the fear network can be described as follows: If a sensory impression (visual, acoustic, olfactory, tactile, gustatory) is recognized as familiar or relevant (again) by the sensory association cortex, this is immediately, but rather "roughly", evaluated with regard to its potential dangerousness by the thalamus. If this is confirmed, this information is passed on very quickly to the amygdala, which is located deep in the temporal lobe area. The amygdala in turn, as the central "fear center", indirectly triggers the psychophysiological mechanisms of the fight-or-flight response via activation of the hypothalamus and brainstem, allowing the individual to respond to the threat in a timely manner. This very fast process, mediated via the so-called *low route*, is extremely important for the immediate reaction to danger, but at the same time it is very susceptible to being triggered due to its "reflexivity", and sensory impressions that are actually harmless can be biologically misjudged as dangerous. In order to avoid or correct a "false alarm", there is an alternative route, which takes almost twice as much

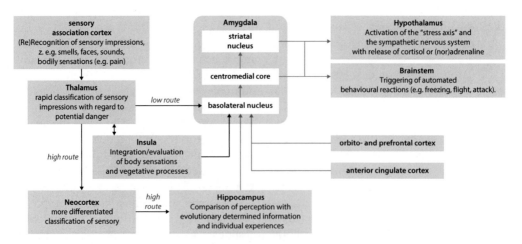

◘ **Fig. 13.2** The fear network. Green arrows: activating effect on the following structure; red arrows: inhibiting effect on the following structure. *High route, low route*: see text

time, but is much more precise. The *high route* leads first from the thalamus to parts of the neocortex and finally to the hippocampus, where perceptions are subjected to a more precise analysis or a comparison with evolutionarily determined information and personal experiences. On the basis of the additional information thus obtained, the hippocampus can either "give the all-clear" and inhibit the activation of the amygdala or confirm the danger and maintain the fear reaction. Amygdala activity is further regulated by (pre)frontal structures, with the prefrontal (PFC), orbitofrontal (OFC), and anterior cingulate cortex (ACC) in particular exerting an inhibitory function on the amygdala. In line also with previous findings incorporated into the RdoC matrix, a different pattern emerges for the relevance of each component of the fear network depending on the type of threat. The basolateral and centromedial nuclei of the amygdala, the hippocampus, the (pre)frontal cortical areas, the hypothalamus, and the insula play a particularly important role in the perception of and response to acute or persistent threat ("fear"), whereas the *bed nucleus of* the stria terminalis plays a more important role in "fear" as a correlate of an anticipated threat.

In the 1990s, the first studies were published that used magnetic resonance imaging (MRI) to investigate the morphology of various components of the fear network in anxiety disorders. In this context, volume increases or decreases in various structures (especially the amygdala) were repeatedly found in panic disorder, agoraphobia, GAD and various specific phobias compared with healthy subjects, the extent of which correlated in part with symptom expression (e.g. Irle et al. 2010; Lai 2011). However, in the following years, this work, which was purely related to structural anatomy, was increasingly complemented by studies that focused on the activity of the individual units of the fear network. With the aid of functional MRI (fMRI), differences in activations within the fear network between patients and healthy subjects are measured, which arise in response to the presentation of disorder-specific, usually visual or acoustic stimuli (e.g. images of crowds in agoraphobia or worry-associated words in GAD). In the meantime, numerous studies have also found differences in activity in different structures of the fear network for different anxiety disorders. Transdiagnostic evidence points in particular to increased activity in the (different core areas of the) amygdala, hippocampus, and insula, as well as decreased activity in frontal structures (especially the PFC or ACC), and correlations between the degree of activity abnormalities and the expression of clinical symptoms have also been observed here (Holzschneider and Mulert 2012). In recent years in particular, the data has been additionally enriched by so-called connectivity analyses. These can be used to determine the extent to which the neuronal connections between the individual parts of the fear network, and thus their activating or inhibitory influence, differ between patients with anxiety disorders and healthy subjects. The idea is that the function of a network is based not only on the function of its individual units, but also, perhaps especially, on their "communication" with each other. Against this background, it has been repeatedly shown that patients with anxiety disorders exhibit connectivity within the fear network that differs from that of healthy individuals. Corresponding studies often found different connectivity patterns depending on the anxiety disorder under investigation and reported both increased and decreased "coupling" between the same structures of the fear network. Overall, however, in the case of panic disorder, GAD and social phobia in particular, there was evidence of deviating connectivity both between the amygdala as the "fear centre" and structures of the frontal cortex (ACC, PFC and OFC) and other (para)limbic regions such as the thalamus, midbrain, insula or precuneus, as well as within the various core areas within the amygdala (Peterson et al. 2014).

Overall, the findings to date clearly indicate that both the activity and the connections within the fear network and within the internal structure of the amygdala are important in the pathogenesis of anxiety disorders. This observation is supported not least by further studies which show that disorder-specific pharmacotherapy and psychotherapy normalise pathological activation patterns. For example, it has been shown with the aid of imaging procedures in patients with various anxiety disorders that treatment with SS(N)RI for several weeks or with a respective disorder-adapted CBT reduces the activity of the amygdala, the striatum, the insula and the thalamus, but increases that of the ventromedial prefrontal cortex, and that these changes often correlate positively with the treatment effect. (e.g. Farb and Ratner 2014; Galvao-de Almeida et al. 2013; Phan et al. 2013).

13.5.3   Neurotransmitters, Neuropeptides and the Stress Axis

The role of monoaminergic neurotransmitter systems in the pathogenesis of anxiety disorders is derived primarily from the clinical and biological effects of antidepressants, which increase the availability of serotonin (5-HT) in the synaptic cleft (▶ Sects. 13.3 and 13.5.2). However, the existing data do not provide a sufficient basis for formulating a "monoamine deficiency hypothesis" by analogy with depression (e.g. Maron et al. 2012). The mechanism by which the SS(N)RI effect on (para)limbic activity is mediated is still largely unknown. However, it is assumed, among other things, that serotonergic drugs, depending on the respective brain area and receptor subtype (5-HT1A, 5-HT1B, 5-HT2C), can alter the availability of 5HT receptors on the cell surface or lead to their (de-)sensitization and thereby indirectly induce corresponding activity changes within the fear network (e.g., Spindelegger et al. 2009). This assumption is indeed supported by studies that used positron emission tomography (PET) to investigate the expression or binding capacity of 5-HT receptors. In particular, patients with panic disorder and social phobia repeatedly showed altered (increased or reduced) receptor availability or binding capacity in different areas of the fear network, such as the amygdala, thalamus, or hippocampus, compared to healthy subjects, some of which returned to normal after successful pharmacotherapy (e.g., Maron et al. 2004; Nash et al. 2008; van der Wee et al. 2008). Against this background, it is discussed that the respective "profile" of the functional changes of the 5-HT receptors differs depending on the respective anxiety disorder and thus could contribute to the discriminative formation of the respective phenotype (Graeff and Zangrossi Jr 2010). Following this assumption, the antidepressant used in each case would then compensate for the respective receptor-mediated serotonergic imbalance in the sense of a "serotonin modulator" and thereby lead transdiagnostically to an improvement in the respective disorder-specific symptomatology. A second neurotransmitter whose relevance is also derived from anxiolytic pharmacotherapy (this time from the benzodiazepines) is γ-aminobutyric acid (GABA). GABA is the most potent inhibitory neurotransmitter in the central nervous system and is critically involved in curbing the excitability of neurons. With respect to inhibitory neurotransmission, changes affecting both transmitter availability and receptor function have also been observed in various anxiety disorders (e.g., panic disorder and GAD) using different imaging techniques. In particular, reduced numbers or binding capacity of the GABA-A receptor within the fear network have been repeatedly found to correlate with symptom severity in some studies (Bandelow et al. 2016). In the context of the GABAergic system, the group of so-called "neuroactive steroids" (e.g., allopregnanolone or allotetrahydrodeoxycorticosterone) have also been the focus of scientific interest in the past. These are steroid hormones that bind as agonists to the GABA-A receptor and thereby increase its binding capacity. While altered concentrations of allopreganolone were already observed in patients with panic disorder some years ago (Ströhle et al. 2002), more recent studies have repeatedly provided evidence that separation anxiety in adulthood may also be associated with reduced steroid synthesis (e.g. Chelli et al. 2008).

In addition to the above-mentioned neurotransmitter systems, in recent decades various substances have also been investigated in the context of anxiety disorders, which are subsumed under the relatively large group of neuropeptides. Neuropeptides represent a heterogeneous group of hormones that are active in the central nervous system and can influence the activity of neurons either directly or by indirectly influencing other neurotransmitters, thereby having an anxiety-enhancing (e.g. cholecystokinin [CCK] and neurokinins [NK] such as substance P) or anxiety-reducing effect (e.g. neuropetide Y, atrial natriuretic peptide [ANP] or oxytocin). In the meantime, numerous studies have found associations between changes within these systems and the presence of various anxiety disorders (see also Bandelow et al. 2016; Plag et al. 2012). For example, patients with panic disorder and specific phobias showed altered NK-receptor binding capacity. Conversely, panic attacks triggered by CCK could be attenuated by ANP administration and a positive correlation between ANP concentration and an anxiolytic effect was observed (e.g. Ströhle et al. 2006). From the corresponding findings, different pharmacotherapeutic approaches were derived (e.g.,

NK1- or CCK-antagonists), but their anxiolytic effects were very heterogeneous in studies and have not reached clinical application to date. Based on the prior knowledge that oxytocin synthesized in the core areas of the hypothalamus affects, among others, systems associated with social cognition and emotional bonding, its effects have been studied, particularly in social phobia. While a reduction in amygdala activity, stronger connectivity between the amygdala and the ACC or PFC, and a reduction in the activity of the neurobiological stress response were indeed observed after (nasal) oxytocin application, the disorder-specific symptomatology in patients with social phobia could not be influenced by this, or only insufficiently (De Cagna et al. 2019). Accordingly, this approach did not reach "market maturity" either.

As in the case of other "stress-reactive diseases", many studies in recent decades have also focused on the activity of the pituitary-hypothalamic-adrenal (HPA) system (the so-called "stress axis") in anxiety disorders as a potentially disease-promoting and -maintaining factor. In contrast to depression or post-traumatic stress disorders, in which hyper- or hypocortisolism are now stably reproducible findings, the data situation for anxiety disorders is considerably more inconsistent (Plag et al. 2013). Against the background of a heterogeneous study situation, no reliable evidence for a general dysfunction of the HPA system has been found so far for panic disorder, GAD and phobic disorders, and individual findings of hypercortisolism could often be explained by the presence of comorbid depression. Also with regard to the responsiveness of the stress axis to various pharmacological or psychological provocation procedures, patients with anxiety disorders did not differ from healthy subjects in most cases.

Overall, there are clear indications of alterations in serotonergic and GABAergic neurotransmission in anxiety disorders, which appear to be caused mainly by changes at the level of transmitter-specific receptors and are addressed by the existing pharmacological treatment options. There are also findings in the field of neuropeptides, from which, however, no therapeutic options could (yet) be derived.

13.5.4 Genetics

Since the last third of the last century, epidemiological studies have pointed to a genetic component in the development of anxiety disorders. This finding, which was initially established in family studies, has been verified in the course of the last decades by twin studies with several thousand subjects in each case, in which a heritability of between 27% (social phobia) and more than 40% (specific phobias) was found for panic disorder, GAD, social phobia and specific phobias. However, in the group of specific phobias, large differences emerged in this regard depending on the specific-phobic subtype; while a heritability of almost 0% was observed for different situational phobias, their proportion was 33% and 45% for the blood-injection-injury type and animal-related fears, respectively (Shimada-Sugimoto et al. 2015; van Houtem et al. 2013). These findings from twin research have been continuously complemented by results from molecular genetic research for several years now. For panic disorder in particular, a relatively large number of linkage studies initially identified numerous chromosomal loci as possible carriers of disease-associated genes. However, most of these findings (each collected in rather small patient groups) could not be replicated, or only inconsistently, in follow-up studies, resulting in meta-analytically weak evidence for the involvement of chromosomes 1, 5, 15, and 16 in the pathogenesis of the disease (Webb et al. 2012). The situation is similar with regard to genetic association studies, which initially investigated specific genes (candidate genes) in isolation with regard to their relevance to disease in anxiety disorders, mainly on the basis of prior knowledge of pathogenetically relevant biological systems. Again, most work in this regard is available for panic disorder. Certain genetic variants were found in genes coding for structures of the monoaminergic (e.g. 5HTTLPR, HTR1A, HTR2A, COMT, MAO) or GABAergic (GABRB3, GABRA5) neurotransmission, the HPA system (e.g. CRHR1) or neurogenesis or neuroprotection (e.g. NPSR1) as well as being associated with the activity of structures of the fear network (e.g. the amygdala). However, even these (individual) findings have so far not withstood meta-analytical scrutiny (Bastiaansen et al. 2014; Howe et al. 2016). Numerically, significantly fewer studies have been conducted in this regard for GAD, social phobia, or specific phobias. Again, associations were found between the respective phenotype and polymorphisms in genes encoding the serotonin transporter or receptor, the enzymatic degradation of monoamines, and neurotrophic factors, or which are thought to influence the function of fear network structures. However, most of these findings remained singular or could not be confirmed by further studies either (e.g., Gottschalk and Domschke 2017; Smoller 2016). Meanwhile, at least one genome-wide association study (GWAS) is also available for many anxiety disorders, in which the entire human genome was searched for disease-associated SNPs (*single nucleotide polymorphisms*). More than five GWAS have now been published for panic disorder, again yielding inconsistent results. Repeatedly, however, two SNPs could be identified within a gene associated with the expression of a protein (132 D) in the neuronal cell membrane. The exact function of 132 D has not been fully elucidated; however, there is evidence that it may be involved in neuronal connectivity within the fear network (Erhardt et al. 2012). In patients with GAD and social phobia, individual GWAS each previously showed SNPs in a wide variety of genes encoding, for example, structures important for intercellular communication or indirectly associated with the metabolism of serotonin (e.g., Stein et al. 2017). In analogy to depression (► Chap. 12) and other mental disorders, a few studies have also investigated the extent to which polymorphisms in certain candidate genes may modulate sensitivity to certain developmental or environmentally associated factors (so-called "gene-environment interactions") and indirectly promote disease development. In this context, individual findings suggest that polymorphisms of the neuropeptide Y or serotonin transporter genes may represent risk factors for the invalidating influences of biographical or developmental stressors (e.g., natural disasters, childhood trauma, low social support) in the development of panic disorder, GAD, or social phobia. However, as in the other areas of genetic research related to anxiety disorder, these findings are also countered by negative results (e.g., Blaya et al. 2010) or have not (yet) been verified by follow-up studies.

In summary, it must be stated that despite clear epidemiological indications of the relevance of hereditary factors in the development of anxiety disorders, molecular genetic research has not yet succeeded in clearly identifying disease-associated polymorphisms for this group of disorders. This is certainly also due to the fact that, comparable to other diseases, a polygenetic pattern of inheritance can also be assumed for anxiety disorders, in which not a single gene, but the interaction of many genes has a risk-increasing effect. However, it is difficult to identify such constellations by looking at individual candidate genes. GWAS, which in principle represent an appropriate instrument for the investigation of polygenetic inheritance, must in turn include a large number of affected individuals in order to obtain reliable results. In our estimation, the GWAS conducted to date in the field of anx-

iety disorders may not yet have reached this size; while studies with more than one hundred thousand participants are available for depression, for example, the corresponding studies in this field have so far been limited to a few thousand individuals each.

However, it seems interesting that gene variants that were only inconsistently found in association with categorical anxiety disorders were in some cases significantly more strongly associated with their developmentally associated risk factors such as behavioral inhibition or neuroticism (Nagel et al. 2018; Smoller 2016). By analogy with findings in depression (► Chap. 12), and consistent with the rationale of RDoC, this suggests that in the context of stress-responsive mental illnesses, gene variants that are associated with a cross-entity increase in stress sensitivity rather than a specific phenotype are particularly important.

13.5.5 Epigenetics

"Epigenetics" primarily refers to the (temporary) influence of the activity of a gene by methylation or acetylation of the DNA/mRNA and the resulting influence on a phenotype. It is assumed that the methylation or acetylation state is indirectly influenced by various environmental influences (e.g. stress) and can thus influence gene expression within an individual as well as transgenerationally. Increased methylation corresponds in each case to reduced transcription of the affected genetic segment and thus to reduced formation or activity of the respective gene product. For the field of anxiety disorders, epigenetic findings are currently available for panic disorder as well as for social phobia (for an overview: Schiele and Domschke 2018). In panic disorder, lower methylation of genes responsible for *monoamine oxidase* (MAO) and *glutamate decarboxylase-1* (GAD-1) expression has been found (Domschke et al. 2012, 2013). These alterations lead to accelerated degradation of monoamines and reduced availability of GABA in the central nervous system, respectively, and thus may promote the development of pathological anxiety. In contrast, patients with social phobia showed hypomethylation of the *oxytocin gene*, the expression of which correlated positively with disease severity as well as cortisol secretion in the context of disorder-specific symptom provocation (Ziegler et al. 2015). This finding contrasts to a certain extent with the lack of clinical effect of oxytocin in this group of patients (▶ Sect. 13.5.3) and thus fits in with the findings, which are not always free of contradictions, that research into anxiety disorders has produced to date.

13.6   A Neurobiological Integrative Approach

The biopsychosocial pathogenesis as a transdiagnostically evident disease model of mental illnesses can be understood particularly well in the example of anxiety disorders. The interplay of biological, psychological and developmental factors increases the susceptibility to the development of the respective syndromes, which is finally "served" by stress as a central symptom-triggering and -maintaining factor. Although the findings are not (yet) universally consistent, the development of anxiety disorders clearly shows the importance of a multi-layered cascade of neurobiological systems, which are often functionally interrelated and ultimately increase the risk of disease: Genetic factors, for example, lead to changes in neurotransmitter systems (e.g. serotonergic neurotransmission), which in turn influence the function of anxiety-associated brain areas in such a way that symptoms develop more quickly under the influence of stress. Biological factors also appear to favor the formation of disease-promoting and stress-

sensitive personality traits such as neuroticism, which in turn may be a risk factor for the formation of dysfunctional conditioning processes and thus disorder-specific psychological learning models. The importance of biology, or the bidirectional link between biological and psychological factors in the development of illness, is underscored not least by the effects of current pharmacological and psychotherapeutic treatment options. Both the use of antidepressants and cognitive behavioral therapy can achieve corresponding changes in the clinical symptomatology and various neurobiological correlates of anxiety disorders; thus, a synergistic effect of both therapeutic options is also emerging at the biological level, which should be exploited in clinical practice (at least above a certain severity of anxiety). Furthermore, the field of epigenetics, which is still relatively under-researched in relation to anxiety disorders, gives rise to hopes for interesting findings in the future. In this context, it seems possible or even to be expected that "stress", which is still predominantly perceived as a purely psychological phenomenon, will become an even more integral part of neurobiological research and thus contribute to the further understanding of intraindividual and transgenerational biological changes in the field of anxiety disorders.

References

American Psychiatric Association (2013) Diagnostic and statistical manual of mental disorders, 5. Aufl. American Psychiatric Association, Arlington

Askew C, Field AP (2007) Vicarious learning and the development of fears in childhood. Behav Res Ther 45:2616–2627

Asmundson GJ, Fetzner MG, Deboer LB et al (2013) Let's get physical: a contemporary review of the anxiolytic effects of exercise for anxiety and its disorders. Depress Anxiety 30:362–373

Bandelow B (1995) Assessing the efficacy of treatments for panic disorder and agoraphobia. II. The panic and agoraphobia scale. Int Clin Psychopharmacol 10:73–81

Bandelow B, Wiltink J, Alpers GW et al. (2014) Deutsche S3-Leitlinie Behandlung von Angststörungen. www.awmf.org/leitlinien.htm

Bandelow B, Baldwin D, Abelli M et al (2016) Biological markers for anxiety disorders, OCD and PTSD—a consensus statement. Part I: Neuroimaging and genetics. World J Biol Psychiatry 17:321–365

Bastiaansen JA, Servaas MN, Marsman JB et al (2014) Filling the gap: relationship between the serotonin-transporter-linked polymorphic region and amygdala activation. Psychol Sci 25:2058–2066

Beck AT (1985) Theoretical perspectives on clinical anxiety. In: Tuma AH, Maser J (eds) Anxiety and the anxiety disorders. Erlbaum, Hillsdale, pp 183–196

Beck JG, Carmin CN, Henninger NJ (1998) The utility of the fear survey schedule-III: an extended replication. J Anxiety Disord 12:177–182

Behar E, DiMarco ID, Hekler EB et al (2009) Current theoretical models of generalized anxiety disorder (GAD): conceptual review and treatment implications. J Anxiety Disord 23:1011–1023

Beutel ME, Scheurich V, Knebel A et al (2013) Implementing panic-focused psychodynamic therapy into clinical practice. Can J Psychiatr 58:326–334

Bienvenu OJ, Onyike CU, Stein MB et al (2006) Agoraphobia in adults: incidence and longitudinal relationship with panic. Br J Psychiatry 188:432–438

Blaya C, Salum GA, Moorjani P et al (2010) Panic disorder and serotonergic genes (SLC6A4, HTR1A and HTR2A): association and interaction with childhood trauma and parenting. Neurosci Lett 485:11–15

Chelli B, Pini S, Abelli M et al (2008) Platelet 18 kDa translocator protein density is reduced in depressed patients with adult separation anxiety. Eur Neuropsychopharmacol 18:249–254

Clark DM, Wells A (1995) A cognitive model of social phobia. In: Heimberg RG, Liebowitz MR, Hope DA, Schneider FR (eds) Social phobia: diagnosis, assessment, treatment. Guildford, New York, pp 69–93

Cuijpers P, Sijbrandij M, Koole SL et al (2014) Adding psychotherapy to antidepressant medication in depression and anxiety disorders: a meta-analysis. World Psychiatry 13(1):56–67

Cyranowski JM, Shear M, Rucci P et al (2002) Adult separation anxiety: psychometric properties of a new structured clinical interview. J Psychiatr Res 36:77–86

De Cagna F, Fusar-Poli L, Damiani S et al (2019) The role of intranasal oxytocin in anxiety and depressive disorders: a systematic review of randomized controlled trials. Clin Psychopharmacol Neurosci 17:1–11

Dilling H, Mombour W, Schmidt MH et al (2015) Internationale Klassifikation psychischer Störungen: ICD-10 Kapitel V (F) klinisch-diagnostische Leitlinien, 10. Aufl. Hogrefe, Bern

Domschke K, Tidow N, Kuithan H et al (2012) Monoamine oxidase A gene DNA hypomethylation—a risk factor for panic disorder? Int J Neuropsychopharmacol 15:1217–1228

Domschke K, Tidow N, Schrempf M et al (2013) Epigenetic signature of panic disorder: a role of glutamate decarboxylase 1 (GAD1) DNA hypomethylation? Prog Neuro-Psychopharmacol Biol Psychiatry 46:189–196

Ehlers A, Margraf J (1989) The psychophysiological model of panic attacks. In: Emmelkamp PMG, Everaerd WTAM, Kraaimaat F, van Son MJM (eds) Fresh perspectives on anxiety disorders. Swets Zeitlinger, Amsterdam, p 1

Erhardt A, Akula N, Schumacher J et al (2012) Replication and meta-analysis of TMEM132D gene variants in panic disorder. Transl Psychiatry 2:e156

Fanselow MS (2018) The role of learning in threat Imminence and defensive behaviors. Curr Opin Behav Sci 24:44–49

Farb DH, Ratner MH (2014) Targeting the modulation of neural circuitry for the treatment of anxiety disorders. Pharmacol Rev 66:1002–1032

Galvao-de Almeida A, Araujo Filho GM, Berberian Ade A et al (2013) The impacts of cognitive-behavioral therapy on the treatment of phobic disorders measured by functional neuroimaging techniques: a systematic review. Rev Bras Psiquiatr 35:279–283

Garcia R (2017) Neurobiology of fear and specific phobias. Learn Mem 16:462–471

Gottschalk MG, Domschke K (2017) Genetics of generalized anxiety disorder and related traits. Dialogues Clin Neurosci 19:159–168

Graeff FG, Zangrossi H Jr (2010) The dual role of serotonin in defense and the mode of action of antidepressants on generalized anxiety and panic disorders. Cent Nerv Syst Agents Med Chem 10:207–217

Hamm AO (2017) Spezifische Phobien. Psych up2date 3:223–236

Harro J (2018) Animals, anxiety, and anxiety disorders: how to measure anxiety in rodents and why. Behav Brain Res 352:81–93

Hayes-Skelton SA, Roemer L, Orsillo SM (2013) A randomized clinical trial comparing an acceptance-based behavior therapy to applied relaxation for generalized anxiety disorder. J Consult Clin Psychol 81:761–773

Heimberg RG, Horner KJ, Juster HR et al (1999) Psychometric properties of the liebowitz social anxiety scale. Psychol Med 29:199–212

Holzschneider K, Mulert C (2012) Neuroimaging in anxiety disorders. Dialogues Clin Neurosci 13:453–461

Howe AS, Buttenschøn HN, Bani-Fatemi A et al (2016) Candidate genes in panic disorder: meta-analyses of 23 common variants in major anxiogenic pathways. Mol Psychiatry 21:665–679

Irle E, Ruhleder M, Lange C et al (2010) Reduced amygdalar and hippocampal size in adults with generalized social phobia. J Psychiatry Neurosci 35:126–131

Jacobi F, Höfler M, Siegert J et al (2014) Twelve-month prevalence, comorbidity and correlates of mental disorders in Germany: the mental health module of the German health interview and examination survey for adults (DEGS1-MH). Int J Methods Psychiatr Res 23:304–319

Kaczkurkin AN, Foa EB (2015) Cognitive-behavioral therapy for anxiety disorders: an update on the empirical evidence. Dialogues Clin Neurosci 17:337–346

Kandel ER, Hawkins RD (1992) Molekulare Grundlagen des Lernens. Spektrum der Wissenschaft 11:66–76

Katzman MA, Bleau P, Blier P et al (2014) Canadian clinical practice guidelines for the management of anxiety, posttraumatic stress and obsessive-compulsive disorders. BMC Psychiatry 14(1):S1

Keefe JR, McCarthy KS, Dinger U et al (2014) A meta-analytic review of psychodynamic therapies for anxiety disorders. Clin Psychol Rev 34:309–323

Kogan CS, Stein DJ, Maj M et al (2016) The classification of anxiety and fear-related disorders in the ICD-11. Depress Anxiety 33:1141–1154

Lahat A, Hong M, Fox NA (2011) Behavioural inhibition: is it a risk factor for anxiety? Int Rev Psychiatry 23:248–257

Lakshmi J, Sudhir PM, Sharma MP et al (2016) Effectiveness of metacognitive therapy in patients with social anxiety disorder: a pilot investigation. Indian J Psychol Med 38:466–471

Lai CH (2011) Gray matter deficits in panic disorder: a pilot study of meta-analysis. J Clin Psychopharmacol 31:287–293

Lang T, Helbig-Lang S, Westphal D et al (2012) Expositionsbasierte Therapie der Panikstörung mit Agoraphobie. Ein Behandlungsmanual. Hogrefe, Göttingen

Leichsenring F, Salzer S, Jaeger U et al (2009) Short-term psychodynamic psychotherapy and cognitive-behavioral therapy in generalized anxiety disorder: a randomized, controlled trial. Am J Psychiatry 166:875–881

Leichsenring F, Salzer S, Beutel ME et al (2013) Psychodynamic therapy and cognitive-behavioral therapy in social anxiety disorder: a multicenter randomized controlled trial. Am J Psychiatry 170:759–767

Lissek S, Powers AS, McClure EB et al (2005) Classical fear conditioning in the anxiety disorders: a meta-analysis. Behav Res Ther 43:1391–1424

Lueken U, Zierhut KC, Hahn T et al (2016) Neurobiological markers predicting treatment response in anxiety disorders: a systematic review and implications for clinical application. Neurosci Biobehav Rev 66:143–162

Maron E, Kuikka JT, Shlik J et al (2004) Reduced brain serotonin transporter binding in patients with panic disorder. Psychiatry Res 132:173–181

Maron E, Nutt D, Shlik J (2012) Neuroimaging of serotonin system in anxiety disorders. Curr Pharm Des 18:5699–5708

Merikangas KR, Swanson SA (2010) Comorbidity in anxiety disorders. Curr Top Behav Neurosci 2:37–59

Meyer TJ, Miller ML, Metzger RL et al (1990) Development and validation of the Penn State Worry Questionnaire. Behav Res Ther 28:487–495

Mowrer OH (1960) Learning theory and behavior. Wiley, New York

Nagel M, Jansen PR, Stringer S et al (2018) Meta-analysis of genome-wide association studies for neuroticism in 449,484 individuals identifies novel genetic loci and pathways. Nat Genet 50:920–927

Nash JR, Sargent PA, Rabiner EA et al (2008) Serotonin 5-HT1A receptor binding in people with panic disorder: positron emission tomography study. Br J Psychiatry 193:229–234

Öhman A, Mineka S (2001) Fears, phobias, and preparedness: toward an evolved module of fear and fear learning. Psychol Rev 108:483–522

Öst LG, Thulin U, Ramnero J (2004) Cognitive behavior therapy vs exposure in vivo in the treatment of panic disorder with agoraphobia (corrected from agrophobia). Behav Res Ther 42:1105–1127

Offidani E, Guidi J, Tomba E et al (2013) Efficacy and tolerability of benzodiazepines versus antidepressants in anxiety disorders: a systematic review and meta-analysis. Psychother Psychosom 82:355–362

Pavlov IP (1927) Conditional reflexes: an investigation of the physiological activity of the cerebral cortex. Oxford University Press, Oxford

Perna G, Alciati A, Riva A et al (2016) Long-term pharmacological treatments of anxiety disorders: an updated systematic review. Curr Psychiatry Rep 18:23

Peterson A, Thome J, Frewen P et al (2014) Resting-state neuroimaging studies: a new way of identifying differences and similarities among the anxiety disorders? Can J Psychiatr 59(6):294–300

13

Phan KL, Coccaro EF, Angstadt M et al (2013) Corticolimbic brain reactivity to social signals of threat before and after sertraline treatment in generalized social phobia. Biol Psychiatry 73:329–336

Plag J, Ströhle A (2012) Pharmakotherapie der Angststörungen. In: Gründer G, Benkert O (eds) Handbuch der psychiatrischen Pharmakotherapie, 2. Aufl. Springer, Heidelberg

Plag J, Hoyer J (2019) Die generalisierte Angststörung—ein update. Psych up2date 13:243–260

Plag J, Siegmund A, Ströhle A (2012) Neue Ansätze in der Angstbehandlung. In: Rupprecht R, Kellner M (eds) Angststörungen. Klinik, Forschung, Therapie. Kohlhammer, Stuttgart

Plag J, Schumacher S, Schmid U et al (2013) Baseline and acute changes in the HPA system in patients with anxiety disorders: current state of research. Neuropsychiatry 3(1–18):1

Reiss S, McNally RJ (1985) Expectancy model of fear. In: Reiss S, Bootzin RR (eds) Theoretical issues in behavior therapy. Academic, New York, pp 107–121

Sanchez-Meca J, Rosa-Alcazar AI, Marın-Martınez F et al (2010) Psychological treatment of panic disorder with or without agoraphobia: a meta-analysis. Clin Psychol Rev 30:37–50

Schiele MA, Domschke K (2018) Epigenetics at the crossroads between genes, environment and resilience in anxiety disorders. Genes Brain Behav 17:e12423

Seligman MEP (1971) Phobias and preparedness. Behav Ther 2:307–320

Shimada-Sugimoto M, Otowa T, Hettema JM (2015) Genetics of anxiety disorders: genetic epidemiological and molecular studies in humans. Psychiatry Clin Neurosci 69:388–401

Skinner BF (1938) The behavior of organisms: an experimental analysis. Appleton-Century, Oxford

Smoller JW (2016) The genetics of stress-related disorders: PTSD, depression, and anxiety disorders. Neuropsychopharmacology 41:297–319

Spindelegger C, Lanzenberger R, Wadsak W et al (2009) Influence of escitalopram treatment on 5-HAT 1A receptor binding in limbic regions in patients with anxiety disorders. Mol Psychiatry 14:1040–1050

Stein MB, Chen CY, Jain S et al (2017) Genetic risk variants for social anxiety. Am J Med Genet B Neuropsychiatr Genet 174:120–131

Steinhausen HC (2010) Fragebogen zur Erfassung des Elektiven Mutismus (FEM). In: Steinhausen HC (ed) Psychische Störungen bei Kindern und Jugendlichen, 7. Aufl. Elsevier, München, pp 558–560

Ströhle A, Romeo E, di Michele F et al (2002) GABA(A) receptor-modulating neuroactive steroid composition in patients with panic disorder before and during paroxetine treatment. Am J Psychiatry 159:145–147

Ströhle A, Feller C, Strasburger CJ et al (2006) Anxiety modulation by the heart? Aerobic exercise and atrial natriuretic peptide. Psychoneuroendocrinology 31:1127–1130

Stubbs B, Vancampfort D, Rosenbaum S et al (2017) An examination of the anxiolytic effects of exercise for people with anxiety and stress-related disorders: a meta-analysis. Psychiatry Res 249:102–108

Thorndike EL (1911) Animal intelligence. Experimental studies. Macmillan, New York

van der Heiden C, Muris P, Van der Molen HT (2012) Randomized controlled trial on the effectiveness of metacognitive therapy and tolerance-of-uncertainty therapy for generalized anxiety disorder. Behav Res Ther 50:100–109

van der Wee NJ, van Veen JF, Stevens H et al (2008) Increased serotonin and dopamine transporter binding in psychotropic medication naive patients with generalized social anxiety disorder shown by 123I-beta-(4-iodophenyl)-tropane SPECT. J Nucl Med 49:757–763

van Houtem CM, Laine ML, Boomsma DI et al (2013) A review and meta-analysis of the heritability of specific phobia subtypes and corresponding fears. J Anxiety Disord 27:379–388

Webb BT, Guo AY, Maher BS et al (2012) Meta-analyses of genome-wide linkage scans of anxiety-related phenotypes. Eur J Hum Genet 20:1078–1084

Wells A, Welford M, King P et al (2010) A pilot randomized trial of metacognitive therapy vs applied relaxation in the treatment of adults with generalized anxiety disorder. Behav Res Ther 48:429–434

Wells A (2011) Metakognitive Therapie bei Angststörungen und Depression. Beltz, Weinheim

Yeganeh R, Beidel DC, Turner SM et al (2003) Clinical distinctions between selective mutism and social phobia: an investigation of childhood psychopathology. J Am Acad Child Adolesc Psychiatry 42:1069–1075

Ziegler C, Dannlowski U, Bräuer D et al (2015) Oxytocin receptor gene methylation: converging multilevel evidence for a role in social anxiety. Neuropsychopharmacology 40:1528–1538

Neuropsychotherapy: Psychotherapy Methods and Their Effect

Nina Romanczuk-Seiferth

Contents

The purpose of this chapter is to provide an introduction to a neurobiological perspective on psychotherapy. For this purpose, the previous development in the interaction of neuroscience and psychotherapy (research) will be considered and neuroscientific methods in psychotherapy research will be introduced. In addition, the major schools of therapy, such as behavior therapy and psychoanalysis, as well as cross-school concepts are considered and related neurobiological findings are presented as examples. Following on from this, implications of the neurobiological perspective for everyday therapeutic practice are explained. Finally, possible future perspectives of this field of research are outlined.

Learning Objectives

The readership, after working through this chapter, should …

… have developed an understanding of the development and interplay of neuroscience and psychotherapy (research).

… have received an overview of the scientific methods in the field.

… have become acquainted with the basic findings on the neurobiological foundations of both school-oriented and cross-school concepts.

… have gained a practical impression of the possible implications of neuroscientific findings for the application of psychotherapeutic methods.

14.1 Neuropsychotherapy: An Integrative Perspective

The following section provides a brief introduction to the topic with a view to the developments to date in the interaction between neuroscience and psychotherapy (research).

While neuroscience uses scientific methods to study the human brain, its structure and functioning, psychotherapy research, as a field of clinical psychology, is concerned with the mode of action of psychotherapeutic procedures, their effectiveness and the further development of psychotherapeutic methods for clinical practice. Thus, while on the one hand humans are viewed and studied as predominantly biological beings, on the other hand humans and their experience and behavior are understood and attempted to be changed predominantly as psychological phenomena. While the cradle of neuroscience is rather to be found in biology and medicine, psychotherapy research originates primarily in empirical psychology, more precisely: the subfield of clinical psychology and psychotherapy. The paths of neuroscience and psychotherapy research would therefore not necessarily have to cross. What both disciplines have in common, however, is that they focus on people and their mental processes and consider their adaptation to a changing environment. Another common feature is that both fields represent multidisciplinary research areas, i.e. scientists from many different disciplines, such as psychology, medicine, biology, computer science or mathematics, are involved and work on common research questions. Thus, it is not surprising that neuroscientific methods have become indispensable to research in clinical psychology and psychotherapy and, conversely, that psychological theories and concepts often inform neuroscientific research questions.

As an integrative concept, the neurobiology of the psyche or the so-called neuropsychotherapy has therefore clearly gained in importance in recent decades. Neuropsychotherapy is concerned with the application of the findings of neuroscience, i.e. from different sub-areas such as neurobiology and neuropsychology, to psychotherapy. The latter is based on empirical findings from psychotherapy research.

The term neuropsychotherapy was coined primarily by the psychotherapist and psychotherapy researcher Klaus Grawe, who used it as the title of a book published in 2004 in which he argued for an integrative view of psychotherapy and for greater mutual promotion of psychotherapy research and the neurosciences (Grawe 2004).

This multidisciplinary and integrative approach is increasingly achieving important results in the field of the foundations of human learning and development processes. The most important example is findings on the neuronal plasticity of the human brain, which had long been considered clearly limited. Neuronal plasticity refers to the enormous capacity of the human brain to change throughout life through various neurobiological processes. This occurs in the context of a person's particular experiences and actions. Findings on lifelong neuronal plasticity therefore also underline the importance of therapeutic interventions for neurobiological changes in the brain of the person being treated. Gaining knowledge about brain changes that are associated with certain diseases or that are related to thera-

peutic interventions in the treatment process therefore represents a great hope for being able to help people with a mental illness as well and effectively as possible. The potential opportunities of looking at neuroscience and psychotherapy (research) from an integrative perspective therefore include several potentially useful aspects (▶ Box 14.1).

From a methodological perspective, findings from e.g. functional imaging add an intermediate step to the previous approach. If psychotherapy research usually wants to describe in more detail and understand how a specific symptom or function associated with a mental illness can be changed by a specific intervention, the neuroscience perspective adds the intermediate step of underlying changes in the brain (◼ Fig. 14.1).

In addition to the justified hopes, there are also justified doubts as to how great the added value of neurobiological findings for psychotherapy can actually be. This is mainly due to the fact that popular neurobiological studies, such as imaging studies, have so far mainly been able to make statements about overall samples. Predictions about clinically relevant parameters, such as diagnoses or expected treatment success, can hardly be made precisely or reliably for the individual with the methods available to date. Concrete translational implementa-

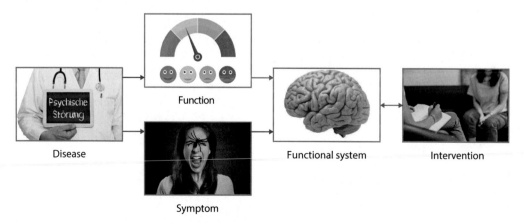

Function

Disease

Functional system

Intervention

Symptom

◼ **Fig. 14.1** Scientific investigation of the effect of therapeutic interventions. (© from left to right: DOC RABE Media, stas111, lassedesignen, danheighton, loreanto, each at ▶ stock.adobe.com)

tions into clinical practice are accordingly still thin on the ground (Won and Kim 2018). It remains to be seen which methodological innovations can remedy this situation (see also ▶ Sect. 14.8).

From a more philosophical point of view, it is equally interesting to note that the study of the brain and its functions in particular, as an explanation for phenomena such as "the mind", evokes astonishing scepticism. David Papineau, a contemporary philosopher, makes the comparison that we have little difficulty describing a clear, odorless, tasteless liquid by the word "water" and at the same time by the chemical structural formula "H_2O" (Papineau 1998). Both represent different approaches to the same phenomenon. It is incomparably more difficult for us humans to capture a feeling of "love" simultaneously by "neuronal activity in area X", even though this is also not a substitution but the addition of a further level of description.

> In this section we have taken a first look at the parallel development of neuroscience and psychotherapy (research). While both initially developed mainly separately from each other, these fields of research are now increasingly interacting with each other.

> **Box 14.1 Opportunities and Possibilities of an Integrative Perspective on Neuroscience and Psychotherapy (Research)**
> - Objectifying description of symptoms, which are by definition subjective in their clinical recording.
> - Broadening the understanding of the development and maintenance of mental illnesses.
> - Optimization of diagnostic procedures.
> - Improving the effectiveness of known interventions.

> - Further and new development of interventions based on neurobiological findings.
> - Expansion of the possibilities of differential indications.
> - Orientation of the therapist's behaviour towards the best possible brain-biological conditions for change.

14.2 Neuroscientific Methods in Psychotherapy Research

In order to be able to see and classify neurobiological psychotherapy research and its significance for the field as a whole, it is first important to explain and discuss the methodological peculiarities in this field of research by way of example. The aim is to create a basic understanding of the opportunities and limitations of the interlocking of neuroscience and psychotherapy research against the background of methodological conditions.

From archaeological finds of early Egypt it is known that man already performed surgical interventions on the human brain about 5000 years ago. Since then, man has tried to expand the understanding of this organ in our head with the possibilities available at the time. While in ancient times a strong interest in the brain and its processes prevailed, this clearly diminished over the Middle Ages, the understanding of the human brain continued to grow steadily in the further course of human history and reached a temporary peak with the great neurobiological findings of the nineteenth and twentieth centuries, such as the discovery of so-called long-term potentiation as a direct indication that experiences can change the activity of nerve cells. In recent decades and currently, this trend has continued, especially in the form of methodological innovations.

14.2.1 Methods of Neuroscientific Psychotherapy Research

The sub-fields of today's modern neuroscience, which deal with the changes in human experience and behaviour in the context of mental illness and their correlates in the brain, now use a wide variety of methods, with different strengths and weaknesses. These methods are also used to investigate the neurobiological basis of psychotherapeutic processes. Classical psychotherapy research focuses primarily on psychological and sociological methods to describe the changes in the individual or the social system, while neuroscience adds various neurobiological oriented approaches. Investigative methods such as electrophysiology, imaging, or so-called stimulation methods belong to macroscopic perspectives on the brain, while cell physiological, molecular biological, or genetic approaches, for example, aim to understand microscopic processes (◻ Fig. 14.2). Ideally, these different approaches complement each other to form an overall picture.

The available neuroscientific methods differ in their applicability to humans. Accordingly, mainly non-invasive methods are used in the field of neuroscientific psychotherapy research. These include various approaches which, with the help of technical devices, allow direct or indirect conclusions to be drawn about the activity of the brain.

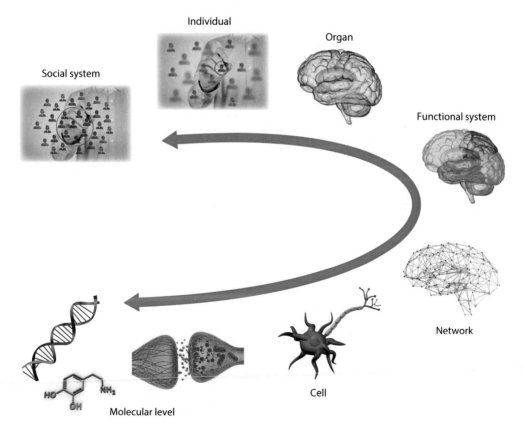

◻ **Fig. 14.2** Different methodological approaches to the (neuro-)scientific study of humans. (© clockwise from top left: Jakub Jirsák (2x), ag visual (2x), tampa-tra, freshidea, vector_factory, dana_c, Alexandr Mitiuc, each at ▶ stock.adobe.com)

One example are the so-called electro-physiological methods, which measure electrical potentials or fields and can map changes in brain activity with temporal accuracy. These include *electroencephalography* (EEG) and *magnetoencephalography* (MEG). The advantage of both methods is that they can measure the activity of large cell clusters with high temporal resolution, whereas the spatial resolution is moderate. Therefore, these methods are particularly suitable for the analysis of successive processing steps in the brain. Methods that are primarily used to identify lesions in the brain, such as *cranial computed tomography* (cCT) or *structural magnetic resonance imaging* (sMRI), play little role in psychotherapy research. In contrast, all imaging methods that can depict functional changes in the brain are particularly relevant, such as *functional magnetic resonance imaging* (fMRI). This method is non-invasive, has good to very good spatial resolution and is widely available in Western countries, but has relatively poor temporal resolution. This group also includes the less widely used methods of *positron emission tomography* (PET) and *single photon emission computed* tomography (SPECT), which can measure molecular correlates of neurotransmission in addition to activation. In addition, *near-infrared spectroscopy* (NIRS) has a good temporal resolution but can only record small parts of the brain. Stimulation methods that can temporarily and locally influence brain activity, such as *transcranial magnetic stimulation* (TMS), *transcranial direct current stimulation* (tDCS) or *deep brain stimulation* (DBS), are also increasingly being used. These methods, some of which are invasive, have so far played only a minimal role in research into psychotherapeutic processes. Newer methods or further developments of the available procedures, on the other hand, are very interesting for psychotherapy research, such as the increasing analysis of fMRI data on the basis of network models and with a view to functional couplings

(functional connectivities) as well as the recording of mechanisms on the basis of mathematical models in so-called *computational neuroscience*.

The use of microbiological or genetic approaches for questions in psychotherapy research has been rarer up to now. However, with regard to the modulation of peripheral physiological parameters by psychotherapy, such as stress-associated parameters, a correspondingly broad repertoire of peripheral physiological methods is used here.

Neuropharmacological approaches are also increasingly emerging in this field, i.e. the use of pharmacological modulation to augment psychotherapeutic effects. For example, the administration of D-CYCLOSERINE to accompany the cognitive behavioral treatment of anxiety disorders is being discussed in order to optimise the effect of psychotherapeutic interventions by modulating the glutamatergic system, but the data on this are still inconclusive (Ori et al. 2015).

14.2.2 Experimental Paradigms in Neuroscientific Psychotherapy Research

Different experimental paradigms are used to study the effects of psychotherapeutic procedures and specific interventions. Three main groups of paradigms can be distinguished:
- Paradigms with symptom provocation,
- paradigms to etiological models of a disease, and
- intervention paradigms.

Also used are so-called *resting-state* paradigms, i.e. measurements of brain activity under resting conditions.

■ **Paradigms with Symptom Provocation**
Paradigms with so-called symptom provocations examine the changes in the brain in

connection with psychotherapy by comparing the changes in brain activity during targeted stimulation of the brain with a symptom-triggering stimulus before and after therapy. Examples include the neural processing of images of fear-inducing objects or situations in phobias (Paquette et al. 2003), auditory presented scripts of traumatic experiences in post-traumatic stress disorder (Lindauer et al. 2008; Malejko et al. 2017), or addiction-related stimuli such as images, smells, or sounds in people with addictive disorders (Owens et al. 2017). As can be seen from the examples, for methodological reasons this approach is particularly suitable when external triggers are present for the symptoms to be stimulated. It is more demanding to implement studies that refer to internal triggers, e.g. to trigger ruminative loops in depressive patients by means of certain thoughts.

Overall, symptom provocation paradigms are now widely used. For example, one of the first imaging studies on this topic (Paquette et al. 2003) showed in people with arachnophobia that the neuronal effects of confrontation with video sequences of spiders had normalized after a successfully implemented cognitive behavioral therapy. While the patients reacted to the spider videos before therapy with increased activity, especially in the dorsolateral prefrontal cortex and the parahippocampal area, which the authors interpreted as memory as well as self-regulatory activity, this was no longer measurable after therapy.

- **Paradigms of Etiological Models of a Disease**

Experimental paradigms on etiological models of a disease use knowledge about the development of a disease to stimulate an etiologically relevant functional system of the brain and to record its altered activity in the context of psychotherapy. For example, this approach can be found in studies on the neural processing of emotional stimuli in depression (Ritchey et al. 2011), on the processing of reinforcers in the brain's motivational system in addiction disorders (Balodis et al. 2016), on action control and its neural correlates in obsessive-compulsive disorders (Nakao et al. 2005), or on changes in prefrontal activity related to working memory functions in schizophrenia (Kumari et al. 2009). The latter study was able to show, for example, that patients with schizophrenia improved significantly with cognitive behavioral therapy (vs. *treatment as usual*) and that this clinical improvement was associated with increased dorsolateral prefrontal activity in a working memory task (Kumari et al. 2009). This example also illustrates well that experimental paradigms for etiological models often do not focus on symptom-related phenomena, such as hallucinations or delusions in schizophrenia, but rather on cognitive or emotional functions, which are regarded as a kind of basic dysfunction within the respective disease.

- **Intervention Paradigms**

So-called *Intervention paradigms*, on the other hand, examine changes in brain function in a therapy-like situation using a targeted intervention. This is found, for example, in the study of effects of clinical hypnosis on pain perception (Rainville et al. 1999) or in the use of neurofeedback (Hohenfeld et al. 2017), i.e., real-time feedback on activity in a specific brain region and its modulation to be learned. As an example of the successful use of Intervention paradigms in neuroscience research, studies have now consistently shown that mindfulness-based interventions directly affect pain perception as well as associated neural activity, particularly in the anterior cingulate, insula, and dorsolateral prefrontal cortex (Bilevicius et al. 2016).

- **Paradigms for the *Resting* Activity of the Brain (*Resting-State*)**

The use of *resting-state paradigms* is based on the assumption that a typical response readiness of the brain can be described in the pattern of brain activity at rest, which can be used diagnostically (Kim and Yoon 2018) or whose change can be measured as a result of psychotherapy. For example, it could be shown by means of PET that metabolic changes in the resting-state activity of the brain occurred in people with depression after cognitive behavioral therapy, particularly in the hippocampus and dorsal cingulate (increase) and in various frontal regions (attenuation), and that this pattern differed from the changes that were observable in connection with the effects of antidepressant medication (Goldapple et al. 2004).

14.2.3 Research Designs in Neuroscientific Psychotherapy Research

Different research designs are used in neuroscientific psychotherapy research to answer the respective questions (◘ Fig. 14.3; Linden 2006).

The classic design comprises a pure pre-post comparison of neurobiological parameters, as shown here, for example, on the basis of an imaging examination before and after an applied psychotherapy (i.e.,

implementation of psychotherapy A in ◘ Fig. 14.3) and a comparison with regard to the respective response or non-response (Lueken et al. 2013). Differential study designs additionally compare psychotherapy with a possible other intervention, e.g. pharmacotherapy, or another form of psychotherapy (psychotherapy A vs. psychotherapy B). Designs that do not look at a form of psychotherapy as a whole, but rather seek to describe the neurobiological correlates of concrete partial aspects, effect factors, or interventions, are also becoming increasingly important. One of the reasons for this is that neuroscience has led to a growing realization that people and their brains are not structured analogously to diagnostic categories, cf. the so-called *Research Domain Criteria* (RDoC) initiative (Insel et al. 2010). Similarly, it is inconceivable to map a treatment according to a specific psychotherapeutic school, such as behavioral therapy or psychoanalytic psychotherapy, precisely in neurobiological terms, since a wide variety of elements and factors are interwoven in each case. Thus, it is only logical to limit questions about mechanisms and modes of action of psychotherapy in particular to concrete interventions or concepts and to consider them

14

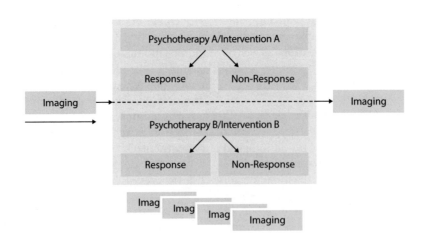

◘ **Fig. 14.3**   Research designs in neuroscientific psychotherapy research

differentially (Intervention A vs. Intervention B; Dörfel et al. 2014). It is also possible to identify neurobiological predictors of therapeutic success in order to use them in the sense of selective indication, e.g. testing of prefrontal functions under stress. Depending on whether a general prefrontal deficit or a situational/stress-related inhibition of functions is more likely to be evident, different intervention profiles can be defined. This type of investigation is also interesting for predicting the general responsiveness to a therapeutic intervention within a population. For example, a correlation between high responsiveness of the amygdala and ventral anterior cingulate to fearful faces before therapy and low response to cognitive behavioral therapy in posttraumatic stress disorder could be shown (◘ Fig. 14.4; Bryant et al. 2008).

Finally, with an appropriate study design, it is also possible to define neurobiological process variables and to collect these, such as via repeated imaging examinations parallel to the therapeutic process (◘ Fig. 14.3), which would make a process-oriented, adaptive indication possible on the basis of continuous feedback on the patient's brain functional changes in the process of therapy.

In this section we have looked at the methodological foundations of neuroscientific psychotherapy research. What people experience, what we perceive, feel, think and do can be scientifically described and investigated on different levels. Today's neuroscientific methods such as functional imaging offer the possibility to add further perspectives to our previous psychological and sociological perspectives and to watch the brain at work, so to speak. To answer the respective questions, neuroscientific psychotherapy research uses different experimental paradigms, including paradigms with symptom provocation, on etiological models of a disorder, intervention paradigms as well as resting-state measurements. These are used in the context of different study designs, which in turn allow various conclusions to be drawn, ranging from simple pre-post comparisons to the prediction of therapy effects.

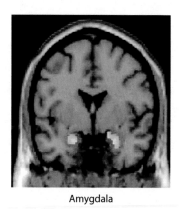

Amygdala

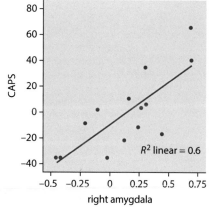

◘ **Fig. 14.4** Amygdalae responsiveness to anxious faces before therapy is related to response to cognitive behavioral therapy for post-traumatic stress disorder (CAPS = Clinician-Administered Post-Traumatic Stress Disorder Scale, scale scores post-pre treatment). (Adapted from Bryant et al. 2008)

14.3 The Different Schools of Psychotherapy

In order to be able to look more closely at the results of neuroscientific psychotherapy research using examples, in the following section we will briefly look at the special features of the various schools of psychotherapy (behavioral therapy, depth psychology-based therapy, psychoanalytic psychotherapy, systemic therapy and client-centered (Rogerian) psychotherapy) and exemplify the most important core assumptions and postulated mechanisms of action.

If we look at the various schools of psychotherapy, it quickly becomes clear that they have developed out of different traditions and along different paths. In modern times, the classical schools of therapy are increasingly converging, yet they differ in their postulated mechanisms of action, which will be briefly presented here in the form of a tabular overview (◘ Table 14.1).

> In this section we have provided an overview of the characteristics of the different schools of psychotherapy (behavioral therapy, depth psychology based therapy, psychoanalytic psychotherapy, systemic therapy and client-centered (Rogerian) psychotherapy).

◘ **Table 14.1** Comparison of the major schools and directions of psychotherapy with a focus on a simplified presentation of the core assumptions and postulated mechanisms of action

Therapy school	Core assumptions and postulated mechanisms of action
Behavioral therapy	• Experience and behaviour have been learned to a large extent and can therefore also be "unlearned" or relearned • Promote understanding of the mechanisms of a problem through problem and condition analysis • Consideration of cognitions, i.e. thought patterns, or emotional and motivational processes that contribute to the problem • Behavioral exercises and alternative experiences help establish new behaviors, confront problematic stimuli (e.g., phobic stimuli), reinforce positive behaviors, and mitigate problematic behaviors
Depth psychology-based psychotherapy	• Focus on conflicts and developmental disorders that arise or are reactivated in the patient's current life situation • Suspected causes of this lie in childhood and adolescence, i.e. looking to the past for better understanding of the present • Interpretation of current statements and conflicts as an expression of an already long-lasting inner conflict in order to uncover dysfunctional coping strategies that are related to the disorder and the current problematic situation • Establishment of more mature processing and manifestations of unconscious conflicts in current life circumstances, especially in current interpersonal relationships
Psychoanalytic psychotherapy	• Unconscious factors influence our thoughts, actions and feelings, which can lead to inner conflicts • Fundamental conflicts acquired in childhood affect the therapeutic interaction (transference/countertransference) • Patients stick (resistance) to the previously used coping strategy (defence mechanism), even if it has led to problems • Unconscious return of the patient to the point of development at which the disturbance arose (regression) in order to bring about a new beginning in the protection of the therapeutic relationship

14

◘ **Table 14.1** (continued)

Therapy school	Core assumptions and postulated mechanisms of action
Systemic therapy	• It is not the patient who is the cause of the problems, but rather a dysfunctional system that is in dysfunctional homeostasis as a result of the index person's symptoms • Relationship processes contribute to the development and maintenance of problems • Metaphorical techniques (e.g. sculptures) reveal family constellations and social relationships; circular questioning and paradoxical interventions break the current constellation and push change process of the whole system
Client-centered (Rogerian) psychotherapy	• Person-centred, humanistic approach, i.e. the focus is on the person, not the problem • A person's hidden or hitherto socially unaccepted abilities are brought to light and patients are enabled to find independent solutions to problems • The positively experienced therapy relationship forms the basis for the development of the patient's potential, i.e. the therapist meets patients with appreciation, without prejudice and with understanding for their life contexts, genuinely and without expert attitude

14.4 Neurobiological Findings on the Effects of the Major Psychotherapeutic Approaches

The following are some concrete examples of previous studies on the neurobiological correlates of school-focused psychotherapies.

Already Sigmund Freud, whom we know as the founder of psychoanalysis and who was a neurologist by training, formulated the assumption in one of his early writings (Freud 1885) that mental states of suffering arise on the basis of changes in brain states. However, he himself did not explicitly pursue these theses during his life. Even the majority of today's psychoanalysts do not necessarily adhere to the vision of a neurobiological description of their activity. At the same time, famous personalities, such as the Nobel Prize winner Eric Kandel, argued that psychoanalysis had to face up to neuroscientific insights if it still wanted to play a productive role in the fields of psychiatry and psychotherapy in the future (Kandel

1998). Comprehensive studies on the effect of psychoanalytic therapies on neurobiological parameters as well as on long-term courses are still largely lacking (Abbass et al. 2014). With a focus on the neurobiological correlates of psychodynamic psychotherapies, for example, a study was conducted on the treatment of chronically depressed patients by experienced psychoanalysts. In addition to extensive clinical diagnostics with psychodynamic psychometric procedures designed to measure, for example, unconscious central conflicts or attachment representations, study participants were treated with a frequency of ≥ 2 therapy sessions weekly and examined over a period of 15 months by means of fMRI and EEG at two measurement time points. Neural processing of personalized attachment-related stimuli was found to change over time in the patients, showing increased activity in regions such as the left anterior hippocampus, amygdala, subgenual cingulate, and prefrontal cortex before treatment, which decreased after treatment and was associated with symptomatic improvement (Buchheim et al. 2012a, b; ◘ Fig. 14.5).

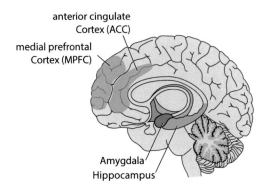

anterior cingulate
Cortex (ACC)

medial prefrontal
Cortex (MPFC)

Amygdala
Hippocampus

☑ **Fig. 14.5** Relevant brain areas in which a change was shown after psychoanalysis during the processing of attachment-relevant stimuli. (According to Buchheim et al. 2012a)

An investigation of the mechanisms of action of psychodynamic therapies is difficult, even with neuroscientific methods, due to the basic concept of Freudian theory, which defies empirical verifiability. Representatives of psychoanalysis argue that evidence of Freud's conceptualized unconscious has now also been provided outside neuropsychoanalysis (Doering and Ruhs 2015), e.g. in the form of the phenomenon of subliminal perception, i.e. perceptions of which we are not aware but which demonstrably exert an influence on our decisions (Berlin 2011). Thus, other neuroscientific findings can also be interpreted in terms of psychodynamic theory. Anderson et al. (2004), for example, saw hyperactivation of the dorsolateral prefrontal cortex combined with hypoactivation of the hippocampus in the process of "active forgetting" as analogous to repression, an important concept in psychoanalysis. Another important concept from the spectrum of methods of the psychodynamic school is free association. Indirectly, studies are also concerned with this, e.g. on the neurobiological basis of creativity and its connection with mental illness. Thus, it is argued that there might be a relationship between the resting-state activity of the brain (*default mode network*), which can be seen as an intrinsic function of

the brain during free thought wandering, and free association as a therapeutic method, which is related to changes in activity in different frontal areas and parieto-temporal regions. Also, creative people showed more activity in association cortices during creative tasks and at the same time showed a higher likelihood of suffering from mental illness (Vellante et al. 2018).

Numerous studies now exist on neurobiological changes in connection with the other major school of psychotherapy, behavioral therapy (Linden 2006; Barsaglini et al. 2014; Lueken and Hahn 2016). This is also due to the fact that behavioral therapy emerged from empirical psychology and is correspondingly science-savvy. The following are some interesting examples of studies in this area: On the question of the neurobiological correlates of successful behavioral therapy, a fivefold repeated fMRI examination of people with borderline disorder—accompanying the course of a 12-week dialectical behavior therapy (DBT)—was able to neurobiologically map, that the affective hyperexcitability existing before therapy decreased during the experimental emotion induction in the MRI in the sense of a reduction of activity in the right anterior cingulate, in temporal and parietal cortices as well as in the left insula (☑ Fig. 14.6; Schnell and Herpertz 2007).

Other researchers investigated, for example, at what point in the treatment of people with obsessive-compulsive disorder exposure therapy would be particularly promising, and also included neurobiological data. Using high-frequency logs of symptoms and repeated fMRI scans over an 8-week period, they were able to show in a single-case analysis that just before symptom reduction in the middle of treatment, the dynamic complexity of the data had increased, which can be interpreted as critical instability of the system that makes change processes more likely. The initially increased activity in the anterior cingulate when provoked with images that usually triggered compulsive

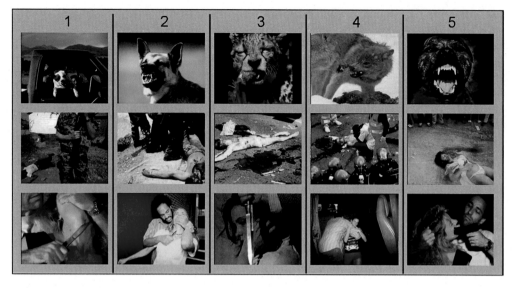

◘ Fig. 14.6 Examples of stimulus materials for the experimental induction of negative emotions. (From Schnell and Herpertz 2007)

actions in the patient (e.g., dirty laundry, etc.) decreased over the course of treatment (Schiepek et al. 2009).

The possibilities of differential comparisons (Furmark et al. 2002; Goldapple et al. 2004; Quidé et al. 2012; Barsaglini et al. 2014) are also interesting, i.e. to investigate in more detail whether psychotherapeutically mediated improvement of psychopathological symptoms is based on similar or distinct mechanisms as effective pharmacotherapy. For example, PET measurements have shown that cognitive behavioral treatment of social phobia, as well as treatment with the selective serotonin reuptake inhibitor citalopram, was associated with reduced responsiveness bilaterally in regions of the amygdala-hippocampal complex during symptom provocation (Furmark et al. 2002). It is possible that such similar effects are based on different neurobiological starting points. A common theory of differential mechanisms of action of different treatments for anxiety disorders and depression is that psychotherapy tends to address regulatory prefrontal functions (so-called "top-down" effects), whereas pharmacotherapy influences the functional networks involved by modulating the activity of subcortical structures (so-called "bottom-up" effects) (◘ Fig. 14.7).

Imaging data have also been successful in identifying predictors of long-term outcome in people with depression. Depressed patients with initially lower activity in the subgenual cingulate and high amygdala activity benefited most from 16 h of cognitive behavioral therapy (Siegle et al. 2006).

In other therapeutic disciplines, the comprehensive use of neurobiological methods is not yet widespread, but is being discussed, for example, in systemic therapy (Schwing 2013). As the social neurosciences continue to grow, insights into the neurobiology of mental illness and its treatment are likely to become increasingly interesting for systemic therapists in the future.

In this section we have looked at exemplary studies of the neurobiological correlates of psychodynamic and behavioural psychotherapies in particular, and their results.

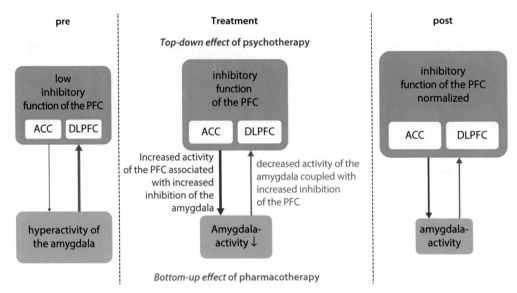

● **Fig. 14.7** Hypothetical model of brain functional changes in anxiety disorders and depression by different treatment approaches. Before treatment (pre), patients show overall increased amygdala activity and reduced inhibitory prefrontal activations. Different treatments lead to differential neurobiological effects in this model (treatment): psychotherapy (blue arrows) leads to an increase in inhibitory prefrontal activity, associated with reduced amygdala activity ("top-down" effect), whereas pharmacotherapy (red arrows) conditions a reduction in amygdala activity, associated with an increase in the inhibitory function of the prefrontal cortex on this structure ("bottom-up" effect). After treatment (post), the structures involved show a corresponding normalization of their function. *PFC* prefrontal cortex; *ACC* anterior cingulate cortex; *DLPFC* dorsolateral prefrontal cortex. (Adapted from Quidé et al. 2012)

14.5 Cross-School Methods and the Concept of General Mechanisms of Change in Psychotherapy

This section will briefly explain integrative concepts of psychotherapy using the example of Grawe's model of general mechanisms of change.

In addition to more school-focused approaches, there are now also cross-school approaches which assume common or similar factors of effect of successful psychotherapies regardless of the respective school. One of the most important models in this area is the concept of general mechanisms of change in psychotherapy according to Klaus Grawe (Grawe et al. 1994). According to this model, successful psychotherapy contains the general factors of effect regardless of the respective school (● Table 14.2):

1. Resource Activation
2. Motivational/meaning clarification
3. Problem actuation
4. Problem solving/coping
5. Therapeutic relationship

Psychotherapies therefore work by improving psychological consistency. High consistency, synonymous with good possibilities for satisfying basic psychological needs (orientation/control, pleasure gain/unpleasure avoidance, attachment, self-esteem enhancement/stabilisation), contributes to protection against the development of psychological disorders. These arise in particular when psychological inconsistencies exist over a long period of time, e.g. because successful satisfaction of needs is blocked or

◻ Table 14.2 General mechanisms of change in psychotherapy. (Grawe et al. 1994)

Impact factor	Explanation
Resource activation	Targeted use of the patient's existing strengths, abilities, interests, values, potentials, social relationships, etc.
Motivational/ meaning clarification	Promotion of insight into individual problematic patterns of experience and behaviour, problem contexts and conditional structures in connection with the illness
Problem actuation	Activation and making the problems in the therapy setting tangible, in order to make them accessible to a new experience in the therapy setting and in everyday life
Problem solving/ coping	Teaching skills for coping with problems and enabling coping experiences in the therapy setting and in everyday life
Therapeutic relationship	Establishment of a therapeutic interpersonal framework that enables a trusting engagement in the therapy and can serve as a basis for change processes

avoidance and approach goals motivationally interfere with each other. This leads to the assumption that therapeutically induced changes become possible when the therapy takes these psychological conditions into account and, if possible, also creates optimal conditions from a neurobiological perspective that enable and support the (re) construction of healthier processes and structures in the brain (Grawe 2000, 2004). For example, it results from this that therapy should be in the service of new, concrete and positive life experiences, i.e. the treatment of a problem should only contain limited interventions around the problem. Rather, it is essential for successful psychological change that important approach goals of the patients are in focus, i.e. that the therapist enables his or her patients as often as possible to experience perceptions that are relevant for the individual motivational goals and thus enable motivational learning, including the associated neuronal changes.

> This section introduced an integrative concept of general mechanisms of change in psychotherapy, which can positively influence the course of psychotherapy independent of the respective orientation of the therapist or the school of therapy.

14.6 Neurobiological Correlates of General Mechanisms of Change in Psychotherapy

The following section presents some exemplary studies on the neurobiological correlates of general mechanisms of change in psychotherapy, such as the therapeutic relationship, resource activation, problem actuation and problem coping.

The physician, neurophysiologist and psychoanalytically trained Eric Kandel, who received a Nobel Prize in 2000 for his work on signal transmission in the nervous system, emphasized early on that we were faced with the fascinating possibility of using imaging methods to monitor the success of psychotherapy in the future. He argued that in the course of psychotherapy there is a change in gene expression in the neurons which, if the treatment is successful, also leads to structural changes in the neurons involved (Kandel 1979).

Basically, this is no surprise when we consider that our brain is an organ specialized in adapting to its constantly changing environment. Psychotherapy is based on these abilities of the brain to adapt to the respective environment and systematically shapes this environment according to psychological principles so that a process of change—i.e. a process of adaptation—comes about. That we do well to understand our brain as an organ of adaptation is now shown by numerous studies. Our brain changes greatly in the course of our life on a macroscopic and microscopic level. As we age, gray matter density decreases nonlinearly in most brain regions, particularly in dorsal-frontal and parietal association cortices. However, white matter volume, that of fiber connections between brain centers, continues to increase into older adulthood (Sowell et al. 2004). Structural imaging studies have also shown that volume in specific brain regions changes locally in response to our experiences. For example, London taxi drivers have higher volumes in the hippocampus, a central structure of human memory (Maguire et al. 2000). People who are newly learning to juggle show an increase in volume in regions of the brain associated with processing complex, visually controlled movements after just 3 months (Draganski et al. 2004). Contrary to earlier doctrines, the brain appears to exhibit a high degree of local structural plasticity in adult humans as well, depending on the demands of their environment. Changes in the organization of functional systems of the brain play an important role in connection with our experiences, which manifest themselves in the form of altered neurotransmission of a cell or synapse, changes in connectivities or synchronicity in a network, etc., and can thus be detected in functional imaging or electrophysiological studies, at least in principle.

The brain can be conceived as a complex network that is in constant adaptation to its environment and may also exhibit features of self-organizing systems (Bassett et al. 2018; Takagi 2018).

Thus, if we assume that therapeutic changes are based on changes in the excitability of certain networks or systems in the brain, these changes take place primarily through processes such as priming/reinforcement, disuse/attenuation, and active inhibition. In this sense, the "extinction" of a fear response would consist in the establishment of an inhibitory or altered neural excitation pattern, whereas extinction in the sense of "forgetting" the relevant excitation patterns is unlikely to be relevant to psychotherapy. It is also likely that the brain can adapt to its environment differently under different conditions. Neuroscientific findings could therefore help to characterize the optimal conditions for change processes in more detail. The focus here is on neurotransmitters such as dopamine or serotonin (Heinz 2017). It is known that the activity of dopaminergic neurons plays an important role for the processing of reinforcers and thus motivational learning (Schultz 2016). Prosocial neuropeptides, such as oxytocin and vasopressin, which are released during sex or breastfeeding, for example, in turn promote dopamine release in the brain's motivational system, thus promoting memory formation and strengthening social relationships (Vargas-Martínez et al. 2014). Neurotransmitters such as serotonin and norepinephrine modulate a person's affective state (Young 2013). It is known that increased or prolonged release of stress hormones such as cortisol and adrenaline can lead to undesirable inhibition of neuronal activity and even to structural changes in the brain in the long term, such as in the context of traumatic experiences (O'Doherty et al. 2015).

At the latest since an improved investigation of emotional and motivational functions of the brain has been possible, the findings of neuroscience have become interesting for a deeper understanding of mechanisms of action in psychotherapies. If we follow Grawe's model of general mechanisms of change, then first of all there is the therapeutic relationship. On the one hand, it represents the basis for effective therapeutic action. On the other hand, the experience of a bond with other people also corresponds to one of our basic psychological needs. Neurobiological research has already shown that, analogous to the different, subjectively perceived qualities of interpersonal relationships, distinct activation patterns can also be mapped in the brain (Zeki 2007), e.g. in maternal vs. romantic love (Bartels and Zeki 2004), mediated, among other things, by differences in dopaminergic neurotransmission (Takahashi et al. 2015). The brain biological correlates of the sensation of unconditional love have also been considered (Beauregard et al. 2009), which is closest to the characteristic of unconditional appreciation in the context of the therapeutic relationship. Although the significance of unfavourable attachment experiences for the development of various mental illnesses is widely undisputed (Bora et al. 2009), detailed studies on the neurobiological correlates of a helpful therapeutic relationship are still lacking.

With regard to the general change factor of resource activation, a number of neurobiological studies can be cited that have examined the effects of approach behaviour, for example. This effect factor is closely related to the basic psychological need for pleasure/unpleasure avoidance. The well-documented negative influence of stress on learning and memory as well as associated neuronal processes (Schwabe 2017) suggests that people in states of subjective well-being are particularly willing and able to learn. The individually optimal conditions of a person's performance lie between respective under- or overload, cf. Yerkes-Dodson law (Yerkes and Dodson 1908). This means that the therapist's first task, also from a neuro-

biological perspective, is to create the most favourable initial conditions possible by understanding the patient's individual preconditions and motivational situation and supporting him or her in the development of his or her potential or enabling needs to be satisfied within the framework of the therapeutic relationship (cf. managing relationships motive-/need-oriented; Stucki and Grawe 2007). Only then would the conditions be optimal from a neurobiological point of view for a process of change. At the same time, these new, so-called corrective experiences within the framework of the therapeutic relationship will lead to an altered neuronal signature, which in turn will condition other expectations and arousal readiness of the brain in future situations. A whole branch of research investigates and substantiates various priming effects, i.e. context effects, and their neurophysiological correlates, both in the basal perceptual psychological domain, e.g. in subliminal priming (Elgendi et al. 2018) as well as with social psychological relevance (Molden 2014). However, the application of neurobiological methods to the psychotherapeutic context is still largely lacking here.

Neurobiological changes of therapeutic interventions that can be linked to the effect factor of motivational clarification could be considered, for example, in studies on the brain functional effects of psychoeducation. Although a positive effect of neurobiologically informed psychoeducation has been documented, especially on a reduction of stigma and help-seeking behavior of patients (Livingston et al. 2012; Han and Chen 2014), a neurobiological understanding of such interventions is still lacking.

With regard to the neurobiological correlates of change factors such as problem actuation and problem solving, it can be said that studies on the processing depth of emotions or on emotion induction on the basis of individual, i.e. personalized, stimuli as well as autobiographical information

(Cabeza and St Jacques 2007) are particularly interesting here, since these bring with them a corresponding emotional depth, liveliness and self-referentiality, which is also reflected in the underlying activation patterns. Commonly used in psychotherapies in this regard are techniques such as affect bridge work or imaginative overwriting (Holmes et al. 2007). In particular, the corresponding coping experience appears relevant for the satisfaction of the basic psychological needs for orientation/control and self-esteem stabilization/increase. In this context, studies on the neurobiological correlates of therapeutic interventions in patients with personality disorders from the so-called cluster B are also interesting. In disorders of this type, such as the so-called narcissistic or borderline personality disorder, the violation of the need for integrity of the self as well as self-esteem plays an important role in life-biographical development. On the psychological level, this manifests itself, for example, in changes in the competencies for affect regulation in interpersonally relevant situations, which therefore represents an important part of a behaviorally oriented psychotherapy for these clinical pictures. For example, in a functional imaging study of the neuronal processing of emotional stimuli, borderline patients initially showed increased responsiveness of the amygdala and insula, as well as increased

orbitofrontal and insular activation compared to healthy individuals during attempts at conscious affect regulation (Schulze et al. 2011). Furthermore, it could be shown that these neuronal abnormalities regress during affect regulation after DBT therapy (Schmitt et al. 2016). In line with this, an increased responsiveness to negative social information in patients with borderline personality disorder was also associated with increased amygdala activity, which could be improved under intranasal therapy with oxytocin (Bertsch et al. 2013). Similarly, for people with narcissistic personality disorder, imaging techniques show that this patient group is particularly sensitive to experiences of exclusion, as evidenced by an attenuated response in regions of the so-called social pain network (anterior insula, dorsal and subgenual anterior cingulate) (◘ Fig. 14.8; Cascio et al. 2015).

In this section, we have embarked on a search for neurobiological studies that allow us to better understand the general mechanisms of change in psychotherapy, such as the therapeutic relationship, resource activation, problem actuation, and problem coping experience. Despite interesting approaches of some studies in this field, further research on affective-motivational factors of therapy and their

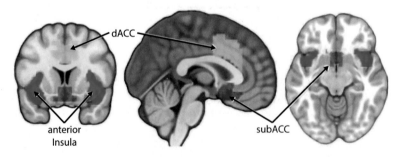

◘ **Fig. 14.8** Illustration of the brain anatomical structures—anterior insula, dorsal (dACC) and subgenual (subACC) anterior cingulate—which are relevant for the so-called social pain network and which appear to be significant, among other things, in experiences of exclusion. (Adapted from Cascio et al. 2015)

neurobiological correlates is necessary to further reveal the mechanisms of change in the therapeutic process and to create the best possible conditions for change also at the neuronal level.

Box 14.2 Implications of a Neurobiological Perspective on Mental Illness for Prac tice
- Complementation of a bio-psycho-social overall perspective.
- Health policy consequences, e.g. due to the extended objectifiability of psychotherapy effects.
- Adaptation of communication with patients.
- Influence on patients' and therapists' expectations of the therapeutic process.
- Promoting the destigmatization of mental illnesses.
- Influencing help-seeking behaviour.
- Stimulation of new approaches, especially for older patients.

14.7 Implications of Neurobiological Findings for the Development of Psychotherapeutic Approaches

In this section, we will present examples of the possible implications of neurobiological research for everyday therapeutic practice. For this purpose, we will look in particular at interventions that take into account and seek to induce changes in the motivational system in the context of mental illnesses such as addiction.

As fascinating and interesting as a more comprehensive knowledge of neurobiological correlates of successful psychotherapies, individual interventions or relevant mecha-

nisms of change may be, this knowledge is only relevant from a therapeutic point of view if it can be translated into therapeutic action and thus serves to improve the care and treatment of people with mental illness. The neurosciences in particular, whose methods are often time-consuming and cost-intensive and in some cases invasive, must accept the question of what added value they provide for clinical practice, what benefit they have for society as a whole and for patients in particular (Walter et al. 2009).

Let us therefore first consider what significance an increasing collaboration between the neurosciences and psychotherapy research could have for society and the development of the field (▶ Box 14.2), before turning to concrete implications for therapists.

Opening up psychotherapy research to neuroscientific methods first of all complements the bio-psycho-social overall concept of the development of mental illnesses that predominates in medicine. Due to the extended objectifiability of psychotherapy effects or as an element of quality assurance, neuroscientific findings could also contribute to a strengthening of the argumentation for a more consistent provision of necessary resources for the psychotherapy of people with mental illnesses at the level of health policy. At the level of therapist or patient behaviour, opportunities lie in adapting communication between therapists and patients, e.g. in the sense of demystifying psychotherapy and promoting transparent information about the underlying mechanisms. Moreover, communication about the neurobiological basis of psychotherapeutic interventions is certainly directly beneficial for patients with a predominantly biological understanding of illness, which is most likely to be due to a change in patients' expectations of effects in the therapeutic process. The dissemination of knowledge about the neurobiological basis of mental illness and therapeutic interventions can also promote the destigmatization of mental illness and thus also have a positive effect on help-seeking behavior (Livingston et al. 2012; Han and Chen 2014). At the same time, it should be noted that a biological perspective on mental illness may also contribute to increased stigmatization by society under certain conditions, e.g., for people perceived as dangerous (Müller and Heinz 2013; Andersson and Harkness 2018). Last but not least, neurobiological findings, such as those on the lifelong high neuronal plasticity of humans, could lead to a further stimulation of therapeutic approaches for people in adulthood or for older patients in particular.

But what does all this mean for professionals in the help system and especially for therapists? As stated before, therapeutic conversations in familiar ways do not lead to long-term changes in the brain. From a neurobiological perspective, this means that the analysis of problems is on the one hand relevant, for example in the form of clinical or functional diagnostics, but on the other hand is only productive if it serves the preparation and targeted implementation of change interventions, i.e. enables new, alternative experiences of the person being treated. In order to achieve changes in experience and behavior, repeated, sustained activations of the specific networks are required, i.e., from a neurobiological perspective, it is helpful to direct the focus to thoughts, feelings, and behaviors that represent an approximation of the therapy goals, such as in the presentation of target states in behavior therapy or the so-called wonder question in systemic therapy. These must be relevant to motivational goals of the patient, otherwise it cannot be assumed that the self-referential network or the motivational network in the brain is activated. The latter, with the attribution of significance, ensures an increased attentional capacity for associated content, paves the way for behavioural impulses in this direction and promotes the memory consolidation of associated experiences (Schultz 2016). Thus, a simultaneity of processing problem constellations and

activating approach goals in psychotherapy should be aimed at in order to stimulate a reconstruction of underlying brain biological systems. Finally, it should be noted that this important question of the concrete usability of the results from the neurosciences for everyday therapeutic practice is increasingly leading to the emergence of new works with a primarily therapeutic target group in addition to corresponding new textbooks and reference books (Eßing 2015).

We take a closer look at the concrete findings and their potential for implementation in practice for the research area of addictions as an example (Romanczuk-Seiferth 2017; cf. also ▶ Chap. 10): In the course of the development of an addiction, an initially positively experienced substance use becomes a habitual, automatic behaviour (Everitt and Robbins 2005; Seiferth and Heinz 2010). It is now well established from studies on the neurobiology of addictive disorders that the learning processes associated with the development of an addictive disorder are accompanied by neurofunctional changes in reward sensitivity and cognitive control. In addicted patients, it has been documented that there is an increased sensitivity of the brain's motivational system, the so-called meso-cortico-limbic reward system, to all addiction-associated contextual stimuli (Schacht et al. 2013), associated with a shift in attentional focus to addiction-relevant stimuli in the environment (Hester and Luijten 2014). At the same time, people suffering from addiction experience a hyposensitivity of the brain, i.e. a reduced responsiveness, especially in the so-called ventral striatum as a central region of the mesolimbic reward system (Wrase et al. 2007; Romanczuk-Seiferth et al. 2015), in the neuronal processing of common reinforcing stimuli, i.e. delicious food, sexuality, social contact, etc.

Furthermore, it is known that in the context of addiction there is also an altered neuronal activation in goal-directed behaviors

(Sebold et al. 2017) or in the cognitive control of behavioral impulses (Bickel et al. 2012). At the same time, patients find it more difficult to adapt their behaviour to changing reward contingencies, i.e. to relearn, which is associated with reduced connectivity between the ventral striatum and the prefrontal cortex (Park et al. 2010).

In principle, the communication of neurobiological knowledge, such as in the context of psychoeducation, therefore appears to be helpful for the treatment of addiction disorders, as it has been shown that shame and (self-)stigmatization are relevant barriers to the initiation of treatment (Wallhed Finn et al. 2014). It was also found that training of medical staff on this topic can contribute to a reduction of stigmatisation of people with addictions (Livingston et al. 2012). In addition, the findings on the impairment of relearning in addiction disorders, for example, can be communicated to patients in a relieving manner and thus serve to maintain motivation throughout the course of treatment.

However, direct therapeutic approaches can also be derived from the neurobiological findings on addictive disorders outlined above. One obvious approach is to reduce the motivational stimulus emanating from a substance or associated stimuli in the environment, as is found in clinical practice in strategies for stimulus control. However, the aim here is not to directly reduce the motivational stimulus, but to reduce the probability of occurrence of addiction-relevant stimuli, i.e. with a high motivational stimulus. It would therefore be interesting to consider newer therapeutic approaches that allow direct modulation of motivational stimuli. Similarly interesting are approaches that aim to weaken automated addictive behaviour. For this purpose, for example, a training was developed that weakens automatic approach tendencies to addiction-associated stimuli. This has been shown to be effective for reducing consumption or maintaining abstinence in smokers and alco-

hol addicted subjects (Kakoschke et al. 2017). On a neurobiological level, this can also be described, patients showed a normalization of the neuronal response to drug-associated stimuli in frontal and limbic areas through such training (Wiers et al. 2015).

Also relevant are measures to strengthen the executive control of addicts, as is already clinically common in rejection training or social skills training. Neurobiologically based therapy components in this sense are also methods that support a weakening of automated addictive behaviour, such as the use of observation tasks, diaries, behavioural analyses and traffic light systems to assess critical situations.

The aforementioned neurostimulation is sometimes used to modulate impulsivity or impulsive decision-making behavior in the context of addictive disorders. It could be shown that after stimulation of the dorsolateral prefrontal cortex via transcranial magnetic stimulation (TMS) in nicotine dependence, a higher ability to postpone reward as an aspect of (lower) impulsivity was detectable. However, this effect did not generalize to reduced cigarette consumption after treatment, so such treatments can so far only be helpful as an adjunct to demonstrably effective psychotherapy, such as cognitive behavioral therapy (Sheffer et al. 2013).

Particularly in view of the comprehensive adaptive abilities of our brain, therapy methods that focus on re-learning in connection with previous substance use and associated behaviour also appear to be useful and helpful, cf. application of exposure training in the field of alcohol/drugs.

Directly derivable from neurobiological findings is also the importance of therapy methods that aim at strengthening protective factors and reactivating the processing of alternative reinforcers, such as contact with people, sensual experiences, etc., as is sometimes found in therapeutic practice in so-called euthymic therapies or pleasure training. By means of exercises for perception with all five senses, the ability to enjoy is (re)strengthened and the brain is sensitized for the processing of these impressions—as an alternative to addictive behavior. It has been shown, for example, that when the motivational system is highly responsive to positive emotional stimuli, such as pictures of laughing people, etc., fewer severe relapses occur in the context of alcohol addiction (Heinz et al. 2007).

> In this section we have looked in more detail at the possible implications of neurobiological research for everyday therapeutic practice. To this end, we have taken a closer look at previous findings on the neurobiology of addictive disorders and discussed their possible significance for therapeutic action in the sense of neurobiologically informed psychotherapies.

14.8 Future Perspectives of Integrative Neuropsychotherapy

The following section will provide a brief concluding explanation of further relevant research perspectives and a general outlook.

For a better understanding of the development of severe mental illnesses, such as schizophrenia, great hopes have been and still are placed in the further development of genetic research. The combination of genetic information with neurobiological characteristics, as is the case with so-called *imaging genetics*, increases the probability of describing relevant endophenotypes of a disease and may continue to play a role as a research branch of human neuroscience in the future (Arslan 2018).

The identification of potential biomarkers of mental illness is of particular interest in terms of their usefulness for the future if they can be used to make predictions about the effectiveness of treatments or their sustainability. While the prediction of relapse seems to be of less prominent importance in other fields, it has traditionally played a major role in research on addictive disorders. Thus, there are now also individual findings on the neurobiological correlates of relapse in addiction disorders, e.g. on the importance of addressing the prefrontal cortex, which could be used to optimise current treatment strategies or for more selective indications (Moeller and Paulus 2018).

In a review article on the results of functional magnetic resonance imaging under *resting-state* conditions in panic disorder and social phobia (Kim and Yoon 2018), researchers were able to show that in people with panic disorder, in addition to changes in the *default mode network* and the medial temporal cortex, interoceptively relevant networks including the sensorimotor cortex were also relevant, while in people with social phobia, changes in connectivity in the so-called salience network were also evident. Such knowledge of overlapping and also distinct changes in brain connectivity at rest could be used to subdivide phenotypically different mental disorders, such as various anxiety disorders, into neurobiologically informed subgroups, which are suitable to better support the search for biomarkers and thus increase the chance of new therapeutic options. Accordingly, this could also advance the development of neurobiologically informed treatment programmes (Knatz et al. 2015). In this context, studies that allow us to directly modulate the brain neurochemically in people with mental disorders also appear to be of increasing interest: for example, it has been shown that intranasal oxytocin administration has a positive effect on the ability to process faces in people with autism (Domes et al. 2013). Overall, the therapeutic potential of neuropeptides and their neurobiological effects appear to be of growing interest (Lefevre et al. 2019).

The potential for translating existing neurobiological knowledge about relevant disease mechanisms into therapeutic action has also not been exhausted everywhere. For example, it can be argued that for eating disorders such as anorexia nervosa and bulimia nervosa, numerous studies on changes in the motivational system of the brain are available, but this knowledge has not yet resulted in concrete interventions that focus on modulating reinforcer processing (Monteleone et al. 2018).

Ultimately, the hope underlying all these approaches to the use of neurobiological knowledge in the therapy of mental illnesses is that, with the help of the knowledge gained, individual therapy decisions can be justified neurobiologically and predictions can be made about the possible prospects of success of a particular intervention on an individual basis (Ball et al. 2014). An important task for the future is therefore certainly to narrow or even close the gap between the findings from basic science and the practical application in everyday therapeutic practice. Individual statistical methods for bridging the gap between population-based research results and clinical practice appear to be very helpful here, e.g. on the basis of the application of machine learning approaches in so-called *"predictive medicine"* (Hahn et al. 2017). This includes methods that are intended to enable the prediction of certain characteristics, such as the presence of a mental illness, from other personal data at the individual level. For example, there are now several findings that can make predictions about the presence of an affective disorder at the individual level on the basis of people's brain-structural data using machine learning approaches and thus nourish the hope of being able to transfer results from neurobiological research directly into everyday clinical practice. At the same time, methodological considerations currently limit the generalizability and thus everyday applicability of these findings (Kim and Na 2018). However, the application of *computa-tional* approaches in psychotherapy (*computational psychotherapy*) is also interesting with a view to expanding our understanding of mechanisms of action: For behavioral therapy approaches in particular, methods of mathematical modeling offer the opportunity to test existing hypotheses about how people make inferences from experience, how this is altered in mental illness, or how it is modified by therapy (Moutoussis et al. 2018).

Such approaches to predicting mental illnesses or the effect of psychotherapy are developing in close connection with the field of so-called *"precision"* or *"personalized"* medicine (Fernandes et al. 2017), which aims to implement treatment and prevention of illnesses taking into account the individual variability of a person with regard to genetics, environmental factors, lifestyle, etc. The aim of these approaches is to develop a new approach to the prediction of mental illnesses and the effect of psychotherapy.

Even though the present chapter focuses on the presentation of the possibilities of a neuropsychotherapeutic perspective, it is important to emphasize that from a scientific-methodological perspective, relevant critical aspects of the findings in this field need to be considered. For example, the results of neurobiological studies are rarely homogeneous; as a rule, there are heterogeneous findings on a specific question, which can be attributed to factors such as different methods and paradigms used, the size and selection of the samples studied, clinical characteristics of the population (comorbidities, duration of illness, etc.) and specifics of the therapy (type, duration, setting, intensity of interventions, etc.), and which limit the generalizability of the results accordingly (Walter and Müller 2011).

With the progress of neuroscience, its opportunities and limitations are also becoming the focus of a debate in society as a whole (Meckel 2018). In particular, the detachment of the techniques generated and the knowledge gained from an application in a clinical context deserve a detailed critical debate. Thus, it will be the task of today's society to decide for itself and future generations how far we want to commit ourselves to an optimization of the healthy brain, as is the goal, for example, in the field of so-called neuroenhancement. Or how far findings about us and our brain should be made usable as a basis for economic interests, as is already the case with so-called neuromarketing.

Last but not least, it would be highly interesting to understand in more detail the neurobiological changes that the therapist's brain undergoes in the course of the numerous therapy processes. As could be shown in experts vs. novices in a compassion mediation, experts show a neuronally enhanced sensitivity to emotionally-averse stimuli (Lutz et al. 2008). This may serve as an example of possible (neural) risks and side effects of being a therapist, which are certainly not yet approximately understood.

> **Overview**
> In this section we have looked at possible future aspects. These include the further development of genetic research in the sense of so-called "imaging genetics", the use of potential biomarkers of mental illness for predicting the effectiveness of various treatments or their sustainability, the description of neurobiologically informed subgroups of mental illness and the development of suitable treatment programmes, and the further development of methods for deriving results that can be used for individual therapy decisions, e.g. in the context of so-called computational approaches in psychotherapy. It remains to be seen whether the hopes placed in these future prospects will be fulfilled. In addition, from a scientific-methodological perspective as well as from a social-ethical point of view,

relevant limitations of the field become apparent, which need to be discussed and taken into account.

Kurt Lewin, the co-founder of modern psychology, said, "If you want to truly understand something, try to change it" ▶ (▶ https://www.verywellmind.com/kurt-lewin-quotes-2795692, accessed 25 June 2019). Probably a joint, fruitful development of neuroscience and psychotherapy (research) has just begun.

References

Abbass AA, Nowoweiski SJ, Bernier D et al (2014) Review of psychodynamic psychotherapy neuroimaging studies. Psychother Psychosom 83:142–147

Anderson MC, Ochsner KN, Kuhl B et al (2004) Neural systems underlying the suppression of unwanted memories. Science 303:232–235

Andersson MA, Harkness SK (2018) When do biological attributions of mental illness reduce stigma? Using qualitative comparative analysis to contextualize attributions. Soc Ment Health 8:175–194

Arslan A (2018) Imaging genetics of schizophrenia in the post-GWAS era. Prog Neuro-Psychopharmacol Biol Psychiatry 80:155–165

Ball TM, Stein MB, Paulus MP (2014) Toward the application of functional neuroimaging to individualized treatment for anxiety and depression. Depress Anxiety 31:920–933

Balodis IM, Kober H, Worhunsky PD et al (2016) Neurofunctional reward processing changes in cocaine dependence during recovery. Neuropsychopharmacology 41:2112–2121

Barsaglini A, Sartori G, Benetti S et al (2014) The effects of psychotherapy on brain function: a systematic and critical review. Prog Neurobiol 114:1–14

Bartels A, Zeki S (2004) The neural correlates of maternal and romantic love. NeuroImage 21:1155–1166

Bassett DS, Zurn P, Gold JI (2018) On the nature and use of models in network neuroscience. Nat Rev Neurosci 19:566–578

Beauregard M, Courtemanche J, Paquette V, St-Pierre ÉL (2009) The neural basis of unconditional love. Psychiatry Res Neuroimaging 172:93–98

Berlin HA (2011) The neural basis of the dynamic unconscious. Neuropsychoanalysis 13:5–31

Bertsch K, Gamer M, Schmidt B et al (2013) Oxytocin and reduction of social threat hypersensitivity in women with borderline personality disorder. Am J Psychiatry 170:1169–1177

Bickel WK, Jarmolowicz DP, Mueller ET et al (2012) Excessive discounting of delayed reinforcers as a trans-disease process contributing to addiction and other disease-related vulnerabilities: emerging evidence. Pharmacol Ther 134:287–297

Bilevicius E, Kolesar T, Kornelsen J (2016) Altered neural activity associated with mindfulness during nociception: a systematic review of functional MRI. Brain Sci 6:14

Bora E, Yucel M, Allen NB (2009) Neurobiology of human affiliative behaviour: implications for psychiatric disorders. Curr Opin Psychiatry 22:320–325

Bryant RA, Felmingham K, Kemp A et al (2008) Amygdala and ventral anterior cingulate activation predicts treatment response to cognitive behaviour therapy for post-traumatic stress disorder. Psychol Med 38:555–561

Buchheim A, Cierpka M, Kächele H, Roth G (2012a) Neuropsychoanalyse: Das Hirn heilt mit. Gehirn Geist 11:50–53

Buchheim A, Viviani R, Kessler H et al (2012b) Changes in prefrontal-limbic function in major depression after 15 months of long-term psychotherapy. PLoS One 7:e33745

Cabeza R, St Jacques P (2007) Functional neuroimaging of autobiographical memory. Trends Cognit Sci 11:219–227

Cascio CN, Konrath SH, Falk EB (2015) Narcissists' social pain seen only in the brain. Soc Cognit Affect Neurosci 10:335–341

Doering S, Ruhs A (2015) Neuropsychoanalyse—Modeerscheinung oder Rückkehr zu den Freud'schen Urkonzepten? Neuropsychiatrie 29:39–42

Domes G, Heinrichs M, Kumbier E et al (2013) Effects of intranasal oxytocin on the neural basis of face processing in autism spectrum disorder. Biol Psychiatry 74:164–171

Dörfel D, Lamke J-P, Hummel F et al (2014) Common and differential neural networks of emotion regulation by detachment, reinterpretation, distraction, and expressive suppression: a comparative fMRI investigation. NeuroImage 101:298–309

Draganski B, Gaser C, Busch V et al (2004) Neuroplasticity: changes in grey matter induced by training. Nature 427:311–312

Elgendi M, Kumar P, Barbic S et al (2018) Subliminal priming-state of the art and future perspectives. Behav Sci 8:e54. (Basel)

Eßing G (2015) Praxis der Neuropsychotherapie: Wie die Psyche das Gehirn formt. Deutscher Psychologen Verlag, Berlin

Everitt BJ, Robbins TW (2005) Neural systems of reinforcement for drug addiction: from actions to habits to compulsion. Nat Neurosci 8:1481–1489

Fernandes BS, Williams LM, Steiner J et al (2017) The new field of "precision psychiatry". BMC Med 15:80

Freud S (1885) Entwurf einer Psychologie. In: Gesammelte Werke—Texte aus den Jahren 1885 bis 1938, pp 375–386

Furmark T, Tillfors M, Marteinsdottir I et al (2002) Common changes in cerebral blood flow in patients with social phobia treated with citalopram or cognitive-behavioral therapy. Arch Gen Psychiatry 59:425

Goldapple K, Segal Z, Garson C et al (2004) Modulation of cortical-limbic pathways in major depression. Arch Gen Psychiatry 61:34

Grawe K (2000) Psychologische Therapie. Hogrefe, Göttingen

Grawe K (2004) Neuropsychotherapie. Hogrefe, Göttingen

Grawe K, Donati R, Bernauer F (1994) Psychotherapie im Wandel: von der Konfession zur Profession. Hogrefe, Göttingen

Hahn T, Nierenberg AA, Whitfield-Gabrieli S (2017) Predictive analytics in mental health: applications, guidelines, challenges and perspectives. Mol Psychiatry 22:37–43

Han D-Y, Chen S-H (2014) Reducing the stigma of depression through neurobiology-based psychoeducation: a randomized controlled trial. Psychiatry Clin Neurosci 68:666–673

Heinz A (2017) A new understanding of mental disorders. Computational models for dimensional psychiatry. MIT Press, Cambridge

Heinz A, Wrase J, Kahnt T et al (2007) Brain activation elicited by affectively positive stimuli is associated with a lower risk of relapse in detoxified alcoholic subjects. Alcohol Clin Exp Res 31:1138–1147

Hester R, Luijten M (2014) Neural correlates of attentional bias in addiction. CNS Spectr 19:231–238

Hohenfeld C, Nellessen N, Dogan I et al (2017) Cognitive improvement and brain changes after real-time functional MRI neurofeedback training in healthy elderly and prodromal Alzheimer's disease. Front Neurol 8:384

Holmes EA, Arntz A, Smucker MR (2007) Imagery rescripting in cognitive behaviour therapy: images, treatment techniques and outcomes. J Behav Ther Exp Psychiatry 38:297–305

Insel T, Cuthbert B, Garvey M et al (2010) Research domain criteria (RDoC): toward a new classification framework for research on mental disorders. Am J Psychiatry 167:748–751

Kakoschke N, Kemps E, Tiggemann M (2017) Approach bias modification training and consumption: a review of the literature. Addict Behav 64:21–28

Kandel ER (1979) Psychotherapy and the single synapse. N Engl J Med 301:1028–1037

Kandel ER (1998) A new intellectual framework for psychiatry. Am J Psychiatry 155:457–469

Kim Y-K, Na K-S (2018) Application of machine learning classification for structural brain MRI in mood disorders: critical review from a clinical perspective. Prog Neuro-Psychopharmacol Biol Psychiatry 80:71–80

Kim Y-K, Yoon H-K (2018) Common and distinct brain networks underlying panic and social anxiety disorders. Prog Neuro-Psychopharmacol Biol Psychiatry 80:115–122

Knatz S, Wierenga CE, Murray SB et al (2015) Neurobiologically informed treatment for adults with anorexia nervosa: a novel approach to a chronic disorder. Dialogues Clin Neurosci 17:229–236

Kumari V, Peters ER, Fannon D et al (2009) Dorsolateral prefrontal cortex activity predicts responsiveness to cognitive-behavioral therapy in schizophrenia. Biol Psychiatry 66:594–602

Lefevre A, Hurlemann R, Grinevich V (2019) Imaging neuropeptide effects on human brain function. Cell Tissue Res 375:279–286

Lindauer RJL, Booij J, Habraken JBA et al (2008) Effects of psychotherapy on regional cerebral blood flow during trauma imagery in patients with post-traumatic stress disorder: a randomized clinical trial. Psychol Med 38:543–554

Linden DEJ (2006) How psychotherapy changes the brain—the contribution of functional neuroimaging. Mol Psychiatry 11:528–538

Livingston JD, Milne T, Fang ML, Amari E (2012) The effectiveness of interventions for reducing stigma related to substance use disorders: a systematic review. Addiction 107:39–50

Lueken U, Hahn T (2016) Functional neuroimaging of psychotherapeutic processes in anxiety and depression: from mechanisms to predictions. Curr Opin Psychiatry 29:25–31

Lueken U, Straube B, Konrad C et al (2013) Neural substrates of treatment response to cognitive-behavioral therapy in panic disorder with agoraphobia. Am J Psychiatry 170:1345–1355

Lutz A, Brefczynski-Lewis J, Johnstone T, Davidson RJ (2008) Regulation of the neural circuitry of emotion by compassion meditation: effects of meditative expertise. PLoS One 3:e1897

Maguire EA, Gadian DG, Johnsrude IS et al (2000) Navigation-related structural change in the hippocampi of taxi drivers. Proc Natl Acad Sci U S A 97:4398–4403

Malejko K, Abler B, Plener PL, Straub J (2017) Neural correlates of psychotherapeutic treatment

of post-traumatic stress disorder: a systematic literature review. Front Psych 8:85

Meckel M (2018) Mein Kopf gehört mir. Eine Reise durch die schöne neue Welt des Brainhacking. Piper, München

Moeller SJ, Paulus MP (2018) Toward biomarkers of the addicted human brain: using neuroimaging to predict relapse and sustained abstinence in substance use disorder. Prog Neuro-Psychopharmacol Biol Psychiatry 80:143–154

Molden DC (2014) Understanding priming effects in social psychology: what is "social priming" and how does it occur? Soc Cognit 32:1–11

Monteleone AM, Castellini G, Volpe U et al (2018) Neuroendocrinology and brain imaging of reward in eating disorders: a possible key to the treatment of anorexia nervosa and bulimia nervosa. Prog Neuro-Psychopharmacol Biol Psychiatry 80:132–142

Moutoussis M, Shahar N, Hauser TU, Dolan RJ (2018) Computation in psychotherapy, or how computational psychiatry can aid learning-based psychological therapies. Comput Psychiatry (Camb Mass) 2:50–73

Müller S, Heinz A (2013) Stigmatisierung oder Entstigmatisierung durch Biologisierung psychischer Krankheiten? Nervenheilkunde 32:955–961

Nakao T, Nakagawa A, Yoshiura T et al (2005) Brain activation of patients with obsessive-compulsive disorder during neuropsychological and symptom provocation tasks before and after symptom improvement: a functional magnetic resonance imaging study. Biol Psychiatry 57:901–910

O'Doherty DCM, Chitty KM, Saddiqui S et al (2015) A systematic review and meta-analysis of magnetic resonance imaging measurement of structural volumes in posttraumatic stress disorder. Psychiatry Res Neuroimaging 232:1–33

Ori R, Amos T, Bergman H et al (2015) Augmentation of cognitive and behavioural therapies (CBT) with d-cycloserine for anxiety and related disorders. Cochrane Database Syst Rev 2015(5):CD007803

Owens MM, MacKillop J, Gray JC et al (2017) Neural correlates of tobacco cue reactivity predict duration to lapse and continuous abstinence in smoking cessation treatment. Addict Biol 23:1189–1199

Papineau D (1998) Mind the gap. Nous 32:373–388

Paquette V, Lévesque J, Mensour B et al (2003) "Change the mind and you change the brain": effects of cognitive-behavioral therapy on the neural correlates of spider phobia. NeuroImage 18:401–409

Park SQ, Kahnt T, Beck A et al (2010) Prefrontal cortex fails to learn from reward prediction errors in alcohol dependence. J Neurosci 30:7749–7753

14

Quidé Y, Witteveen AB, El-Hage W et al (2012) Differences between effects of psychological versus pharmacological treatments on functional and morphological brain alterations in anxiety disorders and major depressive disorder: a systematic review. Neurosci Biobehav Rev 36:626–644

Rainville P, Hofbauer RK, Paus T et al (1999) Cerebral mechanisms of hypnotic induction and suggestion. J Cognit Neurosci 11:110–125

Ritchey M, Dolcos F, Eddington KM et al (2011) Neural correlates of emotional processing in depression: changes with cognitive behavioral therapy and predictors of treatment response. J Psychiatr Res 45:577–587

Romanczuk-Seiferth N (2017) Therapie (ge)braucht das Gehirn. PiD Psychother im Dialog 18:80–83

Romanczuk-Seiferth N, Koehler S, Dreesen C et al (2015) Pathological gambling and alcohol dependence: neural disturbances in reward and loss avoidance processing. Addict Biol 20:557–569

Schacht JP, Anton RF, Myrick H (2013) Functional neuroimaging studies of alcohol cue reactivity: a quantitative meta-analysis and systematic review. Addict Biol 18:121–133

Schiepek G, Tominschek I, Karch S et al (2009) A controlled single case study with repeated fMRI measurements during the treatment of a patient with obsessive-compulsive disorder: testing the nonlinear dynamics approach to psychotherapy. World J Biol Psychiatry 10:658–668

Schmitt R, Winter D, Niedtfeld I et al (2016) Effects of psychotherapy on neuronal correlates of reappraisal in female patients with borderline personality disorder. Biol Psychiatry Cognit Neurosci Neuroimaging 1:548–557

Schnell K, Herpertz SC (2007) Effects of dialectic-behavioral-therapy on the neural correlates of affective hyperarousal in borderline personality disorder. J Psychiatr Res 41:837–847

Schultz W (2016) Reward functions of the basal ganglia. J Neural Transm 123:679–693

Schulze L, Domes G, Krüger A et al (2011) Neuronal correlates of cognitive reappraisal in borderline patients with affective instability. Biol Psychiatry 69:564–573

Schwabe L (2017) Memory under stress: from single systems to network changes. Eur J Neurosci 45:478–489

Schwing R (2013) Spuren des Erfolgs: was lernt die systemische Praxis von der Neurobiologie? In: Syst. Hirngespinste. Vandenhoeck & Ruprecht, Göttingen, pp 63–119

Sebold M, Nebe S, Garbusow M et al (2017) When habits are dangerous: alcohol expectancies and habitual decision making predict relapse in alcohol dependence. Biol Psychiatry 82:847–856

Seiferth N, Heinz A (2010) Neurobiology of substance-related addiction: findings of neuroimaging. In: Del-Ben CM (ed) Neuroimaging. IntechOpen, Rijeka

Sheffer CE, Mennemeier M, Landes RD et al (2013) Neuromodulation of delay discounting, the reflection effect, and cigarette consumption. J Subst Abus Treat 45:206–214

Siegle GJ, Carter CS, Thase ME (2006) Use of fMRI to predict recovery from unipolar depression with cognitive behavior therapy. Am J Psychiatry 163:735–738

Sowell ER, Thompson PM, Toga AW (2004) Mapping changes in the human cortex throughout the span of life. Neuroscientist 10:372–392

Stucki C, Grawe K (2007) Bedürfnis- und Motivorientierte Beziehungsgestaltung. Psychotherapeut 52:16–23

Takagi K (2018) Information-based principle induces small-world topology and self-organized criticality in a large scale brain network. Front Comput Neurosci 12:65

Takahashi K, Mizuno K, Sasaki AT et al (2015) Imaging the passionate stage of romantic love by dopamine dynamics. Front Hum Neurosci 9:191

Vargas-Martínez F, Uvnäs-Moberg K, Petersson M et al (2014) Neuropeptides as neuroprotective agents: oxytocin a forefront developmental player in the mammalian brain. Prog Neurobiol 123:37–78

Vellante F, Sarchione F, Ebisch SJH et al (2018) Creativity and psychiatric illness: a functional perspective beyond chaos. Prog Neuro-Psychopharmacol Biol Psychiatry 80:91–100

Wallhed Finn S, Bakshi A-S, Andréasson S (2014) Alcohol consumption, dependence, and treatment barriers: perceptions among nontreatment seekers with alcohol dependence. Subst Use Misuse 49:762–769

Walter H, Müller S (2011) Neuroethik und Psychotherapie. In: Schiepek G (ed) Neurobiol. der Psychother., 2 Aufl. Schattauer, Stuttgart, pp 646–655

Walter H, Berger M, Schnell K (2009) Neuropsychotherapy: conceptual, empirical and neuroethical issues. Eur Arch Psychiatry Clin Neurosci 259:173–182

Wiers CE, Ludwig VU, Gladwin TE et al (2015) Effects of cognitive bias modification training on neural signatures of alcohol approach tendencies in male alcohol-dependent patients. Addict Biol 20:990–999

Won E, Kim Y-K (2018) Neuroimaging in psychiatry: steps toward the clinical application of brain imaging in psychiatric disorders. Prog Neuro-Psychopharmacol Biol Psychiatry 80:69–70

Wrase J, Schlagenhauf F, Kienast T et al (2007) Dysfunction of reward processing correlates with alcohol craving in detoxified alcoholics. NeuroImage 35:787–794

Yerkes RM, Dodson JD (1908) The relation of strength of stimulus to rapidity of habit-formation. J Comp Neurol Psychol 18:459–482

Young SN (2013) The effect of raising and lowering tryptophan levels on human mood and social behaviour. Philos Trans R Soc Lond Ser B Biol Sci 368:20110375

Zeki S (2007) The neurobiology of love. FEBS Lett 581:2575–2579

14

Psychoneuroscience and Its Relevance for Practice

Andreas Heinz, Gerhard Roth and Henrik Walter

Contents

This book aims to build a bridge between the "psychosciences" (psychology, psychiatry, psychotherapy) and the neurosciences. As shown, such bridging succeeds most easily in areas of perceptual and cognitive psychology, whereas in personality psychology, for example, neuroscience-based concepts are only gradually being applied. In the fields of psychiatry, a dualistic demarcation of the discipline as a "human science" from neurology understood as a natural science is still popular, which at the same time stands in remarkable contrast to an equally widespread belief in pharmacology. Even within psychotherapy, efforts to provide a neuroscientific foundation for the approaches advocated here have not yet progressed far. Psychoanalysis in particular has resisted "neurobiologization" for a very long time—much to its own detriment and contrary to the intentions of its founder, Sigmund Freud (see Roth and Strüber 2018). We hope that our book will help to build bridges across such divides.

But it would be wrong to see this as a purely academic endeavour. Kurt Lewin's famous sentence applies here: "Nothing is more practical than a good theory!" Scientific facts and findings do not interpret themselves, but only acquire meaning through their embedding in an overarching theoretical concept and its practical implications. It is not indifferent for practice whether someone as a psychologist, psychiatrist or psychotherapist assumes that mind or psyche can or even must be understood independently of the material brain in the sense of a dualism, whether one believes in the sense of a reductionism that mental or psychological processes can be completely traced back to neuronal processes, or whether one understands mind-psyche and brain-body as a multiform and at the same time necessary unity. This leads to the question we want to address in our conclusion: What are the theoretical and practical, as well as ethical, implications of the facts presented in this book?

First of all, one thing should have become clear: From the perspective of neuroscience, a dualism that attempts to make categorical distinctions between mental and neurobiologically based diseases of the brain is outdated. For example, in the field of addictive disorders, it makes no sense to distinguish between "physical" and "psychological" addiction. According to such a distinction committed to dualism, physical signs of addiction would be those that can be objectified by means of physically observable phenomena such as sweating and trembling, whereas psychological addiction is said to manifest as strong craving and diminished control over that craving. Even this attempt to assign the key symptoms to a psychological as opposed to a physical range of relevant phenomena, however, can only be constructed at the price of an inadmissible abridgement. For the withdrawal symptoms include not only trembling and sweating but also withdrawal-related disturbances of sleep-wake regulation and mood up to manifest delirium, that is characterized for example by hallucinations, which as disturbances of perception would have to be assigned to the "mental" domain in the case of strict dichotomization.

In addition, neurobiological correlates for the supposedly "physical" as well as "psychological" leading symptoms are found in different neural networks, such as for withdrawal symptoms in the area of noradrenergic regulation of mental functions or with regard to drug craving in the area of dopaminergic neurotransmission (Heinz 2017). From a neuroscientific perspective, the distinction between psychiatric ("mental") and neurological ("organic") disorders is also blurred and in some cases arbitrary (e.g., in dementia). Often "neurological" disorders are accompanied by mental symptoms and vice versa. A disturbance of tonic dopaminergic neurotransmission in the area of the dorsal striatum is associated with Parkinson's syndrome, which is characterized by movement restrictions, but also by depression and other affective symptoms, while a disturbance of phasic dopaminergic neurotransmission in the area of the ventral and associative striatum is understood as a correlate of motivation, drug craving and—in the case of stress-related or chaotic activation—certain psychotic experiences such as delusions (Heinz 2002b).

One of the aims of this book is to break down such outdated dichotomies and to arrive at a *holistic* understanding of those clinical pictures which—so intricate is the language contaminated by dualism—cannot even be referred to the psyche or mind without recourse to traditional dichotomies. This, however, does not mean neglecting the different ways of approaching the relevant phenomena or falling into an inadequate reductionism. The distinctions necessary here overcome a simple opposition of leading symptoms in the external world, which, for example, would have to be counted as objectifiable motor dysfunctions in neurology, and phenomena in the internal world, such as the hearing of voices or the experience of input thoughts, which are then supposed to characterize mental illnesses. Understanding the inner world and thus the experience of people with mental illness is not possible without a "co-world" (Heinz 2014; Plessner 1975), i.e. not without recourse to joint efforts at shared intentionality, articulation and representation of individual experiences.

It is common to speak of "neurobiological correlates" of those phenomena which are considered to be leading symptoms of mental disorders and which can be as diverse in nature as clouding of consciousness in delirium, pain experiences in somatization disorders, hallucinations in the context of epileptic auras or psychotic experiences, and affect-rigid mood in major depression. Putative mental disorders may present with leading motor symptoms, as in so-called catatonia, while conversely psychotic experiences and depressive mood are common in patients with Huntington's disease. Thus, we

argue for a differentiated view of the leading symptoms of the so-called mental disorders, which cannot be forced into a Procrustean bed of supposed mental versus physical phenomena. From this point of view, the historical and interest-driven strict distinction between neurology and psychiatry is obsolete.

Thus, when we speak rather neutrally of "neurobiological correlates" of the leading symptoms of mental disorders, we deliberately leave it open how the relationship of complex cognitive performances to these biological correlates is to be understood, what forms of identity or emergence are thus presupposed here. This has the advantage that our stance is also open to other approaches, even if as psychiatrists, neuroscientists and philosophers we are committed to a materialism that values the beauty and complexity of living beings and their organs (Ernst and Heinz 2013). Abbreviated statements such as "the brain does this or that" that personalize neural processes, while sometimes descriptive, ultimately fail to do justice to this complexity. If we want to dissolve linguistically outdated dualisms and dichotomies in the field of symptoms of mental disorders and their neurobiological correlates, then reductive equations of the acting persons with their brains are not helpful. This is also evident from the fact that in cognitive neuroscience the concept of *situated cognition,* i.e. the assumption that cognitive processes of an organism are always also embodied and embedded in the environment in an active way, is gaining more and more attention in empirical research (Newen et al. 2018; Walter 2018). In addition, we need to be aware that our understanding of the brain and its interaction with the living body is always shaped by contemporary ideas. Thus, in arguments about the history of mental disorders, we find multiple projections of social relations of domination onto the brain as an organ, which then paradigmatically guide its further exploration and lead to circular misconceptions regarding alleged biological foundations of social inequality. This has been exemplified in the construction of schizophrenic disorders (Heinz 2002a). The brain is a wonderfully complex organ whose understanding eludes such ideological approaches (Roth 1996).

What practical and ethical implications arise from the evidence of neurobiological correlates of mental disorders? We would like to elaborate on this at three points, firstly with reference to the question of free will, secondly with reference to the question of predicting the course of diseases, and thirdly with reference to the issue of equal treatment of people with "mental" versus "somatic" disorders. Regarding free will, we as editors hold partly common and partly different positions (cf. Heinz 2014; Roth 2003, 2009; Walter 1999, 2011). We agree that an unconditional freedom, i.e., the possibility to act differently in spite of existing reasons and motives in a situation assumed to be identical, would be arbitrary and cannot establish moral responsibility—the flip side of free will. A conditional freedom, on the other hand, to be able to act differently under similar circumstances on the basis of one's own (personal) reasons, is possible from a neuroscientific point of view. Freedom of will in this sense is a gradual phenomenon and can be impaired in a relevant way. On the other hand, there is a usually very high threshold at which affective-cognitive abilities are so severely impaired that a self-determined decision can no longer be meaningfully attributed on the basis of these pathologically impaired abilities. Such a determination never results from a diagnosis alone; a person's current abilities in the relevant context are decisive. Impairments of life-relevant functional abilities such as consciousness or spatial disorientation in delirium can make such self-determination impossible and thus lead to a "loss of freedom" of the person concerned. Admittedly, we would find it difficult to define this as such if we did not at least grant persons acting in everyday life a conditional freedom of choice.

Overall, we plead for an understanding of mental disorders that refrains from false dichotomies without falling into inappropriate reductionisms. It is precisely respect for matter as the bearer of our life that demands respect for the most complex achievements of biological and thus material organs in their inherent dynamics, without having to resort here to immaterial constructs such as a supposed "free spirit" (Plessner 1975; Scheler 1928). It should only be noted in passing that this is not a technocratic viewpoint of modern times, but rather one that was already articulated in antiquity (cf. Laotse 1921; Lukrez 2014).

The second point concerns the question: Does the prediction of mental disorders and their course from neurobiological correlates run the risk of discouraging patients or depriving them of their options for action? A simple and quick answer based on the current data situation must be "no", because all biomarkers discovered so far allow only a very incomplete and partial prediction of further courses. Although it is possible to a certain extent, with the help of complex examination methods such as functional magnetic resonance imaging, to use images of the preferred drugs to predict which patients are particularly at risk of relapsing on the basis of activation patterns in the central nervous system. But this explains only a very limited part of the variance of future courses (Beck et al. 2012) and cannot (yet) be used for prediction in individual patients or clinical practice.

As a rule, the possibility of predicting the manifestation of a mental illness such as schizophrenic psychosis from genetic risk factors is even more limited. Although it was possible to explain 18.4% of the variance in the occurrence of schizophrenic psychoses in very large data sets in 2014 with knowledge of all occurring genetic SNPs (*single nucleotide polymorphisms*) (Ripke et al. 2014), this is still not clinically relevant for individual questions, and each individual genetic variant (SNP) increases the risk of developing schizophrenia by only a fraction of a percent. Therefore, we consider it important not to speak of "dysfunctional activation patterns" in the case of correlates of frequent genetic polymorphisms with brain structure or activity, for example in the case of allele-specific activation patterns during the performance of certain tasks.

These common genetic variances are widespread in the general population; they are generally not associated with any limitation of performance and contribute to the diversity of human life forms. Mental disorders do not inevitably develop from a particular genetic predisposition; rather, the latter represent vulnerability factors, i.e. they increase the susceptibility of affected individuals to develop a mental illness when certain life events occur (cf. ► Chaps. 5 and 7).

Clinically relevant disorders are defined by the restriction of generally vital functional abilities. These include clouding of consciousness, disturbances of spatial or temporal orientation, disturbances of perception up to complex acoustic hallucinations or impairments of the capacity for the modulation of mood, as found in manias or depressions with rigid mood states. However, even if such key symptoms are present, a clinically relevant disorder is only present if the affected individual suffers from them or is restricted in his or her activities of daily living, which form the basis of social participation. Thus, those who can no longer wash or dress themselves due to their memory disorders in the context of their dementia, or who personally suffer from complex auditory hallucinations, have a clinically relevant illness. In contrast, from a medical point of view, people who value their voices as a source of advice and manage their everyday lives would still exhibit hallucinations and thus disorders of perception, but this would not be associated with clinical relevance (Heinz 2014).

However, pointing to the (still) low informative value of neurobiological changes is not an argument that such prediction might not one day raise ethical problems. An increasingly precise recording of the neurobiological correlates of mental processes could lead to human behaviour becoming more and more predictable and thus also controllable or even manipulable. In addition, *every* psychological phenomenon will have an organic correlate, if only the methods of investigation are fine and precise enough. It is crucial for us to protect human diversity, to respect different ways of life and not to understand disorders as deviations from an arbitrarily set of statistically defined norms. It is thus a question of which functional abilities are relevant to life and thus, from a medical point of view, symptoms of a disease, and of assessing their impact on the lifeworld of the respective individual.

With regard to the third point, the question of treating mentally ill persons on an equal footing with people suffering from so-called somatic disorders, we believe that the description of the neurobiological correlates can serve to destigmatise them. For example, neurobiologically definable mechanisms make it understandable why a person with an addictive disorder may find it particularly difficult to change his or her behaviors; habituated responses to drug stimuli may be difficult to control even in the presence of strong, conscious intentions. A neurobiological understanding can break down outdated dichotomies that classify people as free or unfree, and instead point to the severity of changes in learning mechanisms that make it particularly difficult, but not impossible, for affected individuals to change their behaviors. Ideally, then, neurobiological approaches help to generate understanding of the difficulties experienced by the mentally ill.

However, this is by no means a foregone conclusion. In fact, meta-analyses show that biogenetic explanations of mental illness on the one hand reduce blaming of the sick, but on the other hand can also lead to prognostic pessimism and possibly also to a tendency to view mentally ill people as more likely to be violent (Kvaale et al. 2013). The history of psychiatry is full of stigmatizations, exclusions as well as inhumane ways of dealing with the mentally ill, up to and including the murder of mentally ill people during the National Socialist era. The description of biological mechanisms always runs the risk of labelling those affected as "different" in some way and of excluding them. This should be emphatically countered here: Mental disorders result from the specific position of people in their life-world. All persons can distance themselves from themselves and reflect on themselves and can thus be in the "profoundly human" position of experiencing their own thoughts as "alien" or "entered", as is the case with psychotic ego disorders (Heinz 2014). All persons who can articulate and reflect on their moods are thus able to develop delusions of sin or guilt in addition to depressed mood. A strong desire to consume drugs can also be observed in animal models, but a reduction in control in the sense of drug consumption against conscious intentions to the contrary is only observable in humans, i.e. a being that can make statements about its intentions. This is true not only in psychiatry but also in the field of neurology, because also the clinical examination of aphasia or ataxia is based on the interaction with the afflicted person.

Nevertheless, there is persistent discrimination against the mentally ill in many ways; not only through stigmatization in society and social exclusion, but in Germany through unequal treatment in the health care system. For example, after detoxification, addicts must apply for rehabilitation in order to receive treatment for their underlying disorder, which is addiction to drugs. Imagine if such an imposition were made on patients with affective disorders, and the health insurance were to fund only acute care after a suicide attempt, while treatment of the underlying disease, in this case depression, would only be possible after an application for rehabilitation had been made.

Against the background of such considerations and the struggle for equal treatment of all people with mental as well as somatic illnesses, which is supported by those affected and their relatives, we wanted to describe the neurobiological correlates of mental disorders and contribute to the destigmatisation of them with this book. It is often a long and rocky road from the traditional stigmatization of the mentally ill to inclusion in the enlived environment and society; we hope that our attempt to provide information about the conditions under which mental illnesses develop and about their therapeutic options can contribute to this aim.

References

Beck A, Wüstenberg T, Genauck A, Wrase J, Schlagenhauf F, Smolka M, Heinz A (2012) Effect of brain structure, brain function, and brain connectivity on relapse in alcohol-dependent patients. Arch Gen Psychiatry 69(8):842–852

Ernst G, Heinz A (2013) Die widerspenstige Materie: Neues aus der Naturwissenschaft und Konsequenzen für linke Theorie und Praxis. Schmetterling, Stuttgart

Heinz A (2002a) Anthropologische und evolutionäre Modelle in der Schizophrenieforschung (Das transkulturelle Psychoforum). VWB-Verlag, Bergisch Gladbach

Heinz A (2002b) Dopaminergic dysfunction in alcoholism and schizophrenia—psychopathological and behavioral correlates. Eur Psychiatry 17(1):9–16

Heinz A (2014) Der Begriff der psychischen Krankheit. Suhrkamp, Berlin

Heinz A (2017) A new understanding of mental disorders: computational models for dimensional psychiatry. MIT Press, London

15

Kvaale EP, Haslam N, Gottdiener W (2013) The 'side effects' of medicalization: a meta-analytic review of how biogenetic explanations affect stigma. Clin Psychol Rev 33(2013):782–794

Laotse (1921) Tao Te King. Das Buch des alten vom Sinn und Leben. In: Aus dem chinesischen verdeutscht und erläutert von Richard Wilhelm. Eugen Diedrichs-Verlag, Jena

Lukrez (2014) Über die Natur der Dinge. Galiani, Berlin

Newen A et al (2018) 4E cognition: historical roots, key concepts and central issues. In: Newen A, De Bruin L, Gallagher S (eds) The Oxford handbook of 4E cognition. OUP, Oxford

Plessner H (1975) Die Stufen des Organischen und der Mensch. Einleitung in die philosophische Anthropologie, De Gruyter, Berlin

Ripke S, Schizophrenia Working Group of the Psychiatric Genomics Consortium et al (2014) Biological insights from 108 schizophrenia-associated genetic loci. Nature 511(7510):421–427. (IF 42.3)

Roth G (1996) Das Gehirn und seine Wirklichkeit: Kognitive Neurobiologie und ihre philoso-phischen Konsequenzen. Suhrkamp, Frankfurt a. M

Roth G (2003) Fühlen, Denken, Handeln. Wie das Gehirn unser Verhalten steuert. Suhrkamp, Frankfurt

Roth G (2009) Aus Sicht des Gehirns. Suhrkamp, Frankfurt

Roth G, Strüber N (2018) Wie das Gehirn die Seele macht. Klett-Cotta, Stuttgart

Scheler M (1928) Die Stellung des Menschen im Kosmos. Reichl, Darmstadt

Walter H (1999) Neurophilosophie der Willensfreiheit. Mentis, Paderborn (English edition: Neurophilosophy of free will. MIT, Boston, 2001; Erstveröffentlichung 1998)

Walter H (2011) Contributions of neuroscience to the free will debate: from random movement to intelligible action. In: Kane R (ed) Oxford Handbook of Free Will. Oxford University Press, Oxford, pp 515–529

Walter H (2018) Über das Gehirn hinaus. Aktiver Externalismus und die Natur des Mentalen. Nervenheilkunde 37(7/8):479–486